AUTOMATIC TRANSMISSIONS/ TRANSAXLES

THOMAS W. BIRCH
Yuba College
Marysville, California

GLENCOE

Macmillan/McGraw-Hill

Lake Forest, Illinois
Columbus, Ohio
Mission Hills, California
Peoria, Illinois

Two-Color Insert Technical Art Studio: Fine Line, Inc.
Cover Designer: Larry Didona
Production Services: WordCrafters Editorial Services, Inc.

Library of Congress Cataloging-in-Publication Data

Birch, Thomas, date.
    Automatic transmissions/transaxles / Thomas W. Birch.
      p. cm.
    Includes index.
    ISBN 0-07-005310-3
    1. Automobiles—Transmission devices; Automatic. 2. Automobiles—
Transmission devices, Automatic—Maintenance and repair.
I. Title.
TL263.B57   1991
629.24'46—dc20   90-25119
                                                                           CIP

AUTOMATIC TRANSMISSIONS/TRANSAXLES

Copyright © 1991 by the Glencoe Division of Macmillan/McGraw-Hill
Publishing Company. All rights reserved. Printed in the United States of
America. Except as permitted under the United States Copyright Act of
1976, no part of this publication may be reproduced or distributed in any
form or by any means, or stored in a database or retrieval system,
without the prior written permission of the publisher.

Send all inquiries to:
Glencoe Division, Macmillan/McGraw-Hill
936 Eastwind Drive
Westerville, Ohio 43081

ISBN 0-07-005310-3

1 2 3 4 5 6 7 8 9 0   SEMSEM   9 8 7 6 5 4 3 2 1 0

CONTENTS

PREFACE	v
CHAPTER 1 INTRODUCTION TO TRANSMISSIONS	1
CHAPTER 2 APPLY DEVICES: CLUTCHES AND BRAKES	32
CHAPTER 3 POWER FLOW THROUGH PLANETARY GEAR TRANSMISSIONS	49
CHAPTER 4 HYDRAULIC SYSTEMS: THEORY	104
CHAPTER 5 HYDRAULIC SYSTEM OPERATION	131
CHAPTER 6 ELECTRONIC TRANSMISSION CONTROLS	164
CHAPTER 7 TORQUE CONVERTERS	181
CHAPTER 8 TRANSMISSION DESCRIPTION	195
CHAPTER 9 GENERAL TRANSMISSION SERVICE AND MAINTENANCE	242
CHAPTER 10 PROBLEM SOLVING AND DIAGNOSIS	258
CHAPTER 11 IN-CAR TRANSMISSION REPAIR	293
CHAPTER 12 TRANSMISSION OVERHAUL	314
CHAPTER 13 TORQUE CONVERTER SERVICE	383
CHAPTER 14 AUTOMATIC TRANSMISSION MODIFICATIONS	397
APPENDIXES	
1 English-Metric Conversion Table	409
2 Description of General Motors Automatic Transmissions/Transaxles	410
3 Chronology of General Automatic Transmissions/Transaxles	411

iii

4 Description of Ford Motor Company Automatic Transmissions/Transaxles 412

5 Chronology of Ford Motor Company Automatic Transmissions/Transaxles 413

6 Description of Chrysler Corporation Automatic Transmissions/Transaxles 414

7 Chronology of Chrysler Corporation Automatic Transmissions/Transaxles 415

8 Bolt Torque Tightening Chart 416

GLOSSARY 417

INDEX 421

PREFACE

Along with the modern automobile, the automatic transmission has evolved steadily and gradually. The somewhat bulky and cantankerous transmissions of the 1950s and 1960s have become the relatively lightweight electronically controlled transaxles of the 1990s.

This evolution has had an impact on the transmission repair industry. In the early years, apprentice mechanics could be trained to learn all necessary repair techniques with one of the half-dozen commonly used transmissions by working alongside an experienced mechanic. Today, not only do we have a larger variety of domestic transmissions and transaxles, but also these units are constantly upgraded to meet the demands of consumers, manufacturers, and government agencies. A substantial number and variety of units used in imported cars have also entered the scene. The major areas of change include the addition of final drives and differentials in the development of transaxles and the use of lockup torque converters, overdrive gears, and electronic shift controls. At one time, a mechanic could work on and memorize a particular transmission such as a Powerglide or C4 and be expert in repairing that transmission for a number of years.

The transmission technician of today, however, cannot memorize all the individual transmissions with their updates and electronic circuitry. The number of things that can fail and cause problems has increased dramatically. Problem diagnosis is no longer a process of checking the dipstick and driving around the block. It has become a systematic, many-stepped process involving the use of pressure gauges and electronic testing equipment. Repair has become a procedure that is carefully guided by service manuals with update bulletins.

This textbook explains the operating principles of gearsets, friction devices, hydraulic systems, and torque converters to enable the prospective technician to understand transmission operation. It also discusses the electronic control systems used in most modern transmissions and transaxles. Given this background, the technician will know what should occur for each gear range operation and upshift or downshift.

In-car maintenance and adjustment operations are described so that the student technician will learn what should be done to keep a transmission/transaxle operating correctly. A chapter on problem diagnosis procedures describes all the tests used to locate the cause of transmission/transaxle problems.

Service operations, both in-car and bench, are thoroughly described so that the future technician will know how and why they are performed and what tools and equipment are needed to perform them. The service operations are described in a generic manner because today's transmission technician must also use a service manual while working on a specific transmission/transaxle to ensure that all the required service operations are performed. In addition, a chapter describing various transmission modifications is included to acquaint the student technician with every aspect of transmission work.

The text covers all content areas of the ASE Transmission/Transaxle test. Along with class instruction, it should prepare students to pass this test and obtain certification.

I acknowledge and thank all the firms that granted use of illustrations used in this text; their names appear in the credit lines. Special thanks are in order to Dennis Zeiger, Chrysler Corporation; Wayne Ferrell, Sealed Power Technologies; and J. E. "Corky" Meyers, Torque Converter Rebuilding Systems.

Thomas W. Birch

CHAPTER 1

INTRODUCTION TO TRANSMISSIONS

OBJECTIVES

After completing this chapter, you should:

- Be able to identify the major components of an automatic transmission and transaxle.

- Have an understanding of how different gear ratios can be obtained in a planetary gearset.

- Have a basic understanding of the systems within an automatic transmission and how they relate to each other.

1.1 TRANSMISSIONS: THEIR PURPOSE

Every driver is familiar with the gear shift lever that is moved to control a car's motion. We know that this lever determines gear selection in the transmission, and this determines the driving mode of the car. The transmission provides the various gear ratios for forward and reverse operation.

At one time, most cars mounted the transmission behind the engine and used a drive shaft to transfer power to the rear axle and driving wheels. This is called *rear-wheel drive* (*RWD*) (Fig. 1-1). Most modern cars drive the front wheels (*FWD*). In most cases, the engine is mounted in a *transverse* position, crosswise with the car, with the *transaxle*, a combination of the transmission and final drive axle, attached to it. Two short drive shafts are used to connect the transaxle to each front wheel (Fig. 1-2).

A car cannot operate very well without a transmission, which serves several purposes:

P Park gear in an automatic transmission locks the drive wheels to hold the car stationary.

R Reverse gears allow the car to go backward.

N Neutral allows us to run the engine without moving the car.

D High gears allow the car to go faster while the engine runs slower.

I Intermediate gear prevents high-gear operation.

L Low gears multiply the engine's torque so there will be enough power to move the car.

Before the 1940s every car used a clutch and standard, manual transmission. The first automatic transmission was the Hydra-Matic Drive introduced on the 1940 Oldsmobile; it was an option costing $57. By the 1960s, most of the domestic (produced in America) cars were equipped with automatic transmissions.

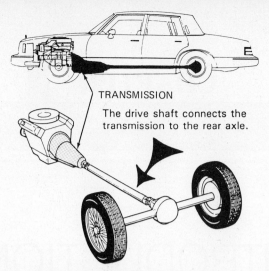

Fig. 1-1 The drive train of a RWD car consists of a clutch and transmission mounted to the rear of the engine, a drive shaft, and a rear-axle assembly. (*Copyrighted material reprinted with permission from Hydra-matic Div., GM Corp.*)

Both types of transmissions serve the same general purpose in providing the needed gear ratios. The difference is that the driver selects the gear and performs each shift while depressing the clutch pedal in a standard transmission. An automatic transmission does not require a manual clutch and can make automatic upshifts and downshifts. Each of these transmission types offer several advantages over the other. Several manufacturers have offered semiautomatic transmissions over the years. Chrysler Corporation's Fluid Drive and Volkswagen's Automatic Stick Shift are examples of semiautomatic transmissions that tried to combine the best features of standard and automatic transmissions. The driver had to perform each of the upshifts and downshifts but did not need to operate a clutch.

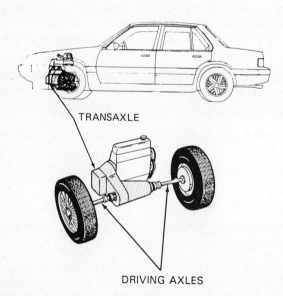

Fig. 1-2 The drive train of a FWD car consists of a transaxle assembly mounted to the engine and a pair of drive axles to the front wheels. (*Copyrighted material reprinted with permission from Hydra-matic Div., GM Corp.*)

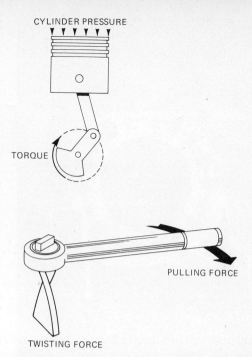

Fig. 1-3 Pressure in the cylinder forces the piston downward, which in turn forces the crankshaft to turn. This produces the rotating torque that turns the transmission gears, drive shaft, and drive wheels. When we pull on a wrench, we produce the torque needed to turn a nut or bolt.

1.2 TORQUE AND HORSEPOWER

Torque is a twisting force (Fig. 1-3). We exert torque on a nut or bolt as we tighten or loosen it. Torque must be exerted at the drive axle in order to turn the wheels and move the car (Fig. 1-4).

Torque is commonly measured in foot-pounds (ft-lb), inch-pounds (in.-lb), or Newton-meters (N-m). A foot-pound of torque is created when we exert a pound of force on a wrench that is one foot long. Force (in pounds) times the wrench length (in feet) equals the amount of torque in foot-pounds. One foot-pound of torque is equal to twelve inch-pounds, or 1.356 N-m.

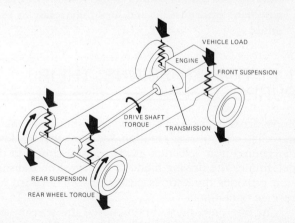

Fig. 1-4 Engineer's view of torque transmission through a drive train. The torque turning the drive wheels is what causes the car to move.

Torque is a form of mechanical energy and, like all other forms of energy, cannot be created or destroyed. It can and is transformed from one form of energy to another. A car's engine converts the potential heat energy of gasoline into torque that rotates the flywheel. The transmission receives the torque and increases or decreases it according to the gear selected. The torque at the drive wheels is transformed into the kinetic energy of the moving car.

Most of us are more familiar with the term *horsepower* than torque. Horsepower is a measurement of the amount of energy developed in the engine. It is a product of torque and engine speed. To determine horsepower, an engineer uses a dynamometer to measure the amount of torque that the engine can produce at various points through its operating range. Then the simple formula, torque times revolutions per minute (rpm) divided by 5252, is used to convert torque at a certain rpm into a horsepower reading. The various readings are then plotted into a curve, as shown in Fig. 1-5. A typical horsepower and torque curve shows us that an engine does not produce very much torque at low rpm, produces the most usable torque in the mid-rpm range, and has a reducing amount of torque with an increasing amount of horsepower at higher rpm.

The amount of torque from an engine can be increased or decreased through the use of gears, belts and pulleys, and chains and sprockets (Fig. 1-6). These are called *simple machines*. Gears are commonly used in transmissions. Transmissions of the future will probably make greater use of a belt with variable-size pulleys; this will be described later in this chapter. Simple machines cannot increase the amount of horsepower or energy. They can modify it by changing the amount of torque or speed. A driver feels the torque increase when he or she steps on the gas, forcing the car to move. A large amount of torque produces faster motion.

The amount of torque that a transmission can handle is referred to as *torque capacity*. Larger engines and heavier cars require larger transmissions with greater torque capacity than lighter cars with smaller engines. A transmission's torque capacity is determined by the diameter of the shafts, the size of the clutch packs and bands, and the size of the gears.

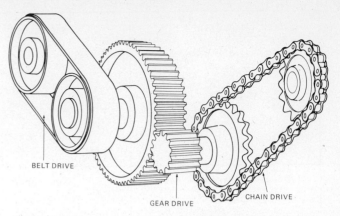

Fig. 1-6 The common methods of transmitting torque are belt, gear, and chain drives.

1.3 GEAR RATIOS

The term *gear ratio* refers to the relative size of two gears. The ratio can be determined using either the diameter or the number of teeth of the two gears. A pair of gears of different size will have different numbers of teeth on them, and the number of teeth is exactly relative to the diameter of the gears (Fig. 1-7). When the driving gear, where the power is put in, is smaller than the driven gear, the ratio between the two gears produces more torque but less speed at the output shaft; this is called *gear reduction*. When the driving gear is larger, an overdrive results. This increases the speed but reduces the torque.

If, for example, one gear has 15 teeth and the gear meshed with it has 30 teeth, the 30-tooth gear is exactly twice the diameter of the 15-tooth gear. If the 15-tooth gear is the input, driving gear, it has to rotate two revolutions for each revolution of the 30-tooth-driven gear.

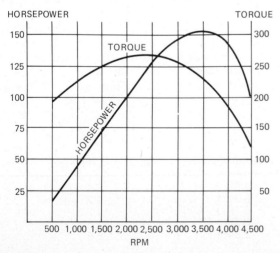

Fig. 1-5 Typical horsepower–torque chart for an automotive engine. Note that the maximum horsepower (155) occurs at about 3400 rpm and that maximum torque (262) occurs at between 2000 and 2500 rpm. This varies depending on the engine size and design.

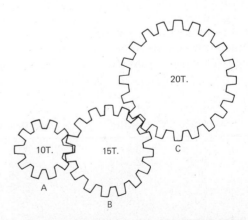

Fig. 1-7 In a matched set of gears, the number of teeth on each gear is related to the diameter. Gear C, which has twice the diameter as gear A, has twice the number of teeth. Gear B, with one and a half times the diameter of gear A, has one and a half times as many teeth.

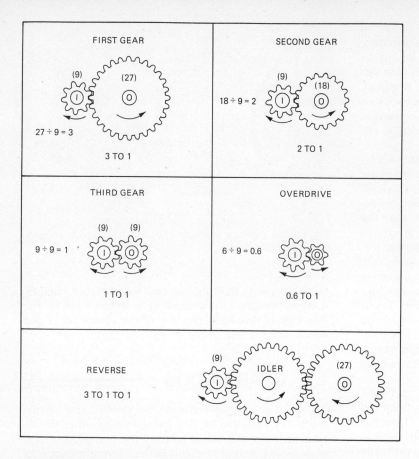

Fig. 1-8 The gear ratio is determined by dividing the number of teeth on the driven (output) gear by the number of teeth on the driving (input) gear.

This gear ratio is 2:1, and the output shaft has twice the torque and half the speed as the input shaft (Fig. 1-8).

When figuring gear ratios on simple gearsets, we always divide the number of teeth on the driven gear by the number of teeth on the driving gear. If the driving gear has 12 teeth and the driven gear 36 teeth, the ratio will be $\frac{36}{12}$, or 3:1—a reduction—and the input shaft will turn three revolutions for each output shaft revolution. If we turn that gearset around, the ratio would be $\frac{12}{36}$ or 0.3:1—an overdrive. In this case, the input shaft would turn 0.3 turns for each revolution of the output shaft.

When power travels through more than one gearset, the overall ratio is determined by multiplying one gear ratio by the next. An example of this is a modern RWD car with a five-speed transmission (shown in Fig. 1-9). A transmission first-gear ratio of 3.97:1 and a rear-axle ratio of 3.55:1 produces an overall ratio of 3.97 × 3.55, or 14.09:1, in first gear. The engine will revolve 14.09 turns for each revolution of the drive axle. An overdrive ratio in fifth gear in this transmission of 0.8:1 times the 3.55 rear-axle ratio produces an overall ratio of 2.84:1. In this gear, the engine will revolve 2.84 turns for each drive axle revolution. This car will have about one-fifth the torque it had in first gear but will be able to go about five times faster in fifth gear for each engine revolution.

Gear	Ratio	Final Drive Ratio	Overall Ratio	mph per 1000 rpm
I	3.97:1	3.55:1	14.09:1	5.4
II	2.34:1	3.55:1	8.31:1	9.2
III	1.46:1	3.55:1	5.18:1	14.7
IV	1:1	3.55:1	3.55:1	21.5
V	0.80:1	3.55:1	2.84:1	26.9

Fig. 1-9 The overall gear ratio can be determined by multiplying the transmission gear ratio by the final drive ratio. If a car has a final drive ratio of 3.55:1 and a first-gear ratio of 3.97:1, the overall ratio will be 14.09:1. All forward-gear ratios for a late-model car with a five-speed transmission are shown here. The mph per 1000 rpm indicates the relative speed for the different gears.

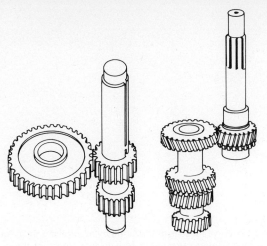

Fig. 1-10 The teeth of a spur gear (left) are cut so they are parallel to the shaft while the teeth of a helical gear (right) are cut at an angle to the shaft.

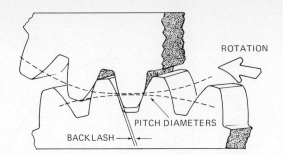

Fig. 1-11 The clearance on the nonloaded side of the gear teeth is called backlash.

Most gears have external teeth with the teeth cut around the outside of the gear. Some gears (e.g., the ring gear in a planetary set) have the teeth cut on the inside of the gear's rim. Most of the gears we will be studying are helical gears as compared to spur gears. The teeth of a spur gear are cut in a straight line, parallel to the axis of the gear and the shaft on which the gear is mounted. The teeth of a helical gear are cut on a spiral helix shape; they would spiral continuously if they were extended (Fig. 1-10). The other major gear type—a bevel gear—is used to transfer power between shafts that are not parallel, and they are made in both a spur bevel and spiral bevel form. A spur gear is less expensive to make, but it is noisier and not as strong as a helical gear. A helical gear tends to be quieter and stronger, but it has a tendency to slide sideways, out of mesh.

Some other important rules to learn about gearsets are:

- The driving gear will always rotate in the opposite direction as the driven gear unless one of them is an internal gear.
- Two gears transferring power push away from each other in an action called gear separation. The amount of gear separation force is proportional to the amount of torque being transferred.
- All gearsets have a certain amount of *backlash* to prevent binding (Fig. 1-11).
- The smaller gear(s) in a gearset is often called a pinion.

1.4 TRACTIVE FORCE

An engineer uses the term *tractive force* to describe the power in an automobile's drive train. It is a product of the engine's torque multiplied by the gear ratio and is plotted in the form shown in Fig. 1-12. Note that a curve is shown for each of the transmission's gears and that first gear produces the most torque and fifth gear the most speed, as we would expect. This curve (b) does not show engine rpm, but in each case in this particular curve, the left end is plotted for 500 rpm and the right end for 4500 rpm.

When discussing vehicle speed, it should be remembered that the tire diameter affects speed. The larger the tire, the faster the car will go at a particular axle rpm and vice versa (Fig. 1-13). You can use the following formula to determine the car speed at a particular engine rpm or vice versa:

$$\text{Engine rpm} = \frac{\text{mph} \times \text{gear ratio} \times 336}{\text{tire diameter}}$$

$$\text{mph} = \frac{\text{rpm} \times \text{tire diameter}}{\text{gear ratio} \times 336}$$

The amount of horsepower required to reach a certain speed is influenced by the weight and shape of the car and on how level the road is. Vehicle motion is resisted by two types of friction: rolling and aerodynamic. Rolling friction is produced by the tires, axles, and other rotating parts of the power train. A heavier car increases the rolling resistance because of the load at the tires and axles. This resistance increases at a constant rate as the car speeds up, and if we double the speed, it requires twice as much power to overcome rolling resistance. Aerodynamic friction or drag is created as we force the body of the car through air. A large, boxy car is much harder to push through air than a small, streamlined car. Aerodynamic drag increases at the square of the car's speed; if we go twice as fast, the aerodynamic drag increases about four times (Fig. 1-14). Combined, these two curves are referred to as *tractive resistance*.

When we place the curve for tractive force over the curve for tractive resistance, we can get an idea of the car's performance (Fig. 1-15). The top speed is where the two curves cross at the right side. We can also see that the lower gears produce much more force than needed to overcome the resistance. This excess force is used to accelerate the car; the greater the amount of excess, the greater the rate of acceleration.

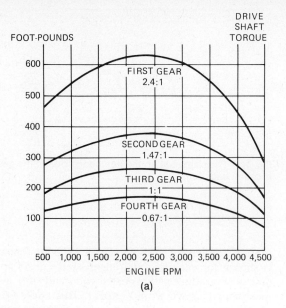

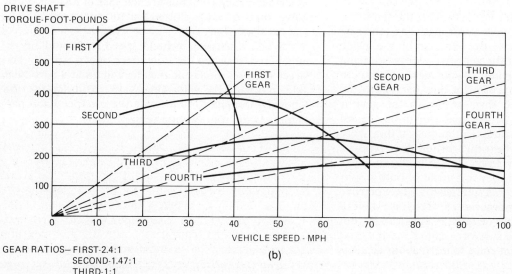

GEAR RATIOS— FIRST-2.4:1
　　　　　　　SECOND-1.47:1
　　　　　　　THIRD-1:1
　　　　　　　FOURTH-0.67:1
FINAL DRIVE RATIO-3.27:1
TIRE SIZE-P215165SR-15

Fig. 1-12 Tractive force is determined by multiplying the engine's torque by the gear ratio, as shown in (a); note that this is the greatest in first gear. This plot is for a typical V8 engine of about 300 in.^{3} (5 L). (b) The same curves as in (a) in relationship with the vehicle speed.

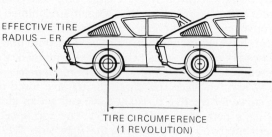

$2 \times ER \times \pi$ = CIRCUMFERENCE
ER = EFFECTIVE RADIUS

Fig. 1-13 The speed of a vehicle is determined by the diameter of the drive tire and how fast we turn that tire.

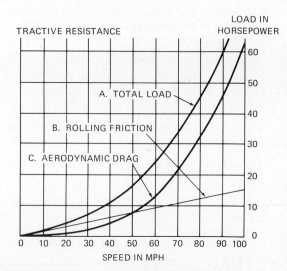

Fig. 1-14 (a) The load on a car increases as it goes faster. It is a combination of (b) the rolling friction, which increases at a constant rate, and (c) aerodynamic drag, which increases at a much more rapid rate. Plot shows the loads on a late-model, domestic coupe with good aerodynamic design.

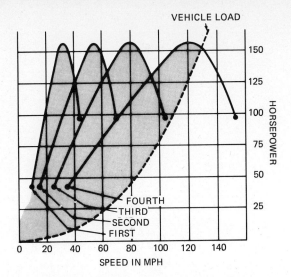

Fig. 1-15 Chart showing the vehicle's road load and the horsepower available. The point where the two intersect, about 128 mph, is the theoretical top speed. The shaded area to the left shows the excess horsepower, which is used for acceleration.

Excess force is also convenient when going up hills because the slope of the road increases the amount of tractive resistance.

1.5 STANDARD TRANSMISSION AND CLUTCH

A standard transmission provides several sets of gears to produce the necessary gear ratios (Fig. 1-16). When a driver selects a particular gear, the power must flow through a particular path as it goes from the input shaft to the output shaft—a 1:1 ratio. A gear reduction or overdrive is produced when the power flows from the main drive gear on the input shaft to the cluster gear and from the cluster gear to a driven gear on the output shaft (Fig. 1-17). The actual shift is made by moving a synchronizer sleeve to couple the desired driven gear to the synchronizer hub and the output shaft (Fig. 1-18).

In a standard transmission, this shift cannot be made while the power is flowing. The power flow must be interrupted or disengaged by depressing the clutch pedal (Fig. 1-19). This allows the gear to be engaged to slow down or speed up to the same speed as the synchronizer sleeve during the shift. This is sometimes called a nonpower shift. The clutch is also needed so the transmission input shaft can come to a stop so we can shift from neutral to low or reverse.

A FWD transaxle is slightly different from a RWD unit in that power is always transferred through three or more shafts that are parallel to each other (Fig. 1-20). The input comes from the clutch; the second shaft transfers power to the pinion gear of the final drive; and the third shaft is the ring gear of the final drive and the differential. The various transmission ratios are the different power paths between the first and second shafts. Like an RWD transmission, a synchronizer assembly is used to make the shifts.

1.6 PLANETARY GEARSETS

With only one exception (the HondaMatic), all automatic transmissions use *planetary gearsets* to provide the different gear ratios. A planetary gearset is a combination of a *sun gear*, two or more *planet gears*, a *planet carrier*, and a *ring gear*. The ring gear, also called an *annulus gear*, is an internal gear; all the other gears are common external gears. The carrier holds the planet gears (also called *pinions*) in position and allows each of these gears to rotate in the carrier (Fig. 1-21). When the gearset is assembled, the sun gear is in the center and meshed with the planet gears, which are around it, somewhat like the planets in our solar system. The ring gear is meshed with the outside of the planet gears.

The three major members of the planetary gearset—the sun gear, ring gear, and planet carrier—have two possible actions: they can rotate or stand still. Depending on what the other members are doing, the planet gears have three possible actions: they can stand still on their shafts and rotate with the carrier; they can rotate on their shafts in a stationary carrier and act like idler gears; or they can rotate on their shafts in a rotating carrier. This last condition is called "walking" (Fig. 1-22). The planet gears walk around the outside of a stationary sun or the inside of a ring gear. The rotating carrier forcing the planet gears around the stationary gear forces the planet gears to rotate on their shafts. This in turn forces the output shaft to rotate.

Planetary gearsets are arranged so power enters one of the members (the sun gear, ring gear, or carrier), leaves through one of the other members, and has the third member held stationary in reaction. One of the basic physical laws states that for every action, there is an equal and opposite reaction. In a gearset, the action is usually an increase in torque; the reaction is an equal amount of torque that tries to turn the gear box in a reverse direction. For example, watch a short-wheelbase truck-tractor pull a heavy trailer from a dead stop. The action needed is to turn the drive wheels and move the truck; the reaction is a lifting of the frame at the left front wheel. The front of the truck lifts in reaction to the drive axle torque, and the left side lifts in reaction to the drive shaft torque. Getting back to the planetary gearset, one of the three members has to be held stationary in reaction to obtain the torque or speed increase in the gear box. There cannot be a ratio change without a reaction member.

Planetary gearsets offer several advantages over conventional gearsets. Because of the multiple planet gears, there is more than one gear transferring power; the torque load is spread over several gear teeth. Because of this, planetary gearsets are quieter and stronger. Also any gear separation forces (as gears transfer power, they tend to push away from each other) are contained within the ring gear, and this load is not transmitted to the case. This allows the transmission case to be thinner and therefore lighter. Another advantage is the small relative size. Conventional gears are normally side by side, and for a 2:1 gear ratio, one gear has to be twice the size of the other. A planetary gearset can easily produce this same ratio in a smaller package. Also, planetary gearsets are in constant mesh; no coupling or uncoupling of mechanical gears is required.

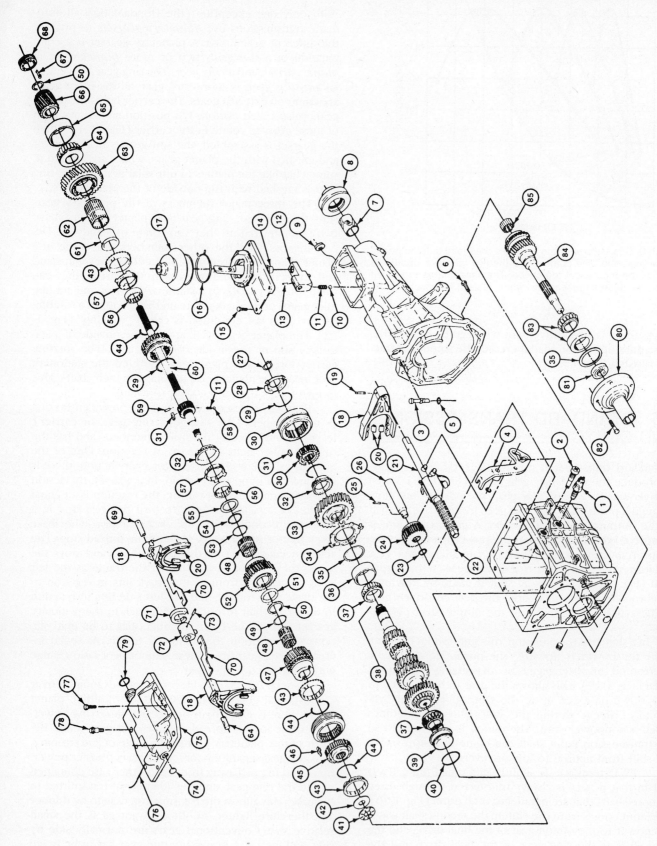

Fig. 1-16 (a) Standard transmission and (b) clutch. Note that a five-speed RWD transmission is shown in an exploded view. (*Courtesy of Ford Motor Company.*)

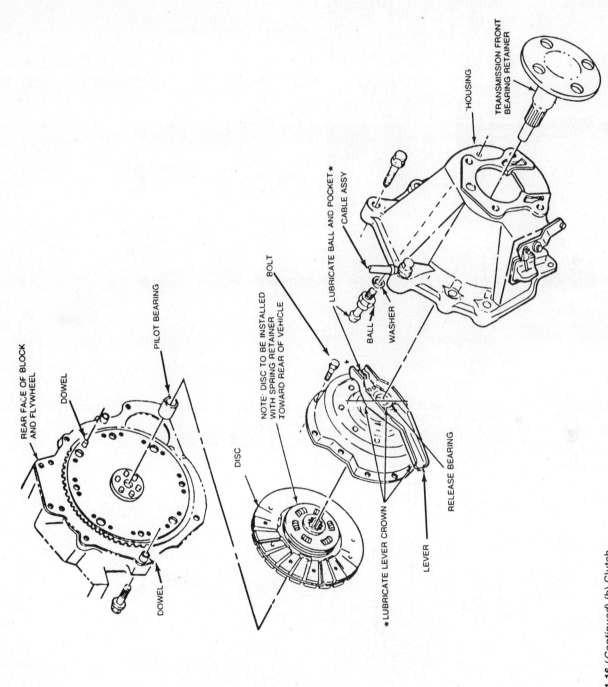

Fig. 1-16 (*Continued*) (b) Clutch.

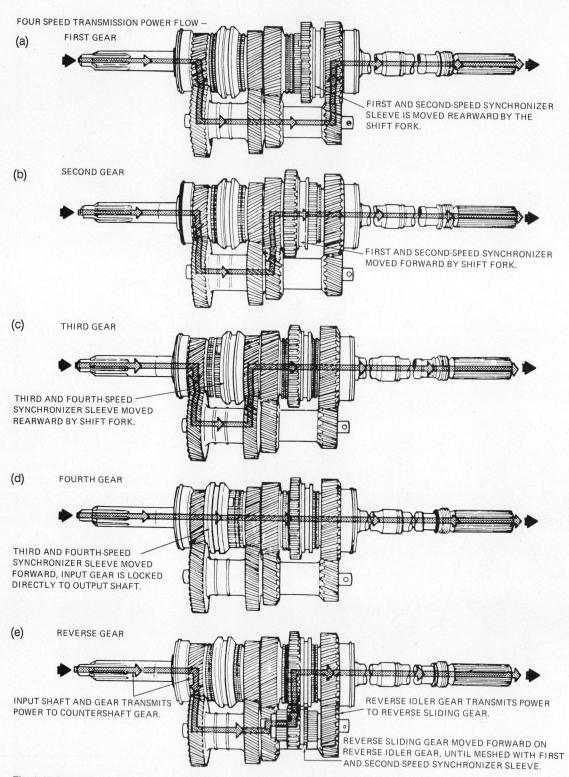

Fig. 1-17 Power flows for first gear of a four-speed standard transmission. The synchronizer sleeves are used to shift the power flow. (*Courtesy of Ford Motor Company.*)

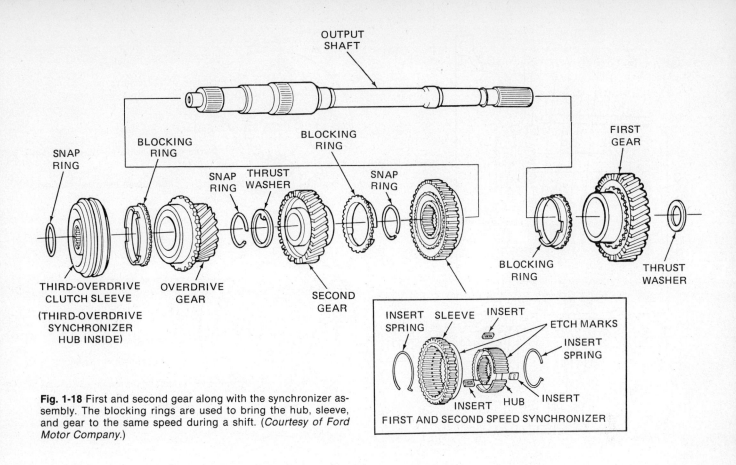

Fig. 1-18 First and second gear along with the synchronizer assembly. The blocking rings are used to bring the hub, sleeve, and gear to the same speed during a shift. (*Courtesy of Ford Motor Company.*)

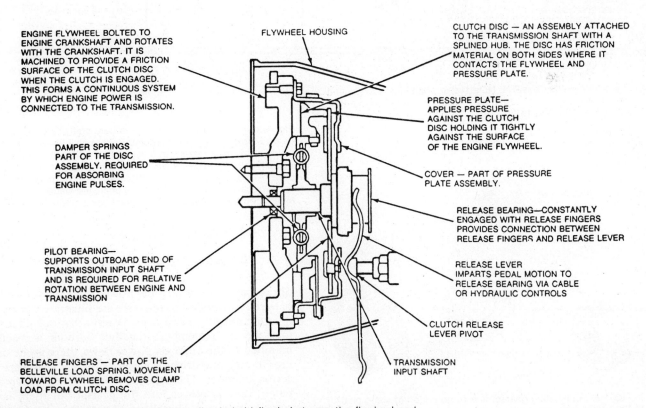

Fig. 1-19 When a clutch is engaged, the disc is held firmly between the flywheel and pressure plate. The pressure plate is moved to the right to release the clutch. (*Courtesy of Ford Motor Company.*)

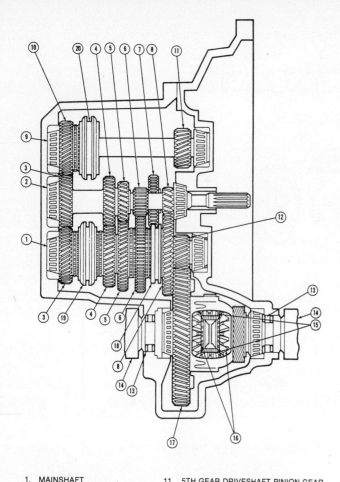

1. MAINSHAFT
2. INPUT CLUSTER GEAR SHAFT
3. 4TH SPEED GEARS
4. 3RD SPEED GEARS
5. 2ND SPEED GEARS
6. REVERSE GEARS
7. REVERSE IDLER GEAR
8. 1ST SPEED GEARS
9. 5TH SPEED GEAR DRIVESHAFT
10. 5TH SPEED GEAR
11. 5TH GEAR DRIVESHAFT PINION GEAR
12. MAINSHAFT PINION GEAR
13. DIFFERENTIAL OIL SEALS
14. CV SHAFTS
15. DIFFERENTIAL PINION GEARS
16. DIFFERENTIAL SIDE GEARS
17. FINAL DRIVE RING GEAR
18. 1ST/2ND SYNCHRONIZER
19. 3RD/4TH SYNCHRONIZER
20. 5TH SYNCHRONIZER

Fig. 1-20 A FWD transaxle. Note that it combines a five-speed transmission with final drive gears (11, 12, and 17) and differential gears (15 and 16). (*Courtesy of Ford Motor Company.*)

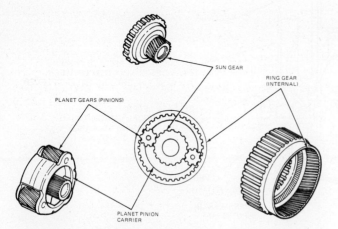

Fig. 1-21 A planetary gearset is a combination of a sun gear in the center, a group of planet gears around it, and a ring gear on the outside. Note that the planet gears are mounted in a carrier. (*Courtesy of Ford Motor Company.*)

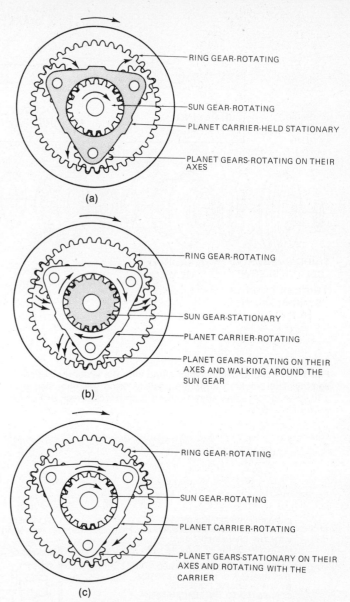

Fig. 1-22 (a) If the planet carrier is held stationary, the planet gears simply rotate in the carrier and act as an idler gear between the sun and ring gears. (b) If the sun or ring gear is held stationary, the planet gears walk around that gear; they rotate on their own shafts as the carrier rotates. (c) If no parts are held and two parts are driven, the planet gears are stationary on their shafts and the whole assembly rotates as a single unit.

1.7 PLANETARY GEAR RATIOS

There can be eight possible power flow conditions in a particular planetary gearset; six are illustrated in Fig. 1-23. One not illustrated is a 1:1 ratio that occurs if you lock any two members together without a reaction member. In the other six power flows, there is one driving (input), one driven (output), and one stationary reaction member. The other condition not illustrated is when there is power into the gearset but there is no reaction member. This allows the members of the gearset to rotate freely, and this effectively stops the power flow and provides a neutral condition.

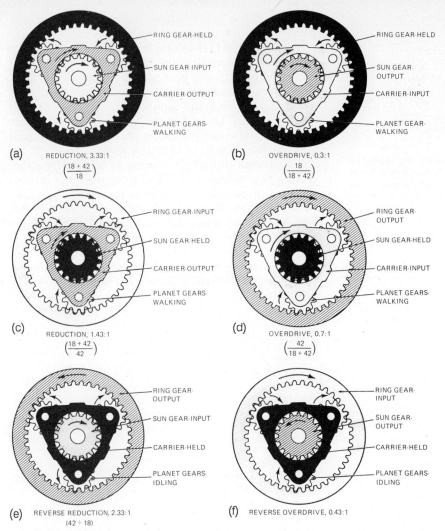

Fig. 1-23 Six possible gear ratios that can be developed in a simple planetary gearset. In each case the input member is shaded, the stationary reaction member is solid black, and the output member is cross hatched. The gear ratios were calculated using a sun gear with 18 teeth and a ring gear with 42 teeth.

Figure 1-23(a) illustrates a reduction where the ring gear is held, the sun gear drives, and the carrier is driven. In this case, the sun gear forces the planet gears to rotate. As they rotate, they must walk around the inside of the stationary ring gear and force the carrier to rotate in the same direction but slower than the sun gear. The gear ratio can be determined by adding the number of teeth on the sun gear and the number of teeth on the ring gear and dividing this by the number of teeth on the sun gear.

Figure 1-23(b) illustrates the reciprocal, or inverse (opposite), of Fig. 1-23(a); what was a reduction in one direction is an overdrive in the opposite direction. Driving the carrier forces the planet gears to walk around the inside of the stationary ring gear, and the rotation of the planet gears forces the sun gear to rotate in the same direction but faster than the carrier. The gear ratio can be determined by adding the number of sun gear teeth and ring gear teeth and dividing this into the number of teeth on the sun gear.

Figure 1-23(c) is commonly used as a second or intermediate gear in many three-speed automatic transmissions. The sun gear is held, and power enters the ring gear. This forces the planet gears to walk around the sun gear, which forces the carrier to rotate at a reduced speed in the same direction. The gear ratio can be determined by adding the number of sun and ring gear teeth and dividing this by the number of ring gear teeth.

Figure 1-23(d) is used for fourth gear in some overdrive transmissions and is the reciprocal of Fig. 1-23(c). In this case the rotating carrier forces the planet gears to walk around the stationary sun gear. The rotating planet gears force the ring gear to turn in the same direction but faster than the carrier. The gear ratio can be determined by adding the number of sun and ring gear teeth and dividing this into the number of ring gear teeth.

Figure 1-23(e) is a reverse gear that is used in many three-speed transmissions. The carrier is held so the planet gears act as idler gears and rotate in a direction that is opposite to the sun gear. Power enters the sun gear and forces the planet gears to rotate in the opposite direction. This in turn forces the ring gear to rotate at a slower speed but in the opposite direction as the sun gear. The gear ratio can be determined by dividing the

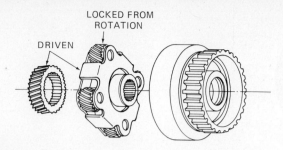

Fig. 1-24 If any two parts of a planetary gearset are driven or locked together, the gearset operates at a 1:1 ratio. In this case, if the carrier and the sun gear are both driven, the planet gears are locked, which forces the ring gear to rotate at the same speed.

number of ring gear teeth by the number of sun gear teeth.

Figure 1-23(f) is an overdrive reverse gear used in some transmissions; it is the reciprocal of Fig. 1-23(e). Power enters the ring gear and forces the planet gears to rotate in the same direction; this in turn forces the sun gear to rotate in the opposite direction as the planet gears at a speed faster than the ring gear. The gear ratio can be determined by dividing the number of sun gear teeth by the number of ring gear teeth.

A straight-through, 1:1 ratio is obtained by either driving two members of a gearset at the same time or locking any two of the members together (Fig. 1-24). At this time, the unit operates as if it were a solid mass, and there is no action within the gearset. The planet gears do not rotate on their shafts.

A technician almost never computes the gear ratios in a planetary gearset, but for those interested in how it is done, the formula is shown in Fig. 1-25.

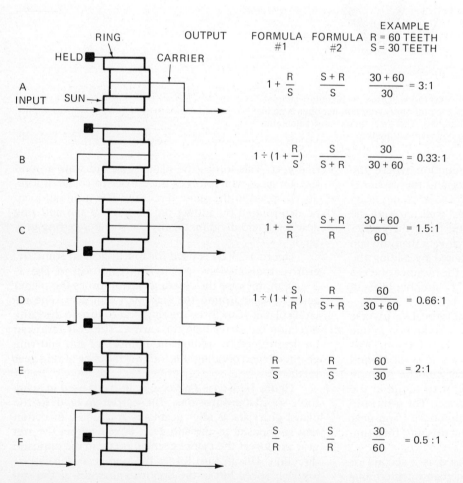

Fig. 1-25 Formula for computing gear ratios for planetary gearset.

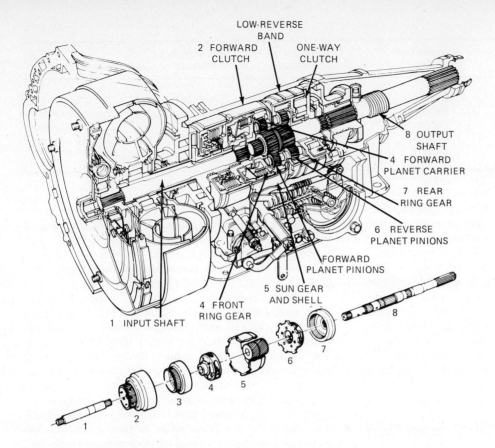

Fig. 1-26 Simpson gear train. Note that the planetary gearset uses a double sun gear with two carriers and ring gears. Also note the two clutches (which can drive either the forward ring gear or the sun gear), the two bands (which can hold either the sun gear or the rear carrier), and the one-way clutch (which can hold the rear carrier). (*Courtesy of Ford Motor Company.*)

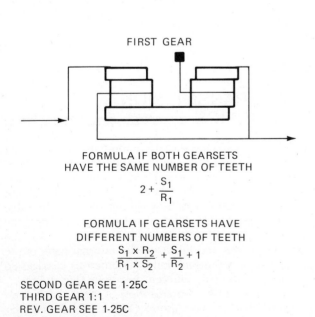

FORMULA IF BOTH GEARSETS
HAVE THE SAME NUMBER OF TEETH

$$2 + \frac{S_1}{R_1}$$

FORMULA IF GEARSETS HAVE
DIFFERENT NUMBERS OF TEETH

$$\frac{S_1 \times R_2}{R_1 \times S_2} + \frac{S_1}{R_2} + 1$$

SECOND GEAR SEE 1-25C
THIRD GEAR 1:1
REV. GEAR SEE 1-25C

Fig. 1-27 Formulas for computing gear ratios for Simpson planetary gearset.

1.8 COMPOUND PLANETARY GEARSETS

Because of the need to be able to shift to different gear ratios, an automatic transmission seldom uses a simple planetary gearset. Most of them use a more complicated *compound gearset* that combines a simple gearset with portions of one or more gearsets in such a way that three or more gear ratios are provided. Occasionally, one simple gearset is used along with a compound set to provide a single additional gear such as overdrive in some modern automatic transmissions.

A common compound gearset used in three-speed transmissions is known as the *Simpson gear train*, after its designer, Howard Simpson. This gearset uses one double sun gear with two ring gears and planet carriers (Fig. 1-26). One ring gear along with the other carrier are attached to the output shaft so they will be the driven or output members. Three inputs are possible: either (1) the other ring gear, (2) the sun gear or (3) both. Also, two reaction members are possible: either the sun gear or the remaining carrier. In newer transmissions, a stepped sun gear of two sizes is used. This allows lower first- and second-gear ratios. Because of the lower ratios, the unit is often called a wide-ratio transmission. If you are interested in how gear ratios are computed for a Simpson planetary gearset, the formulas are given in Fig. 1-27.

TABLE 1.1 Gear Ratios

Gear	Input Ring Gear	Input Sun Gear	Reaction Sun Gear	Reaction Carrier	Action*
Neutral	Released	Released	Released	Released	No power flow
First	Driven	Released	Released	Held	2.5:1 reduction
Second	Driven	Released	Held	Released	1.5:1 reduction
Third	Driven	Driven	Released	Released	1:1 no change
Reverse	Released	Driven	Released	Held	2:1 reversal

* Ratios are approximate.

We will describe the power flow through this gearset as well as other common gearsets in more detail in Chapter 3. For now, in order to get an idea of how an automatic transmission works, let's take a brief look at how shifts are accomplished. With the Simpson gear train there are two possible inputs, two possible reaction members, and an output that can either be the ring gear of one gearset or the carrier of the other. Remember that the front carrier and the rear ring gear will be the output members. By combining the application of the two inputs and two reaction members, there will be a neutral and four different gear ratios as shown in Table 1.1.

1.9 CONTROL DEVICES

There has to be a method of controlling the power flow through the gearset, and this is accomplished by *control devices*—the clutches and bands—also called *apply devices* or *friction members*. A gearset is made with several paths for power to flow through it, and each path is a different gear ratio. The control devices determine which particular path the power must follow at a given time.

There also has to be a way of controlling the control devices, and this is the hydraulic system. By the time we combine the needed systems, the transmission appears rather complex because of all the necessary parts (Fig. 1-28). Hopefully, after we study each of these systems, much of the complexity and mystery will disappear.

When a shift is made in an automatic transmission, some device has to apply force to provide power input or a reaction member. A control device is either a driving or a reaction member. A driving member completes a path so power can transfer from one unit to another. A reaction member stops a portion of the gearset from rotating so the power flow can occur.

Driving devices are nearly always clutches that can connect the transmission's input shaft to the gearset input member. In a Simpson gear train one clutch is used to connect the input shaft to the sun gear and another clutch connects the input shaft to the input ring gear. Multiple-disc clutches applied by hydraulic force are normally used. This type of clutch is made up of several plates that are lined with friction material and several unlined, plain steel plates. The lined plates are connected to one portion, possibly the input shaft, and the unlined plates are connected to the other side, possibly the gearset (Fig. 1-29). These plates are stacked in an alternating fashion. When oil pressure enters behind a hydraulic piston in the clutch assembly, the clutch pack is squeezed together, applying the clutch (Fig. 1-30). When the oil pressure is released, a return spring pushes the piston away from the clutch pack, releasing the clutch.

A reaction member can be a band, a one-way clutch, or a multiple-disc clutch. Some import transmission manufacturers call their reaction members brakes because this is what they act like. A band or a multiple-disc clutch is normally applied by hydraulic pressure and released by a spring. A one-way clutch is a device that allows a rotation in one direction but not the other (Fig. 1-31). One-way clutches can be of the roller or sprag type. The Simpson gear train uses a one-way clutch that allows the reaction carrier to turn in a clockwise direction but blocks it from turning counterclockwise.

A band is lined with friction material and wraps around a drum (Fig. 1-32). It is tightened so that it squeezes the drum to stop it when applied and releases away from the drum to allow rotation. The hydraulic unit that applies a band is commonly called a *servo*. A reaction clutch is built in the same fashion as a driving clutch except one of the sets of plates is connected to the transmission case. When it applies, the gearset member that it connects to will come to a stop.

Because friction-type clutches and bands are used, the power flow does not have to be interrupted during a shift. This allows power shifts to be made, and the drive wheels will have a constant flow of power during a shift.

1.10 TRANSMISSION HYDRAULICS

As soon as the engine starts, a pump—usually driven by the torque converter—sends fluid into the transmission's hydraulic passages. Fluid is pumped from the sump in the transmission pan into the hydraulic circuit where the pressure and flow is controlled by the valve body (Fig. 1-33). This fluid pressure is used to apply the clutches and bands, fill the torque converter so it can transmit power, and lubricate the internal parts of the transmission. A transmission has numerous passages so oil can flow to many different locations (Fig. 1-34).

The valve body contains most of the transmission's control valves in one group (Fig. 1-35). Sometimes, there are one or more valves elsewhere, outside the valve body. The major valves used in a transmission are as follows:

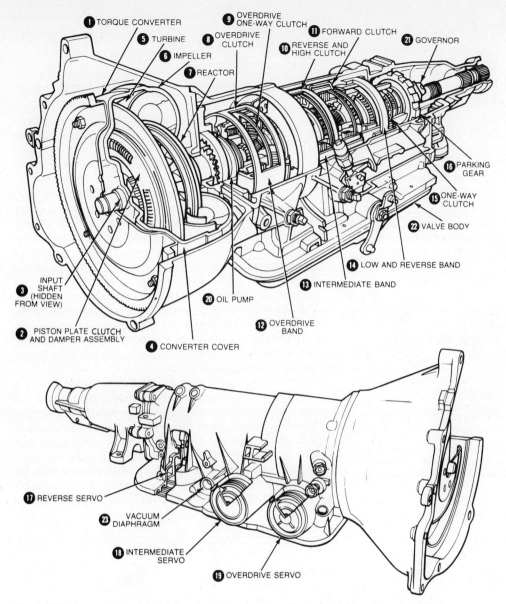

Fig. 1-28 Cutaway view of a modern four-speed automatic transmission. You should be able to identify all of these parts, know what they do, and be able to service them at the completion of this book.

Manual valve Operated by the shift linkage; it selects automatic forward motion, nonautomatic forward motion (manual first, second, etc.), reverse, as well as park and neutral.

Pressure regulator Controls fluid pressure.

Throttle or modulator valve Tailors the pressure regulator to the operation of the transmission and delays upshifts as the throttle is opened.

Governor Operated by the output shaft speed; causes upshifts.

Shift valves One for each upshift; control upshifts and downshifts.

Shift modifier valves Various valves that change the quality of an upshift or downshift to ensure smooth operation.

Torque converter clutch valve Controls the apply and release of the torque converter clutch.

In reality, the hydraulic system is rather complex. We will study it more thoroughly in Chapters 4 and 5. A competent service technician has a good understanding of what the various valves do and what fluid pressures should be in a particular place at a particular time. He or she will often refer to a hydraulic diagram when

17

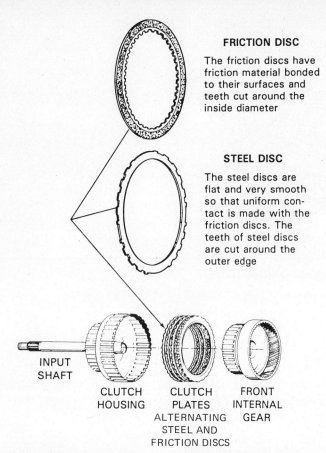

Fig. 1-29 A multiple-disc clutch combines lined friction and unlined steel discs in a clutch assembly. (*Copyrighted material reprinted with permission from Hydra-matic Div., GM Corp.*)

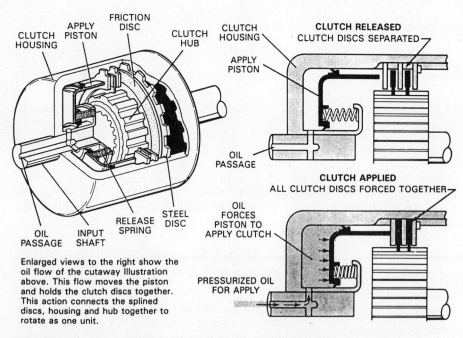

Fig. 1-30 Oil pressure is used to push a piston to force the clutch discs together and apply a clutch. To release a clutch, the oil pressure is released and springs push the piston away from the clutch stack. (*Copyrighted material reprinted with permission from Hydra-matic Div., GM Corp.*)

trying to solve a particular problem. Many automatic transmission problems are caused by a loss of fluid pressure or sticky control valves.

1.11 TORQUE CONVERTERS

Every automatic transmission uses a torque converter to transfer power from the engine to the transmission (Fig. 1-36). It serves as an automatic fluid clutch as well as converts torque (as its name implies). It can multiply torque much like a gearset.

A torque converter is normally made up of three members: a pump or impeller, a turbine, and a stator, and is filled with transmission fluid (Fig. 1-37). The impeller is built into the housing that connects to the engine; when the engine runs, it produces fluid flow inside the converter housing. The turbine is splined onto the transmission input shaft. Oil flow from the impeller causes the turbine to rotate and drive the transmission. The stator is splined onto a shaft extending from the transmission case, and a one-way clutch is used to connect the stator to this stator support. Most modern torque converters also have an internal clutch so torque converter lockup can occur.

At slow engine speeds, fluid flow inside the converter is not strong enough to rotate the turbine and the transmission. As the throttle is opened and engine speed increases, fluid flow and pressure become high enough to rotate the turbine and transmission input shaft moving the car. During slow turbine speeds, the stator redirects and increases the flow effectiveness against the turbine. This has the effect of increasing the torque, producing a ratio of about 2:1. This and other phases of

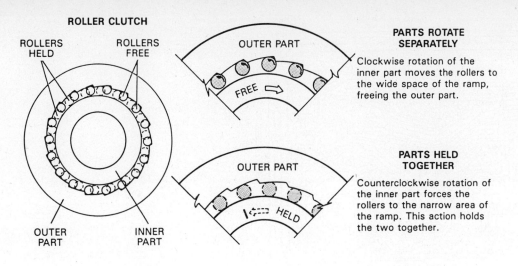
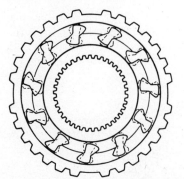

A sprag clutch works under the same principle as the roller clutch. One direction of the inner part acts to hold the sprags "out of the way", the opposite rotation acts to engage the sprags holding the inner and outer parts together.

Fig. 1-31 A one-way clutch, either a roller clutch or a sprag clutch, will let a part (the inner races as shown here) rotate freely in one direction but not the other. (*Copyrighted material reprinted with permission from Hydra-matic Div., GM Corp.*)

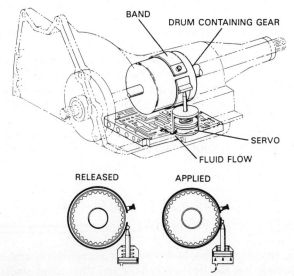

Fig. 1-32 A band is applied when oil pressure pushes a servo piston to squeeze it tightly against a drum. This holds the drum from turning. When the oil pressure is released, the spring pushes the piston back. (*Copyrighted material reprinted with permission from Hydra-matic Div., GM Corp.*)

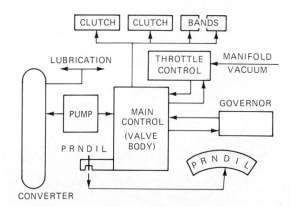

Fig. 1-33 Simplified version of a transmission's hydraulic control circuit. The main control unit (valve body) sends oil pressure to the clutches and bands to control the shifts. It uses signals from the manual shift lever, governor, and throttle valve.

19

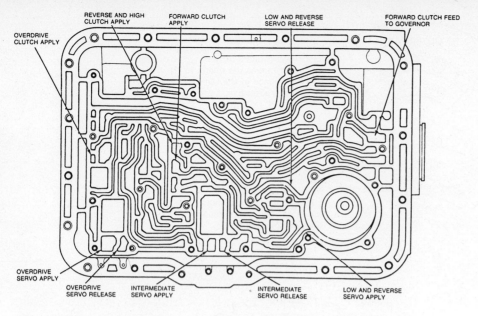

Fig. 1-34 View of the bottom of a transmission case showing most of the fluid passages. These connect with holes drilled through the case to the various parts. (*Courtesy of Ford Motor Company.*)

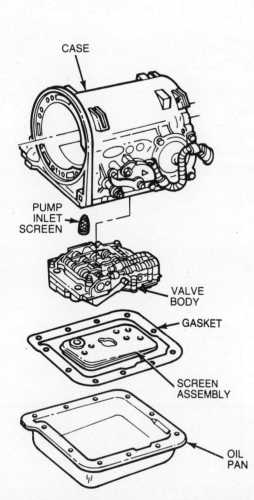

Fig. 1-35 The valve body is usually attached to the bottom of the case, inside the oil pan. (*Courtesy of Ford Motor Company.*)

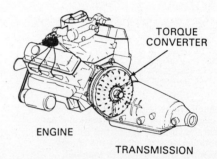

Fig. 1-36 The torque converter transmits power from the engine to the transmission. It acts as an automatic clutch as well as multiplying torque during periods of load. (*Copyrighted material reprinted with permission from Hydra-matic Div., GM Corp.*)

torque converter operation will be described in more detail in Chapter 7.

A plain torque converter is a fluid coupling; all of the power is transferred by fluid. Because of this, it always slips, and this slippage reduces the efficiency and fuel mileage. Modern torque converters contain a mechanical clutch that is applied during cruising conditions to eliminate this slipping (Fig. 1-38). A torque converter clutch improves fuel mileage about 1 to 3 miles per gallon.

1.12 TRANSAXLE FINAL DRIVE AND DIFFERENTIAL

The major differences between a transmission and a transaxle are the final drive unit and the output shafts. A transmission has a single output shaft that connects to the drive shaft and rear-axle assembly. A transaxle has two output shafts, one for each front wheel (Fig. 1-39). Also, the final drive gear reduction and the differential

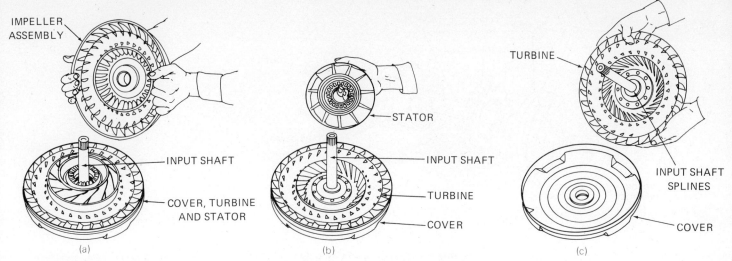

Fig. 1-37 If a converter is disassembled starting from the back, we see (a) the impeller, (b) the stator, (c) the turbine, and the front cover. (*Courtesy of Ford Motor Company.*)

must be built into the area between the transmission's output and the two drive shafts.

The final drive is the gear reduction that determines the overall tractive force of the car as well as the engine operating rpm at cruising speed. In an RWD car, the final drive gears are the ring and drive pinion in the rear axle. Some transaxles use a drive pinion gear mounted on either the transmission mainshaft or a transfer shaft. The ring gear is mounted on the differential case. Helical gears are normally used because the transmission shaft is parallel to the differential (Fig. 1-40). The smaller drive pinion gear driving the larger ring gear provides the amount of desired gear reduction. Some transaxles use a planetary gear final drive. The output shaft of the transmission drives a sun gear that is the input to the final drive; the ring gear is secured into the transmission case so it is the reaction member; and the planet carrier is made into the differential case. This produces the gear reduction described in Fig. 1-23(a) (Fig. 1-41).

A differential is a device that allows two shafts to operate at different speeds. When a car turns a corner, the outer tires must travel farther and therefore must go faster than the inner tires. Most differentials operate through a set of bevel gears, as shown in Fig. 1-42. The differential case is made with a differential pinion shaft running across it; this shaft also passes through a pair of differential pinion gears with the gears free to turn on the shaft. A pair of differential side gears are meshed with both differential pinions. One of these side gears is attached to each drive wheel through a drive shaft or axle.

When the car is going straight ahead, the equal load to drive each side gear and wheel prevents rotation of the pinion gears on their shaft. The entire differential assembly rotates as a unit, and both wheels revolve at the same speed (Fig. 1-43). When the car turns a corner, one side gear rotates slower than the differential case. When this occurs, the pinion gears rotate on their shaft, causing the other side gear to rotate faster than the case. Both drive wheels receive the same amount of torque, but they can operate at different speeds.

1.13 FOUR-WHEEL DRIVE

There are several versions of four-wheel drive (4WD):

- Rear-wheel drive with part-time 4WD.
- Front-wheel drive with transverse transmission and part-time 4WD.
- Front-wheel drive with lengthwise transmission and part-time 4WD.
- Full-time 4WD with lengthwise transmission.
- Full-time 4WD with transverse front transaxle (Fig. 1-44).

Full-time 4WD is also called all-wheel drive (AWD). The concept of driving all of the wheels of a car creates a few problems. One of the greatest is that on corners all of the wheels turn at different speeds (Fig. 1-45). This is because each of the wheels travels around a circle that has a different radius and circumference. Cars normally drive only two wheels, and the simple differential provides for the speed difference between the inner and outer wheels. Vehicles with AWD need three differentials: one between the drive wheels of each axle and another one between the two axles. The center, or interaxle, differential allows the front wheels to travel faster than the rear wheels on corners.

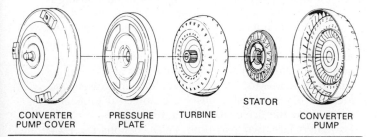

Fig. 1-38 Exploded view of a torque converter with a lockup clutch. During lockup, the pressure plate is pushed forward against the front cover. (*Copyrighted material reprinted with permission from Hydra-matic Div., GM Corp.*)

21

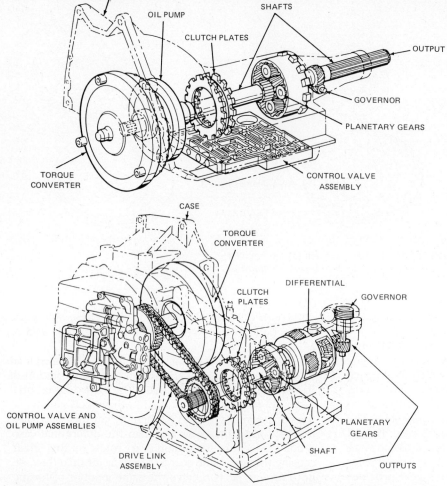

Fig. 1-39 An automatic transaxle (bottom) combines an automatic transmission with final drive gears and a differential. A transmission (top) has a single output. The transaxle uses two outputs, one for each drive shaft. (*Copyrighted material reprinted with permission from Hydra-matic Div., GM Corp.*)

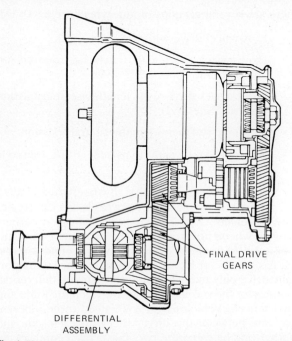

Fig. 1-40 A pair of helical gears are used to provide the final drive gear reduction. The ring gear is usually attached to the differential case. (*Courtesy of Chrysler Corporation.*)

Part-time 4WD normally drives only one axle on a street with good traction. Four-wheel drive is used in poor-traction conditions like mud, sand, snow, or ice, conditions that usually cause enough tire slippage so a center differential is unnecessary. In these conditions, a differential is undesirable because it splits torque equally, and the traction of tires with the poorest traction determines how much torque goes to the other tire on the axle (Fig. 1-46).

The 4WD version of a RWD automatic transmission is basically a same-but-shortened version of the transmission used in two-wheel drive (2WD) cars. Since it must attach to a transfer case, the extension housing is usually omitted, and the output shaft is shortened and designed to connect to a transfer case gear instead of a drive shaft (Fig. 1-47). The transfer case contains the mechanism—often just a dog clutch (one gear slid into mesh with another) and a set of transfer gears or chain and sprockets—to allow engagement and disengagement of 4WD and also offset the universal joint connection for the drive shaft to the front axle.

There are two part-time 4WD versions for FWD vehicles, and these depend on the direction that the transmission shafts run. A few FWD cars use an engine and transmission that are positioned lengthwise in the car. This makes it relatively easy to install a transfer

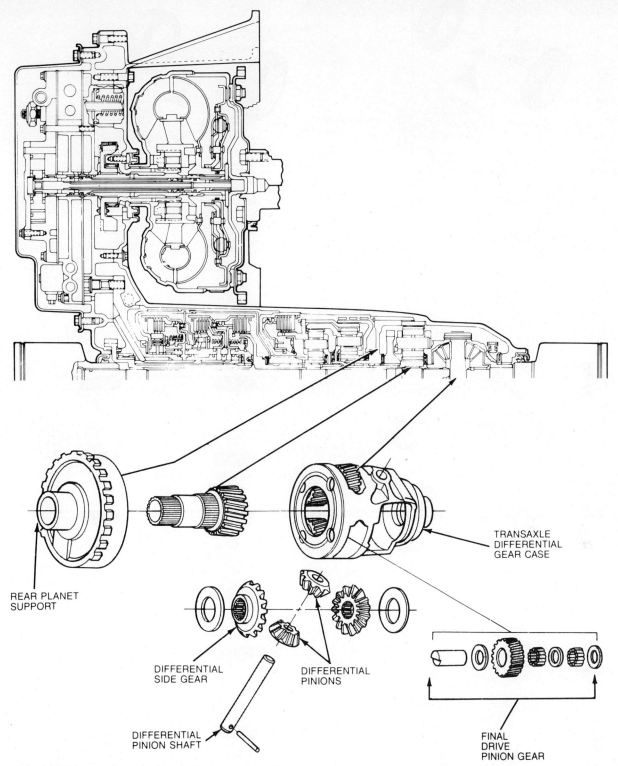

Fig. 1-41 Transaxle that uses a planetary gearset for the final drive reduction. The carrier of this set is made as part of the differential case. (*Courtesy of Ford Motor Company.*)

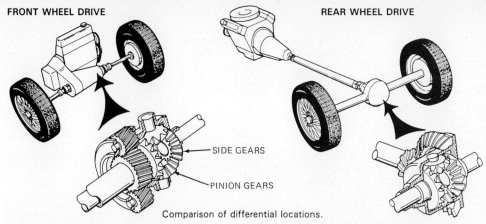

Comparison of differential locations.

Fig. 1-42 A differential usually contains a pair of pinion gears and a pair of side gears. The pinion gears are mounted on a shaft that runs across the case and the side gears are attached to each axle shaft. (*Copyrighted material reprinted with permission from Hydramatic Div., GM Corp.*)

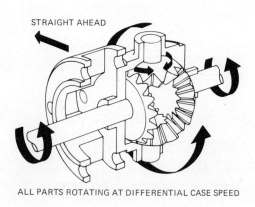

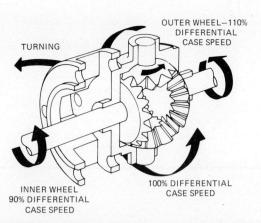

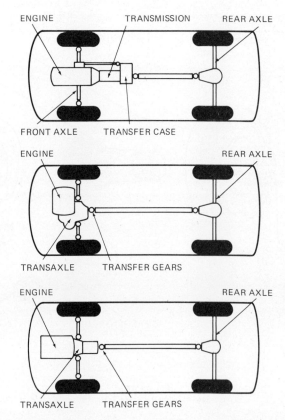

Fig. 1-44 Three major 4WD configurations. The traditional form (top) uses a transfer case to split the torque for the front and rear axles.

Fig. 1-43 When a vehicle is going straight, the differential pinion gears do not turn on their shafts, and the whole differential rotates as one locked-up assembly. When a vehicle turns a corner, the differential pinion gears rotate on their shafts to allow one wheel to speed up and the other to slow down. (*Courtesy of Ford Motor Company.*)

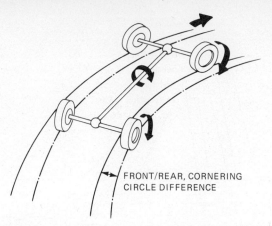

Fig. 1-45 When a vehicle turns a corner, the front wheels must go faster than the rear, and the outside wheels must turn faster than the inside ones.

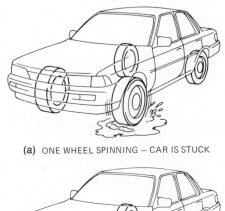

(a) ONE WHEEL SPINNING — CAR IS STUCK

(b) CENTER DIFFERENTIAL LOCKED OUT — FRONT AND REAR WHEELS DRIVING

Fig. 1-46 Because of the differential, a 2WD car can get stuck if one wheel has poor traction. A full-time 4WD can also get stuck under the same condition because of the three differentials. If the center differential is locked, at least one front and one rear wheel will be driven. (*Courtesy of Toyota Motor Sales, U.S.A., Inc.*)

clutch in the rear of the transmission (Fig. 1-48). The clutch can operate an output shaft that runs out the back of the transmission to connect with the drive shaft for the rear axle.

When the engine and transmission are mounted in a transverse position, 4WD becomes more complicated. The power flow must be turned 90° to travel down the car before turning another 90° at the rear axle. The transfer case for a transverse transaxle must include a pair of bevel gears in order to change the direction of power flow (Fig. 1-49). The transfer also includes a clutch to engage and disengage the power flow to the rear axle.

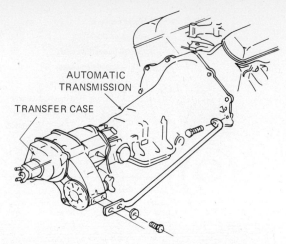

Fig. 1-47 A 4WD transfer case is attached to the rear of this automatic transmission. It provides a means of driving the front axle and wheels. (*Courtesy of Chevrolet Div., GM Corp.*)

Full-time 4WD becomes more complex because the power that leaves the transmission must first go to the center differential and then split to go to the front and rear axles (Fig. 1-50). Some center differentials are torque biasing: they have the ability to split the torque unequally and send a greater amount of torque to the rear axle.

Most center differentials include a lockout feature so both front and rear axles can be driven equally. This ensures that at least one front tire and one rear tire will be driven, and it is only used in very poor traction conditions.

1.14 NONPLANETARY GEAR AUTOMATIC TRANSMISSIONS

As we can well imagine, the automatic transmission described is complicated, has a high number of parts that can fail, is fairly heavy, is fairly large, and is not totally fuel efficient. However, it makes a car that is easy to drive and therefore very popular. In the future, two styles of automatic transmissions in current use may replace the planetary gear transmission: the *continuously variable transmission* (*CVT*) and the *two-clutch transmission*.

Another automatic transmission that is "different" is currently being used in Honda automobiles. This transmission, called the *HondaMatic*, does not use planetary gearsets. The gear train is much like that of a standard transmission with the gears in constant mesh (Fig. 1-51). The fully automatic shifts are made when the multiple-disc clutch for each particular gear is engaged (Fig. 1-52). The shifts are accomplished by hydraulic pressure using a hydraulic system like that of other automatic transmissions. Also like other automatic transmissions, a torque converter is used between the transmission and engine.

The CVT is currently being used in a few compact car models produced in Europe and Japan. It uses a metal V-belt and two variable-size pulleys to produce continuous gear ratios (Fig. 1-53). This transmission does not shift to different specific speeds; the ratio gradually

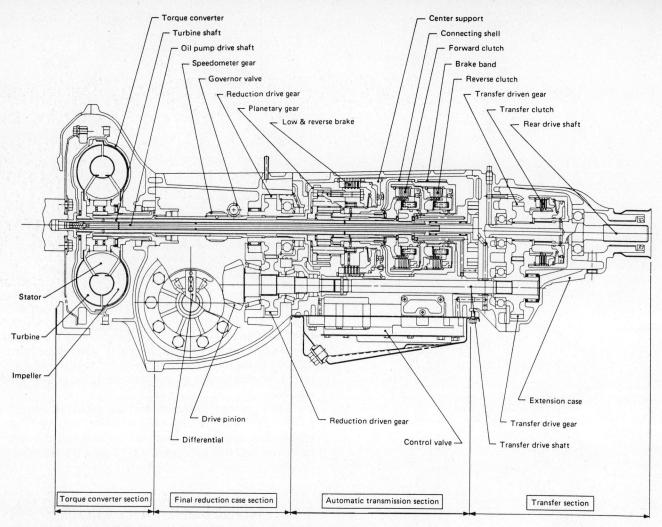

Fig. 1-48 Transaxle with output shaft to drive the rear wheels and clutch in the transfer section to allow the driver to control this power flow. (*Courtesy of Subaru of America.*)

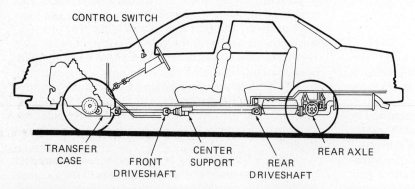

Fig. 1-49 The transfer case, front and rear drive shafts, and rear axle gears of an all wheel drive car. (*Courtesy of Ford Motor Company.*)

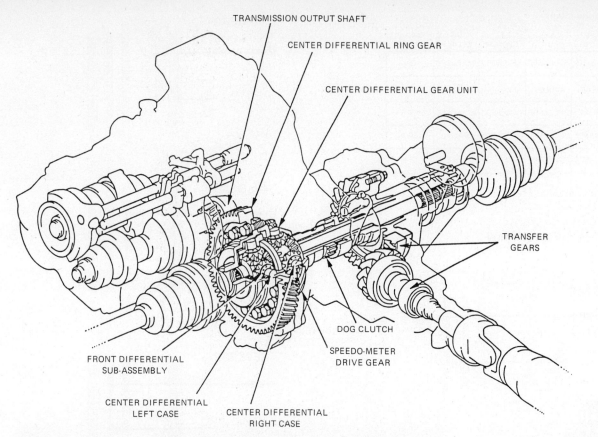

Fig. 1-50 Transaxle with output shafts for the front wheels (left and upper right) and the rear axle (lower right). Note the use of two differentials. The center differential allows the front and rear wheels to rotate at different speeds. (*Courtesy of Toyota Motor Sales, U.S.A., Inc.*)

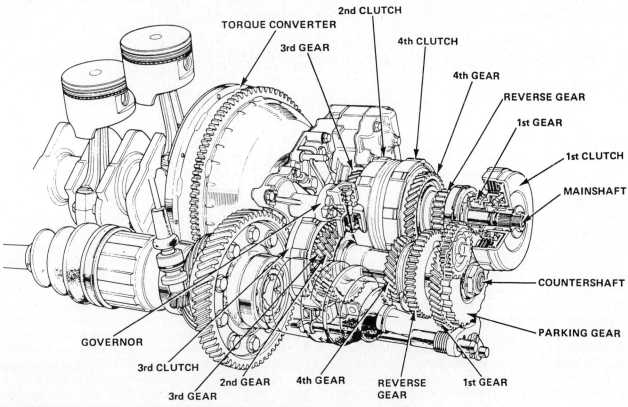

Fig. 1-51 Four-speed HondaMatic transaxle. Note that it combines standard transmission gears and power flow with multiple-disc clutch assemblies and a torque converter. (*Courtesy of American Honda Motor Co.*)

27

CLUTCH APPLICATION CHART

	1st CLUTCH	2nd CLUTCH	3rd CLUTCH	4th CLUTCH	SPRAG CLUTCH
DRIVE 1	APPLIED	RELEASED	RELEASED	RELEASED	ON
DRIVE 2	APPLIED	APPLIED	RELEASED	RELEASED	OFF
DRIVE 3	APPLIED	RELEASED	APPLIED	RELEASED	OFF
DRIVE 4	APPLIED	RELEASED	RELEASED	APPLIED	OFF
MANUAL 2	RELEASED	APPLIED	RELEASED	RELEASED	OFF
REVERSE	RELEASED	RELEASED	RELEASED	APPLIED	OFF

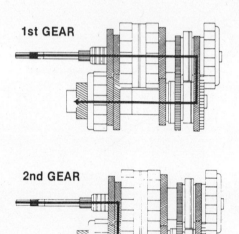
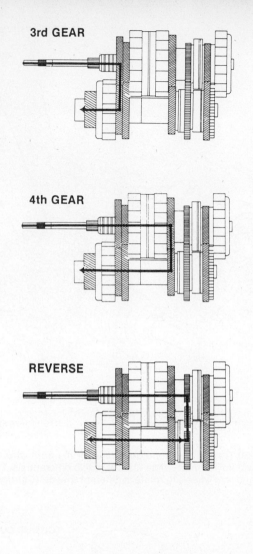

Fig. 1-52 The different power flows through a four-speed HondaMatic are controlled by applying the different clutches. (*Courtesy of American Honda Motor Co.*)

changes from the lowest at a stop to the highest at cruise. While accelerating, the engine runs at almost a constant speed as the transmission varies the ratio. For low gear, a small input pulley drives a large output pulley (Fig. 1-54). As the car speeds up, the diameter of the input pulley becomes larger while the output pulley size becomes smaller. This is accomplished by moving the sides of each pulley together or apart. At cruising speeds (high gear), a large input pulley drives a small output pulley. Hydraulic controls are used to change the pulley diameter in response to throttle position and output shaft speed. A CVT uses an electronic magnetic clutch or a torque converter so the car can be stopped at stop signs and a planetary gearset for reverse so the car can back up.

A *two-clutch transmission* is a combination standard transmission with automatic controls. Power can enter the gearset by one of two paths (the two clutches): a torque converter plus mechanical clutch (the first clutch) for first and third and through a different mechanical clutch (the second clutch) for second and fourth. Synchronizer assemblies like those used in standard transmissions are used to make the shifts, and these are moved by hydraulic servos. When the driver selects "drive," the hydraulic servo shifts to engage first gear; stepping on the gas causes the torque converter or first clutch to transfer power, and the car will move (Fig. 1-55). As the car starts moving, another servo moves to engage second gear, and then the hydraulics will apply the second clutch as it releases the first clutch. At a higher speed, a servo engages third gear, and the first clutch is reapplied as the second clutch is released. For fourth gear, the servo moves to engage fourth gear, the second clutch is reapplied, and the first clutch releases.

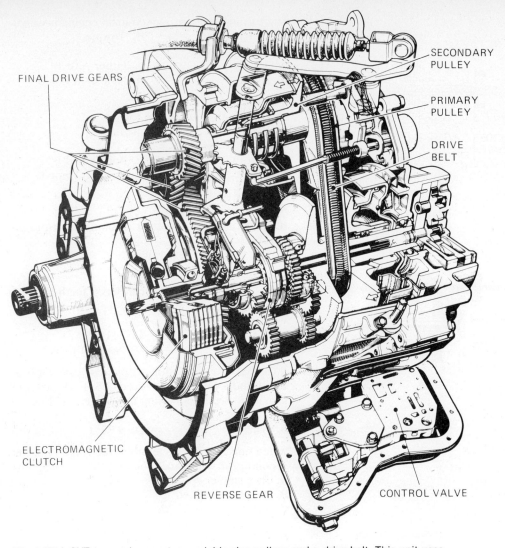

Fig. 1-53 A CVT transaxle uses two variable-size pulleys and a drive belt. This unit uses an electromagnetic clutch that applies when electric current is sent to it. (*Courtesy of Subaru of America.*)

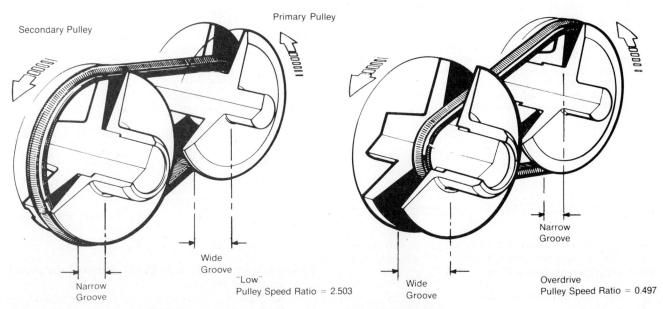

Fig. 1-54 When the driving pulley is as small as possible, the driven pulley is large, and the drive ratio is at its lowest. The drive pulley is made larger and the driven pulley is made smaller to get a higher drive ratio as the car speeds up. (*Courtesy of Subaru of America.*)

29

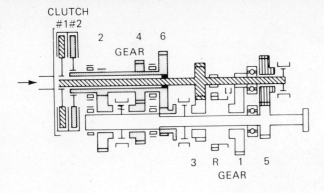
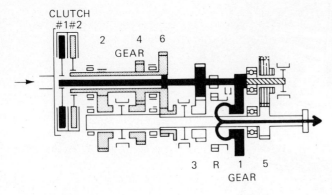
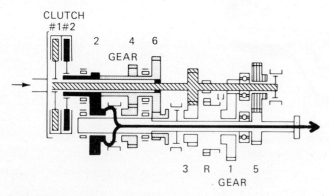

Fig. 1-55 Two-clutch transmission. Clutch 1 drives first and third gears while clutch 2 drives second and fourth gears. The power flow from these gears is controlled by synchronizer assemblies in a manner similar to a standard transmission. In first gear, clutch 1 supplies power to the gearset while the first and third synchronizer sleeve has engaged first gear. In second gear, clutch 2 supplies power while the second and fourth synchronizer sleeve has engaged second gear.

REVIEW QUESTIONS

The questions that follow are provided so you can check the facts you have just learned. Select the response that best completes each statement.

1. The transmission provides
   A. Several gear ratios
   B. A means of backing up
   C. A neutral power path
   D. All of these

2. The turning force produced in an engine is called
   A. Horsepower
   B. Torque
   Which is correct?
   a. A only
   b. B only
   c. Either A or B
   d. Neither A nor B

3. A 2:1 gear ratio will
   A. Double the amount of torque
   B. Reduce the speed by one-half
   Which is correct?
   a. A only
   b. B only
   c. Both A and B
   d. Neither A nor B

4. A ratio of 0.75:1
   A. Is an overdrive ratio
   B. Will offer only a small amount of torque increase
   C. Is a very low gear ratio
   D. All of these

5. If the gear on the input shaft has 9 teeth and the gear on the output shaft has 27 teeth, the gear ratio will be
   A. 2.5:1
   B. 0.33:1
   C. 3:1
   D. None of these

6. A car has a transmission with a low gear ratio of 2.5:1 and a final drive ratio of 3.5:1. The overall gear ratio while operating in first gear will be
   A. 2.5:1
   B. 3.5:1
   C. 6:1
   D. 8.75:1

7. When one gear drives another, the rotation of the second gear will usually be
   A. At a slower speed
   B. At a faster speed
   C. In the opposite direction
   D. All of these

8. The clearance between one gear and another is called
   A. Backlash
   B. Side clearance
   C. Side thrust
   D. End play

9. The speed of a vehicle is determined by
   A. The speed of the engine
   B. The gear ratio
   C. The tire diameter
   D. All of these

10. The amount of torque developed in a vehicle's drive train is called
    A. Power
    B. Tractive force
    C. Torque
    D. All of these

11. A planetary gear set is made up of a
    A. Sun gear
    B. Ring gear
    C. Carrier and planet gears
    D. All of these

12. When the carrier is held, the action of the planet gears will be to
    A. Walk
    B. Be motionless in the carrier
    C. Idle
    D. None of these

13. When the carrier is driven while the sun gear is held, the action of the planet gears will be to
    A. Walk
    B. Be motionless in the carrier
    C. Idle
    D. None of these

14. A Simpson gear train uses
    A. One multiple-disc driving clutch
    B. Two multiple-disc driving clutches
    C. Three multiple-disc driving clutches
    D. Four multiple-disc driving clutches

15. A clutch in an automatic transmission is
    A. Applied by spring pressure
    B. Released by fluid pressure
    Which is correct?
    a. A only   c. Both A and B
    b. B only   d. Neither A nor B

16. A reaction member
    A. Keeps a part of a gearset from rotating
    B. Is always applied by hydraulic pressure
    Which is correct?
    a. A only   c. Both A and B
    b. B only   d. Neither A nor B

17. The transmission's hydraulic circuit
    A. Controls clutch and band operation
    B. Produces upshifts and downshifts
    C. Is rather complex
    D. All of these

18. A lockup torque converter is used in newer transmissions to
    A. Increase first gear torque
    B. Improve fuel mileage
    C. Reduce engine operating speed
    D. All of these

19. When a car is going around a corner, the differential pinion gears are
    A. Rotating on their shafts
    B. In mesh with the axle gears
    Which is correct?
    a. A only   c. Both A and B
    b. B only   d. Neither A nor B

20. The center differential of an AWD vehicle
    A. Allows the front axle to go slower than the rear axle
    B. Forces all four wheels to rotate at the same speed
    C. Is also used in 2WD vehicles
    D. None of these

CHAPTER 2

APPLY DEVICES: CLUTCHES AND BANDS

OBJECTIVES

After completing this chapter, you should:

- **Be able to identify the major portions of a simple and complex planetary gearset.**
- **Understand how the power flow through a planetary gearset is changed.**
- **Be able to identify the portions of a multiple-disc clutch and describe how it operates.**
- **Understand why a reaction member is required for a gear ratio change.**
- **Be able to identify the various types of reaction members and describe how they operate.**

2.1 INTRODUCTION

An automatic transmission is built so there are several paths for power to flow through it; each path provides a different gear ratio. These paths are controlled by the driving and reaction members. The driving members connect the turbine shaft from the torque converter to specific parts of the gear train. The reaction members connect a specific part of the gear train to the transmission case. There are usually two or more driving members and two or more reaction members in a three- or four-speed transmission (Fig. 2-1).

2.2 DRIVING MEMBERS

The driving members are the input to the gearset so they are normally built as part of or splined onto the turbine shaft. A driving member is nearly always a multiple-plate-disc clutch, and in most cases, it will be at the front of the transmission, just behind the pump and next to the converter. A few early transmissions used a fluid coupling inside the transmission (Fig. 2-2). This coupling was filled with fluid to apply and emptied of fluid to release. A good, high-capacity pump was absolutely necessary in order to fill the coupling during a shift. In most cases, the driving member is at the front of the transmission, just behind the front pump, next to the torque converter.

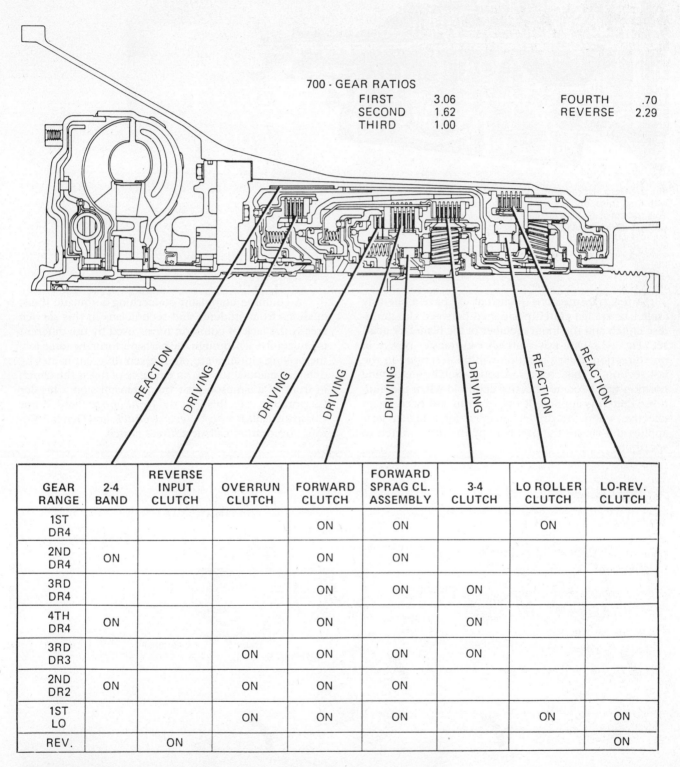

Fig. 2-1 A THM 700 with one band, five multiple-disc clutches, and two one-way clutches. Of these, the reverse input clutch, overrun clutch, forward clutch, forward sprag clutch assembly, and 3–4 clutch are driving members, and the 2–4 band, low roller clutch, and low–reverse clutch are reaction members. *(Copyrighted material reprinted with permission from Hydra-matic Div., GM Corp.)*

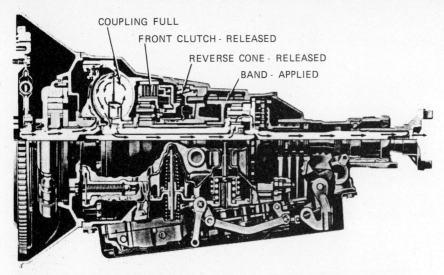

Fig. 2-2 Roto Hydra-matic with a fluid coupling as a driving member. It was filled with first, third, and reverse gears to transfer power and emptied in second gear. This transmission was last used in the early 1960s. *(Copyrighted material reprinted with permission from Hydra-matic Div., GM Corp.)*

A few transmissions in current use have a one-way (roller or sprag) clutch positioned between the multi-disc clutch and the input member of the planetary gearset (Fig. 2-3). One-way clutches can transfer power in one direction; their operation will be described in the last section of this chapter. A roller clutch in a driving position will become effective and hold when the multidisc clutch is applied. It can overrun and become ineffective when the driving member for a higher gear applies and causes the part held by the roller clutch to speed up.

A common complaint concerning automatic transmissions from students and technicians in this service area is the lack of common terms used by the different manufacturers. Although the parts perform the same job, the driving clutches are often given different names by different manufacturers. An example of this is the clutch in front of a Simpson gear train transmission. Chrysler Corporation calls this unit the *front clutch*; Ford Motor Company calls it a *high–reverse clutch*; and General Motors Corporation calls it a *direct clutch*.

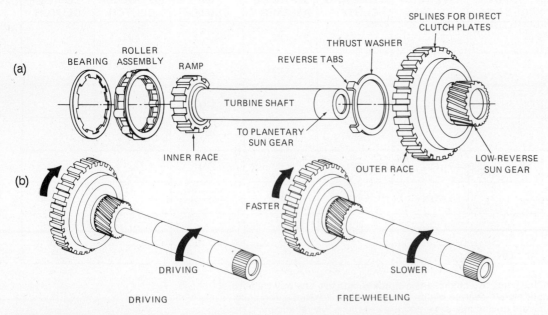

Fig. 2-3 Roller clutch assembly from an ATX transaxle that allows the turbine shaft to drive the low–reverse sun gear as the driving member for first gear. In second and third gears, the roller clutch overruns. *(Courtesy of Ford Motor Company.)*

2.3 MULTIPLE-DISC DRIVING CLUTCHES

A clutch assembly has several major parts: the drum, the hub, the lined plates, the unlined plates, the apply piston, and the piston return springs. The *drum*, also called a *housing*, has a series of splines inside its outer end for the externally lugged plates, usually the unlined plates. The inner diameter of the drum is machined for the apply piston and its inner and outer seals (Fig. 2-4). The drum of the front, or first, clutch is usually built as part of or splined onto the input or turbine shaft. A clutch can be built as a single unit or combined with another drum or hub of a second clutch assembly. For example, the input drum of a THM 700-R4 transmission contains three different clutches as well as the hub for a fourth clutch (Fig. 2-5).

A clutch hub is splined or lugged on its outer diameter to accept the internal lugs of the lined clutch plates. The clutch pack, an alternately stacked series of unlined and lined plates, is normally contained in the drum and is held in place by a thick pressure plate retained by a large retaining ring in the drum. The clutch hub is often built as part of the unit being driven such as a ring gear or a carrier.

The doughnut-shaped apply piston is sealed at both its outer and inner diameter so that fluid entering the

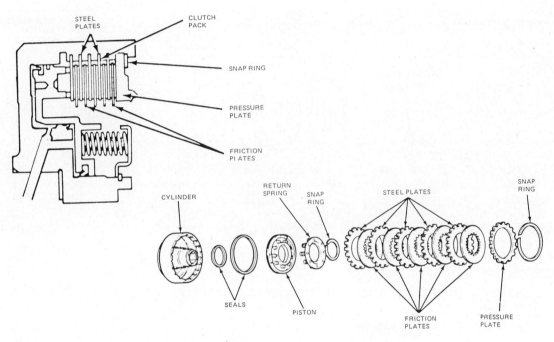

Fig. 2-4 Cutaway and exploded views of a multiple-disc clutch; note the piston used to apply the clutch. *(Courtesy of Ford Motor Company.)*

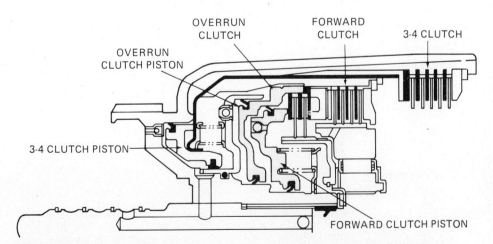

Fig. 2-5 Complex clutch assembly from a THM 700. It has three separate clutch packs and three different pistons, one for each clutch. *(Copyrighted material reprinted with permission from Hydra-matic Div., GM Corp.)*

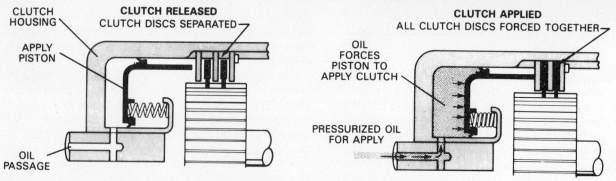

Fig. 2-6 A clutch is released when the springs have moved the apply piston away from the clutch pack (top). Pressurized oil forces the discs together to apply the clutch (bottom). *(Copyrighted material reprinted with permission from Hydra-matic Div., GM Corp.)*

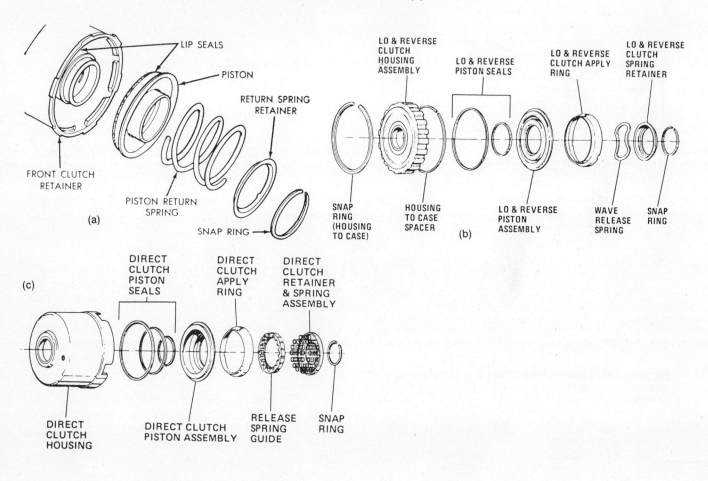

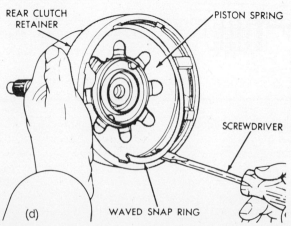

Fig. 2-7 (a) A single, large piston return spring; (b) a single wave return spring; (c) a group of smaller return springs; and (d) a single Belleville return spring. The springs in (c) are the most common. *[(a,d) Courtesy of Chrysler Corporation. (b,c) Copyrighted material reprinted with permission from Hydra-matic Div., GM Corp.]*

cylinder is trapped. During clutch application, the fluid pressure forces the piston to move and squeeze the clutch pack against the pressure plate (Fig. 2-6). When the fluid pressure is released, one or more springs retained in the drum push the piston back to its released position. Most clutch assemblies use a group of small coil springs; some clutches use a single, large coil spring; and some clutches use a single, large Belleville spring (Fig. 2-7). As the clutch is released, the plates can slide away from each other, creating a clearance between each plate.

A *Belleville spring*, also called a *diaphragm spring*, resembles a metal washer that is slightly cone shaped. Spring action occurs when this spring is flattened. This action cushions clutch application. Some clutches position a Belleville spring between the piston and the clutch pack so it acts as an apply lever as well as a return spring. The force increase from this leverage increases the strength of the clutch.

2.4 CLUTCH PLATES

An unlined plate, often called a "steel" or "separator plate," is merely a flat piece of steel stamped into the desired shape. This plate is usually about 0.070 in. (1.78 mm) thick. After being stamped, the plate is carefully flattened; an out-of-flat plate uses up clutch clearance and causes drag while released. Then it is tumbled with other plates in a large drum. This is done to make a series of very small random nicks or dents in the flat friction surface (Fig. 2-8). These recesses hold fluid that lengthens clutch life as well as improves shift quality. Even with these small nicks, a good steel has a surface finish that is smoother than 25 micro-in.; a surface roughness of 12 to 15 micro-in. is desired. An unlined plate usually has lugs on its outer diameter. Steel plates have a secondary purpose of serving as heat sinks to help remove heat from the lined friction plates.

A lined plate is also made from stamped steel, and it has lining material secured to each side. It is often called a *friction plate* or simply a *friction*. The engagement lugs are usually on the inner diameter. The friction material is about 0.015 to 0.030 in. (0.38 to 0.76 mm) thick. This lining material can have a flat or a grooved

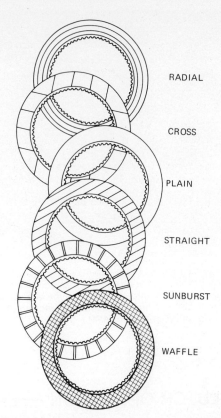

Fig. 2-9 Lined friction plates can be smooth or grooved in one of the patterns shown here.

friction surface, and a grooved plate can have one of several grooving patterns, as shown in Fig. 2-9. The grooves help fluid leave and enter the unlined and lined plates during a shift, and the different grooving patterns have been found to offer different shift characteristics. The oil flow through the grooves and between the plates helps cool the friction surfaces. The different clutch packs of the same transmission sometimes use lined plates with different grooving patterns.

The most common clutch-lining material is paper. Different grades of paper material are blended for particular clutches and types of transmission fluid (Fig. 2-10). The paper materials can be blended with a cotton–asbestos mix, graphite, or other inorganic fillers. The clutches are flooded with transmission fluid while released, and during application this fluid prevents any heat generated by friction from overheating the lining. Some clutches use metallic or semimetallic linings. The metallic material is often sintered bronze or sintered iron.

Manufacturers often provide some means of adjusting the released clearance in a clutch pack. There must be sufficient clearance between the plates to ensure that there is no drag. This clearance should be about 0.010 to 0.015 in. (0.25 to 0.38 mm) between each friction surface—lined and unlined plate. The most common methods of adjusting clutch pack clearance are selective size retaining rings, pressure plates, or steel plates (Fig. 2-11). These units are made in several thicknesses, and the one with the correct width or thickness is selected as a particular clutch is assembled.

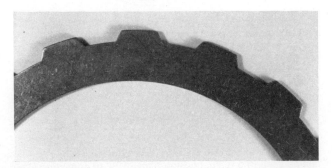

Fig. 2-8 Close-up view of the surface finish on a steel plate. Note the small nicks and grooves that tend to hold the automatic transmission fluid.

37

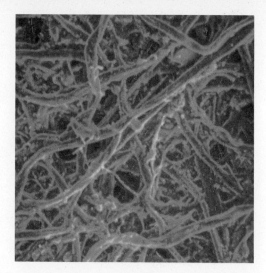

Fig. 2-10 View of the surface of a paper friction material made using a scanning electron microscope. Note the large pores between the paper fibers, which are capable of storing relatively large amounts of oil. *(Courtesy of AFM.)*

2.5 CLUTCH OPERATION

When a clutch is applied, the plates are squeezed together and transfer torque. The amount of torque that can be transferred is determined by the following factors: diameter and width of the friction surfaces, number of friction surfaces (two per lined plate), and the amount of force being applied (hydraulic pressure times the piston area). The greater the plate area, number of plates, piston size, or hydraulic pressure, the greater the torque capacity. The formula used to determine clutch capacity is:

$$T = \frac{N[(D + d)F\mu]}{4}$$

where:

T = torque capacity (in inch-pounds)

N = number of active friction surfaces

D = outside diameter of clutch facings (in inches)

d = inside diameter of clutch facings (in inches)

F = total force on clutch pack (in pounds)

μ = coefficient of friction

A clutch with sufficient capacity will transfer the required amount of torque without slipping. Many transmissions are made with varying torque capacities depending on the engine with which they will be used. The major difference between these strong engine–weaker engine versions is the number of plates in the clutch packs.

When a clutch is released, there must be clearance between the plates (Fig. 2-12). There is often a rather large speed differential between the two groups of plates. For example, during first gear in a Simpson gear train the sun gear revolves at a 2.5:1 ratio in reverse. Imagine the speed difference in a released high-gear

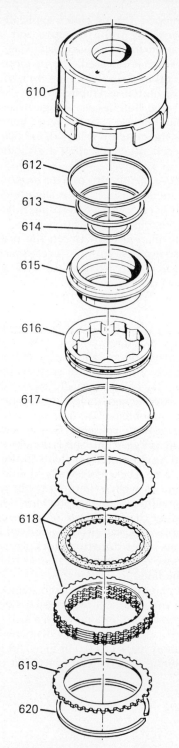

610 HOUSING & DRUM ASM., DIRECT CLUTCH
612 SEAL, DIRECT CLUTCH PISTON (OUTER)
613 SEAL, DIRECT CLUTCH (CENTER)
614 SEAL, DIRECT CLUTCH PISTON (INNER)
615 PISTON, DIRECT CLUTCH
616 APPLY RING & RELEASE SPRING ASM.
617 RING, SNAP
618 PLATE ASM., DIRECT CLUTCH
619 PLATE, CLUTCH BACKING (DIRECT)
620 RING, SNAP

Fig. 2-11 Depending on the particular clutch assembly, the retaining snap ring (620), backing plate (619), or apply ring (616) are of varying thicknesses to allow adjustment of the clutch stack clearance. *(Copyrighted material reprinted with permission from Hydra-matic Div., GM Corp.)*

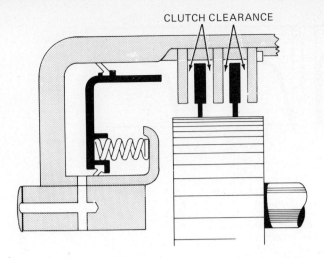

Fig. 2-12 When a clutch is released, there must be sufficient clearance between the plates to prevent any drag and allow a flow of oil to cool the surfaces.

clutch with the input shaft, clutch hub, and lined plates revolving at an engine speed of 3000 rpm in a clockwise direction and the drum and the unlined plates turning 7500 rpm in a counterclockwise direction (Fig. 2-13). Without sufficient clearance and lubrication, these plates would surely drag, create friction, and burn up.

A transmission engineer is concerned with three different friction conditions in a clutch. While the clutch is released, there should be no friction or drag; while applied, there should be sufficient *static friction* to transfer torque without slippage; and while applying there should be the proper *dynamic friction* to get a good, smooth shift. An example of static and dynamic friction can be seen if we place a book on a table. With the book sitting still, the static (stationary) friction be-

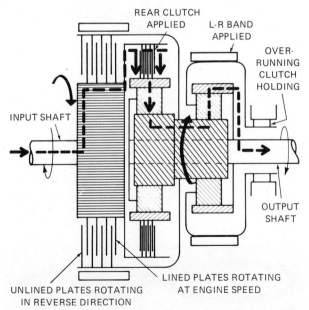

Fig. 2-13 When this transmission is in first gear, the sun gear, clutch housing, and unlined plates of the front clutch rotate in a counterclockwise direction while the hub and lined plates rotate in a clockwise direction. Any drag can cause heat and clutch burnout. (*Courtesy of Chrysler Corporation.*)

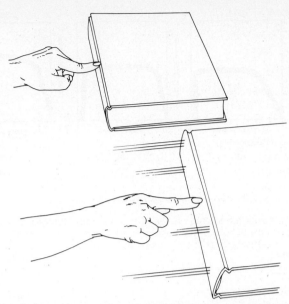

Fig. 2-14 If you push against a stationary book, you will notice a certain amount of static friction that resists the motion. If you push against the same book as it is sliding at a constant speed, you will notice that the dynamic friction is somewhat less.

tween the book and the table holds it in place. Push on the book, and you will notice the amount of force it takes to overcome this static friction (Fig. 2-14). Also notice how much force it takes to keep the book sliding; this is the dynamic or sliding friction. Static friction is always greater than dynamic friction.

Some clutches are power-shifting elements. They are applied under power and have to transfer substantial torque as they are applied (Fig. 2-15). These clutches are often called engaging members and must have a high dynamic coefficient of friction. A first- or reverse-gear clutch needs a high amount of static friction because of the conditions under which it is applied—for example, a garage shift (neutral-to-first or neutral-to-reverse shifts). These clutches have more of a holding function. The shift from neutral to first or reverse is much less severe than the 2–3 shift.

2.6 SHIFT QUALITY

When a Simpson gear train transmission starts a 2–3 (second- to third-gear) shift, the third-gear clutch drum is held stationary by a band and the hub is rotating at engine speed. During this shift, the band is released, and the clutch is applied. Imagine that as the shift begins, the engine is at 3000 rpm and the car is going about 45 mph (72 km/h). In second gear, with a ratio of 1.5:1, the drive shaft will be turning at 2000 rpm. The time duration of the shift has to be gradual enough so that the engine's rpm will be smoothly lowered to the proper speed (in this case, from 3000 to 2000 rpm); a 1:1 third-gear ratio will have the same speed as the drive shaft (Fig. 2-16). While the band is releasing second gear and the clutch is applying third gear, there must be a smooth speed transition. This is referred to as "shift quality."

A shift should be smooth without any unusual noises. In order for this to occur, the clutches and bands must apply smoothly and quietly. Jerks, bumps, or harsh

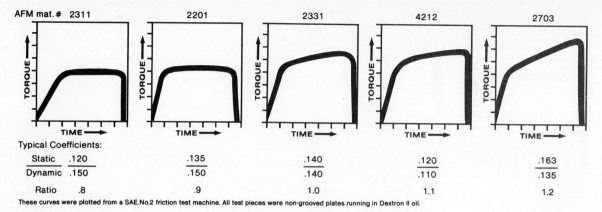

Fig. 2-15 Clutch lockup curves of five different clutches. The dynamic friction is shown as the rate of torque increase at the left side while the static friction is shown as the amount of torque at the right side where the clutch is completely applied. *(Courtesy of AFM.)*

application are considered faults. Improper noises are usually squaks, shrieks, or engine noises if the clutch slips and an rpm flare occurs.

In a clutch, shift quality is controlled by the type of lining material and grooving, the use of wave or Belleville plates, the type of fluid used, and the speed at which fluid enters behind the piston. The last two will be described in Chapter 4.

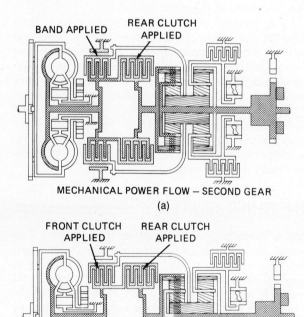

Fig. 2-16 (a) In second gear, the sun gear and clutch drum are held stationary by the band. (b) In third gear, the clutch is applied so the drum is now rotating at engine speed. A driver will feel this change as shift quality. *(Courtesy of Nissan Motor Corporation.)*

A wave plate is an unlined plate that is wavy, not flat. A Belleville plate, like a Belleville spring, is also not flat (Fig. 2-17). These are often called "cushion plates." If either plate is used in a clutch pack, it will be placed between an unlined plate and the pressure plate at the end farthest from the piston. Both plates are designed to compress slightly under pressure during clutch application and slightly prolong the clutch apply time.

Again consider the time just before the 2–3 shift. The high clutch drum and unlined plates are stationary, and the hub and lined plates are revolving at about 3000 rpm. Transmission fluid enters the cylinder behind the clutch piston and starts the piston moving, and this begins reducing the clutch pack clearance. At some point, there will be no clearance, and at that point, there will be a pressure squeezing the plates together. Suddenly, there is dynamic friction between the plates, and a power transfer takes place. If a cushion plate is used, this moment is stretched out and prolonged and a slight slippage occurs, giving a smoother, less severe clutch application.

Many modern transmissions do not use Belleville or wave plates. The hydraulic controls (accumulators and orifices) are designed to produce a better controlled piston motion and the desired shift quality.

2.7 HOLDING MEMBERS

A holding member acts as a brake to hold a portion of the gear train in reaction. Three types of holding members are used: multiple-disc clutches, bands, and one-

Fig. 2-17 Two cushion plates are shown here: a wave plate on the right and a Belleville plate on the left. Note the wavy shape of the wave plate and the dished shape of the Belleville (the outer edge is higher than the inner edge).

way clutches. Some transmissions use a one-way clutch that is connected to a multiple-disc clutch. In the past one transmission used a cone clutch as a holding member (Fig. 2-18). Multiple-disc clutches and bands are applied by hydraulic pressure and therefore, like a driving clutch, are controlled by the valve body. A one-way clutch is self-controlled; it allows rotation in one direction only. In most transmissions, a one-way clutch allows rotation in a clockwise direction but blocks counterclockwise rotation. Applying and holding a one-way clutch is often referred to as being effective, and releasing it is called ineffective or noneffective.

Although reaction members do not rotate, there can be a substantial load on these holding members. Remember that for every action, there is an equal and opposite reaction. A THM 700-R4 transmission uses a 2:1 torque converter ratio and a 3.06:1 first gear. If the engine develops 100 ft-lb (135.6 N-m) of torque, there is 2 × 100, or 200, ft-lb (271 N-m) of torque coming from the torque converter and 200 × 3.06, or 612, ft-lb (829.8 N-m) of torque at the drive shaft going to the rear axle (Fig. 2-19). There will also be the same 612-ft-lb load on the reaction member, trying to revolve it inside the transmission case.

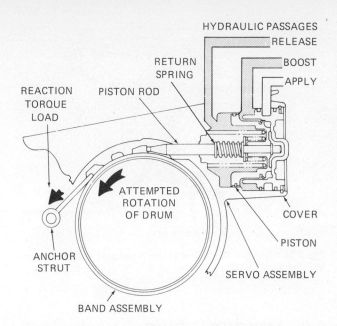

Fig. 2-19 When a band is applied, the reaction load is normally fed back to the transmission case through the anchor strut. *(Courtesy of Ford Motor Company.)*

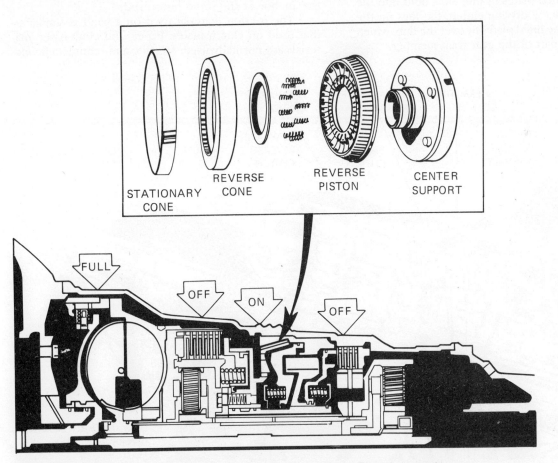

Fig. 2-18 One of the few transmissions to use a cone clutch was the Roto Hydra-matic. The stationary cone is keyed to the transmission case so that when the reverse cone clutch is applied, the front ring gear is held stationary. *(Copyrighted material reprinted with permission from Hydra-matic Div., GM Corp.)*

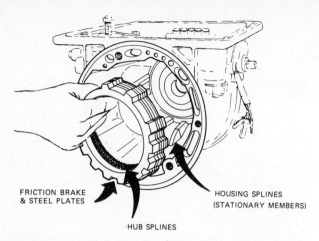

Fig. 2-20 When a multiple-disc clutch is used for a reaction member, the steel plates are splined so they fit into the transmission case. *(Courtesy of Nissan Motor Corporation.)*

2.8 MULTIPLE-DISC HOLDING CLUTCHES

A multiple-disc holding clutch is quite similar to a driving clutch. The major difference is that the drum is the transmission case and this clutch does not rotate. It is called a brake by some manufacturers. The lugs on the outside of the unlined plates fit into slots built into the case (Fig. 2-20). Like a driving clutch, the lugs on the inner diameter of the lined plates fit over the hub, which is usually made as part of the gear train member.

The hydraulic piston is also normally built in the case; it can also be built in the back of the front pump assembly or in a center support (Fig. 2-21). The stationary position of the piston and cylinder make it relatively easy to provide a fluid path into it. Like a driving clutch, the piston is normally returned to a released position by a set of coil springs.

2.9 BANDS

A band is a circular strip of metal that has lining on the inner surface. It wraps around the smooth surface of the drum. Bands come in three basic configurations: a single thick, heavy band; a single thin, light band; and a split, double-wrap, heavy band. The heavy, single-wrap band is also called a *rigid band*, and the thin, single-wrap band is called a *flexible* or *flex band* (Fig. 2-22). A rigid band is strong and provides a good heat sink to absorb some of the friction heat during application. The disadvantage with a rigid band is that it is relatively expensive and doesn't always conform to the shape of the drum. A flex band is less expensive and, because of its flexibility, can easily conform to the shape of the drum. Each band type has end lugs so that the band can be attached to the anchor and the servo. A small link, commonly called a *strut*, is often used to connect the lugs of the band to the anchor or the servo piston rod.

The friction material used on a band is similar to that used on clutch plates. Paper- and cloth-based materials are normally used. Metallic and semimetallic ma-

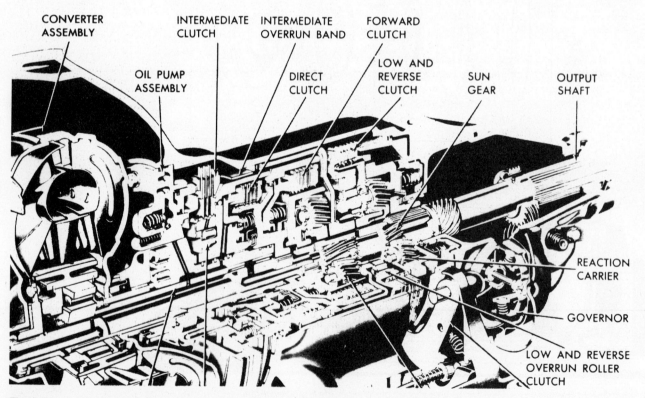

Fig. 2-21 The piston to apply a reaction clutch can be located at the back of the case (upper right) or in the pump assembly (upper left) or in a center support. *(Copyrighted material reprinted with permission from Hydra-matic Div., GM Corp.)*

Fig. 2-22 The three most common band types are a flex band (left), double-wrap band (center), and rigid band (right). Note the lining material on the inner side of each band. This particular flex band is slotted to improve the distribution of the load and increased oil flow away from the band during shifts. *(Courtesy of AFM.)*

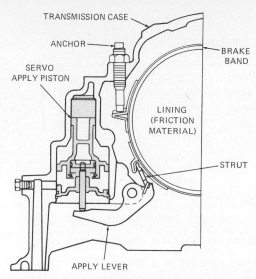

Fig. 2-24 Band with an adjustable anchor that allows the clearance between the band and the drum to be easily adjusted. *(Courtesy of Chrysler Corporation.)*

terials tend to cause severe wear of the drum. The drum must be a smooth cylinder with straight sides in order to have complete contact with the band. The lining surface of the band is often grooved to help control fluid flow during apply and release operations. Similar to a clutch, band friction material and grooving is designed to operate with a particular fluid to ensure good shift quality and long life.

The band is often anchored at the trailing end so drum rotation will tend to pull the band tightly into engagement (Fig. 2-23). A double-wrap band design takes advantage of this wrapping tendency and applies smoother and stronger than a single-wrap band. If possible, the band would rotate with its drum, and the band, anchor, and servo must be designed to absorb these loads. The anchor for the band can be a fixed or adjustable point in the transmission case (Fig. 2-24). An adjustable anchor provides a convenient and easy means of adjusting the clearance between the band and the drum. There must be enough clearance to ensure no band-to-drum contact with the band released but not so much clearance that the band won't completely apply, which might cause slippage.

The servo is the hydraulic piston assembly that applies the band. The piston rod from the servo can push directly on the end of the band or be connected to the band through a lever or linkage attached to the band strut. A band lever usually provides a force increase through the lever ratio. This requires more piston travel to apply the band, but the leverage increases the application force acting on the end of the band (Fig. 2-25). Some manufacturers incorporate an adjustment screw in the apply lever to provide for a band clearance adjustment. Other manufacturters use selective sized servo piston rods for their band adjustment (Fig. 2-26).

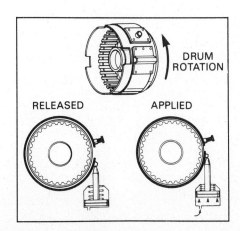

Fig. 2-23 As the band applies, the friction of the lining and the rotation of the drum wraps the band tightly onto the drum. *(Copyrighted material reprinted with permission from Hydra-matic Div., GM Corp.)*

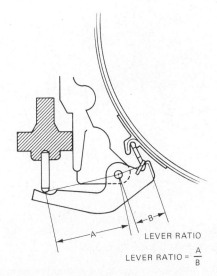

Fig. 2-25 Band with an apply lever that increases the apply force about two and a half times and decreases the apply speed to about one-third of the piston. *(Courtesy of Chrysler Corporation.)*

DIAL INDICATOR TRAVEL		APPLY PIN IDENTIFICATION
.0 - .72mm	(.0" - .029")	1 RING
.72 - 1.44mm	(.029" - .057")	2 RINGS
1.44 - 2.16mm	(.057" - .086")	3 RINGS
2.16 - 2.88mm	(.086" - .114")	WIDE BAND

ILL. NO.	DESCRIPTION
20	RING, SERVO COVER RETAINER
21	COVER, INTERMEDIATE SERVO
22	SEAL, "O" RING (INTERMEDIATE SERVO COVER)
23	RING, OIL SEAL (OUTER)
24	PISTON, INTERMEDIATE SERVO (OUTER)
25	RING, OIL SEAL (INNER)
26	PISTON, INTERMEDIATE SERVO (INNER)
27	RING, OIL SEAL PISTON (INNER)
28	SPRING, INTERMEDIATE SERVO CUSHION
29	RING, SNAP
30	RETAINER, SERVO SPRING
31	SPRING, INTERMEDIATE SERVO (INNER)
32	PIN, INTERMEDIATE BAND APPLY
33	SEAL, INTERMEDIATE BAND APPLY PIN

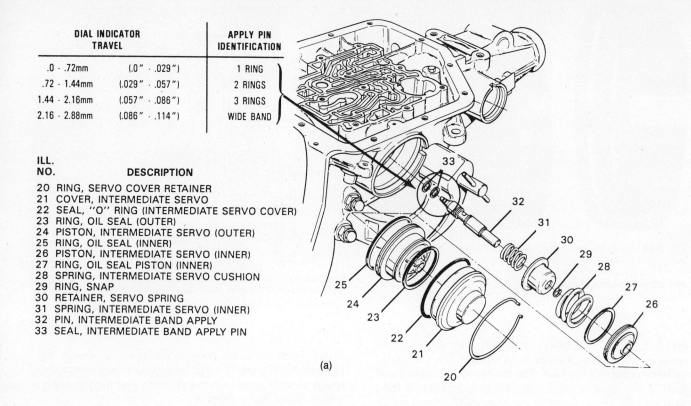

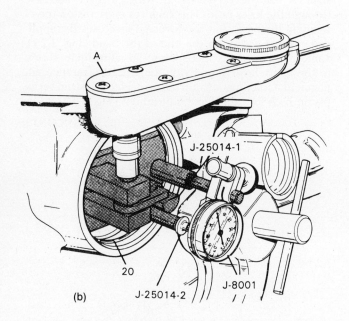

ILL NO.	DESCRIPTION
A	TORQUE WRENCH
20	RING, SERVO COVER RETAINER

Fig. 2-26 The number of rings at the end of the servo pin indicate the relative length of the pin. After the travel of the servo pin is measured using a dial indicator, the proper pin is selected to arrive at the correct band adjustment. *(Copyrighted material reprinted with permission from Hydra-matic Div., GM Corp.)*

2.10 BAND OPERATION

When fluid pressure enters the servo, the servo piston strokes to tighten the band onto the drum. The amount of load that a band can absorb before slipping is determined by the amount of contact area between the band and the drum, the diameter of the drum, and the amount of force squeezing the band onto the drum.

When a band releases, the servo piston backs off, and the springy, elastic nature of the band causes it to move away from the drum. A servo piston is released by two methods. By the first method, when the shift from the gear using the band is to neutral, there is no rush to get the band off, and the release can be fairly slow. Normally only a release spring is used in these servos.

When the shift from the gear using the band is an upshift or a downshift, the band release must be fast and carefully timed as part of the shift. With this method, the band is released by fluid pressure from the shift circuit along with the spring pressure (Fig. 2-27). For example, in a Simpson gear train transmission during a 2–3 shift, the fluid pressure to apply the third-gear clutch is also used to release the second-gear band.

2.11 ONE-WAY CLUTCHES

Two styles of one-way clutches are used in transmissions: *roller clutches* and *sprag clutches*. The roller clutch is more common. A one-way clutch is also called an *overrunning clutch*.

A roller clutch is made up of a hardened, smooth hub or inner race and an outer drum or race, a series of rollers and energizing springs, and a cage or guide to locate the springs (Fig. 2-28). Figure 2-29 shows how each roller fits in a cam section in the drum. An energizing spring pushes each roller so there is a light contact between the roller, the hub, and the drum. A counterclockwise rotation of the hub will wedge the rollers tighter so they lock and block any further rotation in that direction. A clockwise rotation will unwedge the rollers, and each roller will simply rotate, much like a roller bearing. The inner hub will rotate freely or overrun in a clockwise direction.

A sprag clutch uses smooth, round, hardened inner and outer races and a series of sprags in a special cage. A sprag is an odd-shaped device that somewhat resembles an hourglass or fat letter S when viewed from the end. Figure 2-30 shows the two effective diameters of a sprag. The major diameter is greater than the space between the inner and outer races, and the minor diameter is smaller than this distance. The sprags are assembled in a cage that spring loads each sprag in a direction to "stand up" or wedge the major diameter between the two races. A clockwise rotation of the inner race causes the sprags to rotate a little further in the

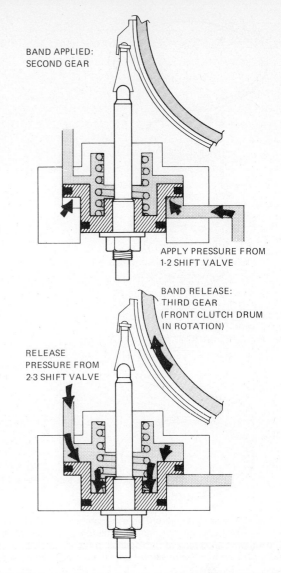

Fig. 2-27 Band applied when the 1–2 shift valve pressure pushes upward on the piston. When the 2–3 shift valve pressure pushes on top of the piston, the band releases. Note that there is a larger piston area on the upper side of the piston. *(Courtesy of Nissan Motor Corporation.)*

stand-up direction, and they wedge firmly between the races and block any further rotation. A counterclockwise rotation of the inner race rotates the sprags in the opposite (lay-down) direction (Fig. 2-31). Each sprag tends to lay down so its minor diameter is between the races, and the inner race rotates freely.

One-way clutches must be thoroughly lubricated. In most transmissions, one or more one-way clutches are overrunning in high gear. The constant motion would quickly cause wear if a good flow of fluid was not available at all times.

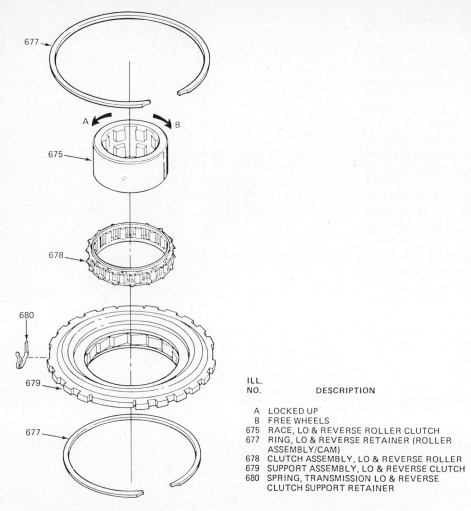

ILL. NO.	DESCRIPTION
A	LOCKED UP
B	FREE WHEELS
675	RACE, LO & REVERSE ROLLER CLUTCH
677	RING, LO & REVERSE RETAINER (ROLLER ASSEMBLY/CAM)
678	CLUTCH ASSEMBLY, LO & REVERSE ROLLER
679	SUPPORT ASSEMBLY, LO & REVERSE CLUTCH
680	SPRING, TRANSMISSION LO & REVERSE CLUTCH SUPPORT RETAINER

Fig. 2-28 One-way roller clutch. Note the ramps/cams at the inner edge of the support assembly (679) and the springs and rollers in the clutch assembly (678). The lugs of the support assembly fit into splines in the transmission case. *(Copyrighted material reprinted with permission from Hydra-matic Div., GM Corp.)*

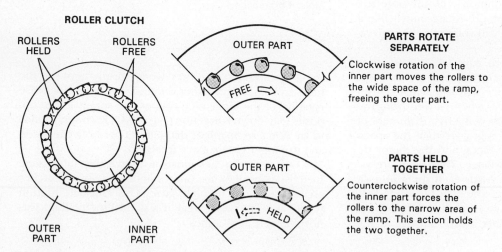

Fig. 2-29 Movement of this particular inner race in a clockwise direction moves the rollers to the wide part of the cams in the outer race, and the clutch freewheels. Movement of the inner race in the opposite direction causes the rollers to wedge tightly between the two races, and the clutch locks. *(Copyrighted material reprinted with permission from Hydra-matic Div., GM Corp.)*

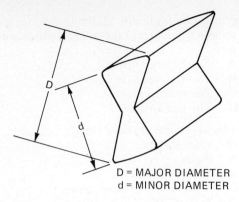

D = MAJOR DIAMETER
d = MINOR DIAMETER

Fig. 2-30 Each sprag has a major and a minor diameter. The major diameter is greater than the distance between the two races, and the minor diameter is smaller.

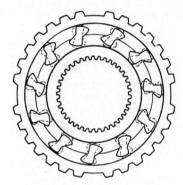

SPRAG CLUTCH

INNER AND OUTER PARTS FREE TO ROTATE

INNER AND OUTER PARTS HELD

A sprag clutch works under the same principle as the roller clutch. One direction of the inner part acts to hold the sprags "out of the way", the opposite rotation acts to engage the sprags holding the inner and outer parts together.

Fig. 2-31 When inner race shown rotates in a counterclockwise direction, the sprags twist so the minor diameter is between the two races, and the unit freewheels. Movement of the race in the opposite direction twists the sprags so the major diameter is wedged between the races, and the clutch locks. *(Copyrighted material reprinted with permission from Hydra-matic Div., GM Corp.)*

REVIEW QUESTIONS

The questions that follow are provided so you can check the facts you have just learned. Select the response that best completes each statement.

1. In order for a planetary gearset to change the ratio and transmit torque, there must be an input member plus
   A. An output member   B. A reaction member
   Which is correct?
   a. A only   c. Both A and B
   b. B only   d. Either A or B

2. Neutral in a planetary gearset can be obtained by
   A. Releasing all of the driving clutches
   B. Not applying any reaction members
   Which is correct?
   a. A only   c. both A and B
   b. B only   d. Either A or B

3. The most commonly used driving member is a
   a. Cone clutch   c. Multiple-disc clutch
   b. One-way clutch   d. Band

4. The plates used in a multiple-disc clutch are
   A. Lined with paper friction material
   B. Unlined steel plates
   Which is correct?
   a. A only   c. Both A and B
   b. B only   d. Neither A nor B

5. The ideal unlined steel plate
   A. Is perfectly flat
   B. Is extremely smooth
   C. Has many very small nicks in its surface
   D. Both A and C

6. The friction plates are lined with
   A. Paper
   B. Cloth
   C. A material like brake lining
   D. Any of these

7. A multiple-disc clutch is applied by
   A. One or more coil springs
   B. Hydraulic pressure pushing against a piston
   C. A piston and air pressure
   D. None of these

8. As a clutch begins applying, there is _____ between the plate surfaces.
   A. Dynamic friction   B. Static friction
   Which is correct?
   a. A only   c. Both A and B
   b. B only   d. Neither A nor B

9. The grooves in the surface of a lined clutch plate
   A. Help cool the friction material
   B. Allow oil to flow between the plates easier
   C. Are cut using different patterns depending on the clutch
   D. All of these

10. The torque-carrying capacity of a clutch is determined by
    A. The number of plates
    B. The amount of lining area on the plates
    C. The amount of pressure squeezing the plates together
    D. All of these

11. Commonly used holding or reaction members are
    A. Bands   C. One-way clutches
    B. Multiple-disc   D. All of these
       clutches

12. A band is
    A. Applied by a hydraulic servo
    B. Released by a spring or hydraulic pressure
Which is correct?
    a. A only   c. Both A and B
    b. B only   d. Either A or B

13. When a multiple-disc clutch is used for a holding member, the lugs of the unlined plates are splined into
    A. A very large drum
    B. The transmission case
    C. The reaction carrier
    D. The output shaft

14. A band is normally lined with _____ on the inner surface.
    A. Paper
    B. Sintered metal
    C. A material like brake lining
    D. Any of these

15. Band clearance can be adjusted using
    A. A threaded adjuster at the band anchor
    B. A threaded adjuster at the servo lever
    C. A selective sized servo piston rod
    D. Any of these depending on the transmission

16. A. The inner and outer races of a sprag clutch are perfectly round and smooth.
    B. One of the races of a roller clutch is perfectly round and smooth while the other one has a series of ramps in it.
    Which is correct?
    a. A only   c. Both A and B
    b. B only   d. Neither A nor B

17. When a sprag clutch locks, the sprags
    A. Try to rotate but wedge between the two races
    B. Roll down the ramps and wedge into place
    C. Rotate into a position that unlocks them
    D. Rotate with the free or rotating race

18. When a roller clutch unlocks, the rollers
    A. Are wedged between the two races
    B. Rotate between the two races like a series of bearing rollers
    C. Rotate to a position where they allow both inner and outer races to rotate
    D. None of these

19. A one-way clutch
    A. Is applied by hydraulic pressure
    B. Is released by a group of springs
    C. Self-applies whenever one of the races tries to rotate in a reverse direction
    D. Any of these

20. The torque load on a reaction member
    A. Is equal to engine torque
    B. Is equal to drive shaft torque
    C. Varies depending on the gear ratio and throttle setting
    D. All of these

48

CHAPTER 3

POWER FLOW THROUGH PLANETARY GEAR TRANSMISSIONS

OBJECTIVES

After completing this chapter, you should:

- Understand how power can be transferred through a planetary gearset to produce the various ratios needed in a transmission.

- Understand the role of the driving and reaction members in producing these different power flows.

- Be familiar with the basic gear train arrangements used in modern transmissions.

- Be familiar with the similarity and dissimilarity of the different transmissions.

3.1 RULES OF POWER TRANSFER

The gearset in an automatic transmission must be able to provide a neutral, one or more gear reductions, a 1:1 or direct-drive ratio, a reverse, and in most newer transmissions, an overdrive. At one time, Chevrolet used a transmission (the Turboglide) that had only a single forward speed and no low gear. There were several transmissions (the Chrysler Powerflite, the Ford two-speed and the General Motors Powerglide, some Dynaflows, and the Hydra-matic 300) that had two speeds: a reduction and a direct drive. Most of the newer transmissions are four-speeds with fourth gear being an overdrive.

To provide these various gear ratios, planetary gearsets are combined in different arrangements, and as mentioned earlier, a particular gear design is often used in more than one make and model of transmission. They can use a complex planetary with more than one of the various parts or two or more simple gearsets can be combined in different manners. A complex or compound gearset combines various parts (e.g., one sun gear and two sets of planet gears) into one unit. Knowledge of an arrangement or type of gearset helps us understand several different transmissions without the need to memorize each one. Each gearset must contain all of the possible gear ratios so the power flows can be changed by applying a particular driving member(s) or a particular holding or reaction member.

The different styles of gearsets all use planetary gears. They just combine them in different fashions. The gearsets can be simple sets, complex sets with two or more planetary sets, or simple sets with portions combined in different ways. Power flow through a simple gearset was described in Chapter 1 if you wish to review the various possibilities. Some rules of power flow through a planetary gearset that we should remember are:

- When there is no driving member or no reaction member, neutral results.
- When the carrier is the output, a reduction occurs.
- When the carrier is the input, an overdrive occurs.
- When the carrier is the reaction member, a reverse occurs.
- When the sun gear is the output, an overdrive occurs.
- When the sun gear is the input, a reduction occurs.
- When the sun gear is the reaction member, an overdrive or a reduction occurs.
- When there are two driving members and no reaction member, direct drive occurs.
- When one external gear drives another, reverse rotation occurs.
- When an external gear drives an internal gear or vice versa, same-direction rotation occurs.

In discussing transmission power flow, the input shaft always turns in a clockwise direction, and in a rear-wheel drive (RWD) transmission, we normally view that direction from the front of transmission looking rearward. With a transaxle, this can become confusing be-

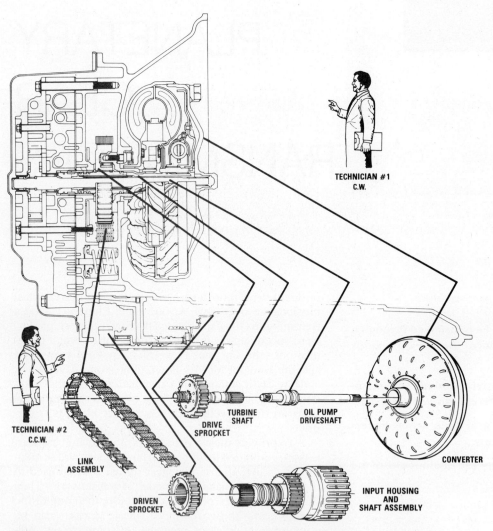

Fig. 3-1. When we view a transmission or a transaxle from the engine end as seen by technician 1, the rotation of the torque converter and turbine shaft is in a clockwise direction. With transmissions and transaxles, which are driven by a chain and sprockets, the rotation of the gear train as seen by technician 2 is counterclockwise. *(Copyrighted material reprinted with permission from Hydra-matic Div., GM Corp.)*

cause the front of the transmission is usually at the right side of the car. When we view the transmission from the right side, the input shaft will turn in a clockwise direction. Some transmissions and transaxles use a drive chain to couple the torque converter, pump, and valve body section to the main gear section. On these units, the first input will be clockwise (viewed from the right), and the second input into the main case will be counterclockwise because it is normally viewed from the left (Fig. 3-1). (Imagine watching a clock's hands rotate from the back side.) Also, most features of a transaxle are the same as a transmission; in the following discussions, we will treat them as the same and only refer specifically to a transmission or a transaxle when necessary.

3.1.1 Park

Every automatic transmission and transaxle includes a *park* range, and most of us realize that this prevents the transmission's output shaft from turning. Park consists of a set of gearlike teeth that are attached to the output member of the planetary gearset, the governor support, or a separate park gear and a park pawl (Fig. 3-2). The park pawl (sometimes called a lever) is mounted on a pivot pin that is mounted in the case (Fig. 3-3).

In all gear positions except park, the park pawl is held away from the park gear teeth by a spring. When the gear selector is moved to PARK, a circular cam on the end of the park-actuating rod pushes the pawl into mesh with the gear teeth, and this will hold the gear and output shaft stationary (Fig. 3-4). The actuating cam is spring loaded so that if the gear teeth are in the wrong position, the gear selector lever can still be shifted into park, and if the output shaft rotates, the spring moves the cam, which in turn moves the pawl into complete engagement.

3.2 THE SIMPSON GEAR TRAIN

As mentioned earlier, the Simpson gear train is a three-speed commonly used in several different transmission/

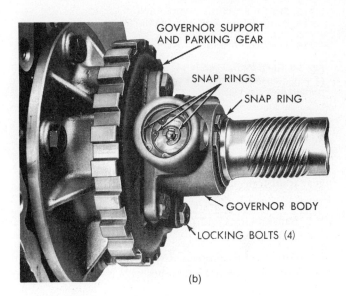

Fig. 3-2 The park gear is often built (a) onto one of the members of the planetary gearset or (b, c) as a separate member attached to the transmission output shaft [(a) is also the output ring gear and (b) also supports the governor.] [(a) Copyrighted material reprinted with permission from Hydra-matic Div., GM Corp. (b) Courtesy of Chrysler Corporation. (c) Courtesy of Ford Motor Company.]

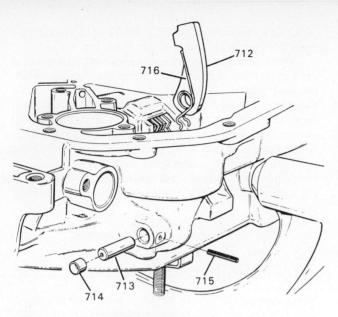

ILL. NO.	DESCRIPTION
712	PAWL, PARKING BRAKE
713	SHAFT, PARKING BRAKE PAWL
714	PLUG, STEEL CUP (.38 DIAMETER)
715	PIN, SLOTTED SPRING
716	SPRING, PARKING PAWL RETURN

Fig. 3-3 The park pawl spring is used to move the park pawl out of engagement with the parking gear. *(Copyrighted material reprinted with permission from Hydra-matic Div., GM Corp.)*

transaxle models. The majority of the automatic transmissions used in domestic cars during the 1960s and 1970s used this gear train. The better known domestic (produced or used in cars made in America) Simpson gear train transmissions are:

Chrysler Corporation: Torqueflite A-727 and A-904 transmissions and the A-404, A-413, A-415, and A-470 transaxles.

Ford Motor Company: Cruisomatic C3, C4, C5, and C6 and Jatco transmissions.

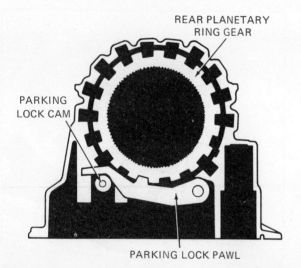

Fig. 3-4 When the control rod is moved, the locking cam pushes the pawl into engagement with the parking gear. *(Courtesy of Chrysler Corporation.)*

General Motors: The THM 200, 250, 350, 375, 400, and 425 transmissions and 125 and 325 transaxles.

Although these transmissions are similar, they are not identical on the inside. They all have two input or driving clutches and a one-way reaction clutch. The power flow paths are all the same, but the reaction members vary (Fig. 3-5). To help understand these different arrangements, we will group them into similar types. Type 1 transmissions use a band for the manual first- and reverse-gear reaction member. Type 2 transmissions use a multiple-disc clutch as a first- and reverse-gear reaction member. Type 1 and Type 2 transmissions use a band to hold the sun gear for the second-gear reaction member. Type 3 and 4 transmissions use a multiple-disc clutch and a one-way clutch for the sun gear reaction member in second gear.

The Hydra-matic THM 400 and 425 use a band for the first- and reverse-gear reaction member and a clutch for the second-gear reaction member so we class them as Type 3 units. Also, the THM 400 and 425 have the gearset turned end for end. In this arrangement, the input ring gear is at the back, and the output ring gear is at the front (Fig. 3-6).

As mentioned earlier, the Simpson gearset consists of one double sun gear that is meshed with the planet gears of two different carriers. One of these carriers (usually the front) is attached to the output shaft, and the other carrier is arranged so it can serve as a reaction member. It is called the *reaction carrier* in some transmissions. The ring gear of one of these gearsets (usually the front) is splined to a driving clutch so it is an input. The other ring gear is attached to the output shaft so it is the output ring gear. The sun gear is attached to a driving clutch through an input shell so it can be an input, but it can also be held stationary by a band or a multidisc holding clutch and can be a reaction member. In most of these transmissions, both ends of the sun gear are the same size. In some transmissions, the rear section has a smaller sun gear so that first gear can have a lower ratio.

3.2.1 Type 1 Gear Train

A Type 1 gear train is a three-speed Simpson gear train that uses bands for reaction members. Figure 3-7 shows a power flow schematic for Type 1 units, and Table 3.1 is a band and clutch application chart for the various gears.

3.2.2 Neutral

Like many other gearsets, neutral occurs by not applying either clutch. Power enters the transmission from the torque converter but only travels as far as the released clutch (Fig. 3-8).

3.2.3 First Gear

This gearset has two slightly different first gears because there are two different methods that can be used to hold the reaction carrier. A one-way clutch is one method, and a band (Type 1) or holding clutch (Type 2) is the

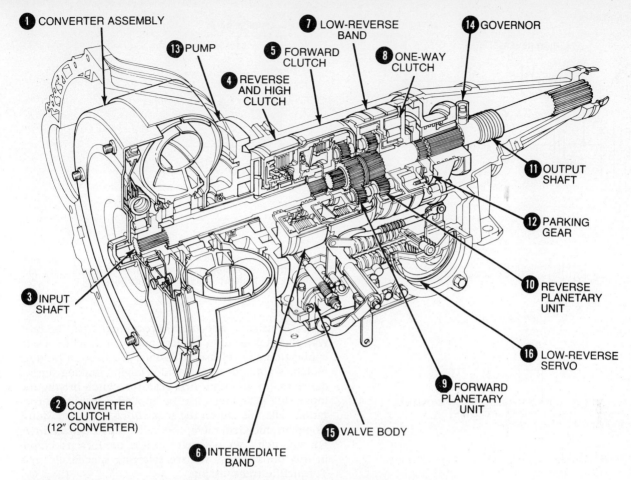

Fig. 3-5 Simpson gear train transmission. Note that there are two gearsets that use the same sun gear. Also note that one carrier and the sun gear can be reaction members, the sun gear and one ring gear can be input members, and the other carrier and ring gear are output members. *(Courtesy of Ford Motor Company.)*

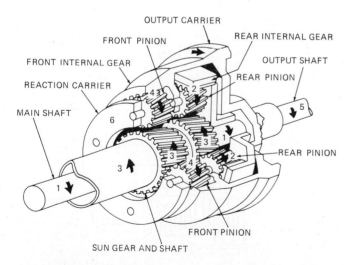

Fig. 3-6 Planetary gearset from a THM 400. Note the reversed position of the gearset. The numbers show the first gear power flow with the mainshaft (1) as input and the reaction carrier (6) held. *(Copyrighted material reprinted with permission from Hydra-matic Div., GM Corp.)*

TYPE 1 GEAR TRAIN

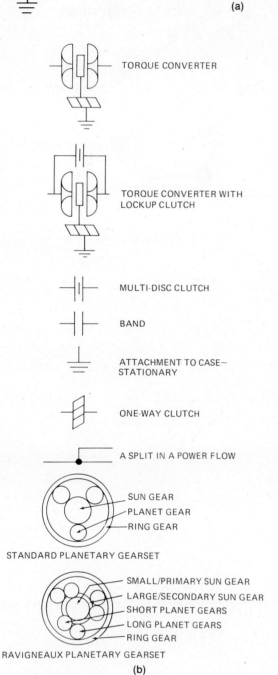

Fig. 3-7 (a) Schematic view of a Type 1 gear train arrangement. The straight lines represent the shafts that interconnect the various parts. (b) Legend explains the other symbols.

other. In drive 1, also called *breakaway first*, the one-way clutch is used; by its very nature, it provides a self-application and release. In manual 1, a band or holding clutch is applied; this provides engine braking during deceleration. In either first gear, the clutch driving the input ring gear must also be applied. Chrysler Corporation calls this clutch the *rear clutch*, and Ford Motor Company and General Motors call it the *forward clutch*. For simplicity, we will refer to it as the forward clutch in this book unless we are referring specifically to a Torqueflite transmission.

When the forward clutch is applied, the input shaft from the torque converter becomes connected to the input ring gear (Fig. 3-9). If there is sufficient engine speed, the turbine in the torque converter will turn clockwise, driving the input shaft and ring gear. At this

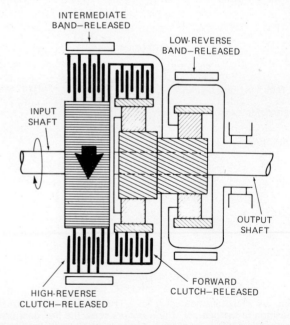

Fig. 3-8 If both the high–reverse and the forward clutches are released, the transmission is in neutral. *(Courtesy of Chrysler Corporation.)*

TABLE 3.1 Band and Clutch Application Chart—Type 1

Gear Range	Forward Clutch	High–Reverse Clutch	Intermediate Band	Low–Reverse Band	One-way Clutch
D1	XXX				XXX
D2	XXX		XXX		
D3	XXX	XXX			
I1	XXX				XXX
I2	XXX		XXX		
L1	XXX			XXX	XXX
R		XXX		XXX	

Note: D = drive, I = intermediate, L = low, and R = reverse. One-way clutches and sprags self-apply; they are shown as applied whenever they can lock up and be effective. In L1, the band is the major reaction member and the one-way clutch the minor.

Type 1 transmissions are the Chrysler Torqueflite A-904, A-998, A-999, and A-727 and the Ford C3, C4, and C5.

Type 1 transaxles are the Chrysler A-404, A-413, A-415, and A-470.

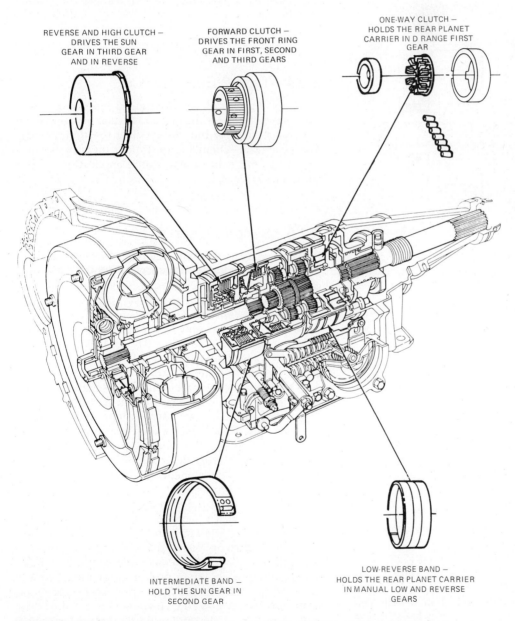

Fig. 3-9 In a Simpson gear train transmission, the reverse and high clutch is arranged so it can drive the sun gear, and the forward clutch is arranged so it can drive the front ring gear. *(Courtesy of Ford Motor Company.)*

time, there is no reaction member in the front gearset other than the output carrier, which is connected to the drive shaft, wheels, and load of the car. Because of this load, the carrier acts as a reaction member and produces a reverse reduction rotation of the sun gear. As the car starts moving, the output carrier becomes a rotating reaction member (Fig. 3-10). The sun gear becomes the input for the rear gear unit, and the reaction carrier, being held from rotating counterclockwise by the one-way clutch, is the reaction member. This arrangement also produces a reverse reduction, and the two reversals become forward rotation. The ratio produced is between 2.45:1 and 2.74:1 depending on the gear sizes in the various transmissions. This power flow produces a gear reduction when the engine drives the car, but it coasts during deceleration because the one-way clutch overruns in a clockwise direction as the power flow is reversed.

When the gear selector is moved to manual 1 (low), the reaction carrier is held by either a band (Type 1) or multiple-disc clutch (Type 2). The Torqueflite transmission family uses a band called the *low and reverse band*. The Ford C3, C4, and C5 and the GM THM 400 and 425 use a band called the *low–reverse band* that is almost the same as the Torqueflite. The Ford C6 and the GM THM 125, 200, 250, 325, and 350 use a multiple-disc clutch called a *low–reverse clutch* (Fig. 3-11). The power flow is exactly the same as drive 1 except that power can be transmitted from the drive shaft to the engine during deceleration. This provides engine braking as the car slows down. This engine braking is easily noticed by comparing the deceleration of a car in drive 1 and manual 1. For simplicity, we will call this holding member either a low–reverse band or a clutch in this text.

3.2.4 Type 2 Gear Train

The Type 2 gear train is a three-speed Simpson gear train that uses a clutch for the low and reverse reaction member and a band for the intermediate reaction member. The gear train schematic for Type 2 units is shown in Fig. 3-12, and a band and clutch application chart is shown in Table 3.2.

3.2.5 Second Gear

For second gear, the sun gear must be held stationary in reaction. This is done by a band in Type 1 and 2 units or a multiple-disc clutch plus a one-way clutch in Type 3 and 4 units. Torqueflite transmissions use a band called a *kickdown band*. The Ford transmissions and the GM THM 125, 200, 250, and 325 transmissions use a band called an *intermediate band*. The GM THM 350, 400, and 425 transmissions use an *intermediate clutch* and an *intermediate roller clutch* (a one-way clutch) (Fig. 3-13). Because of the roller clutch, these units also have two slightly different power flows in second gear: drive 2 and manual 2. Those transmissions using an intermediate clutch and roller clutch also use an intermediate band that is applied during manual 2 to provide engine braking during deceleration. For simplicity, we will call this reaction member either an intermediate band or clutch in this text.

In second gear, the forward clutch stays applied so the input ring gear remains the driving member. When the intermediate band or clutch is applied, the sun gear, which was turning counterclockwise in first gear, is brought to a stop and becomes the reaction member. The planet gears in the front gearset are forced to walk around it, and this forces the output carrier and output shaft to rotate (Fig. 3-14). This produces a reduction of 1.48:1 to 1.57:1 depending on the particular gearset. Although the gears in the rear section of the gearset are rotating, they are not involved in the power flow.

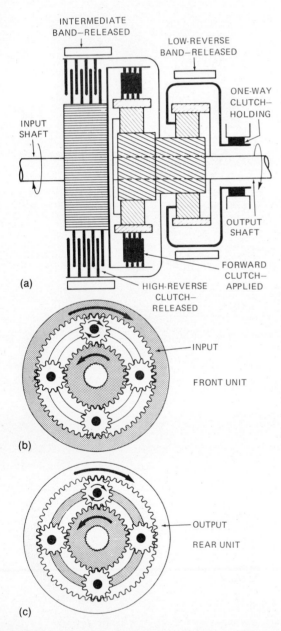

Fig. 3-10 (a) In drive 1, the front ring gear is driven while the rear carrier is held by the one-way clutch. A reverse reduction occurs in (b) the front unit and (b) the rear unit. *[(a) Courtesy of Chrysler Corporation. (b, c) Copyrighted material reprinted with permission from Hydra-matic Div., GM Corp.]*

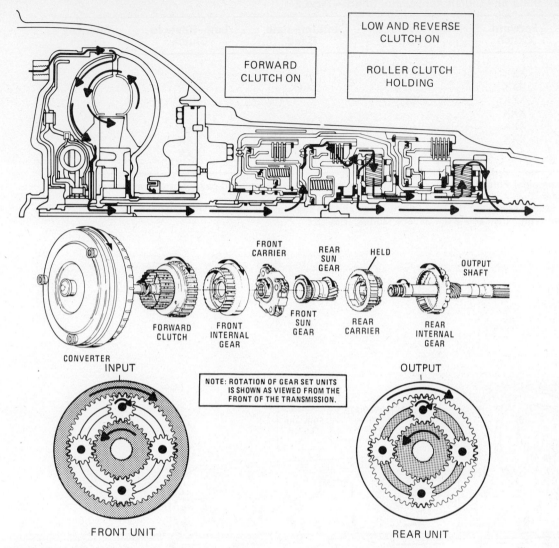

Fig. 3-11 In a Type 2 transmission, the reaction carrier is held by a multiple-disc clutch in manual 1. Otherwise, the power flow is identical to a Type 1 unit. *(Copyrighted material reprinted with permission from Hydra-matic Div., GM Corp.)*

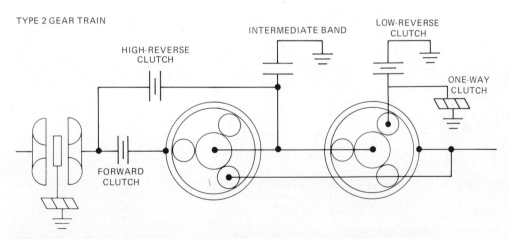

Fig. 3-12 Power flow schematic for a Type 2 transmission. Note that the only difference is the use of a multiple-disc clutch instead of a band to hold the reaction carrier.

TABLE 3.2 Band and Clutch Application Chart—Type 2

Gear Range	Forward Clutch	High–Reverse Clutch	Intermediate Band	Low–Reverse Clutch	One-way Clutch
D1	XXX				XXX
D2	XXX		XXX		
D3	XXX	XXX			
I1	XXX				XXX
I2	XXX		XXX		
L1	XXX			XXX	XXX
R		XXX		XXX	

Note: In some older Ford transmissions, the D2 range will begin with a second-gear start and allow an upshift into third gear.

Type 2 transmissions are the Ford C6 and Jatco and the General Motors THM 200, THM 250, and THM 325. The only Type 2 transaxle is the General Motors THM 125.

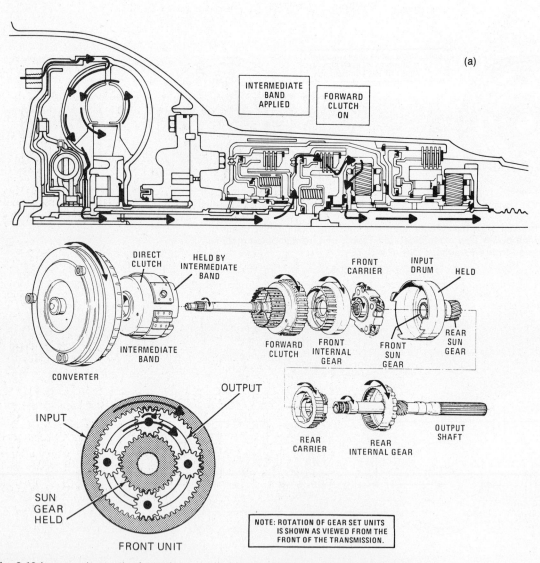

Fig. 3-13 In second gear the front ring gear is driven and the sun gear is held by (a) a band in Type 1 and 2 transmissions or (b) a multi-disc clutch in Type 3 and 4 units. *(Copyrighted material reprinted with permission from Hydra-matic Div., GM Corp.)*

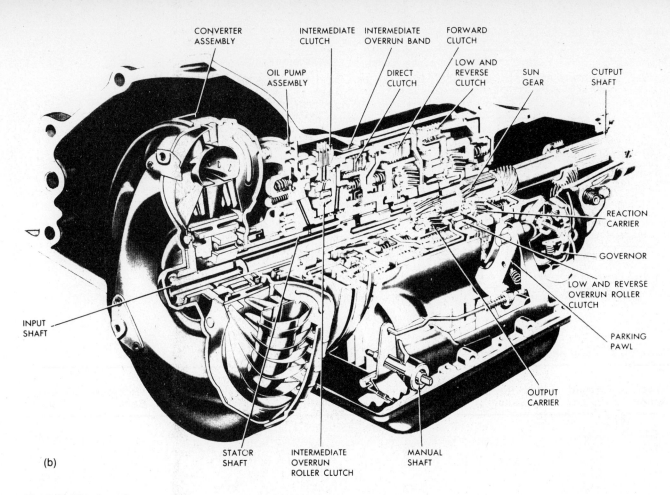

Fig. 3-13 (*Continued*)

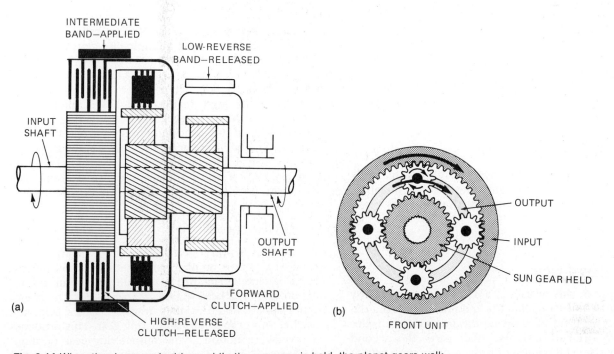

Fig. 3-14 When the ring gear is driven while the sun gear is held, the planet gears walk around the sun gear and force the carrier to revolve at a reduced speed.

59

TYPE 3 GEAR TRAIN

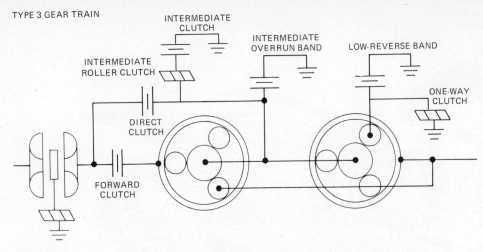

Fig. 3-15 Power flow schematic for a Type 3 transmission. Note that it is the same as a Type 1 unit except for an intermediate clutch and an intermediate one-way clutch.

TABLE 3.3 Band and Clutch Application Chart—Type 3

Gear Range	Forward Clutch	High–Reverse Clutch	Intermediate Clutch	Intermediate One-way Clutch	Intermediate Band	Low–Reverse Band	One-way Clutch
D1	XXX						XXX
D2	XXX		XXX	XXX			
D3	XXX	XXX					
I1	XXX						XXX
I2	XXX		XXX	XXX	XXX		
L1	XXX					XXX	XXX
R		XXX				XXX	

Note: Type 3 transmissions are the General Motors THM 400 and THM 425.

3.2.6 Type 3 Gear Train

We class those transmissions that use an intermediate clutch along with a low–reverse band for reaction members as Type 3 units. The gear train schematic for Type 3 units is shown in Fig. 3-15, and a band and clutch application chart is given in Table 3.3.

3.2.7 Type 4 Gear Train

The THM 350 transmission uses an intermediate clutch along with a low–reverse clutch for reaction members and is classed as a Type 4 unit. The gear train schematic for Type 4 units is shown in Fig. 3-16, and a band and clutch application chart is shown in Table 3.4.

3.2.8 Third Gear

Third gear in this gearset is a 1:1 ratio produced by applying both driving clutches and either releasing the intermediate band or allowing the intermediate roller clutch to overrun. Like the other parts, this second clutch has several names. Chrysler Corporation calls it a *front clutch*, Ford Motor Company calls it a *high–reverse clutch*, and General Motors calls it a *direct clutch*. In all cases, it is a multiple-disc clutch that connects the turbine shaft to the sun gear drive shell. For simplicity, we will call this clutch a high–reverse clutch in this text.

When the high–reverse clutch is applied at the same time as the forward clutch, both the ring gear and the sun gear of the front unit are driven, and this locks the gearset. The planet gears and carrier in the front unit are forced to rotate along with the sun and ring gears (Fig. 3-17). The gears in the entire gearset rotate as a mass with no relative motion between them.

3.2.9 Reverse

Reverse in a Simpson gear train occurs when the high–reverse clutch is applied along with the low–reverse band or clutch. This drives the sun gear while the reaction carrier is held stationary (Fig. 3-18). The planet gears act as idlers as they transfer power from the smaller sun gear to the larger ring gear. This produces a reverse reduction at the output ring gear of about 2.07:1 to 2.22:1. The gears in the front set turn, but they are not involved in the power flow.

TYPE 4 GEAR TRAIN

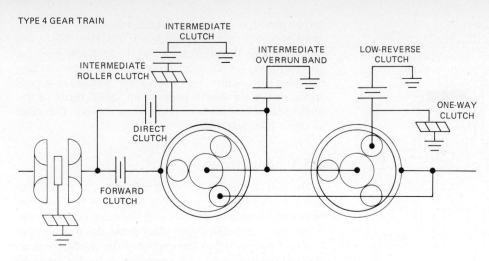

Fig. 3-16 Power flow schematic for a Type 4 transmission. Note that it is the same as a Type 2 unit except for an intermediate clutch and an intermediate one-way clutch.

TABLE 3.4 Band and Clutch Application Chart—Type 4

Gear Range	Forward Clutch	High–Reverse Clutch	Intermediate Clutch	Intermediate One-way Clutch	Intermediate Band	Low–Reverse Clutch	One-way Clutch
D1	XXX						XXX
D2	XXX		XXX	XXX			
D3	XXX	XXX					
I1	XXX						XXX
I2	XXX		XXX	XXX	XXX		
L1	XXX					XXX	XXX
R		XXX				XXX	

Note: The only Type 4 transmission is the General Motors 350.

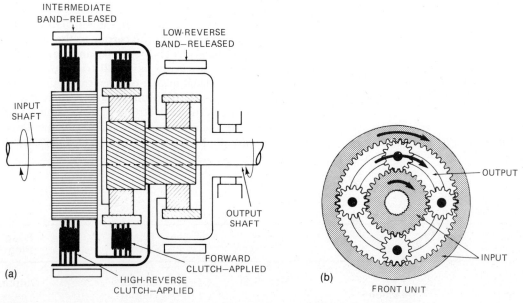

Fig. 3-17 In high gear, both driving clutches are applied so two members (the ring and sun gears) of the same gearset are driven. This locks the gears and produces a 1:1 gear ratio.

61

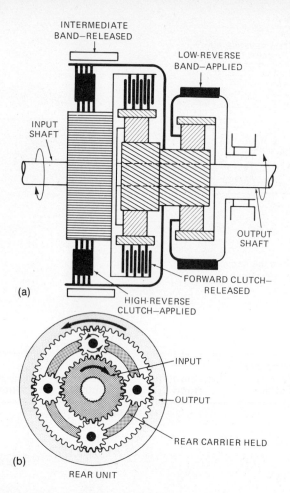

Fig. 3-18 In reverse the sun gear is driven while the rear carrier is held. The planet gears act as idler gears and cause the ring gear to revolve in a reverse direction at a reduced speed.

A summary of the Simpson gear train, input, reaction, and output members used to produce the two reduction, one direct and one reverse, ratios is provided in Table 3.5.

3.3 SIMPSON GEAR TRAIN PLUS AN ADDITIONAL GEARSET

At this time there are five gear train arrangements using the Simpson gear train plus an additional planetary gearset to produce a four-speed transmission or transaxle. Most of these—the Chrysler A-500, the Ford A4LD, the GM THM 200-4R and 325-4L, and the Jeep AW-4—have an overdrive fourth gear. The now-out-of-production Chrysler Torqueflite A-345 has three reduction speeds and a direct for fourth gear. In all cases, three of the four speeds, neutral, and reverse are exactly the same as those just described.

3.3.1 Overdrive

In the A4LD, the THM 200-4R, and the AW-4 transmissions, the overdrive gearset is built into the area between the torque converter and the main gearset (Figs. 3-19 and 3-20). The input shaft from the torque converter is connected to the carrier of the overdrive gearset, and the ring gear of the overdrive gearset is arranged so it becomes the input of the main gearset. In the THM 325-4L, the overdrive unit is built in the rear of the torque converter section so it can also change the speed of the input to the main gear section (Fig. 3-21). The Chrysler A-500 has the overdrive gearset built into the transmission extension housing so it can cause a speed increase between the main gearset and the output shaft (Fig. 3-22). Each of these transmissions uses a different arrangement of bands and clutches so each falls in a different category or type.

The input member for these overdrive gearsets (a simple planetary gearset) is the planet carrier, and the output is the ring gear. The sun gear is arranged so it can be a reaction member and be held stationary by a band or by a multiple-disc clutch. There is also a multiple-disc clutch that can lock the carrier to the sun gear and a one-way clutch. In the A4LD, the one-way clutch is placed so the carrier can drive the ring gear. In the others, the one-way clutch is placed so the sun gear cannot turn faster than the carrier. The different friction members have different names depending on the particular transmission (Fig. 3-23).

In most of these transmissions, power is transmitted through the overdrive gearset at a 1:1 ratio by the one-way clutch in drive 1, 2, and 3. In the A4LD, the power is transferred directly from the carrier to the ring gear; in the 200-4R, the gearset is locked because the sun gear cannot overrun the carrier as it will try to do. The A-500 is a little unusual in that one hydraulic piston and return spring is used for both the direct clutch and the overdrive clutch. The very strong return spring is used to release the overdrive clutch and apply the direct clutch. Hydraulic pressure at the piston releases the di-

TABLE 3.5 Simpson Gear Train Summary

Gear	Front Ring	Sun	Rear Carrier	Front Carrier	Rear Ring
First	Input	—	Reaction	Output	Output
Second	Input	Reaction	—	Output	—
Third	Input	Input	—	Output	—
Reverse	—	Input	Reaction	—	Output

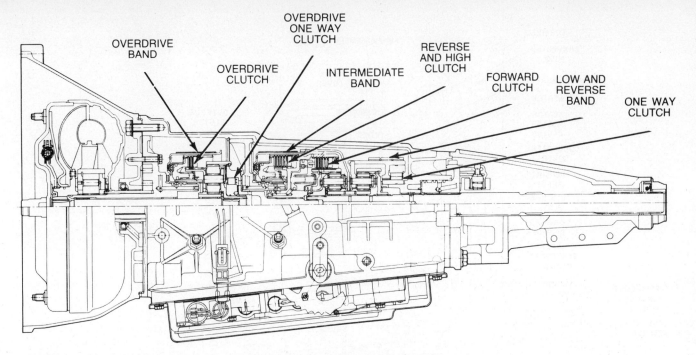

Fig. 3-19 Partially cutaway view of a Ford A4LD transmission. Careful study of this illustration will show, from the left, the torque converter, the overdrive unit, and a Type 1 gearset at the right. *(Courtesy of Ford Motor Company.)*

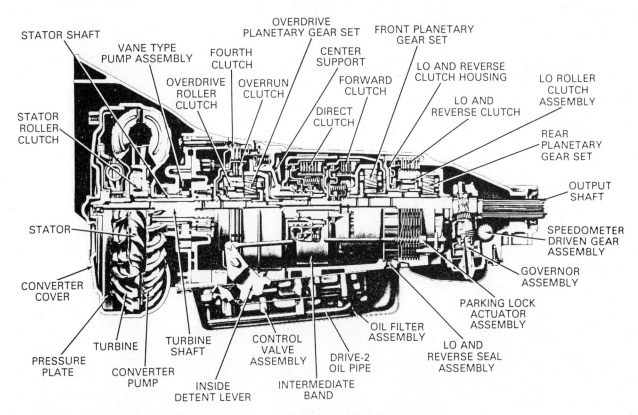

Fig. 3-20 General Motors THM 200-4R transmission. Careful study will show an overdrive unit just in front of a Type 2 transmission. *(Copyrighted material reprinted with permission from Hydra-matic Div., GM Corp.)*

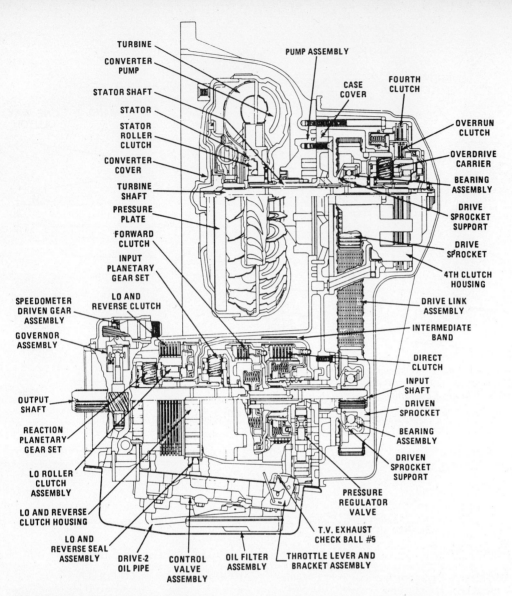

Fig. 3-21 General Motors THM 325-4L transmission. Careful study will show the overdrive assembly at the upper right and a Type 2 gear train at the bottom. *(Copyrighted material reprinted with permission from Hydra-matic Div., GM Corp.)*

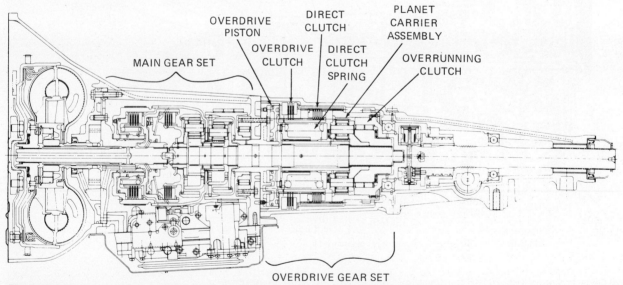

Fig. 3-22 Chrysler A-500 transmission. Careful study will show the overdrive assembly at the right, in back of a Type 1 gearset. *(Courtesy of Chrysler Corporation.)*

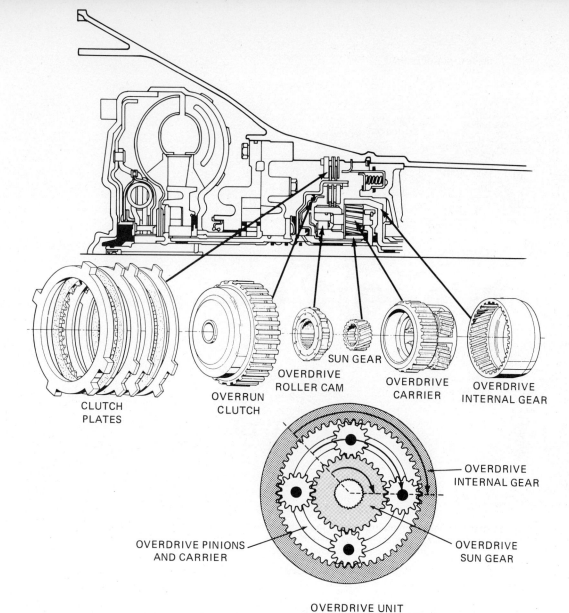

Fig. 3-23 Overdrive gearset. Note that the overdrive roller cam (clutch) can drive the sun gear and that this will lock the gearset, producing a 1:1 ratio. *(Copyrighted material reprinted with permission from Hydra-matic Div., GM Corp.)*

rect clutch and then almost immediately applies the overdrive clutch. With this arrangement, the gearset is locked in either direct drive or overdrive with the overrunning clutch transferring power while the upshift or downshift is made.

Also in these units, fourth gear occurs by applying the band or clutch that holds the sun gear stationary in reaction. The rotation of the carrier forces the planet gears to walk around the sun gear, and this drives the ring gear at a faster ratio of 0.69:1 to 0.75:1 depending on the transmission (Fig. 3-24).

Like other power flows using a one-way clutch, these gearsets overrun during deceleration and do not produce engine braking. To prevent this occurrence in manual 1 or 2, the clutch inside the assembly, overdrive clutch in the A4LD, overrun clutch in the 200-4R, or the OD direct clutch in the AW-4 is applied. This locks the overdrive gear assembly so it operates in direct drive in both directions.

A summary of the input, reaction, and output members of the overdrive gearsets to produce the various ratios is given in Table 3.6.

3.3.2 Type 5 Gear Train

A Type 5 gear train is a Type 1 gear train combined with an overdrive, and the overdrive unit uses two multiple-disc clutches and an overrunning clutch. The gear train schematic for Type 5 units is shown in Fig. 3-25, and a band and clutch application chart is given in Table 3.7.

3.3.3 Type 6 Gear Train

The Type 6 gear train combines a Type 1 gear train with an overdrive unit. In this case the overdrive unit uses one band, one multiple-disc clutch, and a one-way clutch. The gear train schematic for Type 6 units is

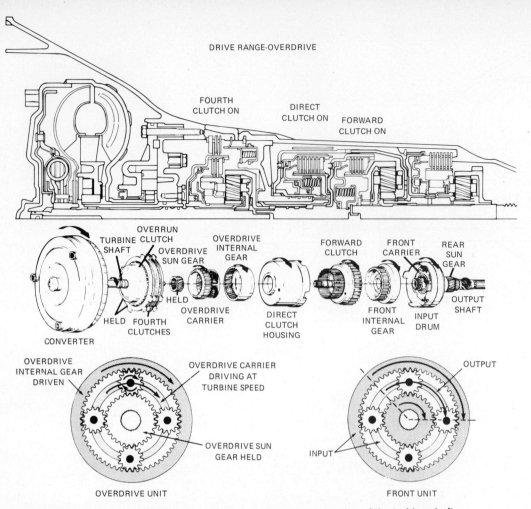

Fig. 3-24 If the fourth clutch is applied so the sun gear is held, rotation of the turbine shaft and carrier forces the planet gears to walk around the sun gear and drive the ring gear at a faster speed. *(Copyrighted material reprinted with permission from Hydra-matic Div., GM Corp.)*

shown in Fig. 3-26, and a band and clutch application chart is given in Table 3.8.

3.3.4 Type 7 Gear Train

The THM 200-4R and 325-4L combine a Type 2 gear train with the overdrive unit, and we class them as Type 7. In this case the overdrive unit uses two multiple-disc clutches and a one-way clutch. The gear train schematic for Type 7 units is shown in Fig. 3-27, and Table 3.9 gives a band and clutch application chart.

TABLE 3.6

Gear	OD Sun Gear	OD Carrier	OD Ring Gear
First	Input	Input	Output
Second	Input	Input	Output
Third	Input	Input	Output
Fourth, OD	Reaction	Input	Output
Reverse	Input	Input	Output

3.3.5 Type 8 Gear Train

The AW-4 combines a Type 4 gear train with the overdrive unit, and we class it as Type 8 (Fig. 3-28). It also uses two multiple-disc clutches and a one-way clutch. The gear train schematic for Type 8 units is shown in Fig. 3-29, and a band and clutch application chart is given in Table 3.10.

3.4 RAVIGNEAUX GEARSETS

The unique feature of the Ravigneaux gearset is that it uses a carrier that has two sets of intermeshed planet gears, two different-size sun gears, and a single ring gear. The pinion gears that mesh with the small sun gear are short and also mesh with the other, long pinion gears. The long pinion gears also mesh with the large sun gear and the ring gear (Fig. 3-30). In some transmissions, the ring gear is in mesh with the short pinions. This gearset is used in some two-speed transmissions (the Chrysler Powerflite, the Ford two-speed, and the General Motors Powerglides, Dynaflows, and THM 300), two three-speed transmissions (the Ford FMX and General Motors

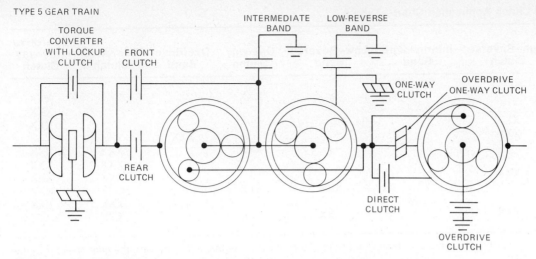

Fig. 3-25 Power flow schematic for a Type 5 gear train. Note that it essentially is a Type 1 unit with an overdrive assembly at the rear.

TABLE 3.7 Band and Clutch Application Chart—Type 5

Gear Range	Forward Clutch	High–Reverse Clutch	Intermediate Band	Low-Reverse Band	One-way Clutch	Direct Clutch	Overdrive One-way Clutch	Overdrive Clutch
D1	XXX				XXX	XXX	XXX	
D2	XXX		XXX			XXX	XXX	
D3	XXX	XXX					XXX	XXX	
D4	XXX	XXX						XXX
I1	XXX				XXX	XXX	XXX	
I2	XXX		XXX			XXX	XXX	
L1	XXX			XXX	XXX	XXX	XXX	
R		XXX		XXX		XXX	XXX	

Note: This transmission uses a six-position gear selector with overdrive selected automatically in drive unless a switch on the instrument panel is moved to switch overdrive off.
The only Type 5 transmission is the A-500.

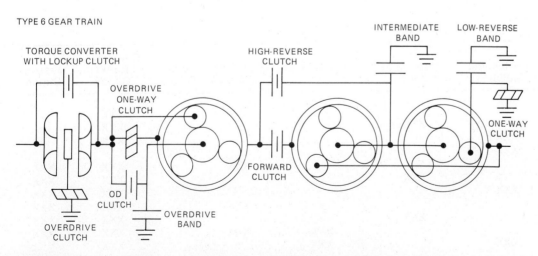

Fig. 3-26 Power flow schematic for a Type 6 gear train. Note that it essentially is a Type 1 unit with an overdrive assembly at the front.

67

TABLE 3.8 Band and Clutch Application Chart—Type 6

Gear Range	Forward Clutch	High–Reverse Clutch	Intermediate Band	Low–Reverse Band	One-way Clutch	Overdrive Band	Overdrive Clutch	Overdrive One-way Clutch
O1	XXX				XXX			XXX
O2	XXX		XXX					XXX
O3	XXX	XXX						XXX
O4	XXX	XXX				XXX		
D1	XXX				XXX		XXX	XXX
D2	XXX		XXX				XXX	XXX
D3	XXX	XXX					XXX	XXX
L1	XXX			XXX	XXX		XXX	XXX
L2	XXX		XXX				XXX	XXX
R		XXX		XXX			XXX	

Note: Gear range O = overdrive; this is often a D with an O around it; Manual 2, L2, is obtained by moving the gear selector to low after a 1–2 shift has occurred.
The only Type 6 transmission is the Ford A4LD.

TYPE 7 GEAR TRAIN

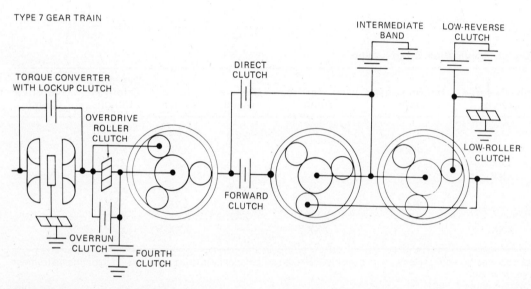

Fig. 3-27 Power flow schematic for a Type 7 gear train. Note that it essentially is a Type 2 unit with an overdrive assembly at the front.

TABLE 3.9 Band and Clutch Application Chart—Type 7

Gear Range	Forward Clutch	High–Reverse Clutch	Intermediate Band	Low–Reverse Clutch	One-way Clutch	Fourth Clutch	Overdrive Roller Clutch	Overrun Clutch
O1	XXX				XXX		XXX	
O2	XXX		XXX				XXX	
O3	XXX	XXX					XXX	
O4	XXX	XXX				XXX		
D1	XXX				XXX		XXX	XXX
D2	XXX		XXX				XXX	XXX
D3	XXX	XXX					XXX	XXX
I1	XXX				XXX		XXX	XXX
I2	XXX		XXX				XXX	XXX
L1	XXX			XXX	XXX		XXX	XXX
R		XXX		XXX			XXX	

Note: Type 7 transmissions are the General Motors THM 200-4R and THM 325-4L.

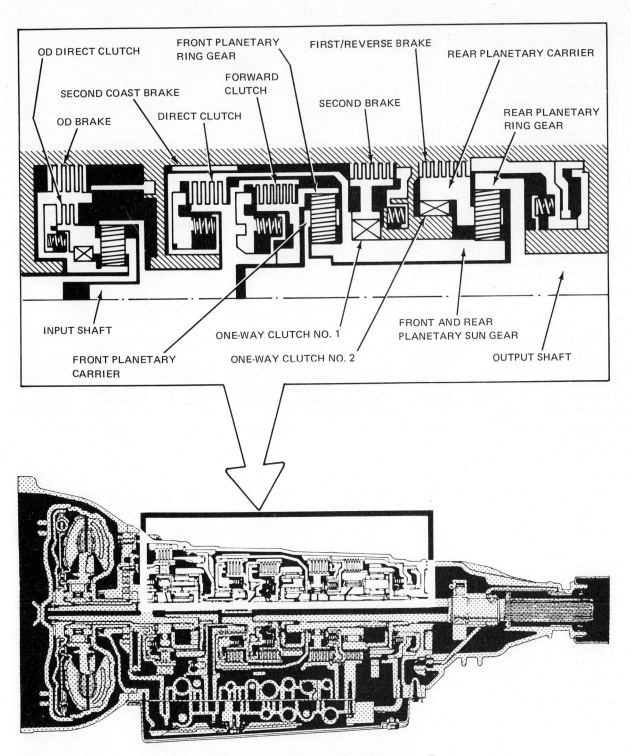

Fig. 3-28 Cutaway view of an AW-4 transmission. Careful study will show that it is a Type 4 transmission with an overdrive assembly in the front part of the case. *(Courtesy of Chrysler Corporation.)*

69

TYPE 8 GEAR TRAIN

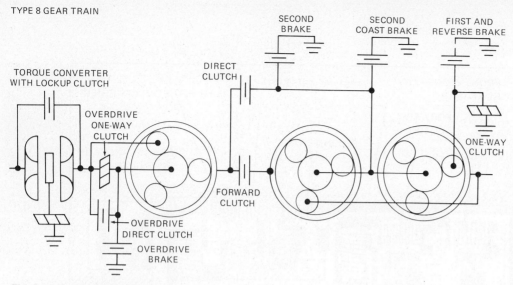

Fig. 3-29 Power flow schematic for a Type 8 gear train. Note that it essentially is a Type 4 unit with an overdrive assembly at the front.

TABLE 3.10 Band and Clutch Application Chart—Type 8

Gear Range	Forward Clutch	Direct Clutch	Intermediate Clutch	Intermediate One-way Clutch	Intermediate Band	First and Reverse Clutch	One-way Clutch	Overdrive Direct Clutch	Overdrive One-way Clutch	Overdrive Brake Clutch
D1	XXX						XXX	XXX	XXX	
D2	XXX		XXX	XXX				XXX	XXX	
D3	XXX	XXX	XXX					XXX	XXX	
D4	XXX	XXX	XXX							XXX
I1	XXX						XXX	XXX	XXX	
I2	XXX		XXX	XXX	XXX			XXX	XXX	
I3	XXX	XXX	XXX					XXX	XXX	
L1	XXX					XXX	XXX	XXX	XXX	
L2	XXX		XXX	XXX	XXX			XXX	XXX	
R		XXX				XXX		XXX	XXX	

Note: The only Type 8 transmission is the AW-4.

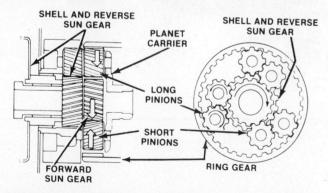

Fig. 3-30 A Ravigneaux gear train uses two sun gears, one carrier with short and long pinions, and one ring gear. Note that the long pinions are in mesh with the larger sun gear, ring gear, and short pinions and that the short pinions are also in mesh with the smaller sun gear. *(Courtesy of Ford Motor Company.)*

THM 180), a three-speed transaxle (the Ford ATX), and a four-speed transmission (the Ford AOD). This gearset is arranged in different manners in these various units.

3.4.1 Two-Speed Arrangement and Operation

Two-speed transmissions were used in the 1960s. They are no longer in production, but so many were produced that they are still encountered. In these transmissions, the small, primary sun gear is attached to the input shaft so it is always an input; the other large sun gear, called the secondary sun gear, is arranged so it can be a reaction member or an input member. The carrier is the output member, and the ring gear can be a reaction member. A driving clutch is placed on the input shaft so the secondary sun gear can be driven, and a band, called a low band, is placed around the clutch drum so the secondary sun gear can be held in reaction. The clutch is often called the *high clutch*, and the band is called a *low band*. The ring gear is held by either a band or a multiple-disc clutch, and this is called either a *reverse clutch* or a *reverse band* (Fig. 3-31).

Neutral occurs in this gearset when none of the driving or reaction members are applied. The primary sun gear turns with the torque converter. This rotates the planet gears, the secondary sun gear, and the ring gear, but the carrier is not driven.

When the low band is applied, the secondary sun gear is held stationary in reaction. At this time, the primary sun gear drives the long pinions in a counterclockwise direction, which in turn drives the short pinions

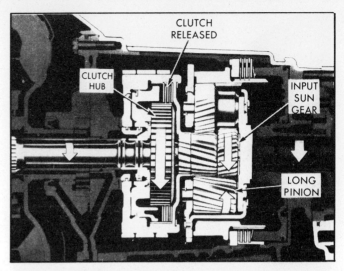

Fig. 3-32 In low gear, the low band is applied to hold the front sun gear stationary; at this time, the long pinions driven by the rear sun gear walk around the front sun gear and drive the carrier. *(Courtesy of Chevrolet Div., GM Corp.)*

in a clockwise direction. As the short pinions walk around the stationary, secondary sun gear, they drive the carrier in a clockwise direction at a reduction ratio of 1.76:1 or 1.82:1 depending on the gearset (Fig. 3-32).

The 1–2, low-to-high shift occurs when the low band is released and the high clutch is applied. At this time, both the primary and secondary sun gears are driven in a clockwise direction, and this motion will try to turn the two sets of planet gears in the same direction.

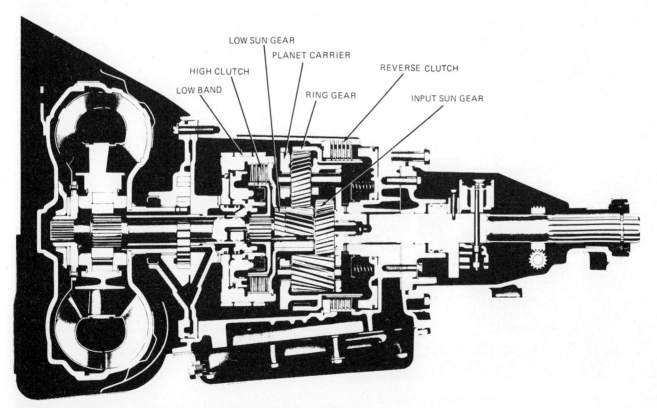

Fig. 3-31 Cutaway view of Powerglide transmission. Note the gear train arrangement and that the larger, rear sun gear is connected directly to the turbine shaft. *(Courtesy of Chevrolet Div., GM Corp.)*

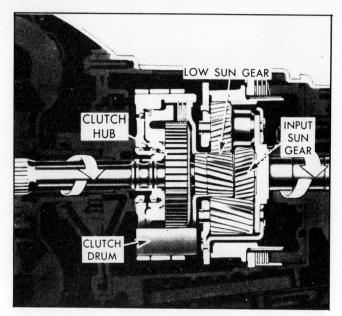

Fig. 3-33 In high gear, the high clutch is applied to drive the front sun gear; this locks the planet gears to produce a 1:1 ratio. *(Courtesy of Chevrolet Div., GM Corp.)*

TABLE 3.11 Gear Ratio Summary

Gear	Large Sun Gear	Small Sun Gear	Ring Gear	Carrier
First, low	Input	Reaction	—	Output
Second, high	Input	Input	—	Output
Reverse	Input	—	Reaction	Output

3.4.2 Type 9 Gear Train

The Type 9 gear train is a two-speed Ravigneaux gearset. We include this transmission in our classifications even though the Powerglide has been out of production for some time because it is still commonly encountered. The gear train schematic for Type 9 units is shown in Fig. 3-35, and a band and clutch application chart is given in Table 3.12.

This causes the gearset to lock and drives the carrier at a direct, 1:1 ratio (Fig. 3-33).

Reverse gear is engaged by applying the reverse band or clutch to hold the ring gear stationary. The power flows from the primary sun gear to the long pinions and on to the short pinions, much like first gear, but in this case, the short pinions are forced to walk around the inside of the ring gear. This forces the carrier to rotate in a counterclockwise, reverse direction at a reduction ratio of 1.76:1 or 1.82:1 (Fig. 3-34).

A summary of the input, reaction, and output members of the various gear ratios is given in Table 3.11.

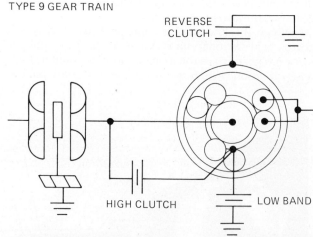

Fig. 3-35 Power flow schematic for Type 9 gear train. It should be noted that the large sun gear is driven directly from the turbine shaft and the small sun gear can be driven by the high clutch or held by the low band.

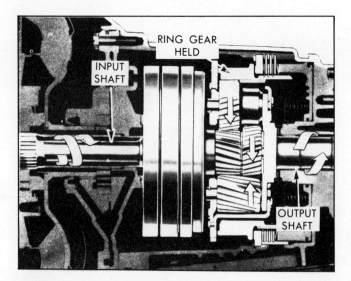

Fig. 3-34 In reverse, the reverse clutch is applied to hold the ring gear stationary; at this time the short pinions walk around the inside of the ring gear and drive the carrier. *(Courtesy of Chevrolet Div., GM Corp.)*

TABLE 3.12 Band and Clutch Application Chart—Type 9

Gear Range	High Clutch	Low Band	Reverse Clutch
D1		XXX	
D2	XXX		
L1		XXX	
R			XXX

Note: Type 9 transmissions are the Chevrolet Powerglide and the General Motors THM 300.

3.4.3 Three-Speed Arrangement and Operation: First Version

There are two three-speed versions of the Ravigneaux gearset. The Ford FMX is the first version. This transmission uses two driving clutches, two bands, and a one-way clutch. The ring gear is the output member, and the carrier with its two sets of pinion gears can be a reaction member held by either a band or the one-way clutch. The one-way clutch race is secured by a stationary transmission part called a *center support*. One sun gear (the smaller, rear, primary one) can be driven by the front, *forward clutch*, and the larger secondary sun gear can be driven by the *high–reverse clutch* or held stationary by the *intermediate band* (Fig. 3-36).

In neutral, both driving clutches are released, and the power flow from the torque converter goes no farther than these clutches.

In first gear, the forward clutch applies to drive the primary sun gear, and the clockwise rotation of the sun gear turns the short pinions in a counterclockwise direction. This may force the carrier in a counterclockwise direction, but it is held by the one-way clutch in drive 1 or by the low and reverse band in manual 1. With the carrier stationary, the rotation of the short pinion gears rotates the long pinions in a clockwise direction, which forces the ring gear to rotate in a clockwise direction at a reduced speed of about 2.4:1 (Fig. 3-37). In manual 1, the carrier is held stationary by the band to provide engine braking during deceleration.

In second gear, the forward clutch stays applied, and the intermediate band is applied. This holds the sec-

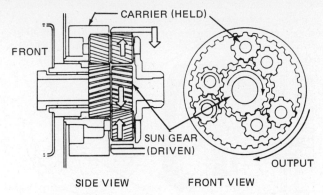

Fig. 3-37 In an FMX, applying the front clutch to drive the small sun gear causes the short pinions to rotate in a counterclockwise direction, which in turn causes the long pinions to rotate in a clockwise direction; holding the carrier stationary forces the ring gear to rotate at a reduced speed. *(Courtesy of Ford Motor Company.)*

ondary sun gear in reaction so the long pinions have to walk around it. The power flow is similar to first gear, from the primary sun gear to the short pinions and then to the long pinions, which will walk around the secondary sun gear. The rotation of the long pinions forces the ring gear to rotate clockwise at a reduced speed of about 1.47:1. As the long pinions walk around the secondary sun gear, the carrier rotates in a clockwise direction and overruns at the one-way clutch (Fig. 3-38).

Third gear is the result of applying both the forward clutch and the high–reverse clutches so both sun gears

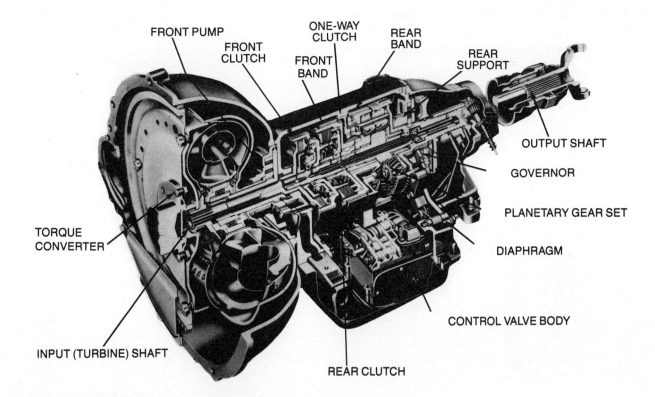

Fig. 3-36 Cutaway view of FMX transmission. Note that the ring gear is connected to the output shaft and that the carrier with its two sets of pinions can be held by either the one-way clutch or the rear band. *(Courtesy of Ford Motor Company.)*

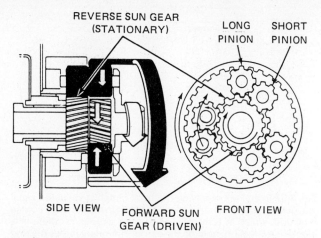

Fig. 3-38 If the large sun gear is held stationary while the small sun gear is driven, the long pinions walk around the large sun gear, and this forces the ring gear to turn and produce second gear. (*Courtesy of Ford Motor Company.*)

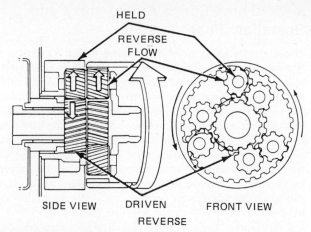

Fig. 3-40 If the carrier is held by the rear band and the small sun gear is driven by the rear clutch, the long pinions rotate in a counterclockwise direction, and this forces the ring gear to also rotate in a counterclockwise direction. (*Courtesy of Ford Motor Company.*)

are driven. Like the two-speed gearset, the two sets of pinion gears try to turn in the same direction, and the gearset locks in direct drive (Fig. 3-39).

To obtain reverse, the high–reverse clutch is applied to drive the secondary sun gear, and the low and reverse band is applied to hold the carrier stationary in reaction. The sun gear forces the long pinions to rotate in a counterclockwise direction, which in turn forces the ring gear to rotate in the same, reverse direction at a reduced speed of about 2:1 (Fig. 3-40).

Table 3.13 provides a summary of the power flows.

TABLE 3.13 Power Flow Summary

Gear	Small Sun Gear	Large Sun Gear	Carrier	Ring Gear
First	Input	—	Reaction	Output
Second	Input	Reaction	—	Output
Third	Input	Input	—	Output
Reverse	—	Input	Reaction	Output

3.4.4 Type 10 Gear Train

A Type 10 gear train is a three-speed Ravigneaux gear train that uses two bands for reaction members and the ring gear as an output member. The gear train schematic for Type 10 units is shown in Fig. 3-41, and a band and clutch application chart is given in Table 3.14.

3.4.5 Three-Speed Arrangement and Operation: Second Version

The General Motors THM 180 transmission and the Ford ATX transaxle use a slightly different arrangement of the Ravigneaux three-speed gearset. Both use two driving

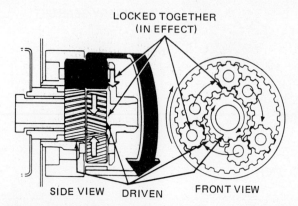

Fig. 3-39 Applying both clutches in an FMX will drive both sun gears; this locks the pinions and produces a 1:1 ratio. (*Courtesy of Ford Motor Company.*)

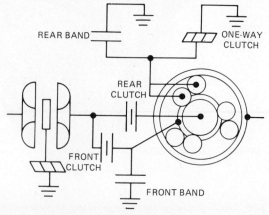

Fig. 3-41 Power flow schematic for a Type 10 gear train. Note that it is similar to a Type 9 gear train.

TABLE 3.14 Band and Clutch Application Chart—Type 10

Gear Range	Forward Clutch	High–Reverse Clutch	Intermediate Band	Low–Reverse Band	One-way Clutch
D1	XXX				XXX
D2	XXX		XXX		
D3	XXX	XXX			
I1	XXX				XXX
I2	XXX		XXX		
L1	XXX			XXX	XXX
R	XXX	XXX		XXX	

Note: The only modern Type 10 transmission is the Ford FMX.

clutches, one holding clutch, a one-way clutch, and a band. The carrier with its two sets of pinion gears is the output member. The ring gear is meshed with the long pinion gearset and is arranged so it can be an input or a reaction member. In both units it can be held stationary by the *reverse clutch*. In the ATX, it can be driven by the *intermediate clutch*, and in the 180, it can be driven by the *second clutch* (Fig. 3-42). The smaller sun gear, called the reverse sun gear in the ATX and the input sun gear in the 180, can be driven by a clutch called the *direct clutch* in the ATX and the *third clutch* in the 180. There is also a one-way clutch in this sun gear drive arrangement that allows the sun gear to over-run the speed of the input shaft. The larger sun gear can be a reaction member. In the ATX, it is called a forward sun gear and can be held by the *low–intermediate band* (Fig. 3-43). In the 180, it is called the reaction sun gear and can be held by the *low band*.

We will use the ATX as an example while describing power flow through this gearset. Remember that power flow through a THM 180 is very similar.

In first gear, the small sun gear is the input, and the large sun gear is the reaction member. The rotation of the small sun gear forces the short pinions to turn in a counterclockwise direction, which in turn causes the long pinions to rotate clockwise. The long pinions then

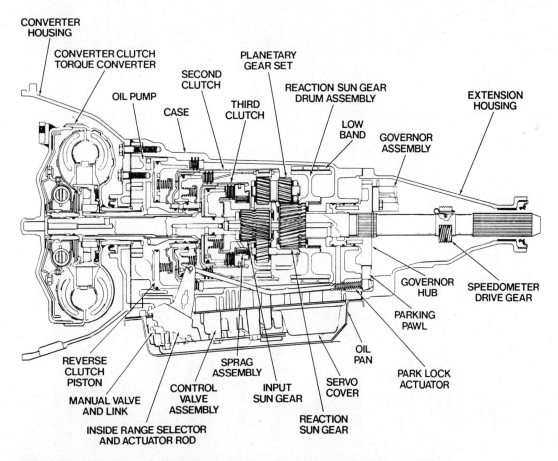

Fig. 3-42 Cutaway view of THM 180C transmission. Note the two sun gears and the long and short pinions in the planetary gearset. Each short pinion is meshed to a long pinion as well as the ring and small sun gears. *(Copyrighted material reprinted with permission from Hydra-matic Div., GM Corp.)*

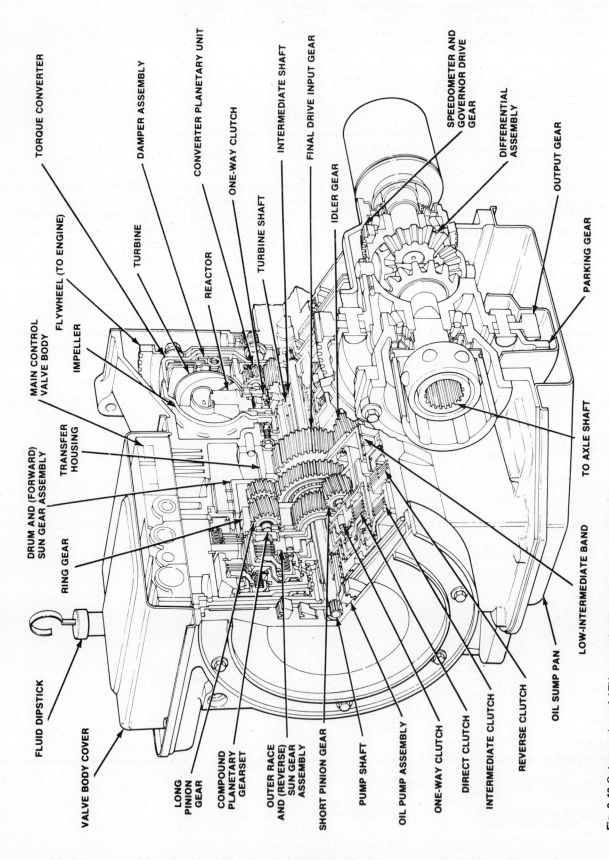

Fig. 3-43 Cutaway view of ATX transaxle. Note the two sun gears and the long and short pinion gears. *(Courtesy of Ford Motor Company.)*

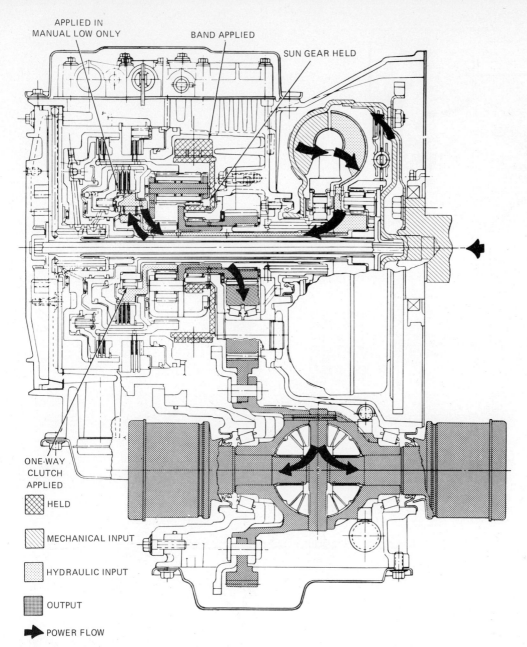

Fig. 3-44 In first gear in an ATX, the large sun gear is held by the band and the small sun gear is driven by the one-way clutch. The short pinions are forced to turn in a counterclockwise direction, which forces the long pinions to turn in a clockwise direction and walk around the large sun gear; this in turn forces the carrier to rotate in a clockwise direction at a reduced speed. *(Courtesy of Ford Motor Company.)*

walk around the stationary, large sun gear and force the carrier to rotate in a clockwise direction at a reduced speed of about 2.4:1 to 2.8:1 (Fig. 3-44).

In second gear, the ring gear is the input and the large sun gear the reaction member. The clockwise rotation of the ring gear forces the long pinions to rotate clockwise and walk around the stationary, large sun gear. This drives the carrier in a clockwise direction at a reduced speed of about 1.5:1 to 1.6:1. At this time, the small sun gear overruns, being driven by the short pinions (Fig. 3-45).

In some versions of the ATX, a split torque converter is used that has an internal planetary gearset. The ring gear of this set is connected to the front of the torque converter through a damper assembly so it will be driven at engine speed; the input sun gear is connected to the turbine shaft; and the carrier is connected to the intermediate shaft. The intermediate shaft (in those transaxles using this torque converter) drives the intermediate clutch in the main gear assembly (Fig. 3-46). This arrangement produces a partial mechanical input of 38 percent in second gear and 94 percent in third gear, which reduces torque converter slippage and improves operating efficiency and fuel mileage.

In third gear, the ring gear and the small sun gear are the inputs, and there is no reaction member. Driving these two members causes the planet gears to lock up and drive the carrier in direct drive (Fig. 3-47).

Reverse occurs when the small sun gear is driven, and the ring gear is the reaction member. The small sun

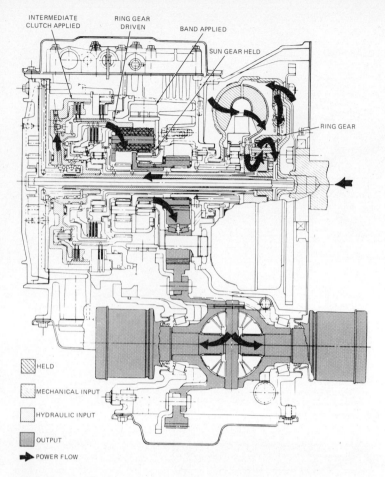

Fig. 3-45 In second gear in an ATX, the ring gear is driven through the intermediate clutch while the large sun gear is held by the band. As the long pinions walk around the sun gear, the carrier is driven at a reduced speed. *(Courtesy of Ford Motor Company.)*

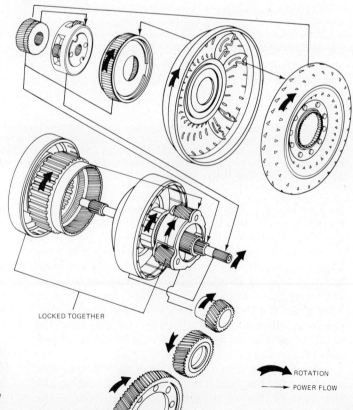

Fig. 3-46 An ATX with a split-torque converter uses one shaft (5) to connect the transmission one-way clutch to the converter turbine/sun gear and another shaft (6) to connect the transmission intermediate clutch to the converter planet carrier. Note that the converter ring gear is driven by the impeller. *(Courtesy of Ford Motor Company.)*

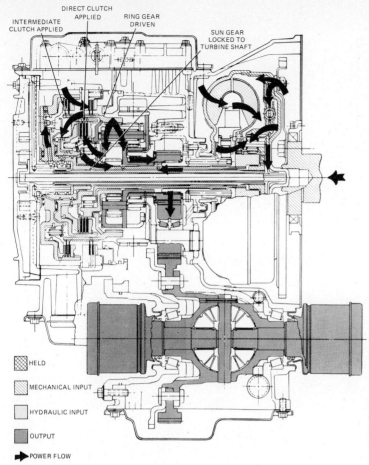

Fig. 3-47 In third gear in an ATX, the ring gear is driven through the intermediate clutch while the small sun gear is driven through the direct clutch. Driving the two members of a gearset locks the gearset and produces a 1:1 ratio. *(Courtesy of Ford Motor Company.)*

gear drives the short pinions in a counterclockwise direction, which in turn drives the long pinions in a clockwise direction, and the pinions walk around the inside of the ring gear. This forces the carrier to rotate in a counterclockwise, reverse direction at a reduced speed of about 1.9:1 to 2:1 (Fig. 3-48).

To summarize these various power flows, the input, reaction, and output members are given in Table 3.15.

3.4.6 Type 11 Gear Train

A Type 11 gear train is a three-speed Ravigneaux gearset that uses the carrier as the output. The gear train schematic for the Ford ATX is shown in Fig. 3-49, and the gear train schematic for the THM 180, which is slightly different, is shown in Fig. 3-50. A band and clutch application chart for Type 11 units is given in Table 3.16.

3.4.7 Four-Speed Arrangement and Operation

At this time, one transmission, the Ford AOD, uses a four-speed version of the Ravigneaux gear train. The gears in the AOD are arranged in a manner similar to the Type 10 Ford FMX described in an earlier section of this chapter. The AOD has an additional input shaft (the direct-drive shaft) and an additional clutch (the direct clutch). These are arranged so the carrier can be an input member in third and fourth gears as well as a reaction member in first and reverse. The direct-drive shaft is driven by a damper assembly at the front of the torque converter so it is a purely mechanical input into the gearset (Fig. 3-51). Other differences between the AOD and the FMX transmissions are: the carrier can be held in reaction by a multiple-disc clutch (called the intermediate clutch) and a one-way clutch (called the intermediate one-way clutch) or a band (called the overdrive band) and the high–reverse clutch of the FMX is used only in reverse in the AOD.

The first-, second-, and reverse-gear power flows in the AOD are exactly the same as those in the FMX with the exception that the intermediate clutch is used to hold the carrier for a reaction member in second gear.

In third gear, the direct-drive clutch is applied to drive the carrier while the forward clutch remains applied to drive the small, forward sun gear. This locks the planet gears and drives the ring gear in direct drive. It will also provide a 60 percent mechanical input through the direct-drive shaft to eliminate most of the torque converter slippage. The intermediate clutch remains applied, but it becomes ineffective because the intermediate one-way clutch overruns (Fig. 3-52).

In fourth gear, the carrier is the input, and the large sun gear is the reaction member. This occurs as the forward clutch releases and the overdrive band applies. At this time, the rotation of the carrier forces the long pinions to walk around the stationary sun gear, which

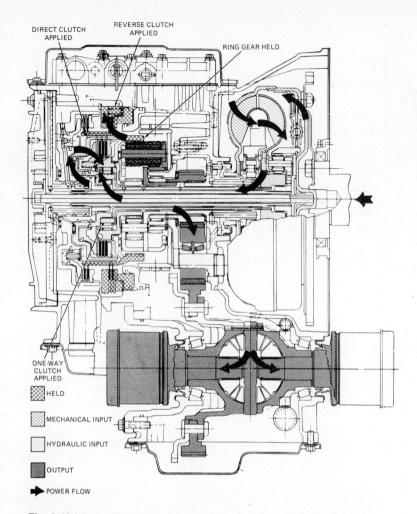

Fig. 3-48 In reverse gear in an ATX, the ring gear is held by the reverse clutch while the small sun gear is driven through the direct clutch. This drives the small pinions counterclockwise and in turn drives the long pinions in a clockwise direction; the long pinions walk around the inside of the stationary ring gear and force the carrier to rotate in a reverse direction at a reduced speed. *(Courtesy of Ford Motor Company.)*

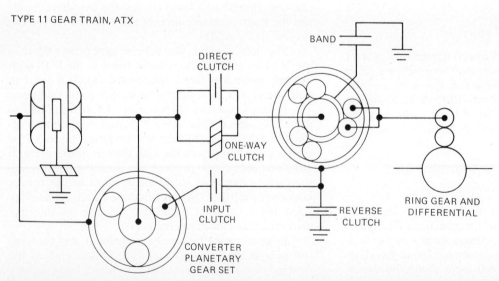

Fig. 3-49 Power flow schematic for a Type 11 ATX gear train. Note the similarity with a Type 9 gear train.

TABLE 3.15 Power Flow Summary

Gear	Small Sun Gear	Large Sun Gear	Ring Gear	Carrier
First	Input	Reaction	—	Output
Second	—	Reaction	Input	Output
Third	Input	—	Input	Output
Reverse	Input	—	Reaction	Output

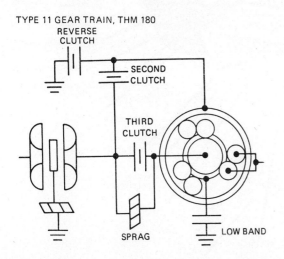

Fig. 3-50 Power flow schematic for a Type 11 THM 180 gear train. Note the similarity and the slight differences with the ATX gear train.

in turn forces the ring gear to rotate at an overdrive ratio of 0.67:1 (Fig. 3-53). The input through the direct drive shaft is 100 percent mechanical, completely bypassing the fluid in the torque converter.

In summary, the input, reaction, and output members for the various power flows are given in Table 3.17.

3.4.8 Type 12 Gear Train

A Type 12 gear train is a four-speed Ravigneaux gear train. The gear train schematic for Type 12 units is shown in Fig. 3-54, and a band and clutch application chart is given in Table 3.18.

3.5 FOUR-SPEED OVERDRIVE GEARSET

A four-speed gearset developed by General Motors and introduced in the THM 700 (Fig. 3-55) is composed of two simple planetary gearsets that have the ring gears of each set interconnected with the carriers of the other set. A second version of this gearset is used in the Chrysler A-604, Ford AXOD, and General Motors THM 440 transaxles and will be described later in this chapter.

The THM 700 uses four multiple-disc clutches plus a one-way clutch as driving members and one multiple-disc clutch, a one-way clutch, and a band for holding members. The driving clutches are arranged so they can drive the sun gear and ring gear in the front gearset, called the input gearset, and the sun gear in the rear gearset, called the reaction gearset. The input housing contains three of the driving clutches and the hub for the fourth. There are two ways that the front sun gear can be driven. One way is through the *forward clutch* and forward one-way clutch, called a *forward sprag*, and the other way is through the *overrun clutch*. The overrun clutch is used in manual 1, 2, and 3 to provide engine braking during deceleration.

The rear carrier (and front ring gear with it) can be held by the one-way clutch or the multiple-disc, *low and reverse clutch*. The rear sun gear can be held by the 2–4 band to serve as a reaction member as well as a driving member.

In neutral, all of the clutches are released so the power gets no farther than the input housing.

In first gear, the forward clutch is applied to drive the front sun gear through the forward sprag. The ro-

TABLE 3.16 Band and Clutch Application Chart—Type 11

Gear Range	One-way Clutch	Intermediate Clutch	Direct Clutch	Reverse Clutch	Intermediate–Low Band
D1	XXX				XXX
D2		XXX	XXX		XXX
D3	XXX	XXX	XXX		
I1	XXX				XXX
I2		XXX			XXX
L1	XXX		XXX		XXX
R	XXX		XXX	XXX	

Note: ATX terminology is used in this chart. In THM 180 terminology a one-way clutch is called an input sprag, intermediate clutch is second clutch, direct clutch is third, reverse has the same name and intermediate–low is low.
The only Type 11 transmission is the General Motors THM 180.
The only Type 11 transaxle is the Ford ATX.

81

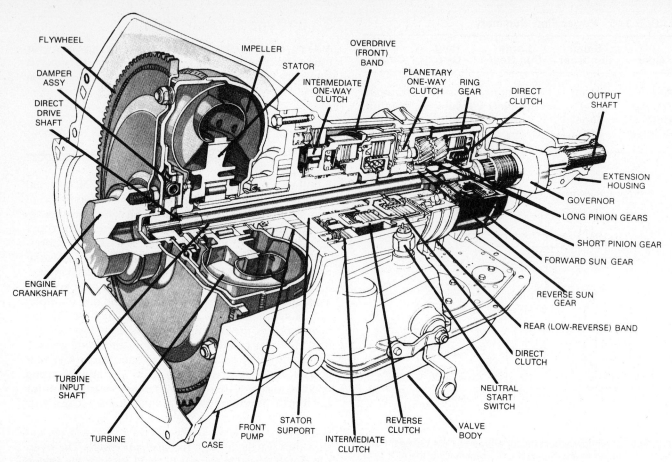

Fig. 3-51 Cutaway view of an automatic overdrive (AOD) transmission. Note the Ravigneaux gear train and the direct-drive shaft coupling the damper assembly at the front of the converter to the direct clutch. *(Courtesy of Ford Motor Company.)*

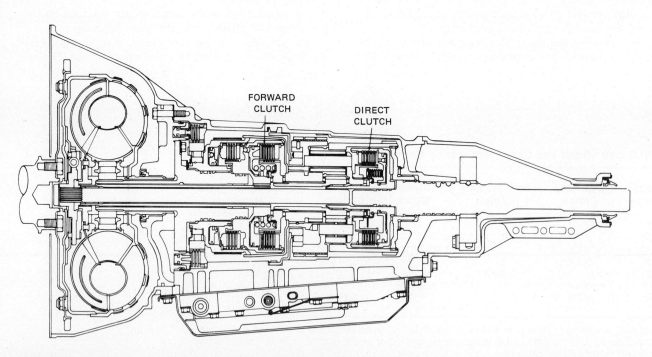

Fig. 3-52 In third gear in an AOD, the forward (small) sun gear is driven through the forward clutch while the planetary unit (carrier) is driven through the direct clutch. This locks the gearset and produces a 1:1 ratio. *(Courtesy of Ford Motor Company.)*

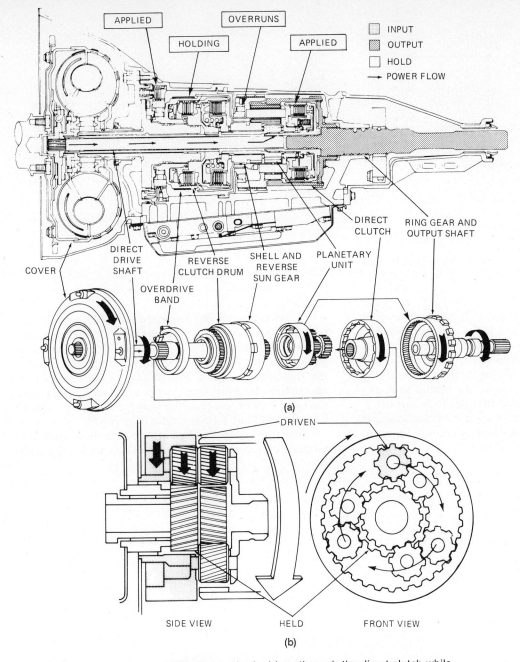

Fig. 3-53 In fourth gear in an AOD, the carrier is driven through the direct clutch while the reverse (large) sun gear is held by the overdrive band. This forces the long pinions to walk around the sun gear and drives the ring gear at an overdrive speed. *(Courtesy of Ford Motor Company.)*

TABLE 3.17 Power Flow Summary

Gear	Small Sun Gear	Large Sun Gear	Carrier	Ring Gear
First	Input	—	Reaction	Output
Second	Input	Reaction	—	Output
Third	Input	—	Input	Output
Fourth	—	Reaction	Input	Output
Reverse	—	Input	Reaction	Output

TYPE 12 GEAR TRAIN

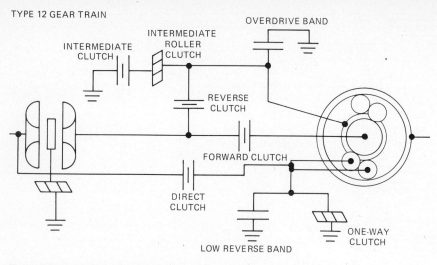

Fig. 3-54 Power flow schematic for a Type 12 gear train. Note the similarity with the Type 10 gear train.

tation of the sun gear causes the pinions in the front gearset to rotate in a counterclockwise direction. The front ring gear also tries to rotate counterclockwise but is held by the one-way clutch at the rear carrier. The pinion gears walk around the inside of the front ring gear and force the carrier to drive the output shaft in a clockwise direction at a ratio of 3.06:1. In manual 1, the rear carrier is locked to the case by the low and reverse clutch to prevent the one-way clutch from overrunning during deceleration (Fig. 3-56).

In second gear, the 2–4 band is applied to hold the rear sun gear in reaction, and the forward clutch remains applied. The power flow is similar to that of first gear except that the planet gears in the rear unit walk around the stationary sun gear. The motion of the rear carrier also drives the front ring gear at a reduced speed, and the front carrier and rear ring gear are driven in a clockwise direction at a 1.63:1 ratio (Fig. 3-57).

For third gear, the 3–4 clutch is applied while the forward clutch stays on (Fig. 3-58). At this time, both

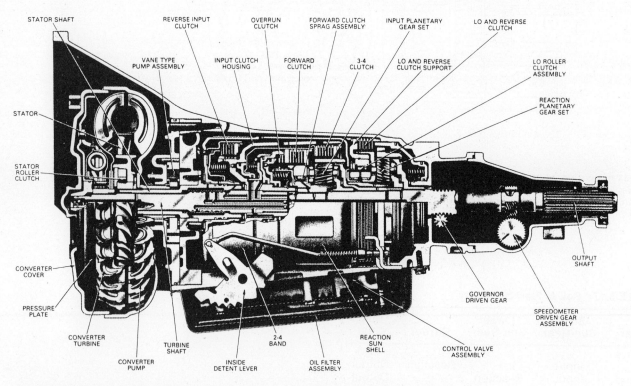

Fig. 3-55 Cutaway view of THM 700 transmission. Note that the carrier for the input gearset and the ring gear for the reaction gearset are splined to the output shaft and that the other carrier and ring gear are coupled together. *(Copyrighted material reprinted with permission from Hydra-matic Div., GM Corp.)*

TABLE 3.18 Band and Clutch Application Chart—Type 12

Gear Range	Forward Clutch	Direct Clutch	Reverse Clutch	OD Band	Intermediate Clutch	Intermediate One-way Clutch	Low–Reverse Band	One-way Clutch
O1	XXX							XXX
O2	XXX				XXX	XXX		
O3	XXX	XXX			XXX			
O4		XXX		XXX	XXX			
D1	XXX							XXX
D2	XXX				XXX	XXX		
D3	XXX	XXX			XXX			
L1	XXX						XXX	XXX
L2	XXX				XXX	XXX	XXX	
R			XXX				XXX	

Note: The only Type 12 transmission is the Ford AOD.

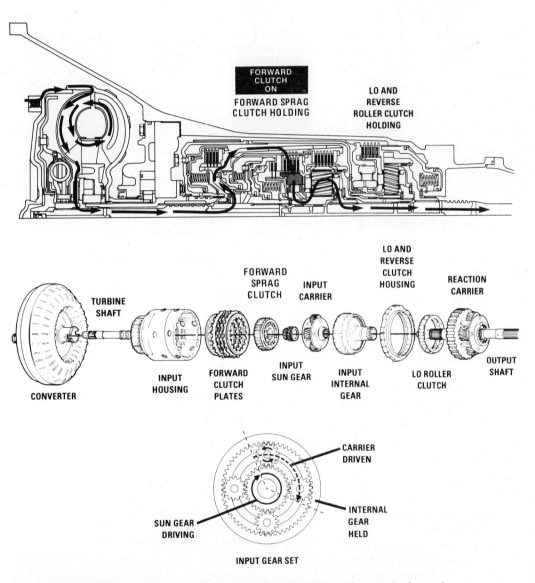

Fig. 3-56 A THM 700 in first gear. The input sun gear is being driven through the forward clutch and forward sprag while the input internal (ring) gear is held by the low and reverse roller clutch. The pinion gears are forced to walk around the inside of the ring gear, and this forces the carrier to rotate at a reduced speed. *(Copyrighted material reprinted with permission from Hydra-matic Div., GM Corp.)*

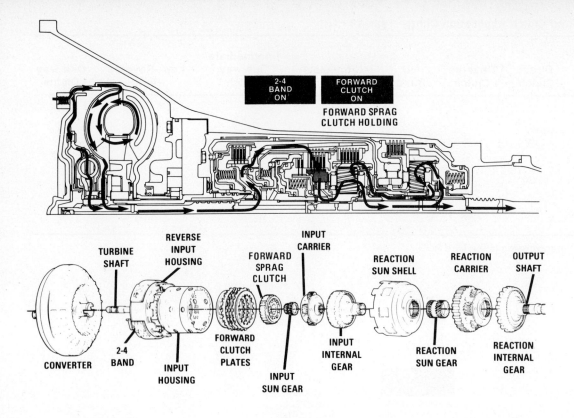

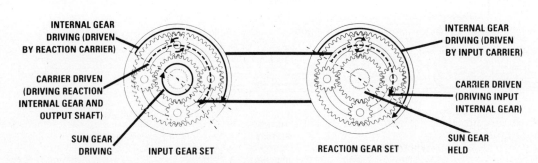

Fig. 3-57 A THM 700 in second gear. The input sun gear is driven through the forward clutch and forward sprag while the reaction sun gear is held by the 2–4 band. This forces the input carrier, output shaft, and reaction internal (ring) gear to rotate, which in turn causes the input internal gear and reaction carrier to rotate, a motion that causes the reaction planet gears to walk around the reaction sun gear. *(Copyrighted material reprinted with permission from Hydra-matic Div., GM Corp.)*

the sun gear and the ring gear in the front gearset are driven, which locks the gearset and produces direct drive.

In fourth gear, the 3–4 clutch stays on to drive the rear carrier (through the front ring gear), and the 2–4 band is applied to hold the rear sun gear in reaction. Rotation of the rear carrier forces the planet gears to walk around the stationary sun gear, and this forces the rear ring gear to rotate clockwise at a 0.70 : 1 ratio (Fig. 3-59). The forward clutch remains applied, and the front sun gear overruns at the front sprag.

In reverse, the rear sun gear is driven through the reverse clutch, and the rear carrier is held by the low and reverse clutch. The rotation of the sun gear forces the planet gears to turn counterclockwise, which in turn forces the ring gear to turn counterclockwise at a 2.30 : 1 ratio (Fig. 3-60).

In summary, the input, reaction, and output members to produce the various power flows are given in Table 3.19.

3.5.1 Type 13 Gear Train

A Type 13 gear train is a four-speed overdrive gearset that uses seven clutches and a band for driving and reaction members and the rear ring gear/front carrier for the output member. The power flow schematic for Type 13 units is shown in Fig. 3-61, and the band and clutch application chart is given in Table 3.20.

3.5.2 Four-Speed Overdrive Gearset: Second Version

The A-604 uses almost the same gearset as the THM 700-R4 with different input and output members and a re-

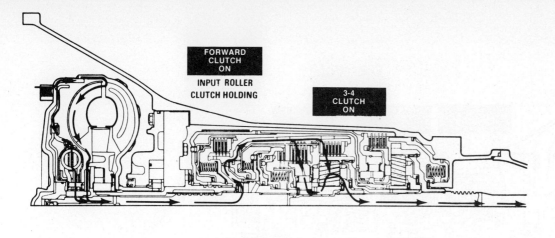
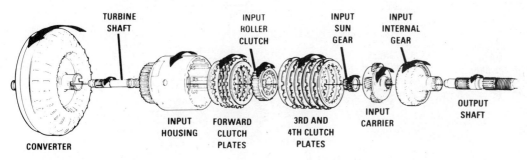
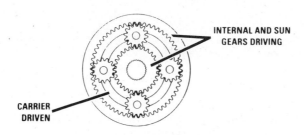

Fig. 3-58 A THM 700 in third gear. The input sun gear is driven through the forward clutch and forward sprag, while the input internal gear is driven through the 3–4 clutch. This locks the input gearset and produces a 1:1 ratio. *(Copyrighted material reprinted with permission from Hydra-matic Div., GM Corp.)*

versal of the gearset so the front ring gear and the rear carrier are the output members (Fig. 3-62). This produces slightly different power flows. The A-604 transmission is also unique in that no bands or one-way clutches are used.

Three of the multiple-disc clutches used in the A-604 are driving members and the other two clutches are holding members (Fig. 3-63). The driving clutches are arranged so they can drive the front sun gear (closest to the engine), the front carrier, or the rear sun gear. Remember that driving the carrier in the front gearset also drives the ring gear in the rear gearset. The holding clutches are arranged so one clutch can hold the sun gear in the front gearset, and the other clutch can hold the ring gear in the rear gearset as well as the carrier in the front set.

Clutch application is controlled by the manual valve and four electric solenoid valves. The solenoid valves are controlled by the transaxle electronic control module, and they are opened and closed to produce the automatic upshifts and downshifts. They are also operated at the exact rate to produce the proper clutch application and release for good shift quality. These valves will be explained in more detail in Chapter 6.

In neutral, the three driving clutches are released. The low–reverse clutch is applied so it will be ready to hold the reaction member in first or reverse gear.

In first gear, the underdrive clutch is applied to drive the rear sun gear, and the low–reverse clutch is applied to hold the rear ring gear. At this time, rotation of the sun gear forces the planet gears to walk around the inside of the stationary ring gear. This forces the carrier to rotate in a clockwise direction at a reduction ratio of 2.84:1 (Fig. 3-64). Because there are no one-way clutches, the power flow in manual 1 is exactly the same as that just described.

In second gear, the low–reverse clutch is released, and the 2–4 clutch is applied to hold the front sun gear. The underdrive clutch stays applied to drive the rear sun gear. Rotation of the rear sun gear turns the planet

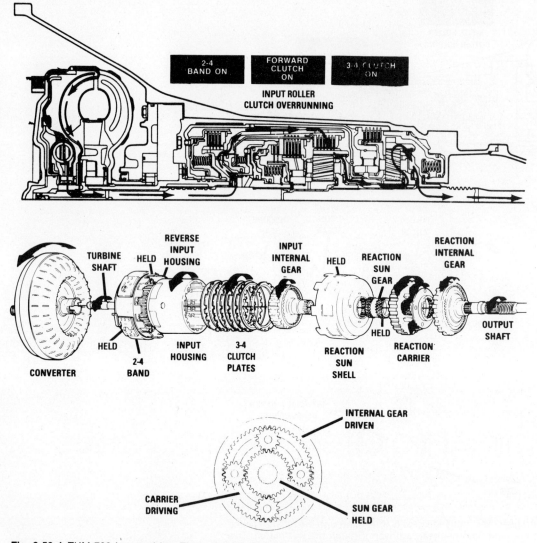

Fig. 3-59 A THM 700 in overdrive. The reaction carrier is driven through the 3–4 clutch while the reaction sun gear is held by the 2–4 band. This causes the reaction planet gears to walk around the sun gear and produces an overdrive ratio. The forward clutch remains applied, but the forward sprag overruns to release the input sun gear. *(Copyrighted material reprinted with permission from Hydra-matic Div., GM Corp.)*

TABLE 3.19 Power Flow Summary

Gear	Front Sun Gear	Front Ring Gear & Rear Carrier	Rear Sun Gear	Front Carrier	& Rear Ring Gear
First	Input	Reaction	—	—	Output
Second	Input	—	Reaction	Output	Output
Third	Input	Input	—	Output	—
Fourth	—	Input	Reaction	—	Output
Reverse	—	Reaction	Input	—	Output

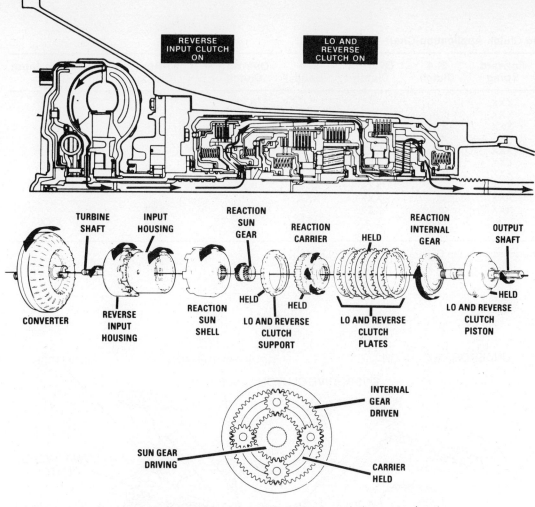

Fig. 3-60 A THM 700 in reverse. The reaction sun gear is driven through the reverse input clutch while the reaction carrier is held by the low and reverse clutch. This forces the reaction planet gears to rotate as idlers and drives the reaction internal gear in a reverse direction. *(Copyrighted material reprinted with permission from Hydra-matic Div., GM Corp.)*

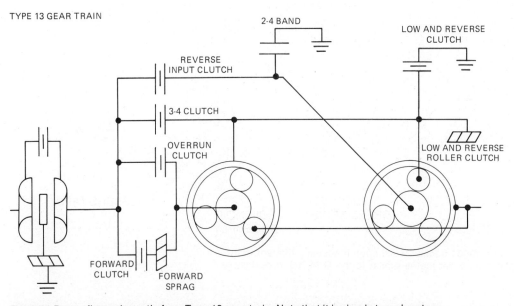

Fig. 3-61 Power flow schematic for a Type 13 gear train. Note that it is simply two planetary gearsets with an interconnection between the two carriers and ring gears.

89

TABLE 3.20 Band and Clutch Application Chart—Type 13

Gear Range	Forward Clutch	Forward Sprag	3-4 Clutch	Overrun Clutch	2-4 Band	One-way Clutch	Low-Reverse Clutch	Reverse Input Clutch
O1	XXX	XXX				XXX		
O2	XXX	XXX			XXX			
O3	XXX	XXX	XXX					
O4	XXX		XXX		XXX			
D1	XXX	XXX		XXX		XXX		
D2	XXX	XXX		XXX	XXX			
D3	XXX	XXX	XXX	XXX				
I1	XXX	XXX		XXX		XXX		
I2	XXX	XXX		XXX	XXX			
L1	XXX	XXX		XXX		XXX	XXX	
R							XXX	XXX

Note: The only Type 13 transmission is the General Motors THM 700-R4.

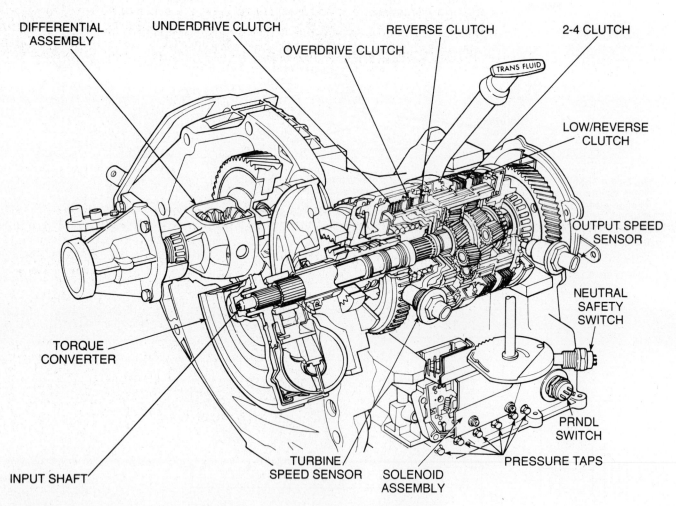

Fig. 3-62 Cutaway view of Chrysler A-604 transaxle. Although it appears different in this view, the gear train is arranged in an order that is similar to the THM 700. *(Courtesy of Chrysler Corporation.)*

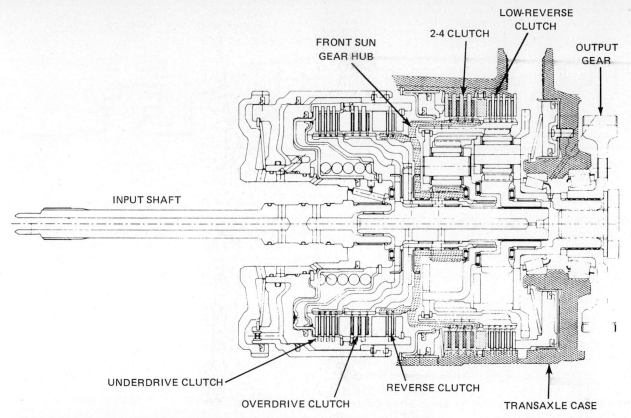

Fig. 3-63 Cutaway view of A-604 transaxle. Note the three driving clutches at the right, the two reaction clutches at the left, and the arrangement of the gear train. *(Courtesy of Chrysler Corporation.)*

gears on the rear carrier, and this causes the rear ring gear and the front carrier along with it to rotate in a clockwise direction. The movement of the front carrier forces the front planet gears to walk around the stationary front sun gear in a clockwise direction (Fig. 3-65). This is turn forces the front ring gear to rotate in a clockwise direction at a reduction ratio of 1.57:1.

In third gear, the 2–4 clutch is released, and the overdrive clutch is applied to drive the front carrier and rear ring gear. The underdrive clutch remains applied to drive the rear sun gear. Driving both the sun gear and the ring gear of the rear planetary set locks it and produces a direct-drive ratio of 1:1 (Fig. 3-66).

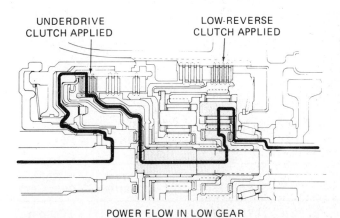

POWER FLOW IN LOW GEAR

Fig. 3-64 An A-604 in first gear. The rear sun gear is driven through the underdrive clutch while the rear carrier is held by the low–reverse clutch. This forces the rear planet gears to walk around the inside of the ring gear and the rear carrier to rotate at a reduced speed. *(Courtesy of Chrysler Corporation.)*

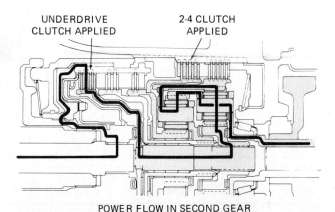

POWER FLOW IN SECOND GEAR

Fig. 3-65 An A-604 in second gear. The rear sun gear is driven through the underdrive clutch while the front sun gear is held by the 2–4 clutch. This forces the rear carrier, output shaft, and front ring gear to rotate, which in turn forces the front planet gears to walk around the stationary front sun gear. *(Courtesy of Chrysler Corporation.)*

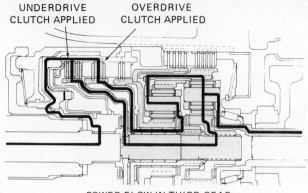

Fig. 3-66 An A-604 in third gear. The rear sun gear is driven through the underdrive clutch while the rear ring gear is driven through the overdrive clutch. This locks the rear gearset and produces a 1:1 ratio. *(Courtesy of Chrysler Corporation.)*

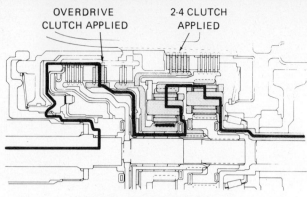

Fig. 3-67 An A-604 in fourth gear. The front carrier is being driven through the overdrive clutch while the front sun gear is held by the 2–4 clutch. This forces the planet gears to walk around the stationary sun gear and drives the ring gear at an overdrive ratio. *(Courtesy of Chrysler Corporation.)*

In fourth gear, the underdrive clutch is released, and the 2–4 clutch is applied to hold the front sun gear. The overdrive clutch stays applied to drive the front carrier. The rotation of the front carrier forces the planet gears to walk around the stationary sun gear, and this motion forces the front ring gear to rotate at an overdrive ratio of 0.69:1 (Fig. 3-67).

In reverse, the reverse clutch is applied to drive the front sun gear, and the low–reverse clutch is applied to hold the front carrier. Rotation of the front sun gear forces the planet gears to rotate as idler gears, and this forces the front ring gear to rotate in a counterclockwise direction at a reduction ratio of 2.21:1 (Fig. 3-68).

In summary, the input, reaction, and output members to produce the various power flows are given in Table 3.21.

3.5.3 Type 14 Gear Train

A Type 14 gear train is an overdrive gearset that uses five driving and holding clutches and the front ring gear and rear carrier as output members. The gear train schematic for Type 14 units is shown in Fig. 3-69, and a clutch application chart is given in Table 3.22.

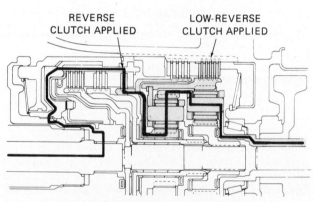

Fig. 3-68 An A-604 in reverse gear. The front sun gear is driven through the reverse clutch while the front carrier is held by the low–reverse clutch. This forces the planet gears to rotate as idlers and drives the ring gear in a reverse ratio. *(Courtesy of Chrysler Corporation.)*

TABLE 3.21 Power Flow Summary

Gear	Front Sun Gear	Front Ring Gear & Rear Carrier	Rear Sun Gear	Front Carrier & Rear Ring Gear
First	—	— Reaction	Input	— Output
Second	Reaction	—	Input	Output Output
Third	—	Input	Input	— Output
Fourth	Reaction	Input	—	Output —
Reverse	Input	Reaction	—	Output —

TYPE 14 GEAR TRAIN

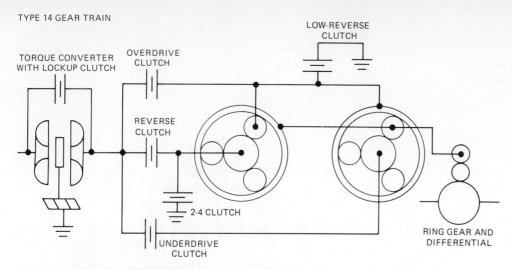

Fig. 3-69 Power flow schematic for a Type 14 gear train. Note the similarity with a Type 13 schematic. If we reverse the position of the two gearsets and change the names of some of the components, they would be the same.

TABLE 3.22 Band and Clutch Application Chart—Type 14

Gear Range	Underdrive Clutch	Overdrive Clutch	Reverse Clutch	2–4 Clutch	Low–Reverse Clutch
O1	XXX				XXX
O2	XXX			XXX	
O3	XXX	XXX			
O4		XXX		XXX	
D1	XXX				XXX
D2	XXX			XXX	
D3	XXX	XXX			
L1	XXX				XXX
L2*	XXX			XXX	
L3*	XXX	XXX			
R			XXX		XXX

* If the accelerator is depressed, at high engine speeds, upshifts to second or third occur.
The only Type 14 transaxle is the A-604.

3.5.4 Four-Speed Overdrive Gearset: Variation of Second Version

The General Motors THM 440 (Fig. 3-70) and the Ford AXOD (Fig. 3-71) transaxles use gearsets that are very similar to the arrangement in the A-604. With these transaxles: The front carrier and the rear ring gears are combined and are the output; the rear carrier and the front ring gear can be a driving member, a reaction, or neither; and the rear sun gear can be a reaction member only. Like the THM 700, the front sun gear can be an input.

In first gear, the front sun gear is driven by the input clutch (THM 440) or the forward clutch (AXOD) and the rear sun gear is held by the 1–2 band (THM 440) or the low–intermediate band (AXOD). The rotation of the front sun gear causes the front ring gear/rear carrier to rotate in the same direction, and this forces the planet gears in the rear unit to walk around the stationary sun gear. The rotation of these pinions drives the rear carrier/front ring gear in the same direction as the input at a 2.92:1 (THM 440) or 2.77:1 (AXOD) ratio (Fig. 3-72).

In second gear, the second clutch (THM 440) or the intermediate clutch (AXOD) is applied to drive the rear ring gear, and the band stays on to hold the rear sun gear in reaction. The rotation of the ring gear forces the planet gears to walk around the stationary sun gear and in turn forces the carrier to rotate at a 1.57:1 (THM 440) or 1.54:1 (AXOD) ratio (Fig. 3-73). The clutch driving the front sun gear stays on, but the gear overruns at the sprag clutch.

In third gear, the band is released, the second/intermediate clutch stays applied, and the third clutch (THM 440) or the direct clutch (AXOD) is applied to lock the front sun gear to the input shaft. At this time, both the sun gear and the carrier in the front gearset are driven, and this locks the gearset into direct drive (Fig. 3-74).

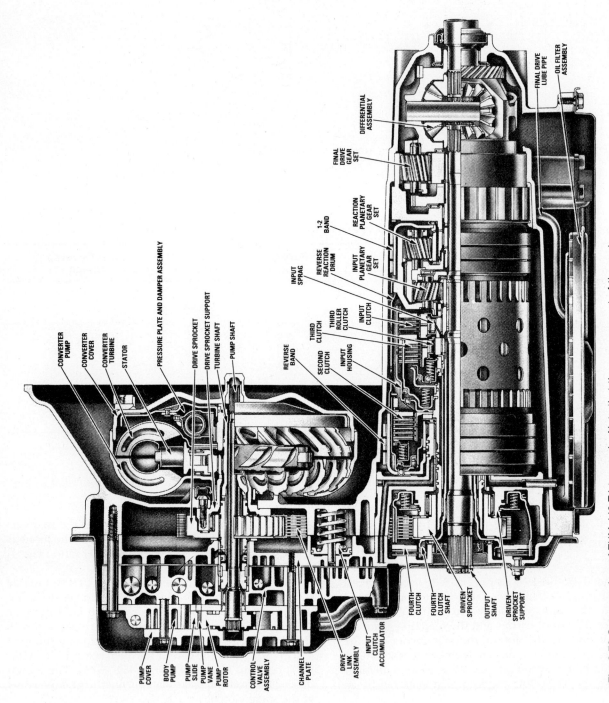

Fig. 3-70 Cutaway view of THM 440-T4 transaxle. Note that the arrangement of the parts is similar to a THM 125. (*Copyrighted material reprinted with permission from Hydra-matic Div., GM Corp.*)

8. **FORWARD CLUTCH** — Locks the driven sprocket to the sun gear of the forward planetary gearset in FIRST gear.

9. **LOW ONE-WAY SPRAG CLUTCH** — Transmits power from the driven sprocket to the sun gear of the forward planetary gearset in FIRST gear and provides engine braking in THIRD gear.

10. **OVERDRIVE BAND** — Holds the sun gear of the forward planetary gearset stationary in FOURTH gear (overdrive).

11. **DIRECT CLUTCH** — Locks the sun gear of the planetary assembly of the forward planetary gearset to the driven sprocket in THIRD gear.

12. **DIRECT ONE-WAY CLUTCH** — Transmits torque from the sun gear to the planetary assembly of the forward planetary gearset in THIRD gear and provides engine braking in MANUAL LOW.

13. **INTERMEDIATE CLUTCH** — Locks the driven sprocket to the planetary assembly of the forward planetary gearset in SECOND and THIRD gear.

14. **REVERSE CLUTCH** — Holds the planetary assembly of the forward planetary gearset, and the ring gear of the rear planetary gearset stationary in REVERSE gear.

15. **PLANETARY GEARS** — Two gearsets are used to provide the four forward speeds, plus REVERSE, dependent upon clutch and/or band applications.

16. **PARKING GEAR** — Allows the output (axle) shaft to be mechanically locked by the parking pawl anchored in the case.

17. **LOW-INTERMEDIATE BAND** — Holds the sun gear of the rear planetary gearset stationary in manual LOW, FIRST and SECOND gears.

18. **FINAL DRIVE SUN GEAR** — Transfers torque from the transmission output to the final drive planetary assembly.

19. **FINAL DRIVE PLANET** — Drives the differential assembly.

20. **DIFFERENTIAL ASSEMBLY** — Drives the front axle shafts and provides the differential action if driving wheels are turning at different speeds.

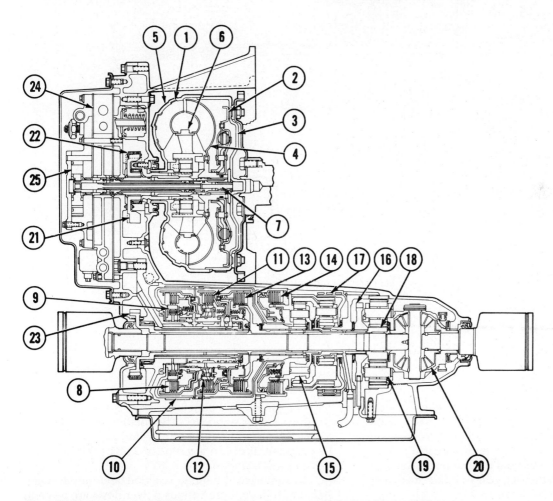

Fig. 3-71 Cutaway view of AXOD transaxle. Note the strong similarity with the THM 440-T4. *(Courtesy of Ford Motor Company.)*

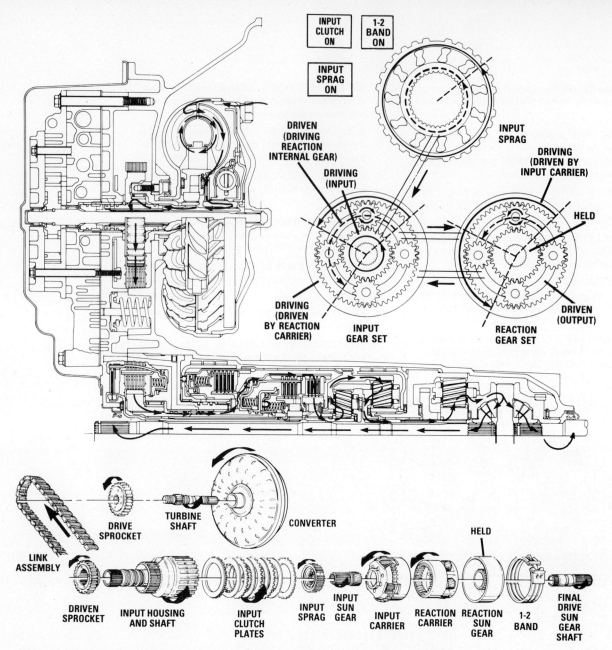

Fig. 3-72 A THM 440 in first gear. The input sun gear is being driven through the input clutch and input sprag while the reaction sun gear is held by the 1–2 band. This forces the input carrier/reaction internal (ring) gear to rotate and the reaction planet gears to walk around the stationary sun gear. This in turn produces a reduction ratio. *(Copyrighted material reprinted with permission from Hydra-matic Div., GM Corp.)*

For fourth gear, the second/intermediate clutch stays applied, and the fourth clutch (THM 440) or the overdrive band (AXOD) is applied to hold the front sun gear stationary. The rotation of the front carrier forces the planet gears to walk around the stationary sun gear, and this in turn forces the ring gear to rotate at an overdrive ratio of about 0.70:1 (Fig. 3-75).

In reverse, the input/forward clutch is applied to drive the front sun gear, and the reverse band (THM 440) or the reverse clutch (AXOD) is applied to hold the front carrier stationary. The rotation of the sun gear forces the planet gears and the front ring gear to rotate in a reverse direction at ratios of 2.381 (THM 440) or 2.26:1 (AXOD) ratio (Fig. 3-76).

In summary, the input, reaction, and output members to produce these various power flows are given in Table 3.23.

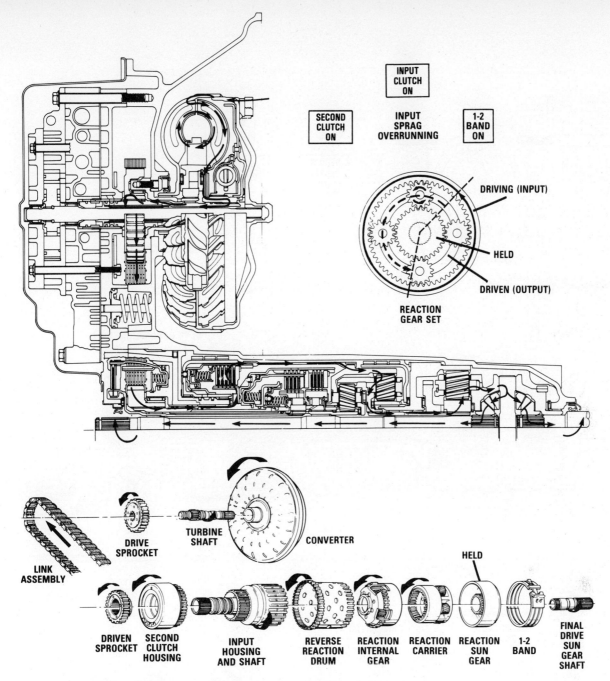

Fig. 3-73 A THM 440 in second gear. The reaction internal gear is driven through the second clutch while the reaction sun gear is held by the 1–2 band. This forces the reaction planet gears to walk around the sun gear and drives the carrier at a reduction ratio. *(Copyrighted material reprinted with permission from Hydra-matic Div., GM Corp.)*

3.5.5 Type 15 Gear Train

A Type 15 gear train is a four-speed overdrive gear train that uses the front ring gear and rear carrier as the output members and a disc clutch to the sun gear in reaction for overdrive. The gear train schematic for Type 15 units is shown in Fig. 3-77, and a band and clutch application chart is given in Table 3.24.

3.5.6 Type 16 Gear Train

Because a slightly different arrangement of bands and clutches is used when compared to the THM 440, we classify the AXOD as a Type 16 unit. The gear train schematic for Type 16 units is shown in Fig. 3-78, and a band and clutch application chart is given in Table 3.25.

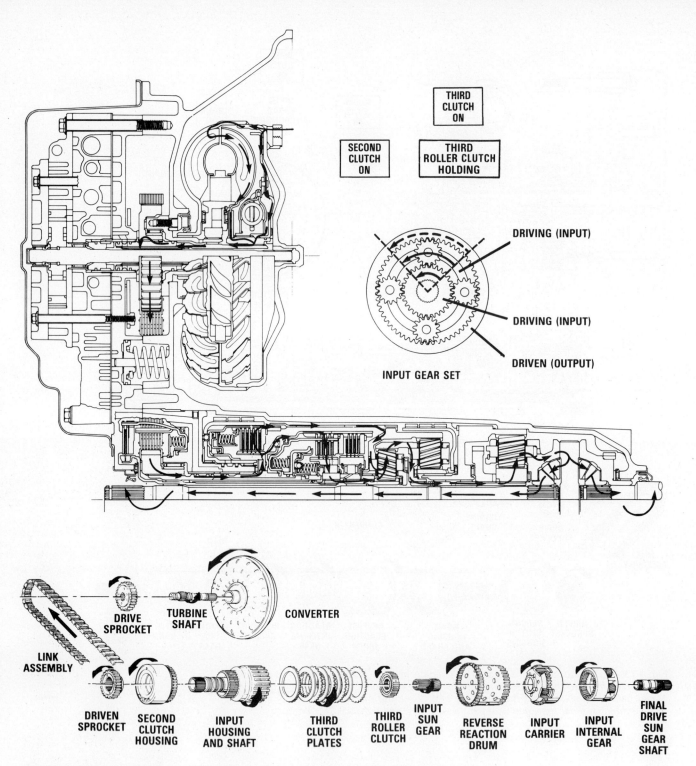

Fig. 3-74 A THM 440 in third gear. The input carrier is driven through the second clutch while the input sun is driven through the third clutch and third roller clutch. This locks the input gearset and drives the ring gear at a 1:1 ratio. *(Copyrighted material reprinted with permission from Hydra-matic Div., GM Corp.)*

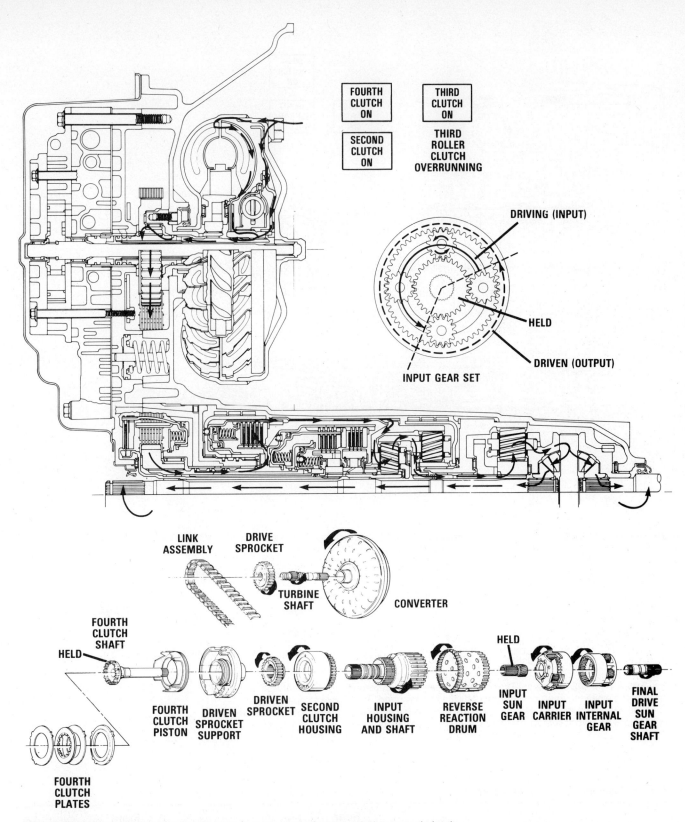

Fig. 3-75 A THM 440 in fourth gear. The input carrier is driven through the second clutch while the input sun is held by the fourth clutch. This forces the planet gears to walk around the sun gear and drives the internal gear in an overdrive ratio. (Copyrighted material reprinted with permission from Hydra-matic Div., GM Corp.)

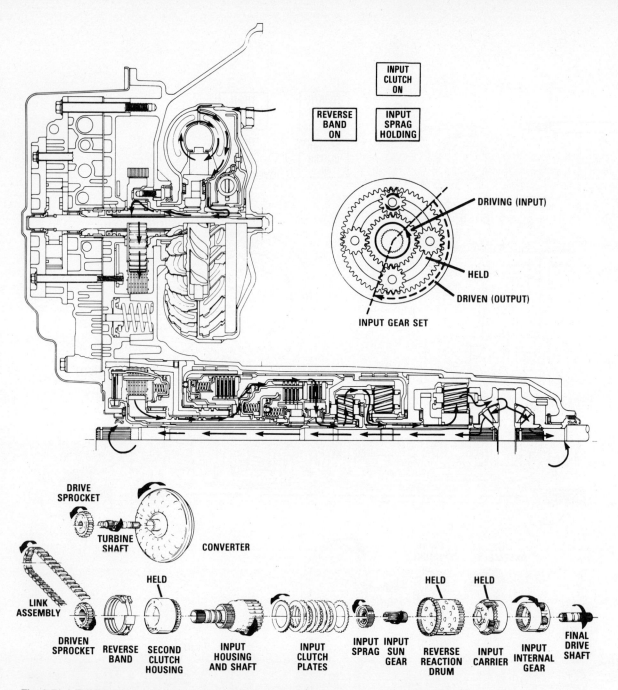

Fig. 3-76 A THM 440 in reverse. The input sun gear is driven through the input clutch and input sprag while the input carrier is held by the reverse band. This forces the planet gears to act as idlers and the internal gear to rotate in a reverse direction. *(Copyrighted material reprinted with permission from Hydra-matic Div., GM Corp.)*

TABLE 3.23 Power Flow Summary

Gear	Front Sun Gear	Front Carrier	Rear Ring Gear	Rear Sun Gear	Front Ring Gear	Rear Carrier
First	Input	—	—	Reaction	Output	Output
Second	—	—	Input	Reaction	—	Output
Third	Input	Input	—	—	Output	—
Fourth	Reaction	Input	—	—	Output	—
Reverse	Input	Reaction	—	—	Output	—

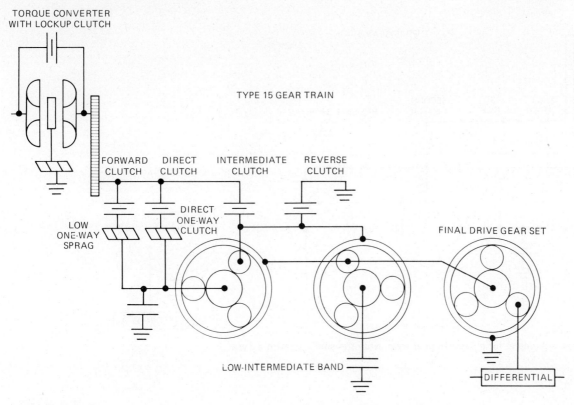

Fig. 3-77 Power flow schematic for a Type 15 gear train. Note the similarity with a Type 14 gear train.

TABLE 3.24 Band and Clutch Application Chart—Type 15

Gear Range	Input Clutch	Input Sprag	Second Clutch	Third Clutch	Third Roller Clutch	1–2 Band	Fourth Clutch	Reverse Band
O1	XXX	XXX				XXX		
O2	XXX		XXX		XXX	XXX		
O3			XXX	XXX	XXX			
O4			XXX	XXX			XXX	
D1	XXX	XXX				XXX		
D2	XXX		XXX			XXX		
D3	XXX	XXX	XXX	XXX	XXX			
I1	XXX	XXX				XXX		
I2	XXX		XXX			XXX		
L	XXX	XXX		XXX	XXX	XXX		
R	XXX	XXX						XXX

Note: The only Type 15 transaxles are the General Motors THM 440-T4, F7, and 4T60-E.

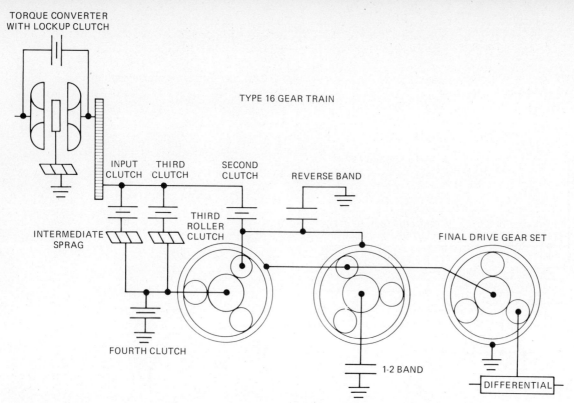

Fig. 3-78 Power flow schematic for a Type 16 gear train. Note the similarity with a Type 15 gear train.

TABLE 3.25 Band and Clutch Application Chart—Type 16

Gear Range	Forward Clutch	Direct Clutch	Direct One-way Clutch	Intermediate Clutch	Overdrive Band	Low–Intermediate Band	Low One-way Clutch	Reverse Clutch
O1	XXX					XXX	XXX	
O2	XXX			XXX		XXX		
O3	XXX	XXX	XXX	XXX				
O4		XXX		XXX	XXX			
D1	XXX					XXX	XXX	
D2	XXX			XXX		XXX		
D3	XXX	XXX	XXX	XXX				
L1	XXX	XXX				XXX	XXX	
L2	XXX							
R	XXX						XXX	XXX

Note: The only Type 16 transaxle is the AXOD.

REVIEW QUESTIONS

The questions that follow are provided so you can check the facts you have learned. Select the response that best completes each statement.

1. In a simple planetary gearset, if the carrier is the input and the sun gear is the reaction member, the ring gear rotates
   A. In a gear reduction
   B. In a reverse direction
   Which is correct?
   a. A only   c. Both A and B
   b. B only   d. Neither A nor B

2. A simple planetary gearset operates in neutral if there is no
   A. Input member
   B. Reaction member
   Which is correct?
   a. A only   c. Both A and B
   b. B only   d. Neither A nor B

3. If the ring gear is driven while the sun gear is held,
   A. The planet gears walk around the sun gear
   B. The ring gear is forced to rotate at a speed slower than the carrier
   Which is correct?
   a. A only   c. Both A and B
   b. B only   d. Neither A nor B

4. If the sun gear is driven while the carrier is held,
   A. The planet gears walk around the inside of the ring gear
   B. The ring gear is forced to rotate at a speed slower than the sun gear
   Which is correct?
   a. A only
   b. B only
   c. Both A and B
   d. Neither A nor B

5. If the ring gear is driven while the sun gear is held, the planet gears
   A. Merely rotate in the carrier
   B. Rotate as they walk around the sun gear
   C. Rotate with the carrier but not on their axes
   D. Are stationary in the carrier.

6. In a Simpson gear train,
   A. One sun gear is an input member while the other is a reaction member
   B. One carrier is an input member while the other is an output member
   Which is correct?
   a. A only
   b. B only
   c. Both A and B
   d. Neither A nor B

7. In a Simpson gear train, if the sun gear is held stationary, the transmission is in
   A. First gear
   B. Second gear
   C. Third gear
   D. Reverse

8. When a Simpson gear train is in first gear, the sun gear is
   A. Rotating clockwise
   B. Rotating counter-clockwise
   C. Stationary
   D. Any of these

9. In a Simpson gear train, third gear occurs when the forward clutch is applied and the
   A. Rear band is applied
   B. Front band is applied
   C. Multiple-disc clutch
   D. None of these

10. In a Simpson gear train, the reaction carrier can be held stationary by a
    A. One-way clutch
    B. Band
    C. Multiple-disc clutch
    D. Any of these

11. When a one-way clutch is used for a reaction member,
    A. The transmission goes into neutral while decelerating in that gear
    B. The clutch self-applies when the throttle is opened
    Which is correct?
    a. A only
    b. B only
    c. Both A and B
    d. Neither A nor B

12. Driving two members of a planetary gearset while there is no reaction member results in
    A. An overdrive
    B. A 1:1 ratio
    C. A reduction
    D. A reverse

13. In an overdrive gearset used with a Simpson gear train,
    A. The sun gear is the input member
    B. The ring is the output member
    Which is correct?
    a. A only
    b. B only
    c. Both A and B
    d. Neither A nor B

14. In third gear (Type 5, 6, 7, or 8)
    A. Both clutches are applied in the main gearset
    B. The reaction member is applied in the overdrive gearset
    Which is correct?
    a. A only
    b. B only
    c. Both A and B
    d. Neither A nor B

15. A Ravigneaux gear train uses
    A. Two sun gears
    B. Two sets of planet pinions
    C. One ring gear
    D. All of these

16. In a two-speed Ravigneaux gear train,
    A. One sun gear is driven whenever the turbine shaft turns
    B. Neutral occurs when there is no reaction member
    Which is correct?
    a. A only
    b. B only
    c. Both A and B
    d. Neither A nor B

17. The Ravigneaux gear train can be arranged so it produces
    A. Two speeds forward
    B. Two or three speeds forward
    C. Two, three, or four speeds forward
    D. Two, three, four, or five speeds forward

18. In a Ravigneaux gear train, the long pinion gears are meshed with the
    A. Large sun gear
    B. Short pinions
    Which is correct?
    a. A only
    b. B only
    c. Both A and B
    d. Neither A nor B

19. In most overdrive gearsets in fourth gear,
    A. The sun gear is driven
    B. The carrier is held stationary
    Which is correct?
    a. A only
    b. B only
    c. Both A and B
    d. Neither A nor B

20. In most gearsets in reverse gear,
    A. The carrier is held stationary
    B. The ring gear is the output member
    Which is correct?
    a. A only
    b. B only
    c. Both A and B
    d. Neither A nor B

CHAPTER 4

HYDRAULIC SYSTEMS: THEORY

OBJECTIVES

After completing this chapter, you should:

- Have a basic understanding of how a hydraulic system operates.
- Be familiar with the parts involved in a hydraulic system.
- Be able to identify the parts of a transmission hydraulic system and understand their purpose.
- Understand the requirements for a hydraulic system.
- Understand the requirements for automatic transmission fluid and the differences between the fluids in use.

4.1 INTRODUCTION

The automatic transmission's hydraulic system has several functions: It is used to apply the clutches and bands and therefore control the power flow in the transmission, transmit sufficient force along with the motion when applying these units to prevent slippage, maintain a flow of fluid through the torque converter for its operation, and maintain a flow of fluid to lubricate the moving parts in the gear train (Fig. 4-1).

4.2 HYDRAULIC PRINCIPLES

Hydraulics, often called *fluid power*, is a method of transmitting motion and/or force. Hydraulics is based on the fact that liquids that are fluid and can flow easily through complicated paths cannot be compressed (squeezed into a smaller volume). Another important feature is that when fluids transmit pressure, that pressure is equal in all directions and acts on all portions of a particular circuit at the same time with the same pressure. This is a simplified version of *Pascal's law* (Fig. 4-2).

If we were to fill a strong container with liquid, we would find it impossible to add more liquid, even by force. The only way for more liquid to enter would be for the container to rupture and leak (Fig. 4-3). Once the container is full, any added force becomes fluid pressure. Pressure is defined as the amount of force pushing on a certain area. In the past in the United States, pressure was measured using *pounds per square inch (psi)*. Ten pounds per square inch means a force of ten pounds on an area of one square inch; a smaller area has a smaller total force on it and vice versa. In Europe and other parts of the world using the metric system, pressure was measured using kilograms per square centimeter (kg/cm^2), or bars (a bar is equal to 14.5 psi). Today, pressure is also measured in *kilopascals (kPa)*; one psi is equal to 6.895 kPa, 0.07 kg/cm^2, or 0.0689 bar. The kilopascal is an international unit for measuring pressure.

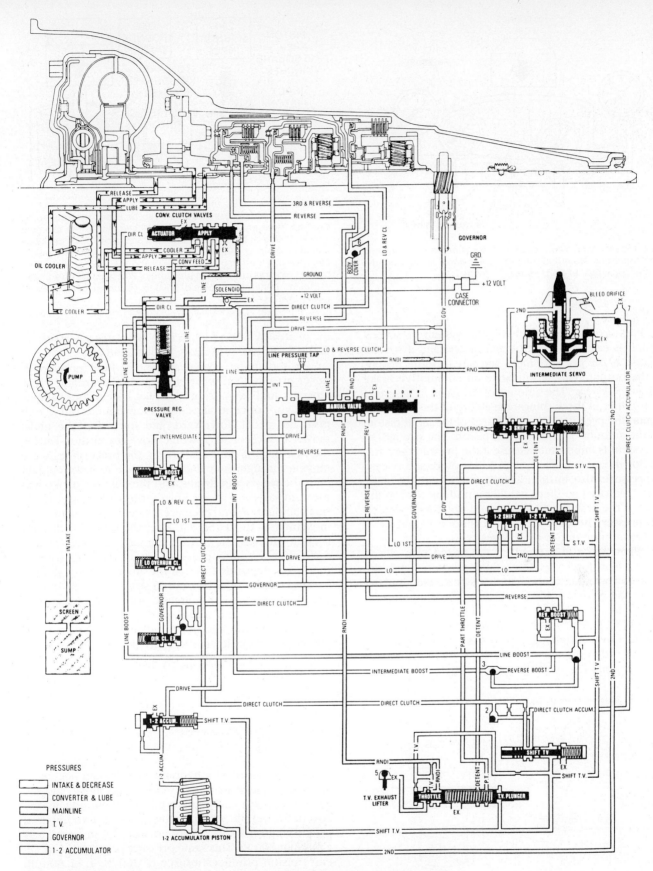

Fig. 4-1 Hydraulic system diagram for three-speed automatic transmission. A diagram such as this is commonly used to determine the relationship of the system components. (*Copyrighted material reprinted with permission from Hydra-matic Div., GM Corp.*)

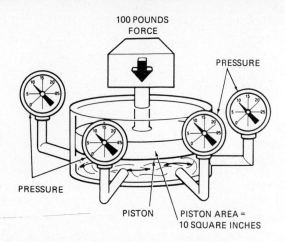

Fig. 4-2 Pressure on a confined fluid is transmitted undiminished in all directions. Note that the pressure is equal throughout the system. (*Courtesy of Ford Motor Company.*)

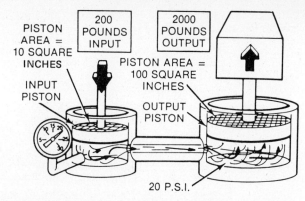

Fig. 4-4 The pressure in a system can be easily determined by dividing force or load on a piston by the area. In the example shown, 200 lb of pressure acting on a piston of 10 in.^{2} develops a fluid pressure of 20 psi. The force being produced by a piston can be easily determined by multiplying the area of the piston by the fluid pressure. A fluid pressure of 20 psi acting on a piston of 100 in.^{2} generates a force of 2000 lb. (*Courtesy of Ford Motor Company.*)

Pressure can enter a hydraulic system in several ways; it is easier to describe and understand if we use a piston as the pressure input and one or more pistons for the output. This is similar to an automotive brake hydraulic system. The amount of pressure in a system is a product of three things: the ability of the system to contain the pressure, the size or area of the input piston, and the amount of force on that piston. The strength of the system is important because if the pressure gets too high for the system, the system will rupture and release the pressure. Also, imagine a brake system with the brake drums removed so there would be nothing to stop the travel of the brake shoes. We would not develop much fluid pressure because the wheel cylinders would move too far and pop out of their bores.

When force is exerted on the piston of a closed system, that force becomes fluid pressure, and the amount of pressure is equal to that force divided by the area of the piston. A 200-lb (90.9-kg) force on a piston that is 1 in.^{2} (6.45 cm^2) in area will generate a force of 200 psi (14 kg/cm^2). This pressure can be converted to 1379 kPa by multiplying 200 by 6.895. The amount of pressure is determined by dividing the force by the area of the piston (Fig. 4-4). This same 200-lb force acting on an area of 0.5 in.^{2} (3.2 cm^2) generates a pressure of 200/0.5, or 400, psi (28 kg/cm^2) (2758 kPa).

It should be remembered that fluid pressure is never greater than what is needed to overcome resistance to fluid flow. Using the brake system, for example, you might notice it takes only a few pounds of force to push the pedal downward, and then after the pedal moves an inch or so, it won't move any farther. During the first part of the pedal travel, the brake shoes were moving through their lining clearance to make contact with the rotors and drums, and the only resistance was the fluid seals and shoe return springs. Only a relatively small amount of fluid pressure is needed to produce this movement—10 to 150 psi. After the shoes touch their friction surfaces, the pistons can move no farther; now the hydraulic system pressure can increase to whatever the driver demands through the amount of pedal force exerted to produce the desired stop.

When discussing hydraulic pistons and computing fluids pressures and forces, it is important to use the area of the piston and not the diameter. The area of a piston or any circle can be easily determined using

$$\pi r^2 \quad \text{or} \quad 0.785 d^2$$

where

$\pi = 3.1416$

$r =$ one-half the diameter

$d =$ diameter

The pressure in a hydraulic system can become a force to produce work and make things move, and the amount of force can be determined by multiplying the area of the output piston by the pressure. Two hundred psi pushing on a piston with an area of 1 in.^{2} produces a force of 200 lb. This same pressure on a 4-in.^{2} (26.17-cm^2) piston produces a force of 200 × 4, or 800, lb (363.6 kg). Application force is multiplied whenever the output piston is larger than the input; force will be divided or made smaller if the input piston is larger. The simple memory triangle shown in Fig. 4-5 can be used as an aid in trying to determine area, force, or pressure.

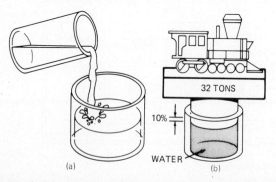

Fig. 4-3 (a) A fluid takes the shape of whatever container it is in and is essentially noncompressible. (b) If a pressure of 32 tons is exerted on 1 in.^{3} of water, the volume only compresses 10 percent. (*Courtesy of Ford Motor Company.*)

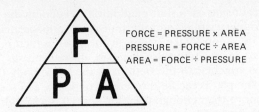

Fig. 4-5 Simple memory triangle. Cover the part you need to determine and the remaining portion of the triangle shows what you need to do to find the answer. For example, if you need to find the pressure, cover the P (pressure); pressure equals F ÷ A, force divided by area).

In a system in which the input is a pumping piston, the output piston motion is also related to the input piston. It should be remembered that a hydraulic system can only transmit the energy—force and motion—that is put into it; it cannot create energy. What goes in one end is all that will come out of the other. However, it is possible to change force to motion or vice versa. Most brake systems multiply and increase force with a loss in motion; other hydraulic systems increase motion.

As the input piston moves, it displaces or pushes fluid through the tubing to the system, and the amount of fluid displaced is equal to the piston area times the length of piston stroke. A 1-in.- (2.54-cm) diameter piston has an area of 0.785 in.^{2} (5.067 cm^2); if this piston strokes 2 in. (5.08 cm), 1.57 in.^{3} (25.7 cm^3) of fluid is displaced (0.785 × 2 in.^{2}) (5.067 × 5.08 cm^2). This much fluid can move a 1-in.-diameter piston 2 in. Given this much fluid, a 3-in.- (7.62-cm) diameter piston that has an area of 7.07 in.^{2} (45.6 cm^2) will travel a distance of 0.22 in.—1.57 in. ÷ 7.07 in.^{2} = 0.22 in. (25.7 cm^3 ÷ 45.6 cm^2, or 0.558 cm) (Fig. 4-6).

4.3 SIMPLE HYDRAULIC SYSTEMS

Many hydraulic systems, including that of an automatic transmission, use an engine- or motor-driven pump as the input. These systems normally consist of the pump, a fluid intake system usually equipped with a filter, a fluid supply, the control valves, and the actuators that provide the system output (Fig. 4-7). In these cases, the amount of pressure is determined by how much power is put into the pump, the strength and condition of the pump, and the strength of the system. Most systems use either a pressure relief valve or a pressure regulator valve to keep the pressure below the point where damage occurs. The amount of flow from the pump is also a product of how much power is put into the pump, the size of the pump, and the amount of fluid available. Pump output is normally rated as the amount of gallons, quarts, or liters that a pump can deliver in a minute at a certain pressure.

The amount of fluid pressure and flow are also related to the amount of energy that is being put into the fluid at the pump. As described in Chapter 1, the mechanical horsepower of an engine is closely related to the torque and rpm of the engine. In a similar fashion, fluid horsepower is related to the pressure and the amount of flow of the fluid. The formula for computing hydraulic, or fluid, horsepower is

$$HP = \frac{psi \times gpm}{1714}$$

where

HP = horsepower

psi = fluid pressure in pounds per square inch

gpm = fluid flow in gallons per minute

From this formula, we can see that the amount of power needed to run the transmission's hydraulic system is de-

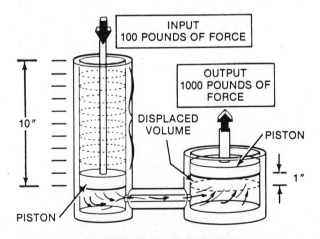

Fig. 4-6 The volume of fluid required to stroke a piston is determined by the diameter of the piston and the length of the stroke. In this case, since the output piston is 10 times as large as the input piston, the input piston has to travel 10 times further to supply the fluid. (*Courtesy of Ford Motor Company.*)

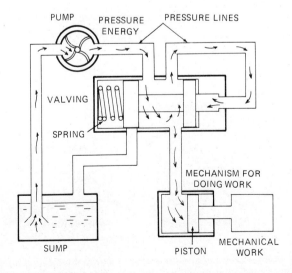

Fig. 4-7 Basic components of a simple hydraulic system. (*Courtesy of Ford Motor Company.*)

termined by how much fluid is being pumped and the pressure at which it is being pumped.

4.4 BASIC AUTOMATIC TRANSMISSION HYDRAULICS

In most automatic transmissions, the pump is built into the bulkhead at the front, engine end of the transmission and is driven by an extension at the rear of the torque converter (Fig. 4-8). In some units, the pump is mounted deeper into the case and is driven by a shaft running from the front of the torque converter. In either case, the pump begins operating and pumping fluid as soon as the engine starts. The pump is sized to produce enough fluid flow to fill the actuators and stroke the pistons during each shift in addition to keeping the torque converter filled and the transmission lubricated.

The pump intake begins at the filter, which is positioned in the transmission's pan; in many cases, the filter is attached to the bottom of the valve body (Fig. 4-9). The pan is the fluid reservoir or sump, and it contains more than enough fluid for normal operation.

All automatic transmissions use a pressure regulator valve that limits the operating pressure as well as provides different hydraulic pressures for different operating conditions. Pressure regulator valves are normally controlled by other valves in the valve body to produce various pressures. The pressure regulator is normally found in the valve body, but it is mounted in the pump body in some transmissions (Fig. 4-10). In addition to the pressure regulator, some transmissions use a pres-

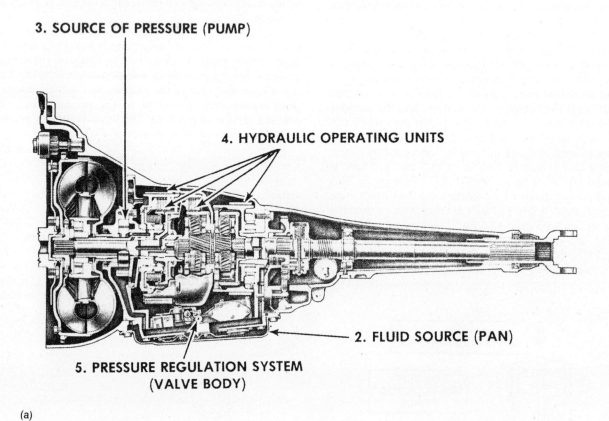

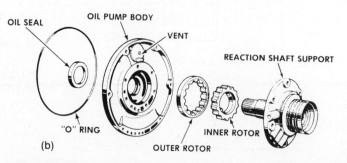

Fig. 4-8 (a) Cutaway and (b) exploded views of typical pump. Note that the pump intake filter is located in the fluid in the pan, with a rather large passage leading to the pump intake, and that the pump is driven off of the back of the torque converter. (*Courtesy of Chrysler Corporation.*)

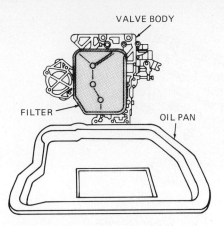

Fig. 4-9 Fluid source for a transaxle (left) and a transmission (right). (*Courtesy of Chrysler Corporation.*)

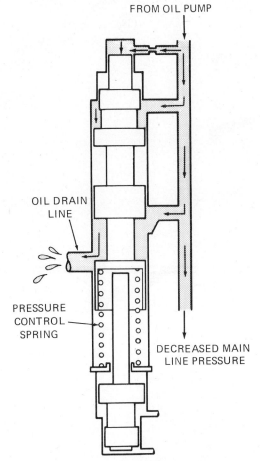

Fig. 4-10 Pressure regulator valve. When the fluid pressure acting on the area at the top end of the valve exceeds the strength of the spring, the valve moves downward to open the drain port and allows the excess pressure to drain off.

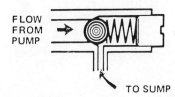

Fig. 4-11 Pressure relief valve. Excessive fluid pressure forces the ball off of its seat to allow the excess fluid to escape. (*Courtesy of Ford Motor Company.*)

sure relief valve to release excess pressure that might occur under some conditions (Fig. 4-11).

Fluid flow through a transmission's hydraulic system is controlled by various valves in the valve body (Fig. 4-12). All transmissions have a manual valve. This valve is connected to the shift lever in the driver's compartment (Fig. 4-13). Nearly all transmissions also have one shift valve for each automatic upshift as well as other valves to tailor the timing, speed, and quality of the upshifts. A two-speed transmission has one shift valve; a four-speed has three (Fig. 4-14).

In many hydraulic systems, the major fluid flow is to the actuator, which is usually a hydraulic ram. The hydraulic actuators of an automatic transmission are the servos that apply and release the bands and the pistons that apply the clutches.

In automatic transmissions, the valves are positioned so that fluid pressure is exerted on the actuator piston all the time that the actuator is applied. When the actuator is released, this fluid pressure is exhausted, usually back at the valve (Fig. 4-15).

4.5 PRODUCING FLUID FLOW AND PRESSURE

Three types of pumps are used to produce the oil flows and pressure within an automatic transmission. They are internal–external gear with crescent (or simply gear pump), gerotor (rotor), and vane pumps (Fig. 4-16). The pumping action in all these is essentially the same in that the inner pumping member (external gear or inner rotor) is driven by the torque converter hub or a drive shaft, and the outer pumping member (the internal gear, outer rotor, or vane housing) is offset or eccentric relative to the inner gear or rotor.

As the inner member rotates, a series of chambers (between the gear teeth, the rotor lobes, or vanes) increase in volume in one area and decrease in another. In the area where the chamber volume increases, a low-pressure area is created, and this area is connected to a passage leading to the filter in the fluid near the bottom of the sump. Atmospheric pressure inside the transmission pushes fluid through the filter, through the intake passage, and into the pump inlet (Fig. 4-17). As the pump rotates, fluid moves in to fill the chambers just as fast as they enlarge.

On the other side of the pump the chambers get smaller, and the outlet port of the pump is positioned in this area. Here, the fluid is forced out of the pump and into the passage leading to the pressure control valve and the rest of the hydraulic system. This fluid flow is often called *supply* or *mainline pressure*.

Most automatic transmissions use a fixed-size, positive-displacement pump. This tells us that the pump will move a certain volume of fluid on every revolution of the pump. Both the gear pump and the rotor pump are of this type. Vane pumps are also positive displacement in that they will pump a certain volume on each revolution, but the displacement, and therefore the amount of fluid pumped, can be changed. This is done by moving the vane housing sideways, reducing the amount of housing offset from the rotor and therefore the amount

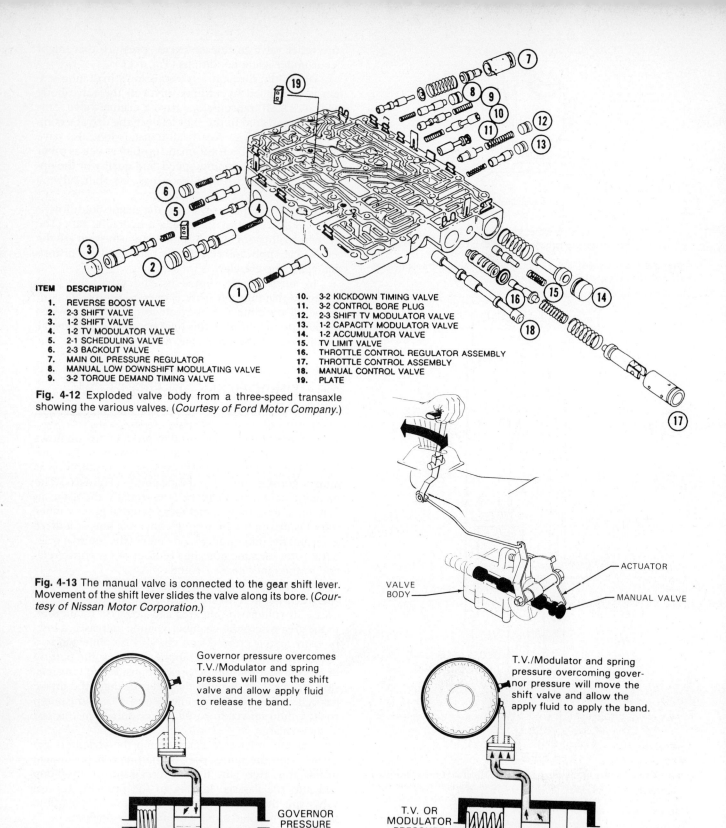

ITEM DESCRIPTION
1. REVERSE BOOST VALVE
2. 2-3 SHIFT VALVE
3. 1-2 SHIFT VALVE
4. 1-2 TV MODULATOR VALVE
5. 2-1 SCHEDULING VALVE
6. 2-3 BACKOUT VALVE
7. MAIN OIL PRESSURE REGULATOR
8. MANUAL LOW DOWNSHIFT MODULATING VALVE
9. 3-2 TORQUE DEMAND TIMING VALVE
10. 3-2 KICKDOWN TIMING VALVE
11. 3-2 CONTROL BORE PLUG
12. 2-3 SHIFT TV MODULATOR VALVE
13. 1-2 CAPACITY MODULATOR VALVE
14. 1-2 ACCUMULATOR VALVE
15. TV LIMIT VALVE
16. THROTTLE CONTROL REGULATOR ASSEMBLY
17. THROTTLE CONTROL ASSEMBLY
18. MANUAL CONTROL VALVE
19. PLATE

Fig. 4-12 Exploded valve body from a three-speed transaxle showing the various valves. (*Courtesy of Ford Motor Company.*)

Fig. 4-13 The manual valve is connected to the gear shift lever. Movement of the shift lever slides the valve along its bore. (*Courtesy of Nissan Motor Corporation.*)

Fig. 4-14 A typical shift valve uses a spring to move the valve to the downshift position. When governor pressure becomes great enough to overcome the strength of the spring plus any throttle pressure, the valve moves to the left and allows fluid to flow from the band servo. (*Copyrighted material reprinted with permission from Hydra-matic Div., GM Corp.*)

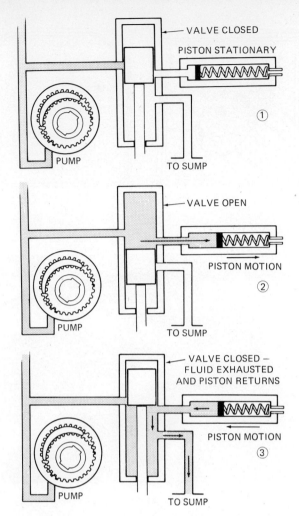

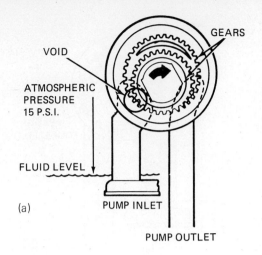

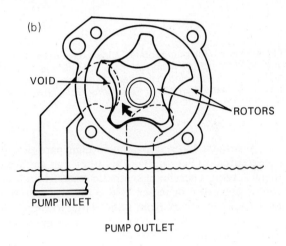

Fig. 4-15 Operation of the valve controls fluid flow into an actuator (1 and 2). The fluid used for apply is usually allowed to return to the sump at the manual valve or shift valve to release the actuator (3). (*Courtesy of Ford Motor Company.*)

of volume change in the pumping chambers (Fig. 4-18). This allows a large pump output to produce the fluid volume needed for shifts and lubrication and a reduced output between shifts, which is most of the time.

The parts in a pump must fit together with only a small amount of clearance to prevent the fluid from leaking across the pump from the high-pressure area to the areas of lower pressure. The fit provides just enough clearance for the parts to move without excess drag.

4.6 PROVIDING CLEAN FLUID

The filter at the pump intake is designed to trap dirt, metal, and any other foreign particles that might cause wear in the pump, bearings and bushings, or gear train or cause sticking of the various valves. Two basic types of filters are used: surface and depth (Fig. 4-19). Both filters offer a potential problem in that as they do

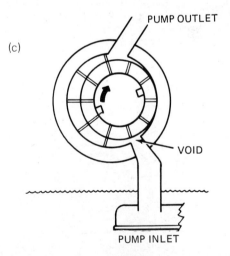

Fig. 4-16 (a) Internal–external gear, (b) gerotor, and (c) vane transmission pumps. [*(a, b). Courtesy of Ford Motor Company. (c) Copyrighted material reprinted with permission from Hydramatic Div., GM Corp.*]

their job, they will eventually plug up. This might cause an excessive pressure drop across the filter and fluid starvation to the pump. A filter must have enough capacity to operate until it is changed.

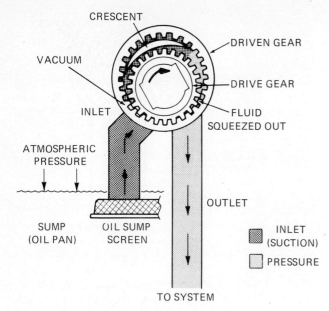

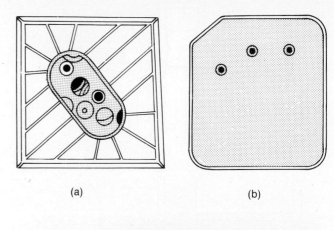

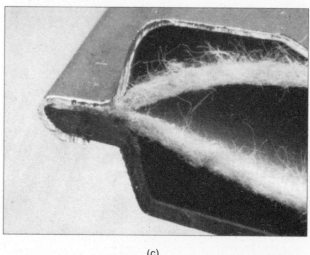

Fig. 4-17 When the pump rotates, a low pressure/vacuum is created as the pumping members move apart in one area while a high pressure is created in the area where the pumping members come together. (*Courtesy of Ford Motor Company.*)

Fig. 4-19 Filters: The major types are (a) surface or screen and (b) depth or felt. (c) Cut-open section of felt filter. (*Courtesy of Sealed Power.*)

A surface filter traps the foreign particles at the outer surface (Fig. 4-20). This filter can be a woven metal screen, a screen of synthetic material such as dacron or polyester, or paper. With a metal screen the size of the openings vary from rather large to rather fine—in the range of 50 to 100 μm. A micrometer is one millionth of a meter or 39 millionths of an inch. A surface filter has the disadvantage that it doesn't have much capacity to trap particles. The mesh openings are the usable area. About one-half of a filter's surface is the fibers or wires that make up the filter with the remainder

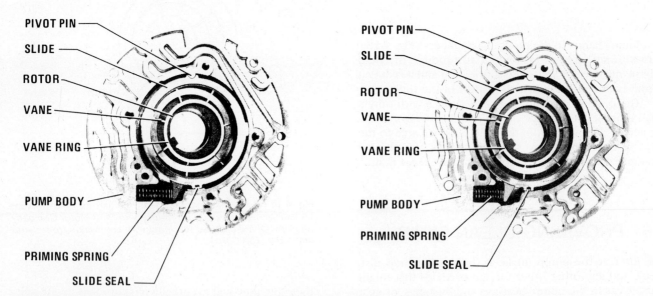

Fig. 4-18 Vane pump in high-output position (left) and low-output position (right). Left: slide is off-center toward the left of the rotor. Right: slide and rotor are almost centered to each other. (*Copyrighted material reprinted with permission from Hydra-matic Div., GM Corp.*)

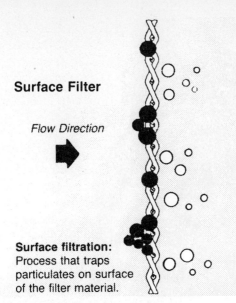

Fig. 4-20 A surface filter traps particles that are too big to pass through the openings in the screen. (*Courtesy of Sealed Power.*)

4.7 CONTROLLING FLUID FLOWS

The fluid flow from the pressure regulator valve to the manual valve and into the control circuit is commonly called *line* or *control pressure*. Fluid flow to and from an actuator is controlled by one or more valves. *Spool valves* sliding in a round bore are the major valves used for this. A spool valve gets its name from its slight resemblance to a spool for thread. This valve has two or more *lands* that fit the valve bore tightly enough that fluid cannot escape along the valve but loosely enough so that the valve can slide freely along the bore (Fig. 4-24). Other valve part names are the *valleys* or *grooves* between the lands and the *faces* or *sides* of the lands. The lands serve as fluid blocks; the valleys serve as fluid passages; and the faces sometime serve as pressure surfaces, called *reaction surfaces*, to produce valve movement. Some valves are relatively long with a series of lands and grooves so fluid flow through two or more passages can be controlled at the same time. The lands of a spool valve often have different diameters in order to provide different-size reaction areas.

A spool valve bore has fluid passages entering from the sides, which connect to grooves that extend clear around the valve. This is done to produce the same pressure entirely around the valve (Fig. 4-25). If pressure were exerted on only one side of the valve, this pressure could force the valve sideways and lock it so the valve would not slide easily along the bore. Only a very few manual valves are designed so as to have a side loading. These valves have one or more grooves running along part of the valve's length. As a spool valve is slid along the bore, the lands open up or close off the side passages and block or allow fluid flow from one place to another or to the sump to allow pressure to be exhausted.

The manual valve (a spool valve) is connected to the shift lever so it is moved mechanically. Most spool

being the openings (Fig. 4-21). The filter material must be strong enough to prevent collapse or tearing of the filter.

A depth filter traps particles as they try to pass through the filter material, which is usually felt or a synthetic material of various thicknesses (Fig. 4-22). The depth of the material allows room to trap particles as well as room for fluid flow. It also offers the ability to trap smaller particles, more capacity to trap them, and the ability to flow better for a longer period of time (Fig. 4-23). Some depth filters trap particles as small as 10 μm.

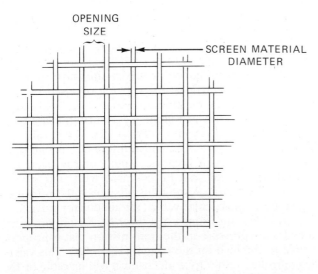

Fig. 4-21 The surface area of a surface filter is reduced somewhat by the material that makes up the screen. The size of the openings between the screen materials determine how fine the filter is.

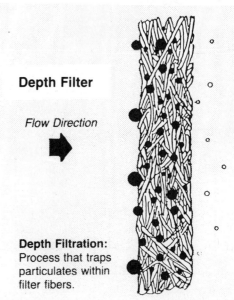

Fig. 4-22 A depth filter is a group of woven fibers of a certain thickness. Foreign particles are trapped as they try to flow through. (*Courtesy of Sealed Power.*)

113

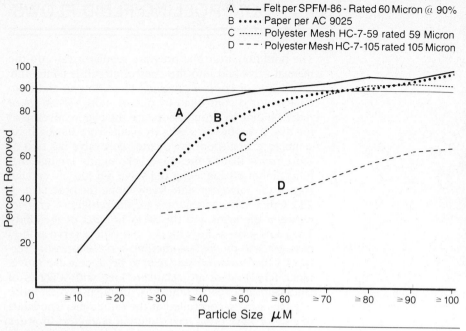

Fig. 4-23 Comparison of the filtering ability of four different filters. (*Courtesy of Sealed Power.*)

valves are positioned with a spring at one end and fluid pressure at the other. The spring tries to hold the valve in one position, and fluid pressure can move the valve to another (Fig. 4-26). Let's imagine a valve with a spring having a 2-lb (0.9-kg) tension pressing on one end and a 0.5-in.- (12.67-mm) diameter land at the other end. The area at that end of the valve is 0.196 in.^{2} (1.26 cm^2), and a fluid pressure of 10.2 psi (70.33 kPa) acting on an area of 0.196 in.^{2} (1.26 cm^2) generates a force of 2 lb. Any greater fluid pressure causes the valve to move, compressing the spring, and changes the fluid flows controlled by the valve. In some cases, pressure between two different-size land faces is used; fluid pressure acting on a larger face develops a force to move the valve in that direction (Fig. 4-27). If the diameter of one land is 0.5 in., the other land is 0.3 in. (7.6 mm), and the shank of the valve is 0.2 in. (5.1 mm), there is a reaction area of 0.165 in.^{2} (1.06 cm^2) on the big land and one of 0.04 in.^{2} (0.26 cm^2) on the small land. Subtracting 0.04 from 0.165 gives us an area difference of 0.125 in.^{2} (0.8 cm^2), and a fluid pressure of 10 psi (68.95 kPa) develops a force of 1.65 lb (0.75 kg) pushing toward the larger land.

Many valves must be kept in one of two positions. For example, a shift valve is spring loaded to be in the downshift position, and when there is sufficient fluid pressure from the governor at the other end, the valve moves to the upshift position. There should be no in-between with a shift valve. Some pressure regulator and throttle valves move to a balanced position to produce a certain, modulated pressure. Some pressure regulators and most pressure relief valves rapidly move back and forth when they operate. Pressure builds up and causes the valve to open and release pressure, and as soon as this happens, the valve closes so the cycle soon repeats.

In the case of governor valves, the valve spool is positioned by centrifugal force being opposed by hydraulic pressure in the governor circuit (Fig. 4-28). The output of this valve is a gradual pressure increase that is proportional to car speed (Fig. 4-29). This is called a *modulated pressure*. A modulated pressure is one that varies to a particular point depending on the signal received. This should not be confused with "modulator pressure," which in some transmissions is the pressure signal from a particular valve (the modulator valve).

Most transmissions use a *throttle valve* to produce a hydraulic pressure in the transmission that is proportional to the load on the engine (Fig. 4-30). This valve is controlled either by a mechanical rod or cable or by a vacuum signal acting on a vacuum diaphragm. The valve is balanced between a mechanical pressure or spring-versus-vacuum-diaphragm pressure at one end and fluid pressure in the modulator circuit at the other.

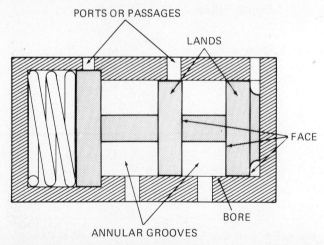

Fig. 4-24 Spool valve and its bore. Note the names of the various parts. (*Courtesy of Chrysler Corporation.*)

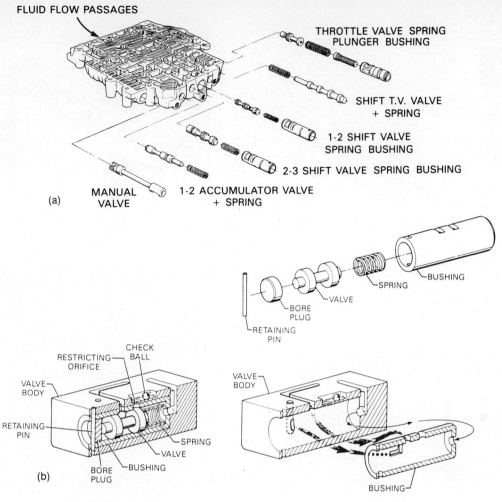

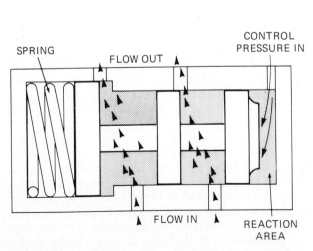

Fig. 4-25 Valve body and valves. (a) Fluid passages that lead to the valves. (b) A bushing is often used to change the flow passage or change the size of the valve and bore. (*Copyrighted material reprinted with permission from Hydra-matic Div., GM Corp.*)

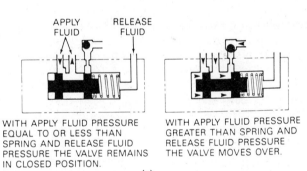

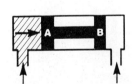

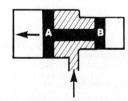

Fig. 4-26 When a valve moves in its bore, the side passages are opened or closed to allow or stop fluid flow. (*Courtesy of Chrysler Corporation.*)

Fig. 4-27 Many valves are moved by fluid pressure acting on the surface area of the valve faces. (*Copyrighted material reprinted with permission from Hydra-matic Div., GM Corp.*)

115

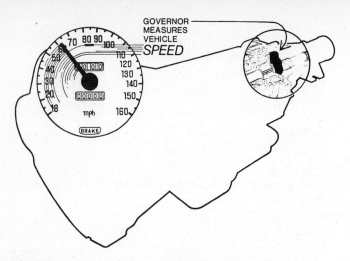

Fig. 4-28 The governor valve produces a fluid pressure that is relative to the speed of the vehicle. (*Courtesy of Nissan Motor Corporation.*)

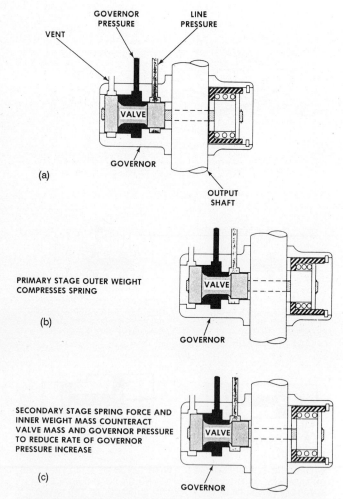

Fig. 4-29 (a) When this vehicle is stopped, fluid pressure at the governor valve moves it to the left where it is vented. (b) As the vehicle starts moving, centrifugal force on the weights moves the weight and valve toward the right to start opening the valve to line pressure. (c) A higher speed moves the valve further to the right to produce still higher governor pressure. (*Courtesy of Chrysler Corporation.*)

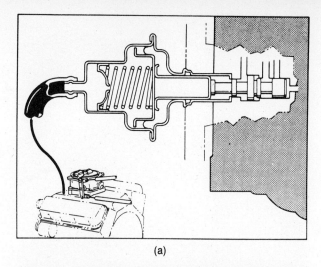

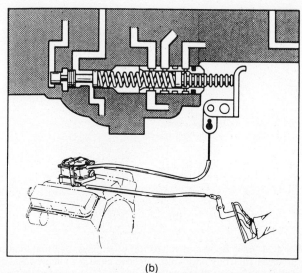

Fig. 4-30 (a) A vacuum-operated throttle valve uses a vacuum line and vacuum modulator to produce an engine-load-sensitive signal at the transmission. (b) A mechanical-controlled throttle valve (in this case cable) does the same thing. (*Copyrighted material reprinted with permission from Hydra-matic Div., GM Corp.*)

In a vacuum motor, the spring is always trying to push against the hydraulic valve, but the engine manifold vacuum reduces the power of the spring. A throttle valve also produces a modulated pressure, and this throttle valve (TV) pressure is very low at idle speed and very high at wide-open throttle (WOT). The maximum pressure that a governor or a throttle valve produces is line pressure.

Still another way to control the position of a spool valve is to use an electric solenoid. The solenoid has the advantage that it can be switched on or off by an electric switch or a computer. The computer can be programmed to inputs from many different sources. Many modern transmissions use a torque converter clutch that is applied and released on command from the engine control module (ECM); some modern transmissions use shift valves controlled by a separate electronic module for the transmission (TCM). Most of these units are arranged so the solenoid can exhaust the fluid pressure

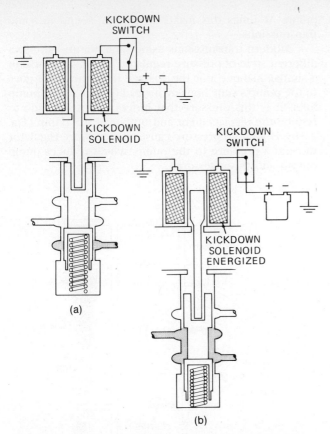

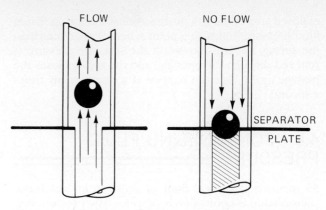

Fig. 4-32 A simple check valve is opened when the fluid flows in one direction and closed when the fluid tries to flow the other way. (*Courtesy of Ford Motor Company.*)

Fig. 4-31 (a) Valve controlled by electric solenoid. When there is no electrical signal at the solenoid, the spring moves the valve to the closed position. (b) The valve opens when the switch closes to energize the solenoid. (*Courtesy of Nissan Motor Corporation.*)

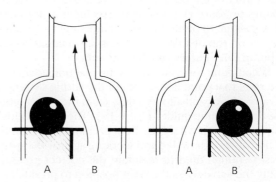

Fig. 4-33 Shuttle valve. Fluid flow from port B to port C will close port A, and fluid flow from port A to port C will close port B.

at one end of a spool valve, and when this happens, a spring or fluid pressure at the other end causes the spool valve to move and change the fluid path through the valve (Fig. 4-31).

Fluid flow can also be controlled by a much simpler means using ball check valves when possible. The ball, which resembles a ball bearing, is positioned above a hole in the *separator plate*, also called a *transfer plate* or *restricter plate*, between the valve body sections or between the valve body and transmission case (Fig. 4-32). An upward fluid flow moves the ball aside and flows around it. A downward fluid flow pushes the ball to seat against the hole and shut off the flow. The flow is in one direction only, upward.

Some transmissions use a *two-way check valve*, also called a *shuttle valve*. The ball valve is positioned above two, side-by-side holes with another flow passage extending upward or to the side. Fluid flow upward through one of the holes in the plate causes the ball to move over and seat against the other hole (Fig. 4-33). Fluid flow upward through the second hole causes the ball to move over and seal the first hole.

A *pressure relief valve* resembles a ball check valve, but it is spring loaded to seat against its seat (Fig. 4-34). Sometimes a flat valve disc or spool valve is used in place of a ball. The fluid pressure required to unseat this valve is determined by the strength of the spring and the size of the opening at the seat. Fluid pressure acting on the

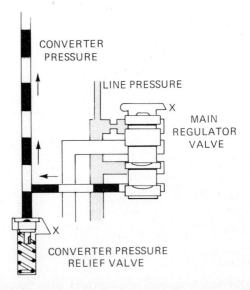

Fig. 4-34 Circuit that supplies fluid through a torque converter. Note the converter pressure relief valve. Excessive fluid pressure forces the valve off of its seat so pressure is exhausted. (*Courtesy of Ford Motor Company.*)

exposed area of the ball pushes against the spring. When fluid pressure builds to a point where it is greater than the spring, the ball moves off the seat; the pressure is relieved through the opening; and the spring closes the passage again. This can happen at a rate of many times a second.

4.8 CONTROLLING FLUID PRESSURE

As mentioned earlier, fluid pressure in an automatic transmission is controlled by a spool-type pressure regulator valve. Fluid pressure must be high enough to apply a clutch or band tightly enough to prevent slippage, but it should not be too high or there will be excessive heat and fluid foaming as well as a higher power drag on the engine. Remember that fluid horsepower is a product of pressure and flow. In most transmissions, the pressure regulator valve is positioned close to the outlet of the pump. This valve is usually arranged so pump pressure is at one and a spring is at the other end. A passage to pump pressure enters the valve at one valley or on top of a land, and a passage leading back to the pump inlet enters an adjacent valley.

When pump pressure times the reaction area of the valve rises to a point above that of the spring's strength, the valve moves and allows a flow from the pump outlet back to the pump inlet (Fig. 4-35). The valve finds a balanced position where the fluid pressure acting on the reaction area of the valve and working against the spring releases enough fluid to maintain the correct pressure. Usually the valve seeks the right position to produce the correct pressure, but a sticky valve that doesn't slide easily in its bore can stick and then overtravel. This can cause the pressure to drop and the valve to move back to its original position, and the cycle soon repeats. A pressure regulator valve can move back and forth quite rapidly, probably several thousand times a second. This movement follows the pressure pulsations from the pump. At times this makes a whining sound in some transmissions.

Modern transmissions using vane pumps can use a different style of pressure regulator. The valve works in a similar manner, but the passage from the valve goes to the pump's vane housing instead of back to the pump inlet. It is this pressure that moves the pump slide to reduce the displacement and output of the pump (Fig. 4-36). High fluid pressure causes the pressure regulator to send a pressure to the pump, which reduces pump output and thus pressure.

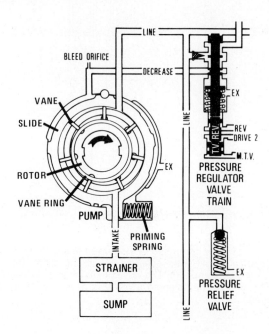

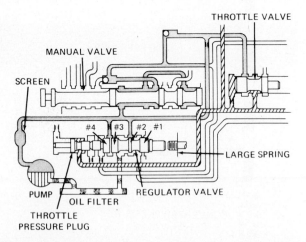

Fig. 4-35 Pressure regulator valve in its bore. When fluid pressure acting on the left side of land 4 becomes high enough, the valve moves toward the right. This lowers the pressure by allowing fluid to flow past land 2 and return to the filter. (*Courtesy of Chrysler Corporation.*)

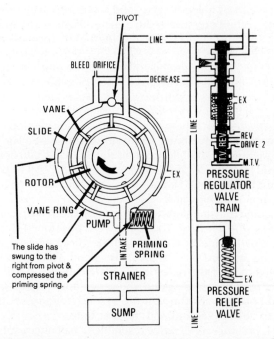

Fig. 4-36 When the pressure in a system using a vane pump gets too high (bottom), the pressure regulator valve opens to send a pressure signal to the pump slide. This moves the pump slide to reduce the pump output and the pressure is reduced as well. (*Copyrighted material reprinted with permission from Hydramatic Div., GM Corp.*)

Regulator valves have additional lands and valleys or another valve in the same bore to provide *boost pressures* for medium- to full-throttle operation or while in reverse and sometimes low and intermediate gear. Because of the probable increased torque requirement, boost pressure is used to provide more holding power at the clutches and bands and a firmer shift while operating in those gears and under wider throttle openings.

4.9 SEALING FLUID PRESSURE

The fluid passages are through the metal valve body, case, metal tubes, and shafts of the transmission. In the valve body, the many passages between the valves look like a bunch of *worm tracks* (as nicknamed by transmission rebuilders) (Fig. 4-37). In the case, the passages are simply holes drilled from one place to another. Metal ball plugs are used to seal unwanted openings, which often result if the passage has to turn a corner. Some transmission shafts are rather busy inside as they may have several different passages for various purposes (Fig. 4-38). Where the fluid must flow from one part to another, seals must be used to keep the pressure from escaping. Seals fall into two categories: *static* and *dynamic*.

A static seal is used to seal the space between two parts that are stationary relative to each other. This seal type includes gaskets and O-rings, which are placed between the two items and squeezed tightly as the parts are fastened together. A static seal must provide a certain amount of give or compression to fill any possible void between the two surfaces (Fig. 4-39). The amount of compression needed for a particular seal depends greatly on the flatness and rigidity of the two surfaces.

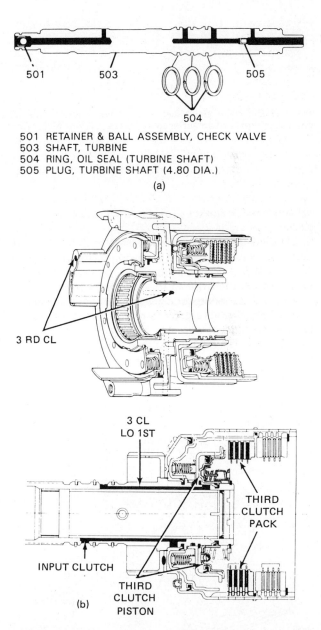

Fig. 4-37 Valve body shown uses an upper and lower section separated by the separator plate. Note how the separator plate closes most of the passages so it becomes a group of ports into the other section. (*Courtesy of Chrysler Corporation.*)

Fig. 4-38 (a) The rather simple passages through a turbine shaft. Note the plug (505) to restrict fluid flow. (b) The fluid passages through the clutch assembly and its support are a little complex. (*Copyrighted material reprinted with permission from Hydramatic Div., GM Corp.*)

119

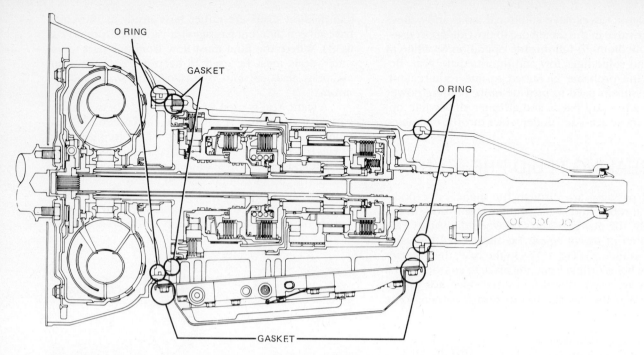

Fig. 4-39 The major static seals used in transmissions are gaskets and O-rings that are squeezed between two surfaces. (*Courtesy of Ford Motor Company.*)

Wavy or weak surfaces need a thick, resilient seal or gasket.

A dynamic seal has a much harder job because one of the surfaces to be sealed is moving relative to the seal. The movement can be a rotation (e.g., the torque converter enters the front of the transmission or the fluid flows from the pump housing into a clutch assembly) or a sliding motion (e.g., the stroking of a clutch piston) (Fig. 4-40).

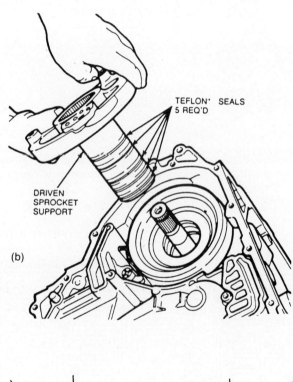

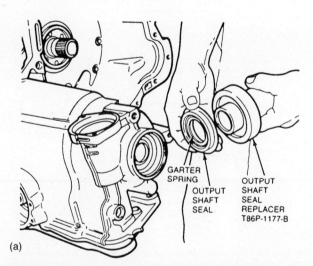

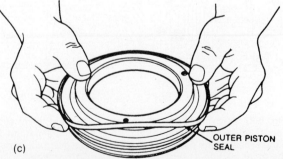

Fig. 4-40 The major dynamic seals are (a) metal-clad lip seals, (b) Teflon or metal seal rings, and (c) rubber piston seal rings. (*Courtesy of Ford Motor Company.*)

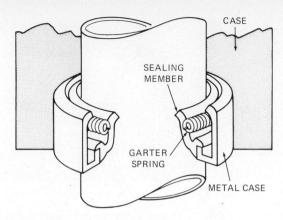

Fig. 4-41 Metal-clad lip seal. The sealing member/lip makes a dynamic seal with the rotating shaft while the metal case forms a static seal with the transmission case.

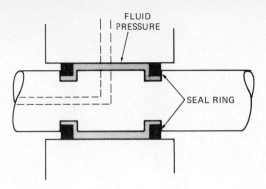

Fig. 4-43 Fluid pressure forces a sealing ring outward in both directions to make firm contact at the side and outer diameter.

At each end of the transmission, a rotating shaft enters or leaves the transmission, and the opening that the shaft runs through must be sealed to keep the fluid in and dirt out. At these openings the torque converter enters the front pump and the output shaft(s) leaves the extension housing. In both cases, a *lip seal* is used. A lip seal has a flexible rubber sealing lip that rubs against the revolving shaft with enough pressure so that fluid cannot flow between the shaft and the seal lip. A garter spring is usually used to increase this sealing pressure (Fig. 4-41). The lip is molded into the seal's case or outer housing, which forms a static seal with the transmission case.

Another style of seal is used to seal the fluid passages when a fluid flow leaves a stationary member and transfers to a rotating member such as the turbine shaft, front clutch assembly, or governor support (Fig. 4-42). This seal is often a metal or Teflon ring that fits tight enough against the bore to make a seal while the side of the seal makes a seal against the side of the groove where it is mounted (Fig. 4-43). A seal with a slight amount of leakage is sometimes desirable to provide lubrication for a bearing area close to the sealing ring. The metal rings can be a full-circle *hook ring* or a ring that has a small gap in it, butt cut. Many Teflon rings are full circle. They can be *scarf cut* or *angle cut*—have the ends cut at an angle so they overlap—butt cut, or uncut (Fig. 4-44). Teflon has the ability to change size, so given the proper tools, this ring can be stretched over a shaft and resized smaller to fit into a groove (Fig. 4-45).

The sliding seals for the clutch pistons are rubber in an *O-ring, lathe-cut seal*, or *lip seal shape*. The seal rings for a band servo piston are a metal ring or a rubber O-ring, square-cut seal, or lip seal. An O-ring is a rubber

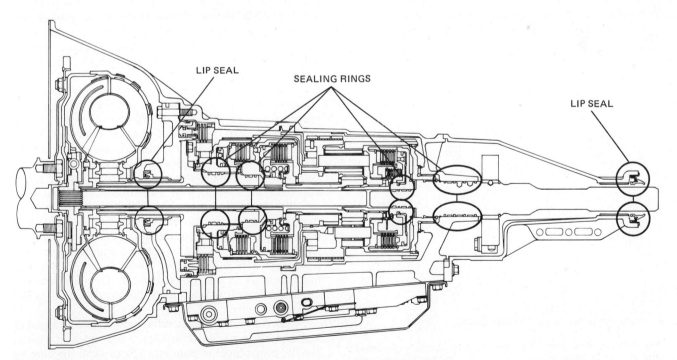

Fig. 4-42 Sealing rings are used to seal the passages between stationary and rotating members. For example, the seal rings at the left keep the fluid flow from the pump to the front clutch from escaping. (*Courtesy of Ford Motor Company.*)

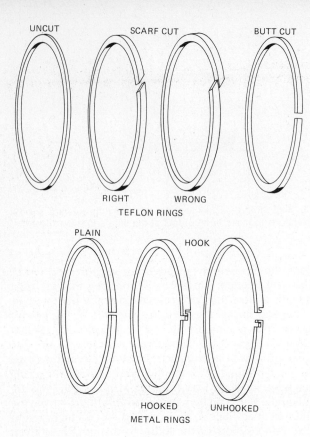

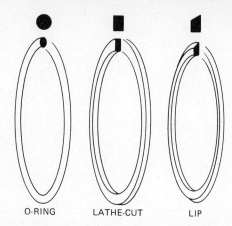

Fig. 4-46 Clutch and servo piston seals are usually O-rings, lathe-cut rings, or lip seals.

Fig. 4-44 Metal seal rings (bottom) have plain ends or hooked ends while Teflon rings are either uncut, butt cut, or scarf cut.

ring with a round cross section; a lathe-cut seal, also called a square-cut seal or quad-ring, is a rubber ring with a square cross section; and a lip seal has a sealing lip much like that described earlier (Fig. 4-46). The type of seal used on a piston is determined by the transmission manufacturer with the groove for the seal being cut to the proper shape and size. By its nature, a lip seal provides the best pressure retention as well as has the ability to adapt to a greater amount of piston-to-bore clearance (Fig. 4-47). The bore in the clutch assembly as well as the center seal area must have straight, smooth sides to provide a good sealing surface and keep the amount of seal wear to a minimum. The outer ends of the bores are normally chamferred to make piston and seal installation easier.

4.9.1 Special Notes on Elastomers

The rubber seal materials used for seals are often called *elastomers* because of the elastic nature of the flexible materials. A dynamic seal must remain flexible and maintain a certain size in order to work properly. Plain rubber is not used in a transmission because it is adversely affected by the heat of and contact with the fluid. Natural rubber has an operating range of −58°F (−50°C) to 212°F (70°C). Higher temperatures cause rubber materials to harden, and exposure to transmission fluid causes it to swell excessively. A hardened seal will probably leak and cause a pressure loss, which in turn will cause slippage.

Synthetic rubbers are blends of various rubberlike compounds. Some of these that are used in automatic transmissions are *Buna N*, or *nitrile rubber* (to 230°F); *polyacrylate* (to 320°F), *ethylene/acrylic* (to 375°F); *fluoroelastomer* (to 400°F); and *silicone* (above 400°F) (Fig. 4-48). Another major difference between these compounds is cost; as the operating temperature range improves, the cost goes up many times. A manufacturer usually selects the lowest cost seal material that will do the job, and we should realize that different seals in a transmission have different working conditions. For example, the high–reverse clutch in a Simpson gear train transmission is subjected to about 30 to 35 times as much activity as the forward clutch. One seal material might work well in one clutch but not stand up to the load of the other.

When a transmission rebuilder purchases a gasket and seal kit to use in rebuilding a transmission, cost is not the only object of concern. The price of the correct seal material can be many times greater than a cheaper Buna N seal. Several methods can be used in the field to identify the quality of a particular seal:

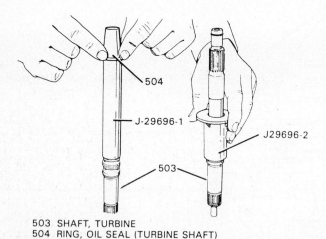

503 SHAFT, TURBINE
504 RING, OIL SEAL (TURBINE SHAFT)

Fig. 4-45 An uncut Teflon seal ring must be carefully stretched and guided to its position on the shaft (left) and then resized down to a smaller diameter (right). Note that special tools are required for these operations. (*Copyrighted material reprinted with permission from Hydra-matic Div., GM Corp.*)

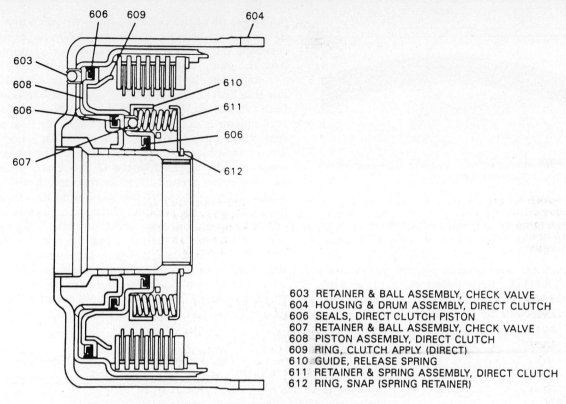

603 RETAINER & BALL ASSEMBLY, CHECK VALVE
604 HOUSING & DRUM ASSEMBLY, DIRECT CLUTCH
606 SEALS, DIRECT CLUTCH PISTON
607 RETAINER & BALL ASSEMBLY, CHECK VALVE
608 PISTON ASSEMBLY, DIRECT CLUTCH
609 RING, CLUTCH APPLY (DIRECT)
610 GUIDE, RELEASE SPRING
611 RETAINER & SPRING ASSEMBLY, DIRECT CLUTCH
612 RING, SNAP (SPRING RETAINER)

Fig. 4-47 The seal forms a fluid-tight fit between the piston (608) and the bore. This clutch assembly used three piston seals (606): an inner one, a middle one, and an outer one. (*Copyrighted material reprinted with permission from Hydra-matic Div., GM Corp.*)

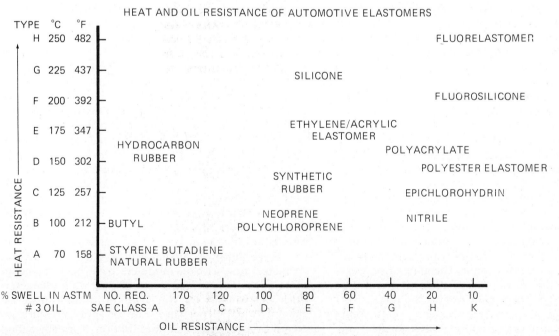

Fig. 4-48 Chart comparing the ability of some of the commonly used seal materials to withstand heat and oil. Note that rubber is very low in both cases.

Fig. 4-49 The code letters cast into the surface of the seal shown identify the manufacturer of the seal.

- Read the letter code molded into the seal surface. The letters are a code registered with the Rubber Manufacturers Association (RMA) to identify the manufacturer. The original equipment manufacturer (OEM) of a transmission seal normally supplies the same OEM seal to the aftermarket (Fig. 4-49). These are the best quality seals.
- Burn a portion of the seal. This destroys the seal, but it gives you a good idea of the material from which it was made. Ignite the seal with a match; let it burn for 2 or 3 sec; and then extinguish the fire. If the smoke odor is like a burning tire, the material is Buna N; if it has a sweet smell, it is polyacrylic material.
- Feel the material. Polyacrylic material is harder and less elastic than rubber.

4.10 MODIFYING FLOW AND PRESSURE

While a shift is occurring, it is usually desirable to provide a relatively gradual pressure increase behind the pistons applying a clutch or band. This improves the *shift quality* and produces a *shift feel* that is desirable for a particular car model. Shift feel is a "seat-of-the-pants" response the driver feels during transmission shifts. As the piston is stroking to take up the clutch or band clearance, the pressures in the circuit are relatively low. But the instant the clearance disappears and the piston stops moving, the pressure rises very rapidly, causing a sudden and possibly very harsh application. There are two commonly used devices to control this phase of band or clutch application: an *orifice* and an *accumulator*. Different versions of a particular transmission often have different orifice sizes and accumulator settings to suit the transmission to the engine and car body weight in which it is to be used.

An orifice is simply a small hole, usually in the separator plate (Fig. 4-50). An orifice produces a resistance to fluid flow and therefore causes a pressure drop as long as fluid is flowing through it. The amount of pressure drop is relative to the size of the orifice and the amount of flow. The smaller the orifice, the greater resistance to flow and the larger the pressure drop. As soon as the flow stops, the orifice no longer has an effect on the flow, and the pressure on both sides is the same (Fig. 4-51). An orifice is often also used to dampen a fluid flow to some of the control valves. We often find one in the passage bringing pump pressure to the pressure regulator valve; this orifice helps keep pressure pulses from the pump, causing the valve to overreact.

An accumulator is usually a piston or valve that is not attached to anything; all it does is stroke down a bore (Fig. 4-52). The pressure required to stroke the accumulator is controlled by a spring or by fluid pressure on the other side of it. An accumulator is tied to a servo piston by a branch of the same fluid passage for apply pressure and is adjusted to stroke at a pressure just above that needed to stroke the servo piston (Fig. 4-53). Just after the servo piston takes up the clutch or band clearance, the accumulator strokes, absorbs a specific amount of fluid flow, and thus causes a lag in the pressure increase at the servo and the band or clutch (Fig. 4-54). The effect is a slightly longer and smoother shift. Some transmissions vary the pressure behind the accumulator to provide different shift characteristics depending on the throttle position; full-throttle shifts are usually firmer than part-throttle shifts (Fig. 4-55).

4.11 AUTOMATIC TRANSMISSION FLUID

Automatic transmission fluid (ATF) is one of the most complex fluids in a car. It has to act as a medium to transfer hydrodynamic energy in the torque converter; to act as a medium to transfer hydrostatic energy at the clutch and servo pistons as well as the valve body; to help transfer sliding friction energy as the clutches and bands apply; to transfer excess heat away from high-temperature locations such as the friction surfaces, gears, and bushings; and to lubricate the various moving parts.

Early automatic transmissions used engine oil for a transmission fluid, but since the internal operating conditions in an engine and automatic transmission are significantly different, a special transmission fluid was developed in the late 1940s. At first the ATF was simply a mineral oil much like motor oil but dyed red. Newer ATF can be a mineral oil or a synthetic lubricant with a number of additives to make it more suited to the needs of a transmission. The first transmission fluid was labeled *Type A* transmission fluid. This fluid type was developed by General Motors. As transmission fluid was improved, Type A was replaced with *Type A, Suffix A; Dexron B; Dextron II-C;* and then *Dexron II-D* (Fig. 4-56). Dexron II-D can be used in older transmissions that specify one of the older types. Some other fluid types were developed by Ford Motor Company; *Types F, G, CJ, H,* and *Mercon* are required for various Ford transmission models. Mercon can be used in place of the various other fluids in most Ford transmissions.

These fluids are similar but have different friction characteristics; friction modifiers have been added to some of the fluids. Transmissions have shift characteristics determined by the type of friction material, amount of apply pressure, rate of apply pressure increase, and coefficient of fluid friction. Normally a fluid has a higher static coefficient of friction than the dynamic coefficient of friction. In a nonmodified fluid, the

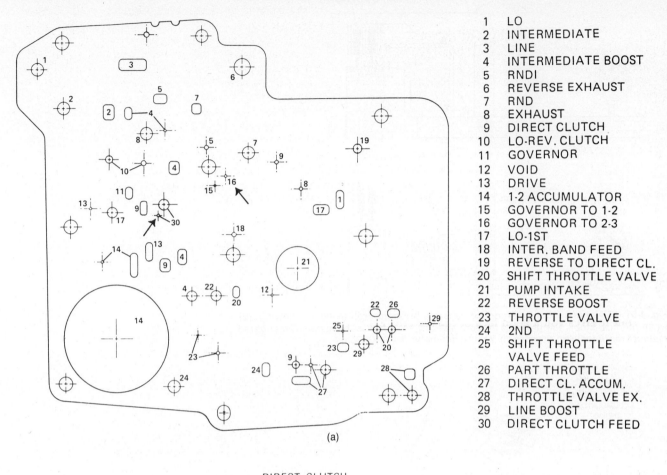

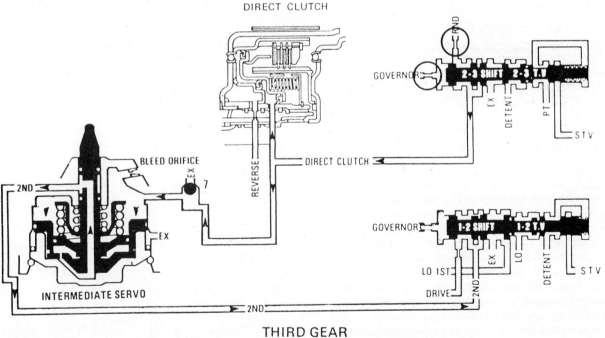

Fig. 4-50 (a) The two orifices in the transfer plate (indicated by arrows) are also shown in the (b) fluid diagram. (*Copyrighted material reprinted with permission from Hydra-matic Div., GM Corp.*)

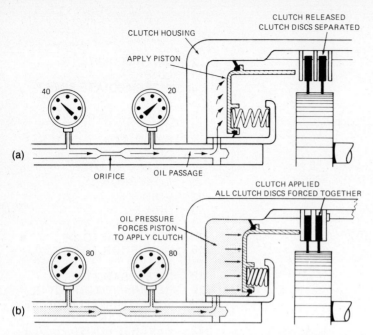

Fig. 4-51 (a) The restriction of an orifice causes a pressure drop as fluid is flowing; (b) when the flow stops, the pressure on both sides of an orifice is the same. (*Copyrighted material reprinted with permission from Hydra-matic Div., GM Corp.*)

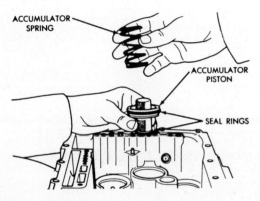

Fig. 4-52 An accumulator is not connected to anything so it does no work. The piston merely strokes against fluid or spring pressure and absorbs fluid. (*Courtesy of Chrysler Corporation.*)

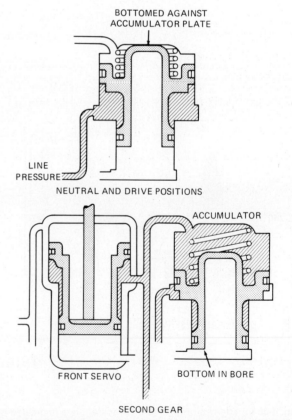

Fig. 4-53 In neutral, the piston shown is moved to the top of its stroke by line pressure; the fluid pressure to apply the front servo in second gear strokes the accumulator piston downward to the bottom of its bore. (*Courtesy of Chrysler Corporation.*)

coefficient of friction increases as a clutch or band locks up. A friction-modified fluid does just the opposite. It has a much lower static coefficient of friction than the dynamic coefficient of friction (Fig. 4-57). A transmission designed for a friction-modified fluid must have more friction area in the clutches and bands to compensate for this lower friction. But the lockup portion of the shift is less harsh, and the shifts are smoother. The fluid specified by the transmission manufacturer or a later version should always be used in a particular transmission. Use of the wrong fluid can produce a harsh shift with higher shock loads to the drive line components in some cases or a slipping shift with a possible transmission burnout in others. A listing of the various fluids is shown in Fig. 4-58.

Transmission fluid is formulated with various additives to produce the favorable operating characteris-

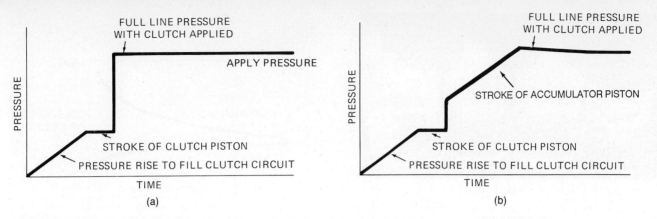

Fig. 4-54 (a) When a clutch is applied fluid pressure increases gradually while the circuit fills with fluid and the piston strokes to take up the clearance. At the point where the piston stops, the rapid increase will probably produce a harsh clutch application. (b) The pressure rise in a circuit that uses an accumulator is less severe. Note that it takes longer for the pressure at the piston to increase to line pressure. This should soften the clutch or band application.

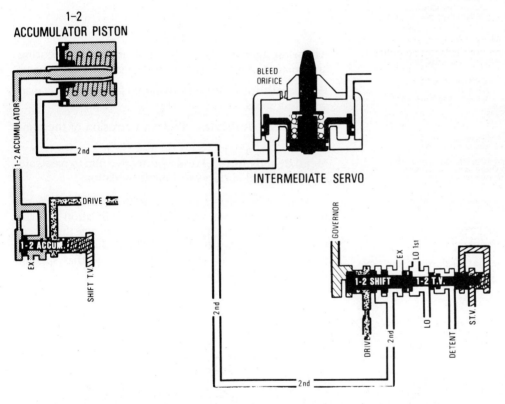

Fig. 4-55 The pressure at the right side of the accumulator piston in this circuit is controlled by the 1–2 accum. valve, which in turn is controlled by shift TV pressure. Therefore this accumulator strokes at a different rate depending on throttle and driving conditions. (*Copyrighted material reprinted with permission from Hydra-matic Div., GM Corp.*)

tics. These additives are chemical compounds, and the reasons for their use are as follows:

Detergents-dispersants Keep the transmission clean and the valves free from sticking by keeping foreign items in suspension until they are removed by the filter or by draining.

Oxidation inhibitors Reduce oxidation and decomposition of the fluid, which can produce varnish and sludge.

Viscosity index improvers Reduce the fluid viscosity change relative to temperature so fluid thickness and shift characteristics remain stable.

Friction modifiers Change the fluid's coefficient of friction.

Foam inhibitors Prevent formation of air bubbles and foam in the fluid.

Seal swellers Produce a slight amount of swelling on the elastomers to compensate for any wear that occurs.

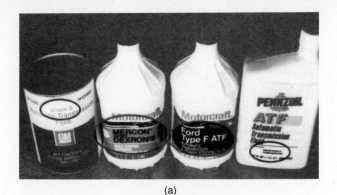

(a)

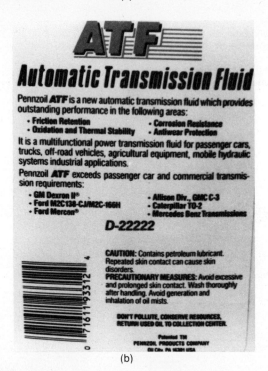

(b)

Fig. 4-56 An ATF container has markings that show the fluid type (circled).

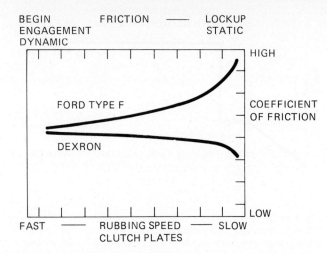

Fig. 4-57 The dynamic coefficient of friction for the two fluids shown is almost the same, but the static friction is very different.

Antiwear agents Reduce friction and prevent scoring and seizure of metal parts running against each other.

Rust inhibitors Prevent rust from forming on the iron and steel parts.

Corrosion inhibitors Prevent corrosion of the nonferrous parts.

Metal deactivators Form a protective film to passivate catalytic metal surfaces to inhibit oxidation.

The transmission fluid that is blended by the many different companies must meet specifications established by the transmission manufacturers and the American Petroleum Institute (API). Fluids that meet a particular specification are allowed to show that specification or type label on the container.

Fluid Type	Year	Friction Modified	Obsolete
A	1949	Yes	X, use newer fluid
A, Suffix A	1957	Yes	
Dexron-B	1967	Yes	X, use Dexron-IID
Dexron-IIC	1973	Yes	
Dexron-IID	1976	Yes	

(a)

Fluid Type	Year	Friction Modified	Obsolete
F, 1P-XXXXXX	1959	No	X, use F, 2P-XXXXXX
F, 2P-XXXXXX	1967	No	
G	1972	No	X, use F, 2P-XXXXXX
CJ	1977	Yes	X, use Mercon
H	1982	Yes	X, use Mercon
Mercon	1988	Yes	

Note: 1P-XXXXXX represents the fluid qualification number, that is, 1P followed by six digits; 2P-XXXXXX is similar.

(b)

Fig. 4-58 Automatic transmission fluids developed by (a) General Motors Corporation and (b) Ford Motor Company.

Probably the greatest problem that transmission fluid faces is heat. Excess heat shortens the fluid's life significantly. Excess temperatures cause the fluid to break down and form gum or varnish. This in turn can cause valve sticking or reduce the amount of fluid flow in certain circuits. All transmissions use a cooler to help remove excess heat. The fluid should be changed more frequently than normal if it operates at temperatures above 175°F (79°C). Adverse driving conditions that tend to produce higher fluid temperatures are trailer towing, driving on hills, and stop-and-go driving such as is done by taxis or delivery vehicles.

REVIEW QUESTIONS

The following questions are provided so you can check the facts you have just learned. Select the response that best completes each statement.

1. Hydraulics is
   A. The process of transferring energy through fluids under pressure
   B. Based on the fact that fluids are compressible
   Which is correct?
   a. A only
   b. B only
   c. Both A and B
   d. Neither A nor B

2. Technician A says that fluid pressure is measured in kilopascals. Technician B says that 1 psi is equal to about 7 kPa. Who is right?
   a. A only
   b. B only
   c. Both A and B
   d. Neither A nor B

3. If a fluid pressure of 50 psi is exerted on a piston area of 10 in.^{2}, it generates a force of
   A. 10 lb
   B. 50 lb
   C. 100 lb
   D. 500 lb

4. If we wanted to increase hydraulic force, we would increase the
   A. Amount of fluid flow
   B. Length of the piston stroke
   Which is correct?
   a. A only
   b. B only
   c. Both A and B
   d. Neither A nor B

5. The amount of fluid pressure in a system is determined by
   A. The pump
   B. The restrictions to flow in the system
   C. The strength of the system
   D. All of these

6. In a hydraulic system, fluid flow and pressure begin at the
   A. Pump
   B. Pressure regulator
   C. Control valve
   D. Filter screen

7. Fluid is forced through the pump intake by
   A. Pump suction
   B. Atmospheric pressure
   C. Inner case pressure
   D. Gravity

8. Automatic transmissions use a pump of the _____ type.
   A. Internal–external gear
   B. Gerotor
   C. Vane
   D. All of these

9. A pressure regulator valve controls fluid pressure by
   A. Letting excess fluid flow back to the pump intake or sump
   B. Reducing the displacement of the pump
   Which is correct?
   a. A only
   b. B only
   c. Both A and B
   d. Neither A nor B

10. Generally speaking, a depth filter _____ than a surface filter.
    A. Can trap smaller particles
    B. Has more filter capacity
    Which is correct?
    a. A only
    b. B only
    c. Both A and B
    d. Neither A nor B

11. When a vane pump is used, mainline fluid pressure is reduced by
    A. Releasing excess fluid to the pump intake or sump
    B. Reducing the displacement of the pump
    Which is correct?
    a. A only
    b. B only
    c. Both A and B
    d. Neither A nor B

12. A major portion of a spool valve is
    A. The land
    B. The valley
    C. The face
    D. All of these

13. The throttle valve is designed to produce a fluid pressure signal that is proportional to
    A. The speed of the car
    B. The gear selector position
    C. The load on the engine
    D. All of these

14. A shift valve is moved to the upshift position by
    A. The throttle valve pressure
    B. The governor pressure
    C. A spring
    D. Any of these

15. The metal balls used in a valve body have the purpose of
    A. Allowing fluid flow that is in only one direction
    B. Closing one passage while fluid is flowing in another
    C. Relieving excess pressure
    D. Any of these

16. A lip seal normally makes a
    A. Dynamic seal with a rotating shaft
    B. Static seal with the case in which it is installed
    Which is correct?
    a. A only
    b. B only
    c. Both A and B
    d. Neither A nor B

17. The elastic seals used in a transmission
    A. Have a lifetime that is affected by transmission temperature
    B. Should all be made from nitrile rubber
    Which is correct?
    a. A only
    b. B only
    c. Both A and B
    d. Neither A nor B

18. An orifice in a fluid passage
    A. Reduces the pressure in a servo while it is in the applied position
    B. Causes a servo to apply faster
    Which is correct?
    a. A only
    b. B only
    c. Both A and B
    d. Neither A nor B

19. An accumulator in a fluid circuit is used to
    A. Increase the fluid flow through the circuit
    B. Cushion a shift by absorbing a portion of the fluid flow
    Which is correct?
    a. A only
    b. B only
    c. Both A and B
    d. Neither A nor B

20. Automatic transmission fluid contains additives designed to
    A. Change the friction characteristics of the fluid
    B. Clean the transmission
    C. Reduce the rate of oxidation
    D. All of these

CHAPTER 5

HYDRAULIC SYSTEM OPERATION

OBJECTIVES

After completing this chapter, you should:

- Have a basic understanding of how the hydraulic system of an automatic transmission operates.

- Have a basic understanding of what valves are used in the average hydraulic system and their function.

- Be able to follow an upshift or downshift sequence of operation through a hydraulic schematic.

- Understand the role of the governor and throttle valves on the automatic shifts.

5.1 INTRODUCTION

The hydraulic system is what makes an automatic transmission automatic. It supplies the force to apply the clutches and bands as well as the valves that control the apply forces. The automatic action of the shift valves produce the upshifts and downshifts. Most of these valves are contained in the valve body, and this is normally a rather complex unit (Fig. 5-1). With experience, you will gain the ability to identify these valves, determine if they are working correctly, and repair any that are not.

The hydraulic system begins at the filter for the pump intake and ends as pressure at the clutch and servo pistons or as lubricating oil at the bushings. In this chap-

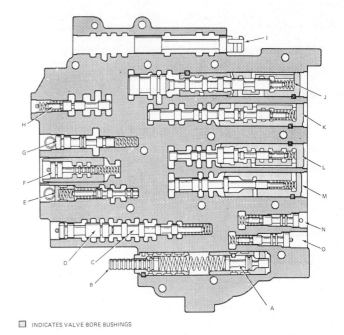

☐ INDICATES VALVE BORE BUSHINGS

ILL. NO.	DESCRIPTION	ILL. NO.	DESCRIPTION
A	THROTTLE VALVE & BUSHING	I	MANUAL VALVE
B	T.V. PLUNGER & BUSHING	J	1-2 SHIFT VALVE TRAIN
C	4-3 SEQUENCE VALVE	K	2-3 SHIFT VALVE TRAIN
D	3-4 RELAY VALVE	L	3-4 SHIFT VALVE TRAIN
E	T.V. LIMIT VALVE TRAIN	M	CONVERTER CLUTCH SHIFT VALVE TRAIN
F	ACCUMULATOR VALVE TRAIN	N	M.T.V. UP VALVE TRAIN
G	LINE BIAS VALVE TRAIN	O	M.T.V. DOWN VALVE TRAIN
H	3-2 CONTROL VALVE TRAIN		

Fig. 5-1 Cutaway view of valve body for modern four-speed transmission. Note that the pressure regulator valve is in the pump assembly and that about half of the valves use one or two bushings. *(Copyrighted material reprinted with permission from Hydra-matic Div., GM Corp.)*

131

TABLE 5.1 Hydraulic System Subcircuits

Name of Circuit	Purpose
1. Supply	Provides fluid to pressure regulator valve and transmission.
2. Main control pressure	Supplies fluid at a controlled pressure to transmission and converter when the engine is running.
3. Converter and cooler	Regulates converter pressure; provides gear train lubrication and cooling of oil.
4. Drive	Applies forward clutch and supplies control systems in forward-gear ranges.
5. Manual 1	Applies low and reverse band and locks out second gear.
6. Manual 2	Prevents upshift from second gear range.
7. Reverse	Increases control pressure and supplies reverse–high clutch and low–reverse band.
8. Shift valve trains	Switch pressure and exhaust pressure to servos and clutches for upshifts and downshifts.
9. Kickdown (downshift, detent)	Forces downshifts by overriding governor and/or throttle valve control of shift valves.
10. Governor	Provides a pressure proportional to road speed for transmission shift speed and shift "quality" control.
11. Throttle valve	Provides a pressure proportional to engine vacuum for shift speed and shift quality control.
12. Boosted throttle	Compensates for lower rate of vacuum change at 50% or more throttle opening.
13. Modulated throttle	Adjusts throttle pressure for control of shift valves.
14. Accumulator	Cushions shift by softening clutch or band apply.
15. Converter clutch	Controls the converter clutch to time clutch lockup.

ter, we trace the fluid flows through a hydraulic system diagram. Later, you will use a diagram as a tool to locate fluid routes.

The hydraulic system will be easier to understand if it is broken down and divided into subcircuits, as shown in Table 5.1. A control valve is the dividing point between the different circuits. Also note that the main feature for each circuit is to serve a particular purpose.

5.2 PRESSURE DEVELOPMENT AND CONTROL

As soon as the engine starts, the torque converter drives the transmission pump to produce a fluid flow that is called *line, mainline, supply,* or *control pressure.* Some sources separate supply pressure as that portion between the pump and the pressure regulator valve. Line pressure usually refers to regulated transmission pressure. As soon as the supply passages fill with fluid, the fluid pressure acts on one end or one large land of the regulator valve acting in opposition to a spring positioned at the other end. When pressure becomes strong enough, the regulator valve moves and opens a passage that either diverts a portion of the oil flow back to the pump intake or sends a pressure signal to the housing of a vane pump (Figs. 5-2 and 5-3). During start-up, the pressure should build up fairly rapidly because the manual valve is blocking the flows to the clutches and bands, and the only flow is to the regulator valve, the throttle valve, and the torque converter. It should be noted that some transmissions do not build fluid pressure in park because the manual valve opens line pressure to exhaust; fluid flow is through the manual valve and back to the sump (Fig. 5-4). The term *exhaust* refers to a fluid flow back to the oil pan to empty or drain a circuit.

Pressure regulator valves balance the line pressure against the strength of the spring. The spring has a predetermined strength so as to maintain a fluid pressure of about 50 to 60 psi (345 to 414 kPa) at light-throttle openings. In some transmissions, this spring is adjustable by turning a screw; in others it is adjustable by adding or removing a shim (Fig. 5-5).

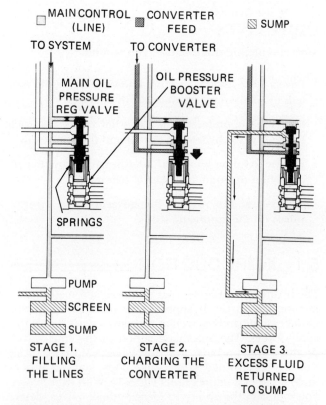

Fig. 5-2 The pressure regulator valve used with most gear pumps applies line pressure to one end of the valve with a spring at the other. When the engine first starts, pressure opens the valve slightly to charge/fill the converter. When line pressure tries to build too high, the valve moves farther so the excess fluid returns to the pump intake or sump. (*Courtesy of Ford Motor Company.*)

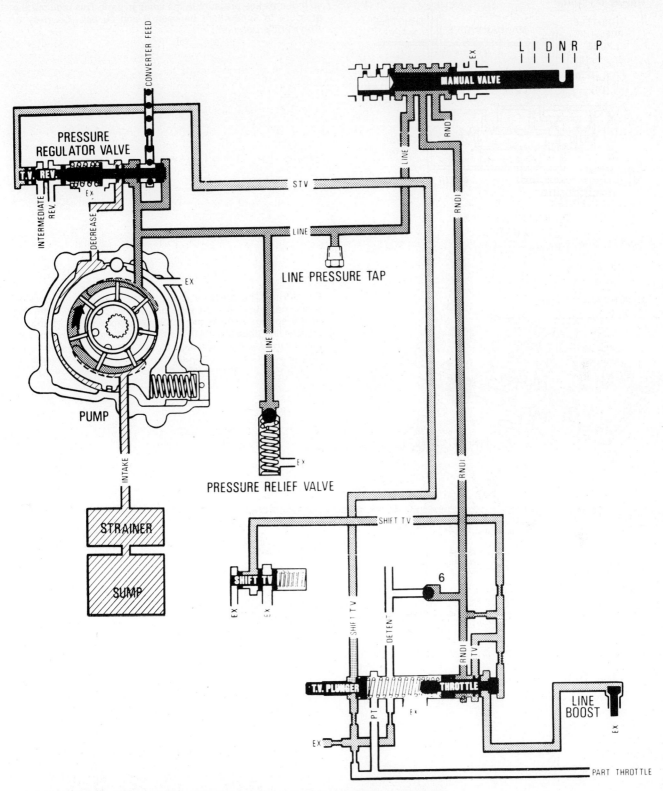

Fig. 5-3 The pressure regulator valve used with most vane pumps is positioned between line pressure and a spring in a manner similar to other pressure regulator valves. When line pressure tries to go too high, a "decrease" pressure signal is sent to the pump, which repositions the pump slide to decrease the pump displacement and therefore the pressure. *(Copyrighted material reprinted with permission from Hydra-matic Div., GM Corp.)*

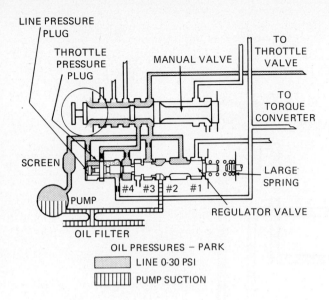

Fig. 5-4 The manual valve of the transmission shown is arranged so it dumps line pressure to the sump in park. This reduces line pressure to somewhere between 0 and 30 psi. *(Courtesy of Chrysler Corporation.)*

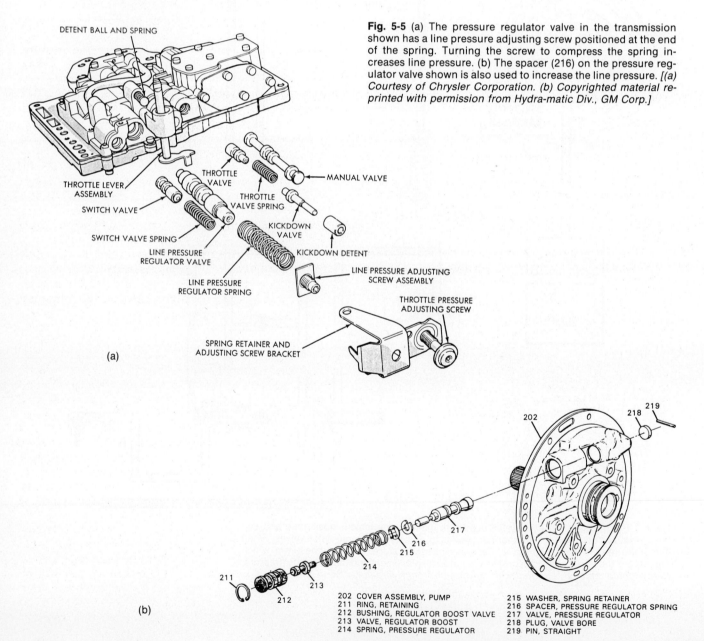

Fig. 5-5 (a) The pressure regulator valve in the transmission shown has a line pressure adjusting screw positioned at the end of the spring. Turning the screw to compress the spring increases line pressure. (b) The spacer (216) on the pressure regulator valve shown is also used to increase the line pressure. *[(a) Courtesy of Chrysler Corporation. (b) Copyrighted material reprinted with permission from Hydra-matic Div., GM Corp.]*

Fluid pressure during mid to full throttle, reverse, and in some other cases low and intermediate gears is boosted by one or more boost areas of the regulator valve (Fig. 5-6). As we will see later, the throttle valve, TV, circuit produces a pressure signal relative to the throttle opening or engine load (Fig. 5-7) TV pressure will be somewhere between zero and line pressure depending on throttle position, and it will act on the regulator valve in the same direction as the spring. In a way, this increases spring pressure and causes the regulator valve to balance to a higher pressure. Throttle boost is used to produce firmer shifts, improve shift feel, and increase clutch and band holding power at higher throttle openings. Full throttle in forward gears causes line pressure to increase to about 90 psi (621 kPa).

Line pressure is increased in reverse because the reverse passage at the manual valve sends a pressure signal to the reverse boost valve at the pressure regulator valve (Fig. 5-8). This pressure, like that of the throttle valve, acts with the spring. Reverse boost produces line pressures of between 150 and 300 psi (1034 and 2068 kPa) depending on the particular transmission and the throttle opening. The purpose for boosting pressure in reverse is to ensure sufficient force at the band and clutch to prevent slipping.

Some transmissions also boost pressure in manual 1 and sometimes manual 2 to ensure strong clutch and band application while operating in these gears. When used, another boost area of the regulator valve is pressurized by the manual valve in a manner similar to reverse boost. When used, low-gear boost produces line pressures of about 150 psi (1034 kPa).

5.3 TORQUE CONVERTER, OIL COOLER, AND LUBRICATION CIRCUIT

As soon as the supply circuit begins to develop pressure, the regulator valve moves slightly and opens a passage to the torque converter. This fluid flow serves several purposes: it ensures that the torque converter is filled so it can transmit torque from the engine to the transmission's input shaft; it helps control fluid temperature; and it provides lubrication for the moving parts inside the transmission (Fig. 5-9).

It should be noted that this feature shuts off the fluid flow to the torque converter, cooler, and lubrication circuit whenever the fluid pressure drops below the regulated pressure point. There are several possibilities where a transmission with a slightly worn pump running at idle or very low engine speeds under conditions calling for high line pressures (full throttle, low vacuum, reverse gear, etc.) can be operated without converter, cooler, and lubrication flow. This can cause severe thrust washers, bushing, and gear train wear if it is too long of a time period.

5.3.1 Torque Converter Pressure Control

A potential problem with torque converters is drain-down or a partial emptying while the engine is off. This causes slippage and delay in power flow until the converter is refilled. The possibility of drain-down is reduced by shutting the fluid passage to the converter at the regulator valve as soon as the pressure drops when the engine is shut (Fig. 5-10). Some transmissions use an additional valve in the passage, leaving the converter for this same purpose, to prevent drain-down. The simple, spring-loaded check valve closes off the passage as soon as fluid flow through the converter stops.

The fluid pressure inside the converter is controlled by how fast the fluid enters and how fast it can leave, and this is controlled by the sizes and the restrictions in the two passages. Since there is not much restriction on a converter's outlet passages, normal internal pressures do not get too high [about 15 to 75 psi (103 to 517 kPa)] (Fig. 5-11). Another potential con-

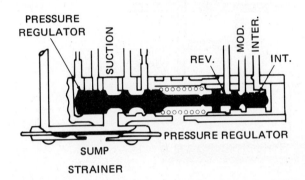

Fig. 5-6 The pressure regulator valve shown will increase/boost line pressure on demand from the modulator and in intermediate and reverse gears. Pressure in any of the ports at the right side of the pressure regulator valve will cause the reverse and/or intermediate valves to move toward the left and increase line pressure. *(Copyrighted material reprinted with permission from Hydra-matic Div., GM Corp.)*

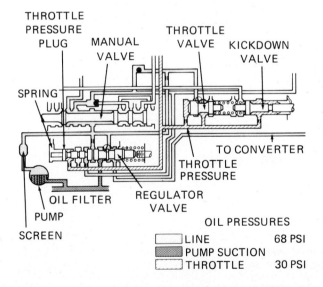

Fig. 5-7 Normally the spring at the left of the throttle pressure plug shown moves the regulator valve toward the right to lower line pressure. However, throttle pressure acting on the throttle pressure plug resists this action and produces higher line pressures. *(Courtesy of Chrysler Corporation.)*

135

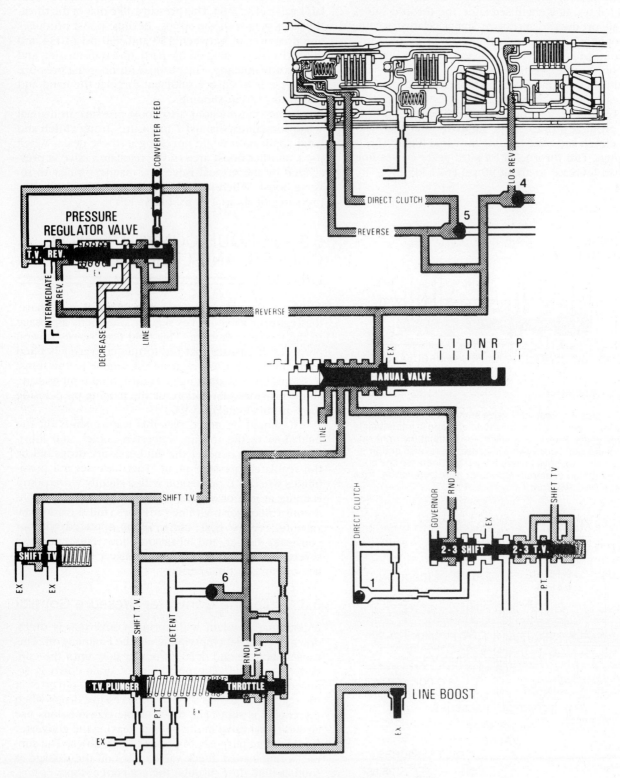

Fig. 5-8 In the transmission shown, shifting into reverse sends pressure to the direct clutch, low and reverse clutch, and reverse boost valve at the pressure regulator. The reverse boost valve moves toward the right, which increases the pressure on the pressure regulator valve spring, which in turn increases line pressure. *(Copyrighted material reprinted with permission from Hydra-matic Div., GM Corp.)*

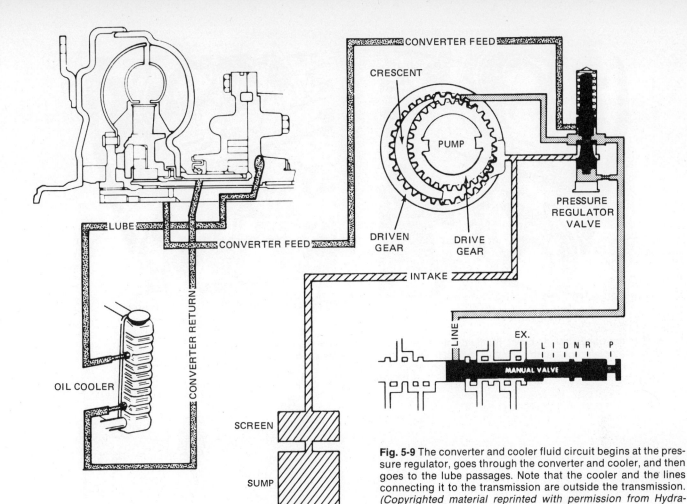

Fig. 5-9 The converter and cooler fluid circuit begins at the pressure regulator, goes through the converter and cooler, and then goes to the lube passages. Note that the cooler and the lines connecting it to the transmission are outside the transmission. *(Copyrighted material reprinted with permission from Hydra-matic Div., GM Corp.)*

verter problem is ballooning. This is a swelling of the outer shell of the converter that can result from excessive internal pressure. Many early transmissions used a pressure relief valve or a torque converter pressure regulator valve in the converter passage to limit the pressure to about 100 psi (689 kPa) maximum (Fig. 5-12).

5.3.2 Torque Converter Clutch Control

Most modern transmissions use a torque converter clutch that is applied by reversing the fluid flow inside the converter. Depending on the transmission, the valve that controls this is called a converter clutch control valve, a switch valve, or a converter clutch apply valve (Fig. 5-13). When this valve moves, the fluid flow through the converter is reversed. In some transmissions, this valve is controlled completely by fluid flows in the transmissions; in others, it is controlled by electronic controls as well as the fluid flows. Electronic control is usually contained in the engine control module (ECM) used with most modern cars to control spark timing and fuel mixture. The ECM uses input signals from the coolant temperature sensor, throttle position sensor, vehicle speed sensor, engine speed sensor, and manifold pressure sensor to determine the correct engine settings, and these sensors plus the brake switch determine if the torque converter clutch should be applied (Fig. 5-14).

In older Chrysler Corporation transmissions, the switch valve is operated by fluid pressure from the drive circuit acting on one end of the valve and a spring at the other end that tries to hold the valve in a nonlockup position. The valve moves to the lockup position after the lockup valve (moved by pressure in the governor

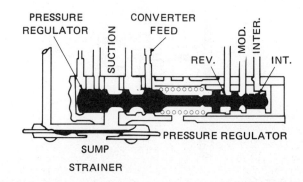

Fig. 5-10 The converter feed port is shut off by the regulator valve spring when the engine is not running. This helps prevent converter drain-down. *(Copyrighted material reprinted with permission from Hydra-matic Div., GM Corp.)*

137

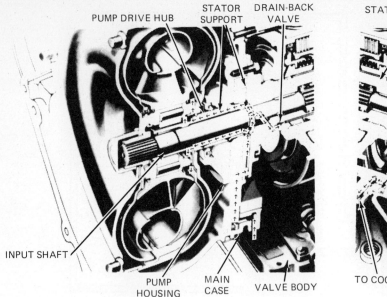
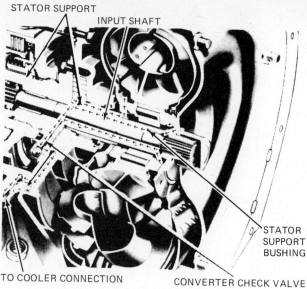

Fig. 5-11 In the transmission shown, the fluid enters the converter through the area between the stator support and the pump drive hub. It leaves through the area between the input shaft and the stator support. *(Courtesy of Ford Motor Company.)*

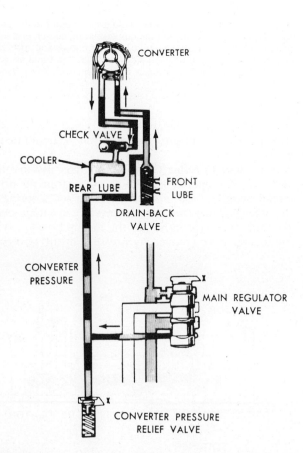

Fig. 5-12 Converter and cooler circuit including a pressure relief valve to prevent converter pressures from going too high and a check valve and drain-back valve that closes to trap the fluid in the converter when the engine is shut off. *(Courtesy of Ford Motor Company.)*

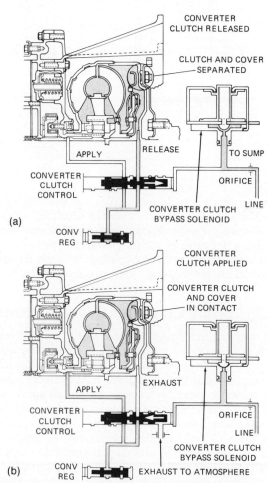

Fig. 5-13 (a) A torque converter clutch is released when the fluid flows from the front of the converter to separate the clutch from the cover. (b) When the converter clutch control valve moves toward the left, fluid flow tries to reverse, and fluid pressure from the left forces the clutch against the cover to lock the converter clutch. Note how the solenoid can control the pressure at the right end of the converter clutch control valve. *(Courtesy of Ford Motor Company.)*

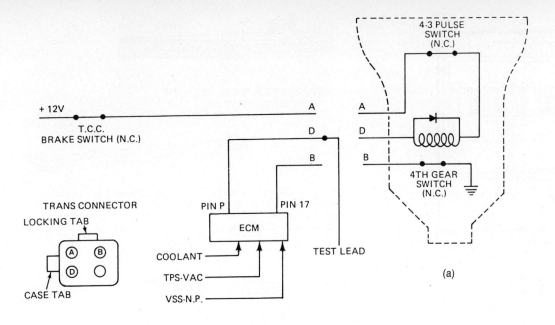

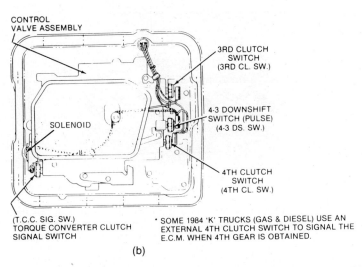

Fig. 5-14 (a) Transmission with two transmission switches and an ECM with its sensors to control converter clutch solenoid. (b) Locations of transmission switches and solenoid. *(Copyrighted material reprinted with permission from Hydra-matic Div., GM Corp.)*

circuit) and fail-safe valve (moved by pressure from the third-gear circuit) move to upshift positions. Torque converter clutch lockup only occurs above a certain speed and after the transmission has shifted into third gear (Fig. 5-15). It releases below that same speed, when the transmission downshifts out of third gear or when there is enough throttle pressure to move the fail-safe valve to a nonlockup position.

5.3.3 Cooler Flow

The fluid leaving the torque converter is routed out of the transmission case and into a steel line to pass on to the cooler. The cooler is positioned in the colder tank of the radiator (Fig. 5-16). Another steel line is used to bring the fluid from the cooler back to the transmission.

A cooler is often called a heat exchanger because it moves heat from one location to another—from the transmission fluid to the engine coolant. It is either a flat or a tubular-shaped plate-type cooler that is simply two outer metal surfaces with a *turbulator* between them (Fig. 5-17). The outer surfaces are in contact with the relatively cool coolant in the radiator tank, and the transmission fluid flows between them. The screenlike turbulator causes a turbulence in the fluid flow to ensure constant mixing and thorough cooling of the fluid.

A plain-tube cooler is not very effective when cooling oil because oil tends to increase its viscosity and slow down as it cools. The cooler oil then tends to become stationary on the outer, cooler areas of the cooler while the hotter, more fluid oil flows through the center (Fig. 5-18). A well-designed oil cooler should contain a turbulator to cause a constant mixing of the oil. The

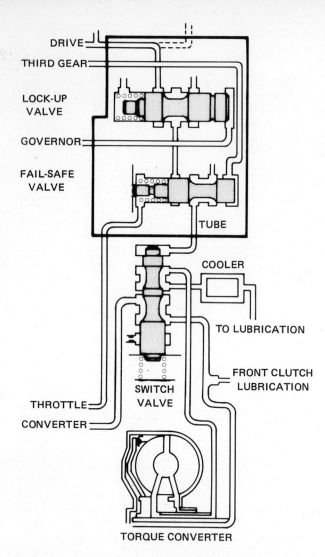

Fig. 5-15 Transmission with lockup valve and fail-safe valve to control switch valve, which in turn controls fluid flow through converter and therefore converter lockup. *(Courtesy of Chrysler Corporation.)*

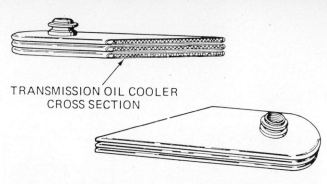

Fig. 5-17 An oil cooler is a large, flat passage from its inlet to its outlet. This area is almost filled with a screenlike turbulator. *(Copyrighted material reprinted with permission from Hydramatic Div., GM Corp.)*

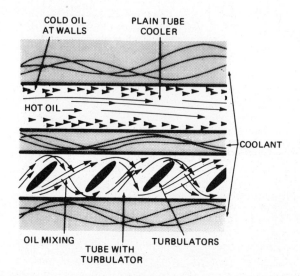

Fig. 5-18 An oil cooler should have a turbulator, a device that causes oil turbulence, to promote oil mixing so the cool oil does not cling to the tubing walls while the hot oil flows through the center.

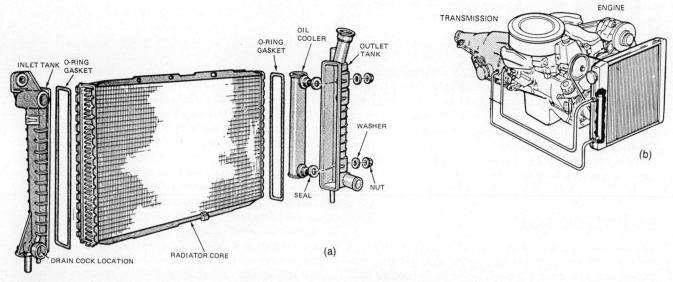

Fig. 5-16 (a) The transmission oil cooler is normally mounted in the colder, outlet tank of the radiator. (b) Line connections for transmission. *(Courtesy of Ford Motor Company.)*

140

turbulator causes some restriction to the fluid flow, but most coolers are sized properly to allow for this.

In an automatic transmission, the cooler tends to become a filter screen that traps foreign particles and can become plugged if there is too much debris. This is especially true when the fluid is extremely dirty or contains metal particles or there is torque converter clutch failure. The procedure for checking cooler flow and cleaning of the cooler will be described in Chapter 9. At least one aftermarket manufacturer markets a filter that can be installed in the transmission-to-cooler line. This filter provides added protection by removing foreign particles from the fluid and preventing cooler blockage.

Supplementary coolers are also available that can be placed in series with the original equipment manufacturer (OEM) cooler to ensure adequate cooling ability. These are oil-to-air coolers (the OEM cooler is normally an oil-to-water cooler). The heat in the supplementary cooler flows directly to the air passing through the cooler. The supplementary cooler is normally positioned where there is an adequate air flow, and this is often in front of the radiator or air-conditioning condenser (Fig. 5-19). This is not the best location because the new cooler tends to block the air flow through the other heat exchangers and increases their operating temperatures. The supplementary cooler is connected to one of the transmission's oil lines and then to the OEM cooler. This author recommends that the hot oil from the transmission be routed through the supplementary cooler first, then through the OEM cooler, and then back to the transmission (Fig. 5-20). Running the oil through in this manner improves the efficiency of the supplementary cooler by operating it at the highest possible temperature, and when operating in very cold temperatures, the OEM cooler reduces the chance of overcooling the fluid. It raises the fluid's operating temperature and helps ensure an adequate flow.

5.3.4 Lubrication Flow

As mentioned earlier, the fluid returning from the torque converter is used to lubricate the moving parts inside the transmission. The fluid flows from the steel line and enters the lubrication passages at the front or rear of the

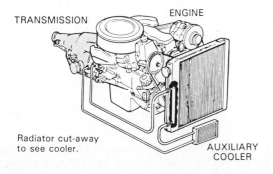

Fig. 5-19 Normally fluid flows from the transmission, through the cooler, and then back to the transmission (top). An auxiliary cooler is mounted so the fluid flows through it and then through the standard cooler in the radiator. *(Copyrighted material reprinted with permission from Hydra-matic Div., GM Corp.)*

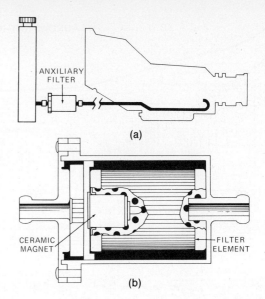

Fig. 5-20 (a) An auxiliary filter can be installed in one of the cooler lines to help remove contaminants. (b) Cutaway view of filter in (a) showing the internal magnet and filter element. *(Courtesy of Sealed Power.)*

case. It flows through holes drilled in the case to the main shaft bushings where it passes into holes drilled in the output shaft (Fig. 5-21). From here, it flows down the shaft to side holes that align with support bushings, thrust washers, planetary gearsets, and clutch packs. The final drive gears and differential of a transaxle are also lubricated by this circuit; because of this added requirement, some transaxles use an oversize pump.

A car with an automatic transmission should not be towed or pushed very far without the engine running because there will be no lubrication oil. The gearsets and bushings will run dry, wear, and overheat or burn out without a constant flow of oil (Fig. 5-22). Most manufacturers recommend towing only when absolutely necessary, and it should be limited to a few miles with a maximum speed of 20 to 25 mph (32 to 40 km/h). If possible, the drive wheels should be lifted off the ground or the drive shaft removed from a rear-wheel drive (RWD) car.

5.4 THROTTLE PRESSURE

In most transmissions, a circuit carrying a throttle pressure signal is always available while the engine is running. Throttle pressure provides a pressure signal that is proportionate to the load on the engine or, similarly, to the amount of throttle opening (Fig. 5-23). This controlled, variable pressure is also called a modulated pressure. The throttle signal can be brought to the transmission by two commonly used methods—mechanical or vacuum—or one that promises to be of more use in the future—electrical.

The transmission control system uses throttle pressure, *TV* or *TP*, to reprogram or reschedule several areas of operation:

141

THE ARROWS IN THE CROSS-SECTION INDICATE THE OIL FLOW DIRECTION.

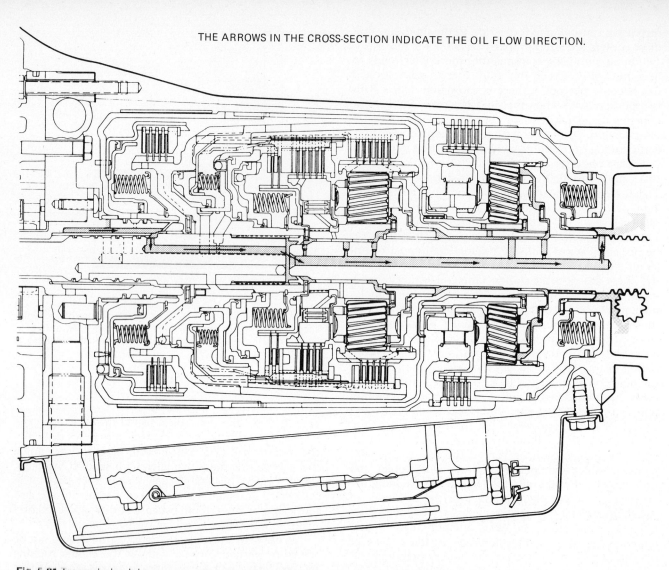

Fig. 5-21 Transmission lube passages. Note how the passage leads to the gears and bearings. *(Copyrighted material reprinted with permission from Hydra-matic Div., GM Corp.)*

Fig. 5-22 Planetary gearset that probably burned out from lack of lubrication oil; note excess clearance and dark coloration at the planet gear shafts.

- *Clutch and band application* The friction devices should apply more firmly while under full throttle to handle the increased torque. This is accomplished by boosting line pressure.

- *Shift timing* A full-throttle upshift should occur at a higher speed than a part-throttle shift. This is accomplished by throttle pressure working against governor pressure at the shift valves (Fig. 5-24).

- *Shift feel* A full-throttle shift should have a firmer, faster quality to reduce the amount of slippage during a shift. This is accomplished by changing the rate of accumulator operation used to cushion a shift (Fig. 5-25).

- *Torque converter clutch control* The torque converter clutch should be released during full-throttle operation so the converter is able to multiply the torque needed for acceleration and power. This is accomplished by the throttle pressure acting on the torque converter clutch control valve.

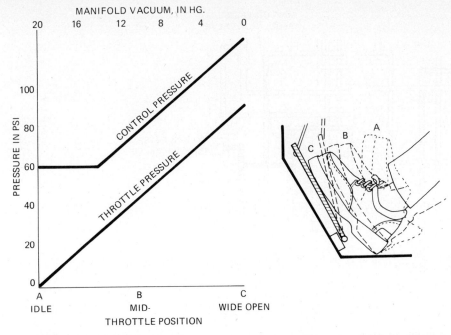

Fig. 5-23 The throttle valve produces a pressure signal that is directly related to throttle opening or engine load. This pressure is used to change line pressure so that line pressure increases as the throttle is opened.

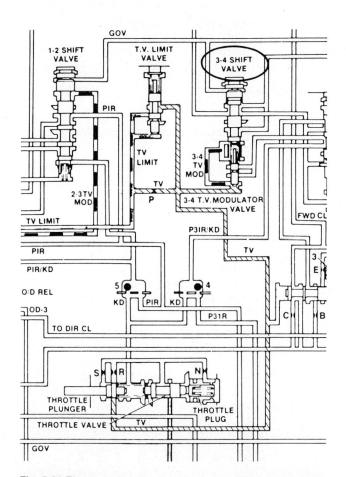

Fig. 5-24 The typical shift valve shown has TV pressure at the lower end that is opposing governor pressure at the upper end. If TV pressure is increased by opening the throttle, the upshifts occur at a higher speed. *(Courtesy of Ford Motor Company.)*

Most transmissions use either a double throttle valve (mechanical operation) or two separate units (vacuum and mechanical or electrical) (Fig. 5-26). The second throttle valve is called a *detent, downshift,* or *kickdown valve,* and it signals full or nearly full throttle operation. Some newer transmissions use a mechanical throttle and detent valve to control shift timing and a vacuum unit to control line pressure and shift feel.

Some transmissions modify the TV pressure at one or more of the valves in order to achieve precise shift timing. The 3–4 modified throttle valve (MTV) and 4–3 MTV act to reduce TV pressure at low throttle openings.

5.4.1 Mechanical Throttle Valves

Most Chrysler transmissions use a mechanical throttle valve and a detent valve, called a kickdown valve. These valves are in the same bore, with the kickdown valve positioned by a spring to rest against the throttle lever assembly (Fig. 5-27). The throttle lever is connected to the carburetor linkage by a metal rod; the lever moves as the throttle is opened or closed. The throttle valve is positioned at the other end of the kickdown valve spring; the spring pushes the two valves in opposite directions. The throttle valve is also positioned by pressure in the throttle pressure passages, but as throttle pressure moves the valve against the spring, the valve moves to cut off throttle pressure. The mechanical force of the throttle linkage tends to increase throttle pressure, but throttle pressure tries to reduce itself. The throttle valve will be balanced between the amount of mechanical throttle pressure—the strength of the spring controlled by the kickdown valve position—and the amount of fluid throttle pressure. At the wide-open throttle (WOT) position, the throttle valve is pushed to

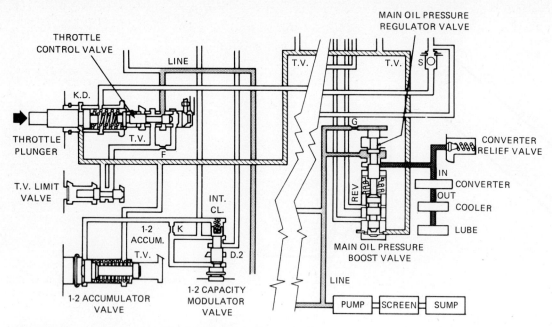

Fig. 5-25 TV pressure acts on the lower end of this pressure regulator valve so that an increase in TV pressure produces an increase in line pressure. *(Courtesy of Ford Motor Company.)*

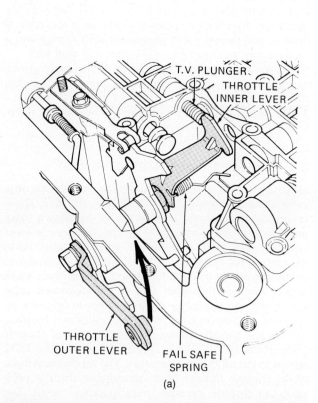

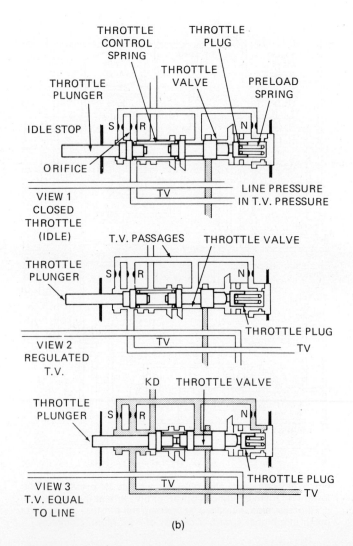

Fig. 5-26 (a) Transmission is upside-down with the pan removed. (b) Movement of the throttle outer lever (arrow) causes the throttle inner lever to push inward on the TV plunger. The TV plunger produces a TV pressure that is proportionate to throttle opening and, at wide-open throttle, a higher pressure in the KD/kickdown passage. *(Courtesy of Ford Motor Company.)*

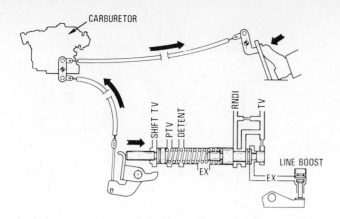

Fig. 5-27 Throttle valve connected to throttle by a cable. Movement of the accelerator to full throttle moves the throttle valve plunger toward the right. *(Copyrighted material reprinted with permission from Hydra-matic Div., GM Corp.)*

the end of the bore by an extension of the kickdown valve. At this time, throttle pressure increases to equal line pressure, and the kickdown valve opens the kickdown passages to allow fluid pressure to enter (Fig. 5-28).

5.4.2 Vacuum Throttle Valves

Many Ford and General Motors transmissions use a throttle valve operated by a vacuum unit called a *vacuum modulator* that is secured to the outside of the transmission case and is connected to the engine's intake manifold by a metal or rubber tube (Fig. 5-29). The vacuum unit contains a flexible diaphragm and a spring. The spring normally pushes the diaphragm and its extension toward the transmission while sufficient manifold vacuum acts on the diaphragm, compresses the spring, and moves the diaphragm and its extension away from the

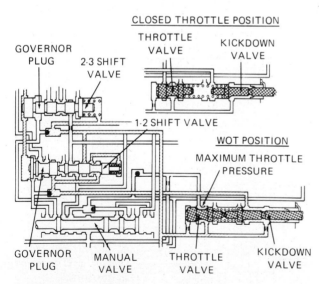

Fig. 5-28 When throttle is opened completely, the kickdown valve pushes the throttle valve against the end of the bore. At this position, throttle pressure is equal to line pressure and there is line pressure in the passage leaving the kickdown valve. *(Courtesy of Chrysler Corporation.)*

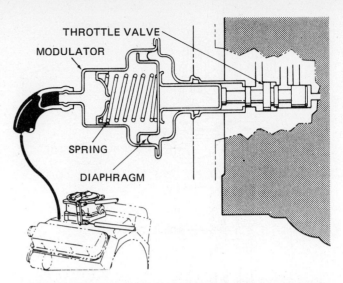

Fig. 5-29 A metal and rubber tube bring a vacuum signal from the intake manifold to the vacuum modulator. A high manifold vacuum pulls the diaphragm toward the left and compresses the spring in the modulator. *(Copyrighted material reprinted with permission from Hydra-matic Div., GM Corp.)*

transmission. As with other vacuum devices, a vacuum is really a pressure that is lower than atmospheric, and atmospheric pressure pushes against the opposite side of the diaphragm and the spring (Fig. 5-30).

The output of the vacuum modulator is a product of the area of the diaphragm, the pressure differential across the diaphragm (atmospheric pressure minus the lower vacuum pressure), and the strength of the spring (Fig. 5-31). Vacuum modulators are made with different sizes of diaphragms and with springs of different strengths to suit the operating conditions in cars of different sizes or engines with different torque outputs. A stronger spring or smaller diaphragm area produces a stronger outward force while a weaker spring or a larger diaphragm produces a weaker force. The stem of the vacuum modulator pushes against one end of the throttle valve while throttle pressure acts against the opposite end, much like in a mechanical throttle valve.

Vacuum-operated throttle valves produce a throttle pressure signal that is more relative to engine load than a mechanically operated valve. Engine manifold vacuum is a product of throttle position, engine load, and engine speed. At idle speed and while operating at light throttle, intake manifold vacuum is relatively high, about 18 to 20 inches of mercury (in. Hg). At this time, the pumping action of the pistons pulls air out of the manifold faster than it can flow in past the closed throttle plates (Fig. 5-32). At WOT, manifold vacuum is usually lower than 1 in. Hg because air can easily flow past the open-throttle plates. During cruise conditions, the throttle is partially open, and manifold vacuum is somewhere in the neighborhood of 10 in. Hg. Cruise vacuum will vary greatly depending on engine size, vehicle weight, and gear ratio. During cruise conditions, an increase in engine load such as a head wind or going up an incline slows down the engine relative to the throttle opening and thus causes a drop in vacuum, from 10 to 6 to 8 in. Hg, for example.

145

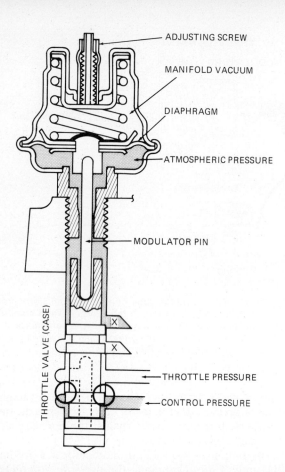

Fig. 5-30 If manifold vacuum drops, the spring in the modulator moves the modulator diaphragm and operating rod toward the left. This moves the throttle valve toward the left, which allows fluid to flow past the circled areas of the valve to become throttle pressure. Note that throttle pressure can also flow through the valve so it can apply pressure at the left end. *(Courtesy of Ford Motor Company.)*

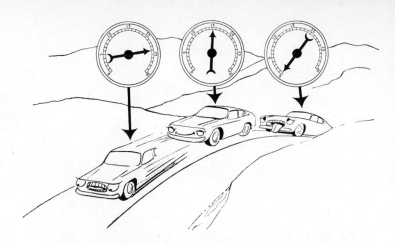

Fig. 5-32 Manifold vacuum varies with load. A high load such as going uphill often produces zero vacuum while a closed-throttle, downhill descent produces a vacuum of about 21 to 24 in. Hg. *(Courtesy of Nissan Motor Corporation.)*

During the 1970s, as the various pollution control devices were added to engine control systems, vacuum became a less accurate indicator of engine load. For example, the exhaust gas recirculation (EGR) valve allows exhaust gases to enter the intake manifold vacuum during cruise conditions. This results in lower manifold vacuum, which signals the vacuum modulator that the engine is under a greater load than it really is. One manufacturer used a double-diaphragm vacuum modulator on some of these transmissions. When the EGR valve begins to operate, the second diaphragm receives a vacuum signal that helps produce more accurate transmission operating pressures.

Diaphragm Diameter (in.)	Diaphragm Area (SI)	Manifold Vacuum			(1.76 SI Diaphragm) Force (lb)
		in. Hg.	psia	Pressure Difference	
$1\frac{1}{4}$	1.227	0	14.7	0	0
$1\frac{9}{32}$	1.289	5	12.2	2.5	4.4
$1\frac{1}{2}$	1.767	10	9.7	5	8.8
$1\frac{9}{16}$	1.917	15	7.2	7.5	13.25
$1\frac{11}{16}$	2.23	20	4.7	10	17.7
2	3.1416				

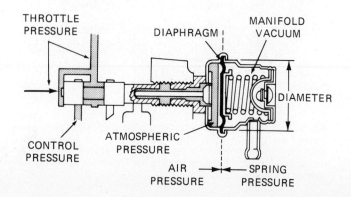

Fig. 5-31 Some modulators are available with different diameter diaphragms (left). A larger diaphragm provides a greater area on which vacuum and atmospheric pressure can act. The spring pressure always tries to force the diaphragm in one direction, the pressure difference on the two sides of the diaphragm tries to force the diaphragm in the other.

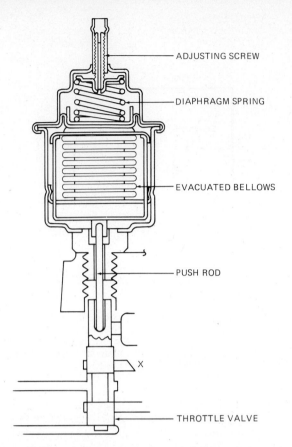

Fig. 5-33 The evacuated bellows in modulator shown grows longer at higher altitudes, which in turn produces lower throttle pressure. *(Courtesy of Ford Motor Company.)*

Some vacuum modulators contain an evacuated bellows assembly. This unit acts against the spring to produce more consistent shift response at differing altitudes. As a vehicle is operated at higher altitudes, the air is thinner, and atmospheric pressure is lower. This produces less torque in the engine, and the transmission pressures should be lowered slightly to prevent harsh shifts. At this time, the reduced atmospheric pressure acting on the bellows allows the bellows to lengthen and increase the pressure acting against the diaphragm spring. This has the effect of reducing the spring's strength, which reduces the force acting on the throttle valve and therefore the throttle pressure (Fig. 5-33).

5.4.3 Wide-Open Throttle Kickdown Valve

A vacuum modulator cannot accurately indicate the exact amount of throttle opening; another signal is needed. The WOT signal can be brought to the transmission by a mechanical linkage, a rod or a cable, or an electric current. A WOT kickdown valve is also called a detent valve. A mechanical detent valve resembles the mechanical throttle and detent valve combination described in an earlier section. This action is illustrated in Fig. 5-28. Throttle motion is transmitted to the detent valve so it can produce a full pressure in the detent passages at WOT.

An electric detent valve is controlled by a switch in the throttle linkage and a solenoid at the valve body. When the throttle is opened to a point close to WOT, the switch is closed so that an electrical signal, usually battery voltage, is sent to the solenoid. This energizes the solenoid and causes it to stroke a metal rod or move a metal plate that either moves a valve or opens a passage (Fig. 5-34). In either case, fluid pressure is sent to the detent passages.

5.5 MANUAL VALVE

The manual valve controls the oil flow to the band and clutch apply pistons and the shift valves for the various forward and reverse driving modes. This valve is also called a selector valve. It receives the major flow from the pump at regulated line pressure. The manual valve is moved by the shift selector in the passenger compartment and is positioned by the detent cam at the valve body (Fig. 5-35). This detent is a spring-loaded roller or ball that drops into notches in the cam to position the manual valve properly in the bore. Because of its shape, the detent cam is often called a *rooster comb* by many transmission technicians.

When the selector is in neutral or, in most transmissions, in park, fluid flow through the manual valve is blocked by a land or trapped between two lands (Fig. 5-36). Many transmissions send line pressure oil to the throttle valve in neutral so the TV circuit is ready to operate as soon as the neutral–drive or neutral–reverse shift is made.

In the other four or five gear selector positions (drive, intermediate, low, and reverse) (in some cases, park, neutral, overdrive, drive, intermediate, and low), depending on the particular transmission, the valve is moved to allow a flow through the valve to various locations. Most three- and four-speed transmissions have the same number of gear selector positions. Some four-speed transmissions have one more position that allows fourth-gear operation in overdrive and limits the transmission to first- through third-gear operation in drive. For simplicity, the following descriptions are for a three-speed, Simpson gear train transmission. In general, the fluid flows are as follows:

Drive Fluid is directed to the forward clutch, the 1–2 shift valve, and the governor (Fig. 5-37).

Intermediate Same as drive but the 2–3 shift valve is blocked from moving or not fed an input flow (Fig. 5-38).

Low Fluid is directed to the forward clutch, the low–reverse clutch or band, and the governor, and the 1–2 shift valve is blocked from moving. In many transmissions, a pressure signal is sent to the pressure regulator to boost line pressure (Fig. 5-39).

Reverse Fluid is directed to the high–reverse clutch, the low–reverse clutch or band, and the pressure regulator to boost the fluid pressures (Fig. 5-40).

The various fluid flows leaving the manual valve are labeled drive oil, manual 2, manual 1, and reverse.

147

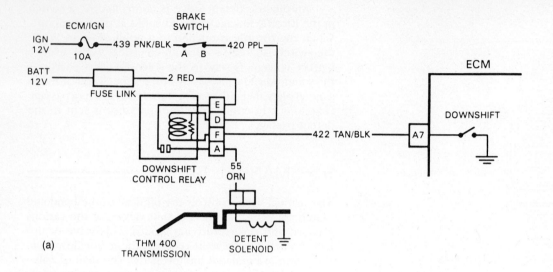

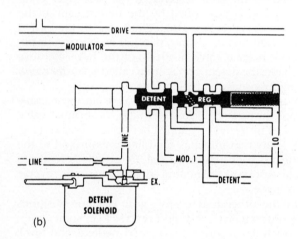

Fig. 5-34 (a) Electrical circuit of an electric downshift solenoid. (b) When the solenoid is energized, line pressure trapped at the left end of the valve moves the valve toward the right to try to produce a downshift. *(Copyrighted material reprinted with permission from Hydra-matic Div., GM Corp.)*

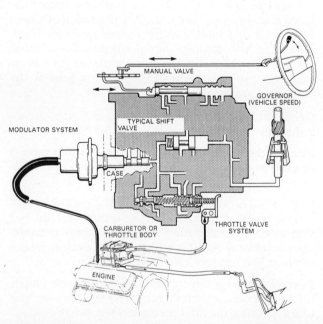

Fig. 5-35 When the gear selector is moved, the manual valve slides along its bore in the valve body. *(Copyrighted material reprinted with permission from Hydra-matic Div., GM Corp.)*

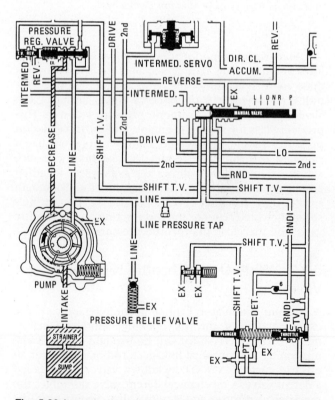

Fig. 5-36 In park, the manual valve blocks line pressure and opens the passages from the clutches and band servos to exhaust. In neutral, line pressure is sent through the RNDI passage to the throttle valve but not to the clutches and bands. *(Copyrighted material reprinted with permission from Hydra-matic Div., GM Corp.)*

DRIVE RANGE-FIRST GEAR

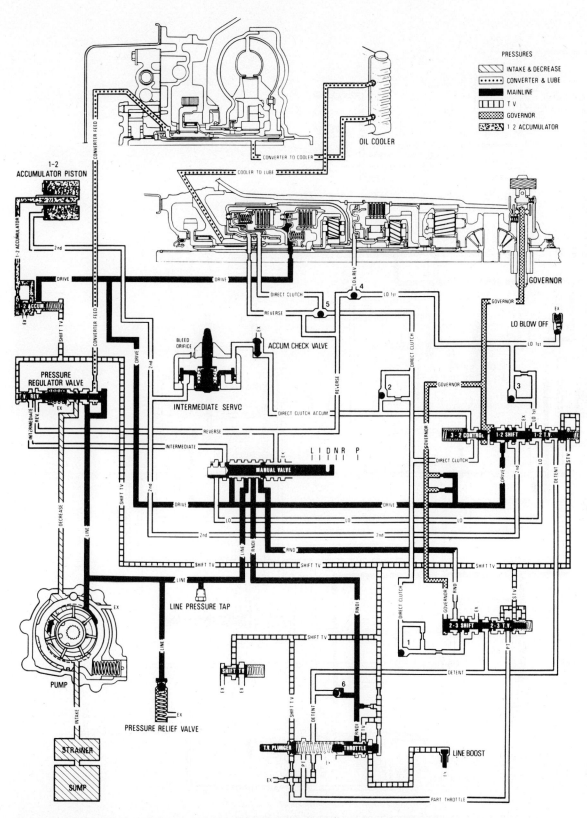

Fig. 5-37 Moving the gear selector to drive repositions the manual valve to send oil into the drive passages to those points shown by dashed lines. *(Copyrighted material reprinted with permission from Hydra-matic Div., GM Corp.)*

INTERMEDIATE RANGE

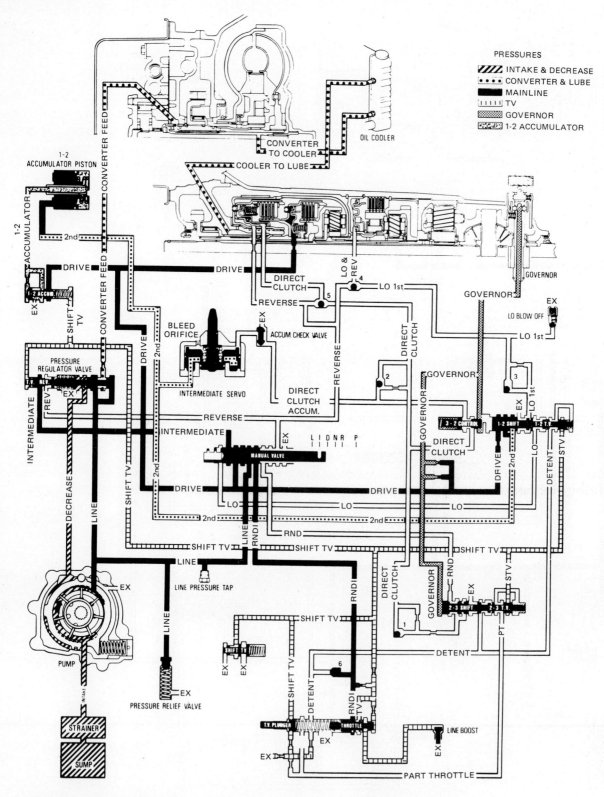

Fig. 5-38 Moving the gear selector to intermediate repositions the manual valve to send oil into the intermediate passages (dashed lines). *(Copyrighted material reprinted with permission from Hydra-matic Div., GM Corp.)*

LO RANGE

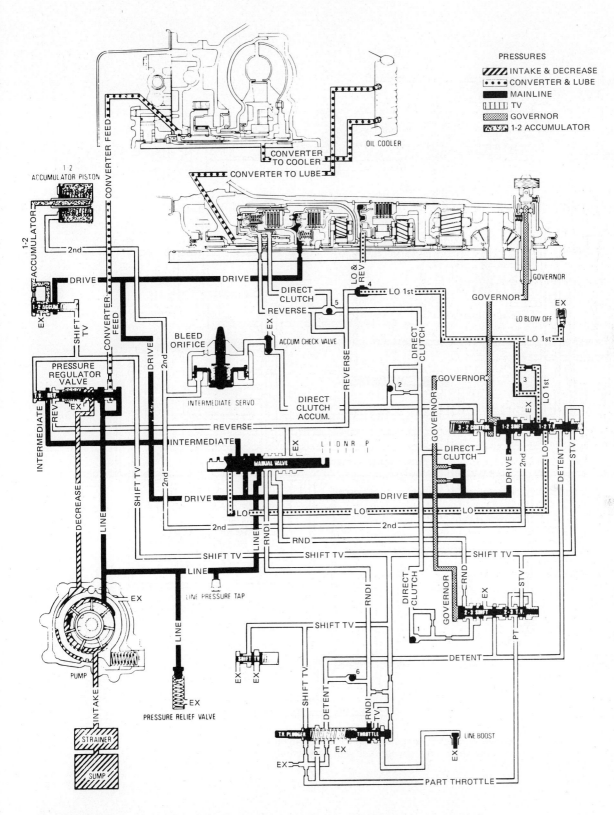

Fig. 5-39 Moving the gear selector to low repositions the manual valve to send oil into the LO passages (dashed lines). *(Copyrighted material reprinted with permission from Hydra-matic Div., GM Corp.)*

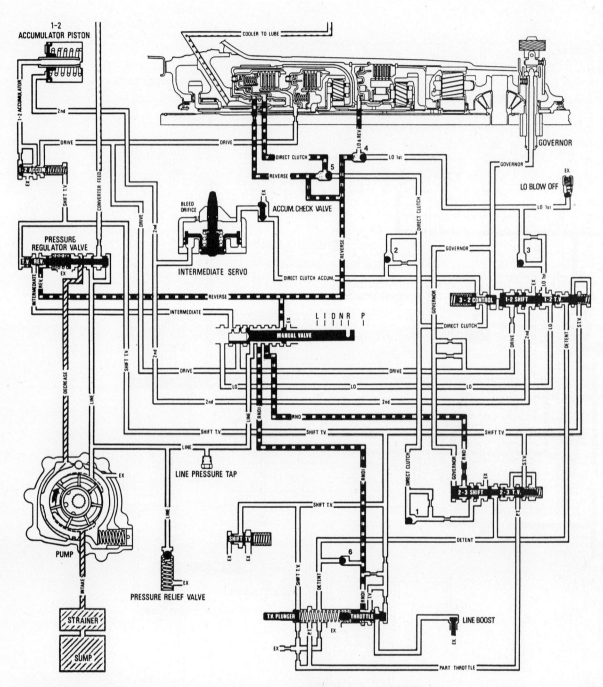

Fig. 5-40 Moving the gear selector to reverse repositions the manual valve to send oil into the reverse passages (dashed lines). *(Copyrighted material reprinted with permission from Hydra-matic Div., GM Corp.)*

5.6 GOVERNOR VALVE

With the selector lever moved to drive, intermediate, or low, the transmission can begin transferring power and driving the output shaft. When the output shaft starts revolving, the governor spins with it. The governor is either shaft mounted (attached directly onto the output shaft) or case mounted (driven by a gear on the output shaft) (Fig. 5-41).

The governor valve produces a pressure signal that is proportionate to the vehicle speed used to move the shift valves to an upshift position. With the vehicle at rest, the governor pressure should be zero. When the vehicle reaches a speed above where the highest speed upshift should occur, governor output pressure should be equal to line pressure. There should be a steady pressure increase between these two points (Fig. 5-42). Governor pressure is routed to one end of each shift valve. It is also sent to any other valve that should have a speed-related response, such as a torque converter clutch control valve or a downshift inhibitor valve.

A governor uses the centrifugal force acting on a pair of weights to measure speed. In most governors, centrifugal force moves the valve to allow fluid flow between the drive and the governor passages, and somewhat like a throttle valve, governor pressure acts on a reaction area and tries to move the valve in the opposite

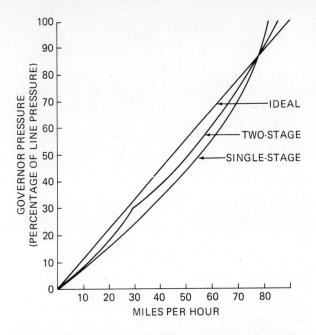

Fig. 5-42 The ideal governor pressure increases in exact proportion to vehicle speed, but a simple governor produces a pressure curve that is too low at intermediate speeds or too high at high speeds. The pressure curve of a two-stage governor comes closer to matching the ideal.

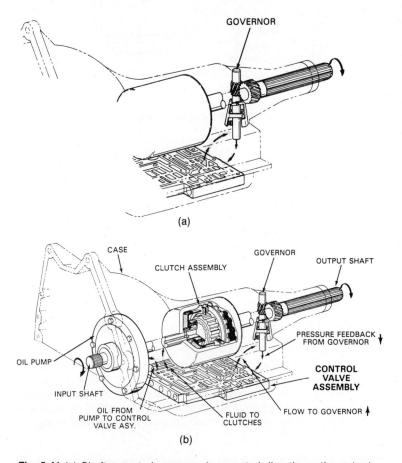

Fig. 5-41 (a) Shaft-mounted governor is mounted directly on the output shaft. (b) Case-mounted governor is mounted in the case and driven by a gear on the output shaft. *(Copyrighted material reprinted with permission from Hydra-matic Div., GM Corp.)*

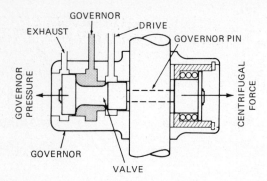

Fig. 5-43 As the vehicle speeds up, centrifugal force acting on the weight tries to move the weight and the valve toward the right to increase governor pressure. This is opposed by pressure at the valve, which tries to move it toward the left and reduce governor pressure. (*Courtesy of Chrysler Corporation.*)

direction. This movement tries to reposition the valve and exhaust governor pressure (Fig. 5-43). Governor pressure is balanced between output shaft rpm and governor pressure.

All governor assemblies share the same basic engineering problem: centrifugal force increases at an ever-increasing rate. Pressure more than doubles as rpm doubles. This produces an output that is not properly aligned with speed. Most governor assemblies use two or more stages to produce a more straight-line pressure increase.

Governor valve assemblies are manufactured in four basic styles. Two are shaft mounted, and two are case mounted:

- *Shaft Mounted with the Weight(s) Opposing the Valve*: The valve is on one side of the shaft and the weight(s) on the other, and they are connected by a pin that passes through the shaft (Fig. 5-44).

- *Shaft Mounted with a Primary and Secondary Valve*: These two valves, which also act as the weights, are mounted so the output of one valve acts to help control the other (Fig. 5-45).

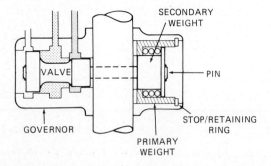

Fig. 5-44 The two weights are connected to the valve by a pin that passes through the transmission output shaft. The primary weight acts against a spring while the secondary weight acts directly on the pin. At a certain speed, the primary weight bottoms out against the stop and has no more effect on the valve. (*Courtesy of Chrysler Corporation.*)

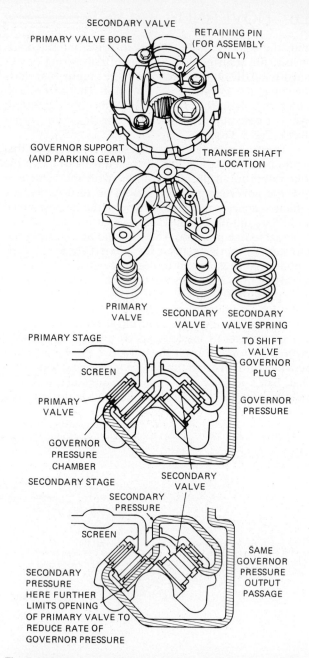

Fig. 5-45 Shaft-mounted governor that uses a primary valve and a secondary valve. (*Courtesy of Chrysler Corporation.*)

- *Case Mounted with a Pair of Primary and Secondary Weights*: The valve is operated by a pair of levers extending from the weights (Fig. 5-46).

- *Case-Mounted Bleed-Off System*: Two governor pressure balls bleed off pressure to act as the valve. With the vehicle at rest, pressure escapes past these balls. They are seated by a pair of levers extending from the weights as speed increases to cause an increase in governor pressure (Fig. 5-47).

In all these governor assemblies, governor pressure acts on the valve(s) or balls in such a way as to move the valve(s) (balls) so the pressure can escape or bleed off to exhaust. This occurs as the vehicle slows to a stop.

154

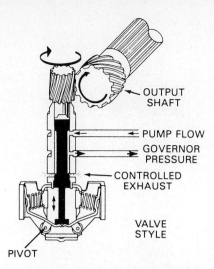

Fig. 5-46 Case-mounted governor. Centrifugal force moves the weights outward, which in turn moves the valve (black) upward to increase governor pressure. *(Copyrighted material reprinted with permission from Hydra-matic Div., GM Corp.)*

As the vehicle speed increases, the centrifugal force acting on the weights causes the exhaust flow to stop, and in the valve assemblies, a flow is opened between the drive and governor passages. Governor pressure is a balance between governor pressure trying to open the valve to exhaust and centrifugal force trying to open the valve to drive oil as it closes the exhaust port (Fig. 5-48). The primary governor weight is heavier so that centrifugal force begins to act on it as soon as the car starts moving. This weight is the major control for slow-speed operation. The primary weight is designed so it reaches a limit and bottoms out on a stop at middle operating speeds, and the stop absorbs any further centrifugal force from that point. From this speed and higher, the secondary weight becomes the controlling factor in the governor.

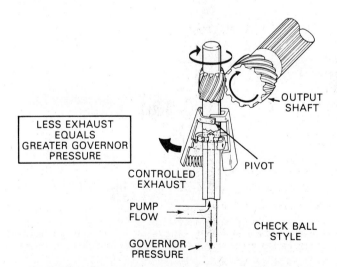

Fig. 5-47 At low speeds, governor pressure can leak past the two balls and escape. As the vehicle speeds up, centrifugal force moves the weights outward, which pushes the balls inward, reduces this flow, and increases governor pressure. *(Copyrighted material reprinted with permission from Hydra-matic Div., GM Corp.)*

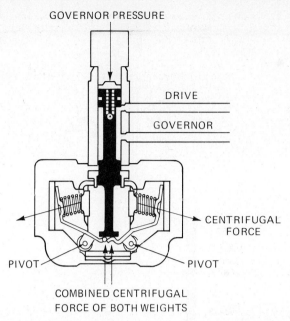

Fig. 5-48 Centrifugal force produces an upward pressure on the valve that increases governor pressure if the valve moves upward. A passage allowing governor pressure to act on the top end of the valve opposes this action. The valve is positioned between these two forces. *(Copyrighted material reprinted with permission from Hydra-matic Div., GM Corp.)*

The governor valve is the least reliable valve in the transmission because of its tendency to stick in its bore and not respond to produce accurate increases and decreases. This is basically a result of the very gradual movement of the valve in normal operation; most of the other valves "snap" from one end of their bore to the other. A governor valve has to move freely to respond to every change in speed. A very fine screen is normally placed in the fluid passage leading to the governor to trap particles that might cause sticking.

5.7 SHIFT VALVES

An upshift or downshift occurs when the shift valve moves. Shift valves are balanced between the governor pressure trying to move the valve to cause an upshift and the spring plus throttle pressure trying to resist an upshift (Fig. 5-49). A typical shift valve has a land for the governor reaction area at one end and a spring at the other. Throttle pressure enters from the side, crosses through a valve valley, and reenters at the end with the spring. The controlled flow to the next gear's apply device also enters from the side and is blocked by a land or a closed valley while the passage to the apply device is open to exhaust.

As the car is accelerating, governor pressure gradually increases until the force at that end of the valve becomes stronger than the spring plus the force created by throttle pressure. At this time, the shift valve should move, but we want the valve to snap from one end to the other to ensure a complete application of the controlled apply device. This happens in most shift valves because the first valve movement cuts off the throttle

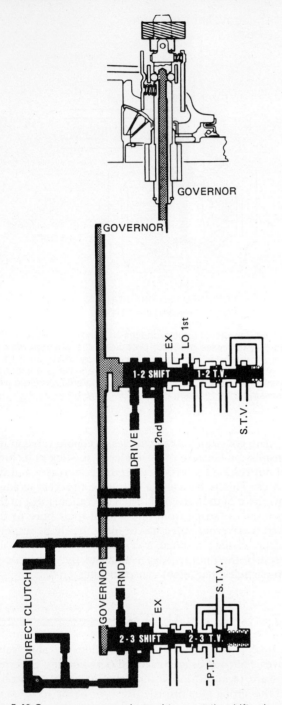

Fig. 5-49 Governor pressure is used to move the shift valves to the upshift position. It is normally opposed by a spring plus TV pressure at the other end of the valve. *(Copyrighted material reprinted with permission from Hydra-matic Div., GM Corp.)*

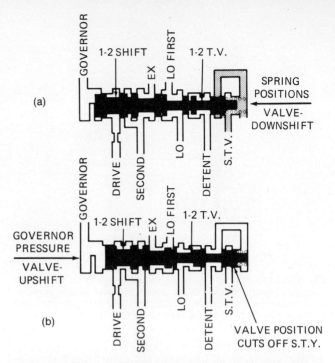

Fig. 5-50 When the shift valve shown moves from (a) the downshift position to (b) the upshift position, the STV pressure at the right end of the valve is cut. This prevents the valve from shifting right back to a downshift position. *(Copyrighted material reprinted with permission from Hydra-matic Div., GM Corp.)*

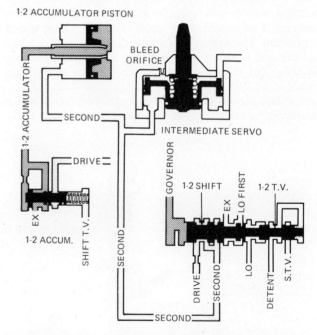

Fig. 5-51 When the 1–2 shift valve moves to upshift, drive pressure flows into the second passage to apply the band through the intermediate servo and to stroke the 1–2 accumulator piston. *(Copyrighted material reprinted with permission from Hydra-matic Div., GM Corp.)*

(TV or STV) pressure at the end of the valve and opens this chamber to exhaust (Fig. 5-50). As a result, the valve is no longer balanced and quickly completes the movement to an upshift position.

The governor land on the 1–2 shift valve is usually larger than the 2–3 shift valve governor land to guarantee that the 1–2 valve moves before the 2–3 valve. When the 1–2 shift valve moves to upshift, it opens a passage so that drive oil can flow through the valve and onto the intermediate band or clutch and apply it (Fig. 5-51). When the 2–3 shift valve moves to upshift, it opens a passage so that pressure is sent to the high–reverse clutch and the intermediate band. This pressure should release the band and apply the clutch (Fig. 5-52).

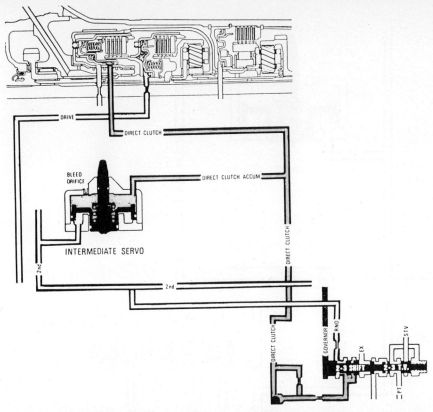

Fig. 5-52 When the 2–3 shift valve moves to upshift, RND (Reverse/Neutral/Drive) pressure flows into the direct-clutch passages where it releases the intermediate servo and band and applies the direct clutch. *(Copyrighted material reprinted with permission from Hydra-matic Div., GM Corp.)*

5.7.1 Shift Overlap

Most upshifts and downshifts require the application of one device—a clutch or a band—at the same time that another releases. For example, in second gear in many transmissions the forward clutch and the intermediate band are applied, and during a 2–3 shift, the intermediate band releases while the high–reverse clutch applies (Fig. 5-53). The band must be released an instant before the clutch is applied. An early application of the clutch or a late release of the band produces a *"fight"* between these two parts, and this produces a harsh shift and probable damage to the clutch or band. A late application of the clutch or an early release of the band produces an engine overspeed—a *"flare"* or *"buzz-up"*—as the transmission falls back to first gear before the shift to third is completed.

Most of the overlap control—the timing of these two devices—occurs by using the upshift fluid pressures from the shift valve to apply the clutch and also release the band (Fig. 5-54). In some units, band release fluid flow is used as an accumulator control for the clutch apply.

During a 3–2 downshift, the clutch must release as the band reapplies, and this action must be carefully coordinated, just like an upshift. An additional problem occurs at this time because centrifugal force in the spinning clutch tends to inhibit the oil from flowing back to the entrance port at the center. The clutch exhaust check ball is designed to allow the oil to leave from the outer area of the clutch chamber (Fig. 5-55). When the clutch applies, fluid pressure seats the check ball to prevent a fluid pressure loss. When the clutch releases and the pressure drops, centrifugal force moves the ball outward to an unseated position. This allows the fluid in the clutch assembly to escape so the clutch can release quickly and completely.

5.7.2 Boosted Throttle Pressure

The same amount of throttle pressure does not always produce the correct result at all of the valves using it. Some transmissions increase or modify the rate of fluid

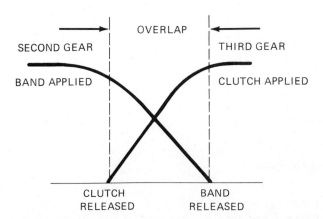

Fig. 5-53 Shift overlap occurs when the band releases to release second gear and the direct clutch applies to engage third gear.

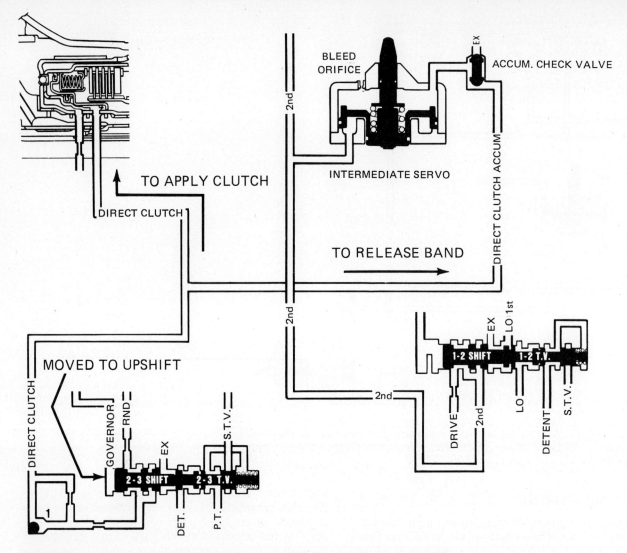

Fig. 5-54 Much of the 2–3 shift overlap timing is accomplished by using direct-clutch apply oil pressure to release the intermediate band servo. *(Copyrighted material reprinted with permission from Hydra-matic Div., GM Corp.)*

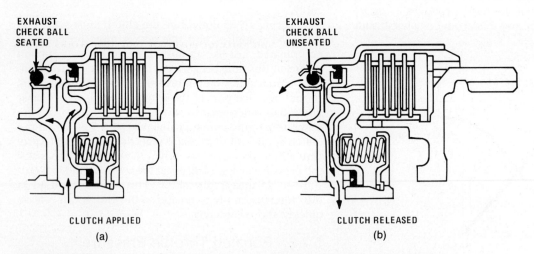

Fig. 5-55 (a) When a clutch is applied, fluid pressure keeps the exhaust check ball seated. (b) When the pressure is released, centrifugal force moves the check ball off its seat and allows fluid to flow out. *(Copyrighted material reprinted with permission from Hydra-matic Div., GM Corp.)*

pressure to produce the proper operation at different throttle-operated valves. Some of these valves merely delay throttle pressure, whereas others produce a pressure signal that is greater than the original throttle pressure.

5.7.3 Downshifts

The shift valves move to downshift position under four operating conditions:

Coasting downshift What normally occurs as a car is brought to a stop. The spring moves the valve as governor pressure drops off (Fig. 5-56).

Part-throttle downshift When the throttle is opened slightly to accelerate slowly and the rise in throttle pressure plus the spring can overcome governor pressure (Fig. 5-57).

Detent downshift Full-throttle position operates the detent valve and sends pressure through the detent passages. This pressure plus the spring must overcome governor pressure (Fig. 5-58).

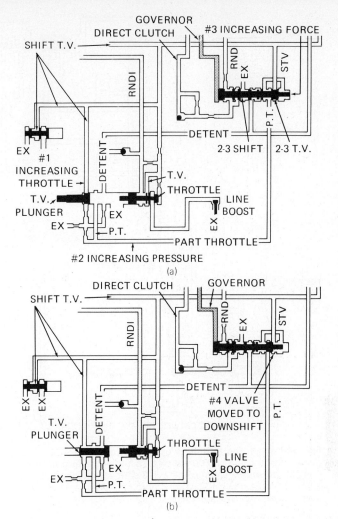

Fig. 5-57 When the throttle is opened slightly, STV (shift TV) oil pressure can become P.T. (part-throttle) oil pressure at the throttle valve. This pressure can cause the 2–3 shift valve to move toward the left and produce a downshift. *(Copyrighted material reprinted with permission from Hydra-matic Div., GM Corp.)*

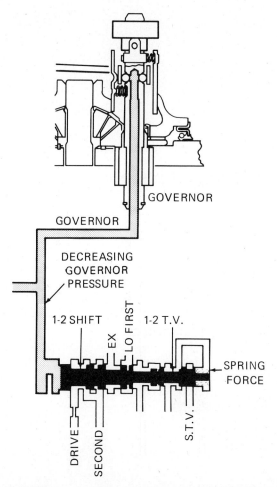

Fig. 5-56 When a car comes to a stop, the amount of governor pressure decreases and the spring force moves the shift to the left and produces a downshift. *(Copyrighted material reprinted with permission from Hydra-matic Div., GM Corp.)*

Manual downshift Movement of the shift lever to the manual 1 or 2 position sends pressure to a reaction area on the shift valve or governor plug that works against governor pressure (Fig. 5-59).

In most transmissions, coasting downshift is 3–1, not 3–2 and then 2–1. A 3–1 downshift cannot be felt because the one-way clutch (the first-gear reaction member) overruns. Under most driving conditions, we don't really need the deceleration that a 3–2 shift provides because there is adequate power in the car's braking system. A 3–2 shift causes more wear in the transmission and an unnecessary downshift "bump" in the operation.

If the shift selector is moved to manual 1 at speeds above 30 mph (48 km/h), the 1–2 shift valve normally stays in the upshift position until the speed drops off. A downshift at too high of a speed causes the engine to overrev and possibly causes damage. This delay in a manual downshift is obtained by the relative sizes of the governor and manual 1 reaction areas.

159

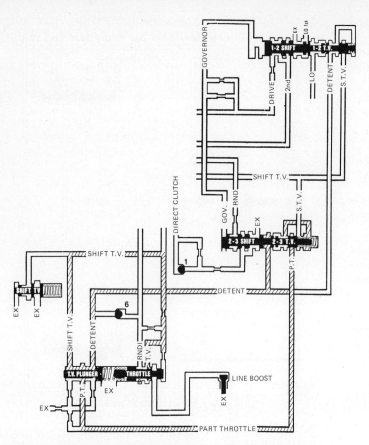

Fig. 5-58 When the throttle is opened completely, STV oil becomes part throttle and detent oil pressure at the throttle valve. The two pressures can move the 2–3 shift valve toward the left and produce a downshift. *(Copyrighted material reprinted with permission from Hydra-matic Div., GM Corp.)*

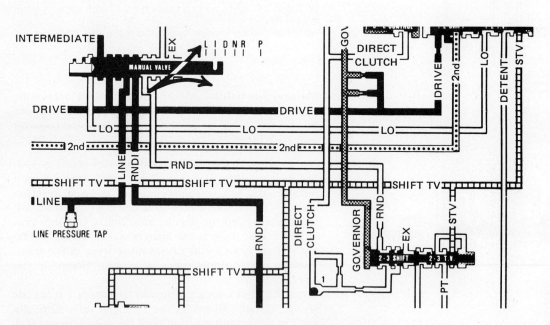

Fig. 5-59 When the gear selector is moved from drive to intermediate, RND oil flows to exhaust at the manual valve. This pressure released the intermediate band and applied the direct clutch; when it is exhausted, the transmission downshifts into second gear. *(Copyrighted material reprinted with permission from Hydra-matic Div., GM Corp.)*

5.8 SHIFT MODIFIERS

Most automatic transmissions contain some valves for the purpose of changing the quality—speed and firmness—of a shift. A light-throttle shift is made with an engine that is not producing much torque; the shifts can be made fairly slow and at light pressures. At heavy throttle, the engine is producing more torque, and higher pressures should be used for the shift in order to lock the band or clutch being applied. Too slow of a shift at pressures that are too low causes slippage and burning of the friction material while too fast of a shift at pressures that are too high produces an aggressive, harsh shift. Several methods are used to tailor a shift to the amount of throttle opening or speed of the vehicle.

Many transmissions use an accumulator tied hydraulically to the clutch or band servo. As described earlier, an accumulator absorbs fluid during the pressure buildup stage of a band or clutch apply, and this has the effect of reducing the pressure and lengthening the time it takes for the friction device to lock up. As clutch or band apply pressure is entering the apply side of the accumulator, fluid must leave the other, exhaust side of the piston, and the pressure on the apply side and rate of stroke can be controlled by how easily the fluid leaves the exhaust side. An accumulator valve or shift control valve is often placed in the accumulator exhaust passage. The accumulator valve is usually balanced between throttle pressure and accumulator pressure in such a way that high throttle pressure closes down the accumulator exhaust flow and produces higher accumulator pressure and therefore a firmer, quicker shift (Fig. 5-60).

Another method of changing upshift speed relative to downshift speed is to use an orifice alongside a one-way check ball. During upshifts, the ball seats so the flow passes through the orifice, and the diameter of the orifice controls the apply rate. During a downshift, the flow unseats the ball so the fluid flows quickly and allows a fast release (Fig. 5-61).

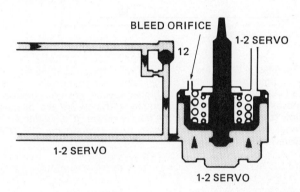

Fig. 5-61 Fluid flowing into the accumulator shown to stroke it upward has to flow through the orifice because the flow seats the check ball. Outward flow is faster because it moves the ball off its seat and flow is around the ball. *(Copyrighted material reprinted with permission from Hydra-matic Div., GM Corp.)*

Still another method of changing shift speed is to position a valve that can open a bypass circuit alongside an orifice. Under high throttle pressures, the valve opens the bypass so that a higher flow and pressure reaches the clutch or band apply piston and produces a faster shift. Low throttle pressures close the bypass valve so the apply fluid must pass through the orifice and be controlled by the small size of the opening (Fig. 5-62).

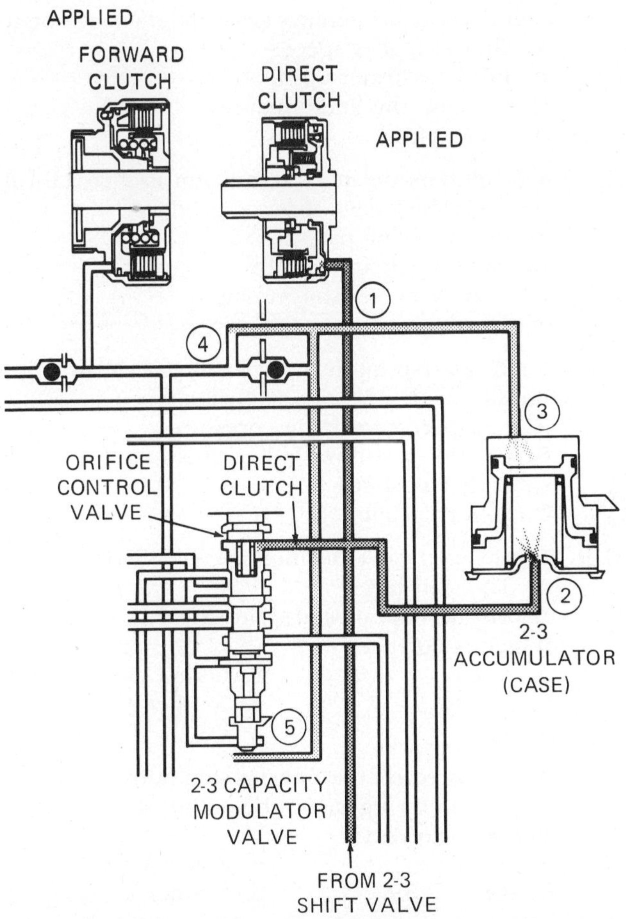

Fig. 5-60 A 2–3 shift applies the direct clutch (1), and the same pressure strokes the 2–3 accumulator upward (2). As the accumulator strokes, fluid at the top of the accumulator must leave past the orifice next to the check ball (4), which closes, or through the 2–3 capacity modulator valve (5), which controls the flow rate. *(Courtesy of Ford Motor Company.)*

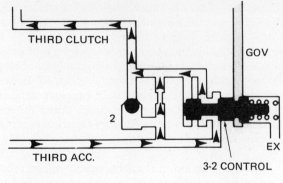

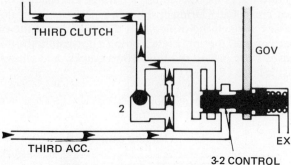

Fig. 5-62 (a) When the 3–2 control valve shown is at the left, 3rd Acc. (third-accumulator) oil can flow through the valve rapidly to become 3rd clutch oil. At higher speeds, governor pressure moves the control valve toward the right; now 3rd acc. oil must flow through an orifice so the flow is slower. *(Copyrighted material reprinted with permission from Hydra-matic Div., GM Corp.)*

REVIEW QUESTIONS

The following questions are provided so you can check the facts you have just learned. Select the response that best completes each statement.

1. Hydraulic pressure is created by
   A. The pump
   B. The pressure regulator valve
   C. Restrictions to fluid flow
   D. All of these

2. Most pressure regulator valves are arranged so line pressure tries to move the valve in one direction while _____ tries to move the valve in the other.
   A. A spring   B. Boost valve pressure
   Which is correct?
   a. A only   c. Both A and B
   b. B only   d. Neither A nor B

3. Normal unboosted line pressure is about
   A. 10 to 15 psi   C. 50 to 60 psi
   B. 25 to 35 psi   D. 125 to 135 psi

4. After the fluid passes through the torque converter, it goes to the
   a. Cooler   c. Sump
   b. Lube passages   d. a, b, and then c

5. In many transmissions, torque converter clutch lockup occurs when the
   A. Fluid flow through the converter is reversed
   B. Fluid flow to the cooler is shut off
   C. Converter pressure is raised to 125 psi
   D. All of these

6. The transmission cooler is usually located in the
   A. Oil pan
   B. Coolest tank of the radiator
   C. Area in front of the air-conditioning condenser
   D. Rear of the car, under the trunk

7. An automatic transmission can be damaged by towing the car because
   A. It won't be lubricated unless the engine is running
   B. The torque converter will seize up
   C. The clutches and bands won't apply
   D. All of these

8. The transmission receives the signal indicating the amount of throttle opening through a
   A. Vacuum line   B. Metal rod or cable
   Which is correct?
   a. A only   c. either A or B
   b. B only   d. Neither A nor B

9. Higher throttle openings cause the transmission to
   A. Shift at higher speeds
   B. Produce firmer upshifts
   C. Increase the line pressure
   D. All of these

10. A vacuum modulator transmission also has a throttle-controlled valve that can produce
    A. Reduced line pressure
    B. Full-throttle downshifts
    C. Slightly earlier shift timing
    D. All of these

11. A stronger spring in the vacuum modulator produces
    A. Less part-throttle line pressure
    B. Greater part-throttle line pressure
    C. Earlier upshifts
    D. Softer upshifts

12. The manual valve position is controlled by
    A. The shift lever
    B. The detent cam and spring-loaded ball or roller
    Which is correct?
    a. A only   c. Both A and B
    b. B only   d. Neither A nor B

13. The governor is
    A. Mounted on the transmission output shaft
    B. Driven by a gear on the output shaft
    Which is correct?
    a. A only   c. Either A or B
    b. B only   d. Neither A nor B

14. The pressure signal from the governor is used to
    A. Produce higher line pressure at higher speeds
    B. Move the shift valves to the upshift position
    C. Produce firmer shifts at higher speeds
    D. Move the shift valves with the vehicle stopped

15. As the speed increases, centrifugal force acting on the governor weights will
    A. Move the valve to increase governor pressure
    B. Balance the output shaft to prevent vibrations
    Which is correct?
    a. A only
    b. B only
    c. Both A and B
    d. Neither A nor B

16. Most shift valves use _____ to move them to a downshift position when the vehicle stops.
    A. Throttle pressure
    B. Governor pressure
    C. Spring pressure
    D. Any of these

17. When a shift valve moves against the spring pressure, the transmission will
    A. Upshift
    B. Downshift
    C. Increase the line pressure
    D. Shift to neutral

18. In a Simpson gear train transmission, when the governor forces the 2–3 shift valve to move, fluid pressure will go to the
    A. High–reverse clutch
    B. Intermediate band
    Which is correct?
    a. A only
    b. B only
    c. Both A and B
    d. Neither A nor B

19. An _____ is used to reduce fluid pressure in a servo that is applying to produce a softer upshift.
    A. Accumulator
    B. Orifice
    Which is correct?
    a. A only
    b. B only
    c. Both A and B
    d. Neither A nor B

20. An orifice will
    A. Cause a fluid pressure drop when fluid is flowing through it
    B. Have no effect on fluid pressure if there is no flow
    Which is correct?
    a. A only
    b. B only
    c. Both A and B
    d. Neither A nor B

CHAPTER 6

ELECTRONIC TRANSMISSION CONTROLS

OBJECTIVES

After completing this chapter, you should:

- Have a basic understanding of how electronic controls are used to control transmission shifts.

- Be able to identify the major electronic control units and describe what they do.

- Have a working knowledge of the basic and common electrical terms.

6.1 INTRODUCTION

More and more modern transmissions are using solid-state electronics to control the automatic functions of the transmission. Many modern transmissions use electronics to control the shifts and the operation of the torque converter clutch. The domestic units using electronic shift controls include Chrysler Corporation's A-604 and A-500; Ford Motor Company's E40D; and General Motors' F7, 4L30-E, and 4T60-E. Electronic control modules can give a more accurate shift timing, can easily control and tailor shift quality, and can integrate the operation of the transmission with the engine. One manufacturer is controlling shift feel by electronically altering the spark timing of the engine as the shift occurs.

The major electronic controls are used to replace the hydraulic governor and throttle valves and either control the shift valve(s) by an electric solenoid(s) or replace the shift valves entirely with solenoids and to control the lockup of the torque converter.

With the introduction of electronic transmission controls, the automatic transmission technician needs a working knowledge of electricity and basic electronics as well as hydraulics. In the past, diagnosing and repairing the simple circuits that were used for detent, neutral start switch, and backup light circuits could often be accomplished with a limited knowledge of electrical circuits and a simple test light. With the introduction of modern solid-state computerized electronics in the torque converter clutch and some shift circuits, there is a greater need to understand the system and electrical test procedures. The ability to measure voltage and resistance and interpret these measurements has become very important.

Solid-state electronics is at the heart of computerized circuits. It includes transistors, diodes, and integrated microchips. These control and sensing devices are quite fragile relative to other automotive electrical devices. They are also relatively trouble free and usually have a long life. Nothing moves in a solid-state circuit except electrons so nothing should wear out.

A course in basic automotive electronics is necessary to thoroughly understand electricity and how to measure it. The description that follows is a brief review.

6.2 ELECTRICAL BASICS

A technician is normally concerned with three measureable aspects of electricity: *volts*, *amperes* (or *amps*), and *ohms* (Fig. 6-1). Voltage can be called electrical pressure because it is the push that forces an electrical current to flow through a wire or component. This current flow is known as amperes. In a car, the voltage is supplied by a source of electrical power—either the battery or the alternator. The total amount of electrical power being used is called *watts*, and this is determined by multiplying amperage by voltage.

The complete electrical path that allows amperes to flow from the power source through the electrical components and back to the power source is called a *circuit* (Fig. 6-2). This path is composed of wires, commonly called conductors, usually one or more switches, and the electrical component(s). A switch is a unit that is used to complete a circuit and allow a current flow or open a circuit and stop a current flow. Some switches are normally open and then closed by an outside force

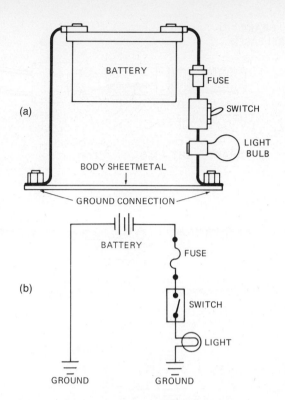

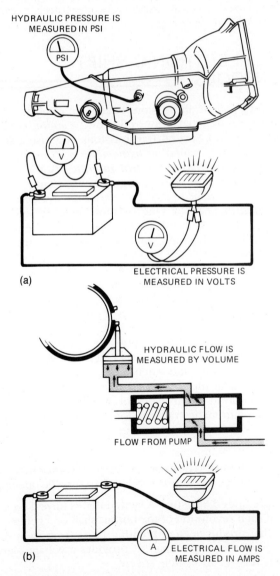

Fig. 6-2 The electrical path through a circuit is shown in (a) picture form and (b) a wiring diagram or schematic that uses symbols to represent the different components. Schematics are normally used in repair shops to trace circuits.

Fig. 6-1 If we compare electricity with hydraulics, (a) voltage and pressure and (b) current flow and fluid flow are very similar.

or normally closed and then opened by an outside force (Fig. 6-3). Some automatic transmissions have one or more internal switches that are opened or closed by fluid pressure from a particular hydraulic circuit (Fig. 6-4). A *ground circuit* is used to conduct electricity from the electrical component back to the source of power. This portion of the circuit uses the metal of the car body, frame, engine block, transmission case, and so on, as an electrical conductor. In modern cars, the negative ($-$) terminal of the battery and the alternator are connected to ground, and the positive ($+$) side is insulated (Fig. 6-4). All cars use direct current (dc) in the major electrical circuits. Direct current always travels in one direction, from negative to positive. Normal house current is alternating current (ac). Alternating current travels in one direction and then reverses to flow in the opposite direction many times a second.

Electrical resistance is measured in units called ohms (Ω). The amount of resistance in a circuit determines the amount of current flow. A large amount of resistance stops or severely limits current flow, and a small amount of resistance allows a large current flow (Fig. 6-5).

Many technicians see a strong resemblance between hydraulics and electricity. A strong comparison can be made between hydraulic pressure and voltage (both forces cause something to move), fluid flow and amperage (both flow), a hydraulic circuit and an electrical circuit (both are flow paths), a valve and a switch (both control flow), a hydraulic orifice and an electrical resistor (both reduce flow), a hydraulic check valve and a diode (both allow flow in only one direction), and an

165

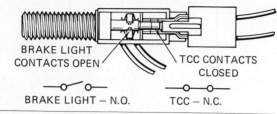

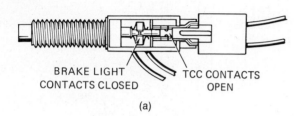

(a)

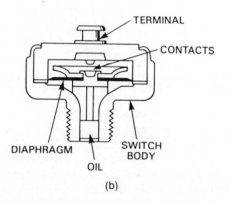

(b)

Fig. 6-3 (a) A switch can be normally open and then closed by an action like the brake light contacts in the brake switch shown when the brakes are applied or normally closed and opened by an action like the torque converter clutch (TCC) contacts in the switch. (b) Switches are also commonly found in single- and double-terminal configurations. *(Copyrighted material reprinted with permission from Hydra-matic Div., GM Corp.)*

accumulator and an electrical capacitor (both absorb a given amount of flow). We should remember that both hydraulics and electricity are methods of transferring energy or power from one place to another.

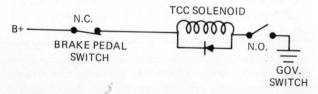

Fig. 6-4 Circuit that uses a normally closed, two-contact TCC switch at the brake pedal and a normally open, hydraulic pressure switch in the transmission. Both switches must be closed to energize the TCC solenoid and apply the TCC. *(Copyrighted material reprinted with permission from Hydra-matic Div., GM Corp.)*

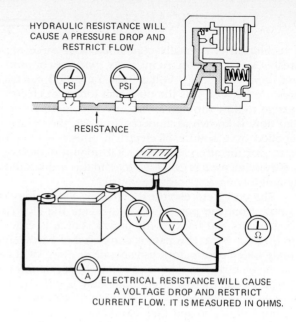

Fig. 6-5 Electrical resistance and a resistance to fluid flow like an orifice are very similar. Resistors are often placed in electric circuits to limit current flow.

6.3 ELECTRONIC SHIFT CONTROLS

For years, many transmissions have been using a torque converter clutch that is controlled by the *engine control module* (*ECM*). Several transmissions now use a separate computer, called a *transmission control unit* (*TCU*), a *transmission control module* (*TCM*), or the ECM to control the shift timing. For consistency, in this book, we will refer to it as the *electronic transmission control unit* (*ETCU*). One or more electric solenoid valves at the valve body are used to control the movement of hydraulic fluid, and these solenoids are controlled by the ETCU (Fig. 6-6). An electronic-shifted transmission operates the same as a standard unit. The major difference is that the shift valves are moved by hydraulic pressure from the governor in a standard unit; with an electronic-shifted unit, the upshift oil is controlled by the electric solenoid(s).

To put it as simply as possible, a computer is an electronic device that receives electronic signals from a group of sensors, and when these signals match a program stored in the computer's memory, the computer sends an output signal to one or more actuators (Fig. 6-7). In an automatic transmission, the actuators are the solenoids, and the input is a group of sensors such as the following:

Throttle Position Sensor A variable resistor that provides a voltage signal that is relative to throttle opening.

Vehicle Speed Sensor A switch that is turned on and off on each revolution of the output shaft or a generator unit that produces an alternating current on each revolution of the output shaft. It provides a voltage signal that is relative to vehicle speed.

Neutral Start Switch A multicontact and multiterminal switch that provides the gear range position signal.

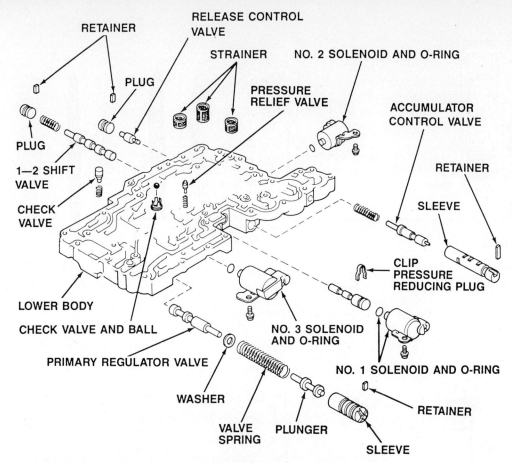

Fig. 6-6 Valve body that uses three different solenoids to control shift valve operation. Additional valves are contained in the upper valve body. *(Courtesy of Chrysler Corporation.)*

Brake Switch A switch mounted at the brake pedal that provides a signal when the brake pedal is depressed.

Power–Economy Switch Allows the driver to raise (power) or lower (economy) the shift points.

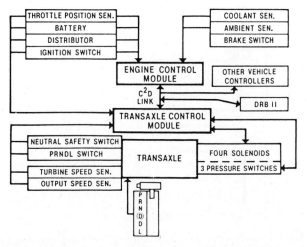

Fig. 6-7 Transaxle controlled by four solenoids that receives data from two transmission switches, three transmission pressure switches, two transmission speed sensors, and the ECM plus other vehicle sensors. *(Courtesy of Chrysler Corporation.)*

6.3.1 Shift Solenoids

Shift solenoids are much like the solenoids used for torque converter clutch control and are also similar to the detent downshift solenoids used in some transmissions. They open or close an oil passage so that the pressure will either be trapped and increase, drain and escape, or pass through to apply a clutch or servo. A solenoid is essentially a coil of wire that becomes an electromagnet when current flows through it, and it loses its magnetism when the current flow is shut off. An iron plunger or plate at the solenoid is spring loaded to one position, and when the solenoid is energized, the plunger is moved to the other position (Fig. 6-8). A solenoid can be normally closed and then opened when activated by the electrical signal or normally opened and then closed by the electrical signal.

Depending on the transmission, the solenoid either directly controls the flow of oil to the apply device or controls a shift valve that in turn controls the flow to the apply device. The solenoid shown in Fig. 6-9 has direct control. When it is energized, the check ball is held on its seat, and the fluid passage is closed. When the solenoid is deenergized, the ball can leave the seat, and oil flows to apply the clutch. This arrangement also allows the solenoid to be cycled on and off to control how fast the pressure builds up in the circuit. This in turn controls how fast the clutch applies and therefore controls the shift quality.

167

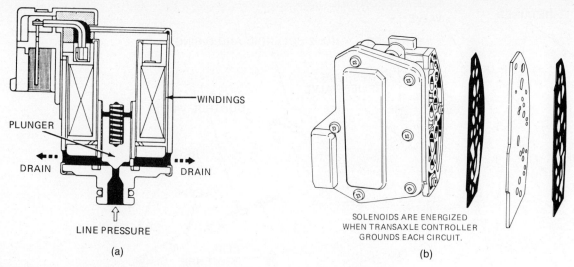

Fig. 6-8 (a) When current flows through the windings of the solenoid, the plunger is lifted and line pressure dumped to the drain openings. Solenoids can be mounted individually or in a group. (b) The solenoid assembly contains four solenoids. *(Courtesy of Chrysler Corporation.)*

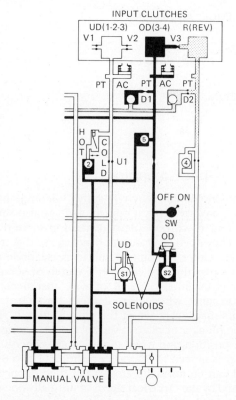

Fig. 6-9 Hydraulic circuit. The UD (underdrive) solenoid is energized to hold the S1 ball on its seat, which blocks fluid flow to the UD clutch. The OD (overdrive) solenoid is also energized, but it holds the S2 ball off its seat to allow the OD clutch to apply. *(Courtesy of Chrysler Corporation.)*

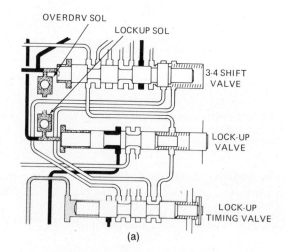

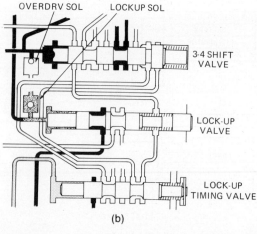

Fig. 6-10 (a) The overdrive solenoid and the lockup solenoid are not energized so fluid pressure is vented at the solenoids. (b) When the overdrive solenoid is energized, it closes the vent, and pressure moves the 3–4 shift valve to upshift. The lockup solenoid controls the lockup valve in the same manner. *(Courtesy of Chrysler Corporation.)*

The solenoids are also mounted so they open or close a passage that is fed through a small orifice from the line or drive pressure. When the solenoid is open, fluid pressure drains from this line, and when it is closed, pressure builds up (Fig. 6-10). Because of the orifice, there will not be a great loss of pressure from the main circuits when the solenoid is open. The pressure controlled by the solenoid acts on one end of one or two shift valve(s) while a spring acts on the other end. With this arrangement, we can easily control the shift valve position.

The example shown in Fig. 6-11 shows a normally closed solenoid valve (1) that is on while the normally closed solenoid (2) is off. This produces a pressure at the end of the 1–2 shift valve that compresses the spring and keeps the valve downshifted. This combination also allows the springs to hold the 2–3 and the 3–4 shift valves downshifted.

The 1–2, 2–3, and 3–4 shifts in this transmission occur as follows:

- *1–2 Shift* Activate solenoid 2 to drain shift pressure; the spring moves the 1–2 shift valve to upshift (Fig. 6-12).

- *2–3 Shift* Turn solenoid 1 off. The shift pressure moves the 2–3 shift valve to upshift (Fig. 6-13).

- *3–4 Shift* Turn solenoid 2 off. The shift pressure moves the 3–4 shift valve to upshift (Fig. 6-14).

A summary of the relationship of the transmission gear and the solenoids for this particular transmission is as follows:

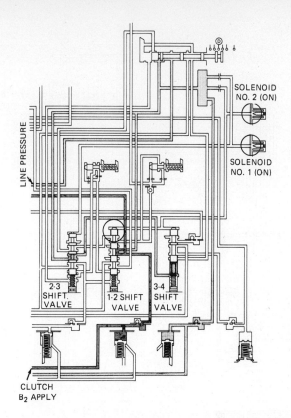

Fig. 6-12 Transmission in second gear. Note that both solenoids are on. Brake B2 is applied by pressure from the 1–2 shift valve, which is controlled by solenoid 2. *(Courtesy of Toyota Motor Sales, U.S.A., Inc.)*

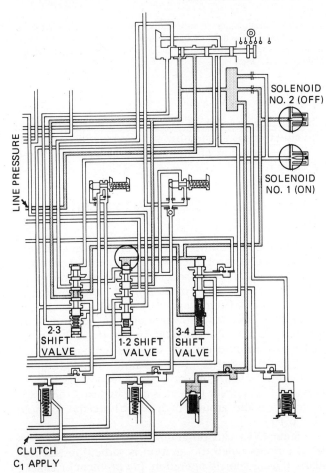

Fig. 6-11 Transmission in first gear. Note that solenoid 1 is on and solenoid 2 is off. Clutch C1 is applied by pressure from the manual valve. *(Courtesy of Toyota Motor Sales, U.S.A., Inc.)*

Gear	Solenoid 1	Solenoid 2
First	On	Off
Second	On	On
Third	Off	On
Fourth	Off	Off

Downshifts occur in exactly the reverse manner as upshifts. For example, a 4–3 kickdown occurs from fourth gear if solenoid 2 is turned on. It should also be noted that the transmission will make a second-gear start if solenoid 1 fails; make a 1–4 shift and skip second and third gears if solenoid 2 fails; and start out in fourth gear if both solenoids fail.

The Chrysler A-604 transaxle does not use shift valves. The relatively simple valve body contains only a pressure regulator valve, manual valve, solenoid switch valve, lockup (converter clutch) switch valve, and the torque converter (T/C) control valve (Fig. 6-15). The gear ranges are controlled by the manual valve and the operation of four solenoids.

Each solenoid operates a check ball, and two also operate a vent valve (Fig. 6-16). The low–reverse/lockup and the overdrive solenoids are *normally vented*. When they are not energized, the vent is open,

169

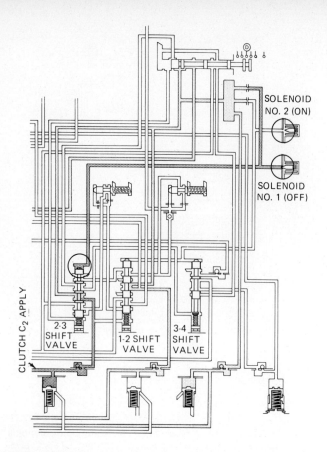

Fig. 6-13 Transmission in third gear. Note that solenoid 1 is off and solenoid 2 is on. Clutch C2 is applied by pressure from the 2-3 shift valve, which is controlled by solenoid 1. *(Courtesy of Toyota Motor Sales, U.S.A., Inc.)*

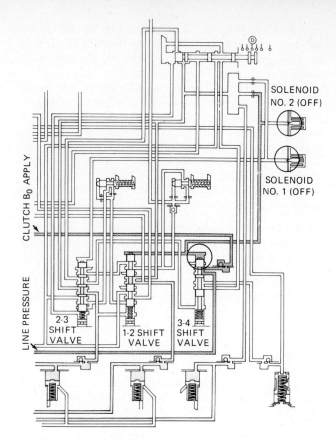

Fig. 6-14 Transmission in fourth gear. Note that both solenoids are off. Brake B0 is applied by pressure from the 3–4 shift valve, which is controlled by solenoid 2. *(Courtesy of Toyota Motor Sales, U.S.A., Inc.)*

and the check ball is allowed to operate in a normal manner (Fig. 6-17). When they are energized, the check ball is held open—off its seat—and the vent is closed. The 2–4/low–reverse and the underdrive solenoids are *normally applied*. When they are not energized, the check balls are allowed to operate in a normal manner (Fig. 6-18). When they are energized, the check balls are held closed—on their seats. These solenoids are pulsed—rapidly cycled on and off—at a high frequency to control the rate of pressure rise and clutch apply. In this way, the ETCU can control clutch application and therefore shift feel. Three pressure switches are used in the circuits controlled by the solenoid valves to allow the ETCU to tell when the solenoid valve actually opens or closes.

Depending on the position of the solenoid switch valve, the low–reverse/lockup solenoid controls the operation of either the low–reverse clutch (all gear ranges except second, third, or fourth) or the torque converter clutch (TCC) (second, third, or fourth gear). The low–reverse clutch or the TCC applies when this solenoid is energized. The overdrive solenoid controls the operation of the overdrive clutch; the clutch applies when this solenoid is operated. The 2–4/low–reverse solenoid controls the low–reverse clutch when the manual valve is in reverse (the clutch applies when the solenoid is energized) and the 2–4 clutch in forward-gear ranges (the clutch applies when the solenoid is *not* energized).

The underdrive solenoid controls the underdrive clutch; the clutch applies when the solenoid is *not* energized.

The operation of these various solenoids is given in Table 6.1.

6.3.2 Electronic Transmission Control Unit

The ETCU controls the operation of the shift solenoids and the TCC control solenoid by switching them off or on at the correct time. A small computer notes the speed of the vehicle, position of the throttle, and position of the hydraulic switches and various other sensors and then either provides a power source, B^+, or completes a ground path to the solenoid. In one transmission, 12 different sensors are used. In some units, the ETCU switches on power B^+ to a solenoid that is grounded in the transmission; in others, the solenoids receive power B^+ whenever the car is operated, and the ETCU switches on the ground path to activate the solenoid (Fig. 6-19).

Some ETCUs are programmed to monitor their electrical circuits when the ignition is turned on and regularly during vehicle operation. If a problem or fault in the operation is found, a code indicating the nature of the problem is stored in the ETCU's memory. If the fault is serious enough to damage the transmission, the ETCU

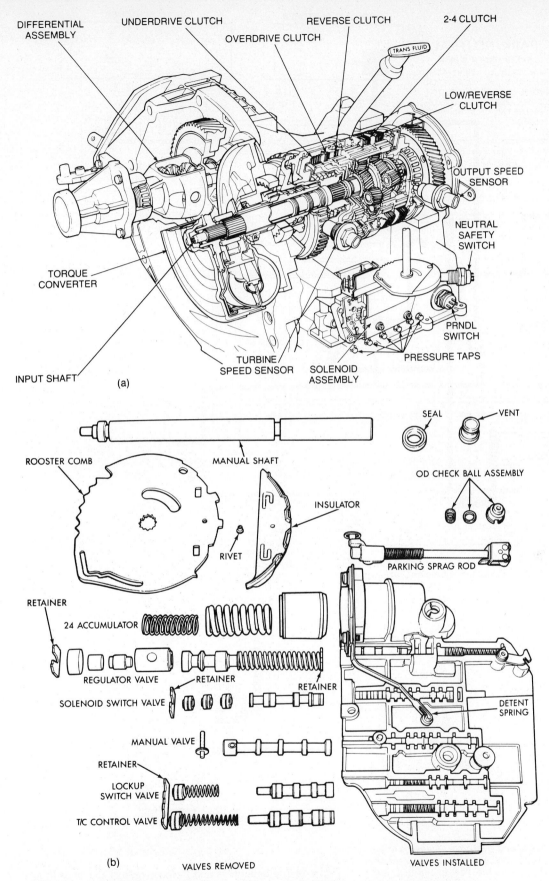

Fig. 6-15 (a) An A-604 transmission with electronic shift controls. Note the turbine and output speed sensors, the neutral safety and PRNDL switches, and the solenoid assembly. (b) The valve body for this transmission is much simpler than that for a hydraulic-controlled transmission. *(Courtesy of Chrysler Corporation.)*

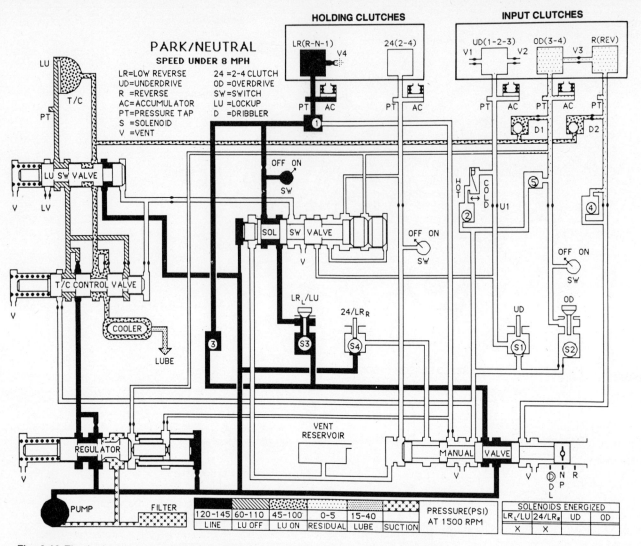

Fig. 6-16 The hydraulic diagram for an A-604 transmission is not complicated by shift valves, accumulators, or other shift control devices. Shift quality is controlled by cycling the solenoids during the shifts. *(Courtesy of Chrysler Corporation.)*

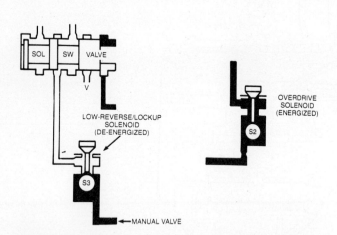

Fig. 6-17 The low–reverse/lockup and overdrive solenoids in an A-604 are normally closed and normally vented. In the deenergized position shown, ball S3 blocks the flow to the SOL SW (solenoid switch) valve while the passage above the ball vents any pressure off. When the solenoid is energized (right), the ball is moved to open the valve while the solenoid stem closes the vent. *(Courtesy of Chrysler Corporation.)*

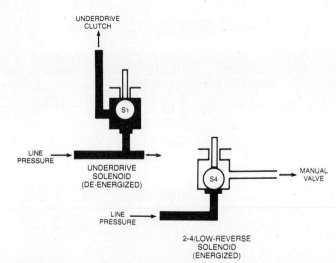

Fig. 6-18 The underdrive and 2-4/low–reverse solenoids in an A-604 are normally applied (open); ball S1 (underdrive solenoid) is off its seat. When it is energized, the ball (right) is moved against its seat and the vent above the ball is opened. *(Courtesy of Chrysler Corporation.)*

172

TABLE 6.1 Solenoid Operation

Gear	Low–Reverse/Lockup	2–4/Low–Reverse	Underdrive	Overdrive
Park–neutral	On	On	Off	Off
Reverse	Off	Off	Off	Off
First	On	On	Off	Off
Second	Off	Off	Off	Off
Third	Off	On	Off	On
Fourth	Off	Off	On	On
TCC	On (in second, third, or fourth gear)			

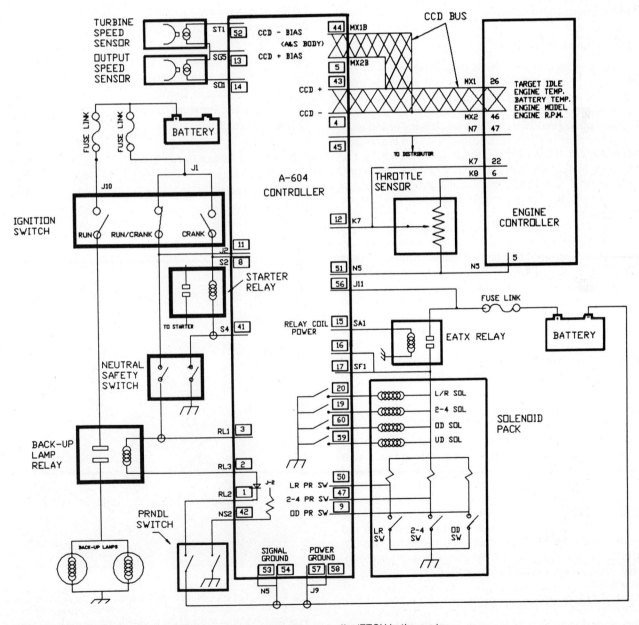

Fig. 6-19 Electrical diagram for A-604 transmission. Note the controller/ETCU in the center and the many connections to the various devices. *(Courtesy of Chrysler Corporation.)*

shuts down the electronic operation, and the transmission's speed is limited to a *default* or *limp-in* mode. It operates in one gear only (usually second) in all of the forward-gear positions. There will be no upshifts or downshifts, but the driver will not be stranded. A technician using the proper equipment and procedure can read the fault code as an aid in determining what is wrong with the unit (Fig. 6-20).

The ETCU, like other automotive computers, is somewhat fragile and fairly expensive. It is normally mounted in a relatively clean and cool location where it is protected from possible damage. A fairly standard location is behind the instrument panel above the glove box (Fig. 6-21).

The electrical circuit for the ETCU is fairly complex. It must be connected to the power source, B^+, and a good ground. The power feed is usually through the ignition switch and a fuse for protection. The ETCU must also have one or more connections to each of the sensors and to the solenoids.

6.3.3 Speed Sensor

The speed sensor is driven off the output shaft in the same manner as a governor. The speed sensor is used in place of the governor in a hydraulically controlled transmission. One style of speed sensor consists of a normally open reed switch. The switch is mounted next to a rotor that has a magnet built into it (Fig. 6-22). When the car starts moving, the magnet passes by the reed switch and pulls the contacts closed for a moment. By "counting" how often the speed sensor switches on and off, the ETCU can accurately determine how fast the car is going.

Another style of speed sensor operates by using a coil of wire that is wrapped around a magnetic core. This sensor is mounted next to a gearlike toothed wheel (Fig. 6-23). As the toothed wheel revolves, an alternating voltage is produced in the sensor, and the rate or frequency of this pulsating voltage is used by the ETCU to determine transmission speed.

The Chrysler A-604 uses two speed sensors: one for the turbine shaft speed (input) and one for the output shaft speed. The relative speed of the two sensors is used to determine how fast a clutch is being applied and therefore shift quality.

6.3.4 Throttle Position Sensor

The *throttle position sensor* (*TPS*) is attached to the throttle shaft (Fig. 6-24). It is a variable resistor that changes resistance value as the throttle is opened and closed. A voltage signal passing through the TPS has the voltage modified relative to resistance, which is a result of the throttle position. By measuring the voltage, the ETCU can accurately determine the throttle position and produce correctly timed upshifts or downshifts. The TPS circuitry takes the place of the throttle/vacuum modulator valve in hydraulically controlled transmissions.

6.3.5 Neutral Start Switch

The neutral start switch is mounted at the transmission and is operated by the shaft connecting the gear selector

FAULT	CODE	LIMP-IN
Internal A-604 controller	11	Yes
Battery was disconnected	12	No
Internal A-604 Controller	13	Yes
Relay output always on	14	Yes
Relay output always off	15	Yes
Internal A-604 Controller	16	Yes
Internal A-604 Controller	17	Yes
Engine Speed Sensor circuit	18	Yes
Bus communication with SMEC	19	No
Switched Battery	20	Yes
OD Pressure switch circuit	21	Yes
2-4 Pressure switch circuit	22	Yes
2-4/OD Pressure switch circuit	23	Yes
LR Pressure switch circuit	24	Yes
LR/OD Pressure switch circuit	25	Yes
LR/2-4 Pressure switch circuit	26	Yes
All pressure switch circuits	27	Yes
Illegal shifter positions	28	No
Throttle position signal	29	No
OD Hydraulic Pressure Switch	31	Yes
2-4 Hydraulic Pressure Switch	32	Yes
OD/2-4 Hydraulic Pressure Switch	33	Yes
Solenoid Switch Valve	37	No
Lockup Control	38	No
Turbine/Trans Output Speed circuit	39	Yes
LR Solenoid circuit	41	Yes
2-4 Solenoid circuit	42	Yes
OD Solenoid circuit	43	Yes
UD Solenoid circuit	44	Yes

(a)

(b)

Fig. 6-20 (a) If a transmission has a problem, the A-604 displays one of the fault codes shown, which indicate the nature of the problem. Note that some of the problems cause the transmission to go into the limp-in mode. (b) A special scanner is required to read these codes. *(Courtesy of Chrysler Corporation.)*

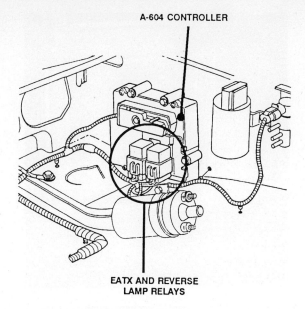

Fig. 6-21 The controller/ETCU is normally mounted in a location that is fairly well protected from heat, cold, moisture and so on. *(Courtesy of Chrysler Corporation.)*

C, Y, A BODY

C = F.W.D. NEW YORKER, DYNASTY
Y = F.W.D. IMPERIAL
A = SPIRIT, ACCLAIM

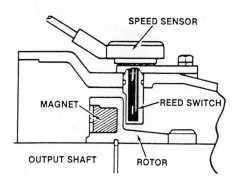

Fig. 6-22 Sensor shown is a reed switch that is opened momentarily as a magnet mounted on the output shaft moves past it. It makes a pulse in the circuit on each revolution of the output shaft. *(Courtesy of Chrysler Corporation.)*

Fig. 6-23 Sensors shown use a coil of wire wrapped around a magnetized iron core. As the toothed wheel moves past the end, the wires generate electrical pulses that allow the ETCU to determine the speed of the toothed wheel. *(Courtesy of Chrysler Corporation.)*

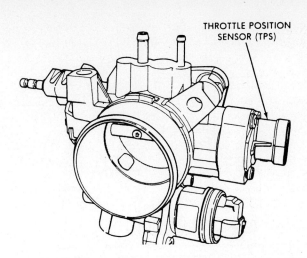

Fig. 6-24 The TPS is mounted on the throttle body. It provides an electrical signal so the ETCU knows if the throttle is closed or opened and how far it is opened. *(Courtesy of Chrysler Corporation.)*

and manual linkage to the manual valve. It is a multi-position switch. This switch also has multiple connectors leading to the ETCU, and as the gear range selector is moved, the neutral start switch makes a variety of switch connections for each gear range (Fig. 6-25). By noting which connections are completed, the ETCU can determine which gear range has been selected.

The signals from the neutral start switch produce the following responses from the ETCU:

- Keep the engine from starting in any gear position except park or neutral.
- Allow a 1–2–3–4 shift sequence in drive.
- Allow a 1–2–3 or cause a 4–3 shift sequence in third.
- Allow only first gear or a 1–2 upshift or a 3–2–1 shift sequence in first.
- Operate the backup lights in reverse.

6.3.6 Comfort–Power Switch

Some vehicles use a comfort–power switch mounted at the instrument panel (Fig. 6-26). This switch allows the driver to select one of two operating modes programmed into the ETCU. In the "comfort" mode, the upshifts and downshifts are made at normal shift points producing quiet, economical operation. This is also called the economy mode in some units. In the "power" mode the upshifts are made later, at higher engine rpm and at wider throttle openings, and the downshifts occur at higher speeds.

6.3.7 Brake Switch

The brake switch is mounted at the brake pedal and can be part of the stop light and cruise control disconnect switch (Fig. 6-27). The major function of this switch is to release the TCC to prevent engine stalling as the car

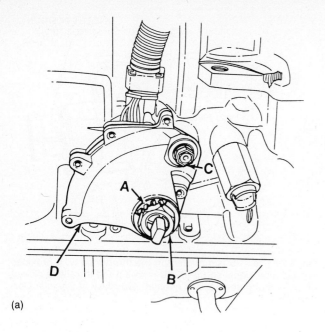

Fig. 6-26 When used, a comfort–power switch mounted on the instrument panel changes the upshift and downshift timing. *(Courtesy of Chrysler Corporation.)*

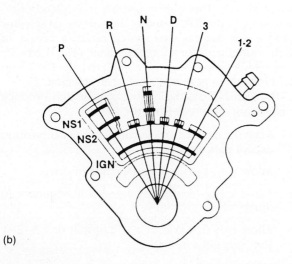

Fig. 6-25 (a) Neutral start switch mounted at the transmission's manual shift linkage. It serves several other functions besides being just a neutral switch. (b) Internal switch circuitry. *(Courtesy of Chrysler Corporation.)*

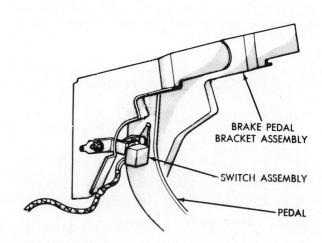

Fig. 6-27 The brake switch is mounted at the brake pedal; when the brakes are applied, it causes the TCC to release. *(Courtesy of Chrysler Corporation.)*

comes to a stop or in case the drive wheels lock up during a hard stop.

6.3.8 Engine Coolant Temperature

This sensor tells the ETCU when the engine is warmed up and operating normally (Fig. 6-28). This signal is used to prevent TCC lockup while the engine is cold to prevent driveability problems. In some transmissions, this signal is used to produce earlier than normal converter clutch lockup if the engine temperature rises too high. In some transmissions, a cold engine temperature delays upshifts so they occur at a slightly higher speed.

6.4 ELECTRICAL CIRCUIT PROBLEMS

The common types of electrical problems are *open*, *shorted*, or *grounded* circuits. These problems are fairly easy to check except for some solid-state units. An unwanted open circuit is a broken, incomplete circuit through which no current flows (Fig. 6-29). A switch opens a circuit intentionally when it is turned off. Open circuits are usually the result of a broken wire or a burned-out fuse or fusible link. A high-resistance, partially open circuit is usually caused by a loose or dirty connection. Source voltage will usually be available up to the point where the circuit is open. An open circuit

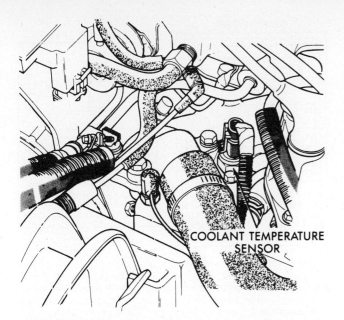

Fig. 6-28 An engine coolant temperature sensor is mounted near the coolant outlet of the engine. When the engine is cold, the ETCU does not allow converter clutch lockup and may delay upshifts. When the engine is too hot, the ETCU locks the converter clutch earlier than normal. *(Courtesy of Chrysler Corporation.)*

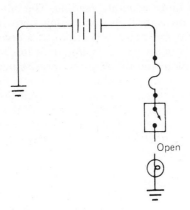

Fig. 6-29 An open circuit is caused by any break in the circuit that stops the current flow. It can occur at any point in the circuit.

is somewhat like a leak in a hydraulic system, although a hydraulic leak causes a pressure loss while an open circuit stops the passage of voltage.

When the conductor of a current-carrying wire or component touches ground, the bare metal of the car body, a grounded circuit occurs. Normally, the wire's insulation prevents a grounded circuit, but if it wears through and allows the metal conductor to touch the ground metal, a grounded circuit results (Fig. 6-30). A grounded circuit provides an unwanted, low-resistance path for current flow, and the current flow increases because of the reduced resistance. The result is usually a burned-out fuse, tripped circuit breaker, or burned-up wire(s). A grounded circuit usually produces an open circuit when it burns out a fuse, fusible link, or wire. Another name for a grounded circuit is a *short to ground*; it is sometimes referred to as a copper-to-iron contact.

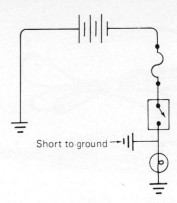

Fig. 6-30 A grounded circuit results if the conductor contacts ground to complete an unwanted path to ground. It can occur anywhere between the battery positive terminal and ground connection of the electrical component.

A short circuit can occur in electrical units that use coils of wire. If the insulation between the coils fails so they make electrical contact with each other, a shorter-than-normal electrical path is created. A short lowers the resistance of the component, which in turn increases the current flow. It also reduces the operating efficiency of the unit (Fig. 6-31). It is sometimes called *copper-to-copper contact*.

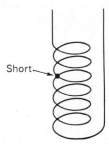

Fig. 6-31 A short circuit occurs in a coil if the insulation is damaged so the conductors touch and make a shorter than normal circuit.

6.5 MEASURING ELECTRICAL VALUES

A test light or a volt-ohmmeter is used when troubleshooting an electrical circuit. A test light is a simple, inexpensive device that can quickly indicate if a circuit has voltage (Fig. 6-32). The amount of voltage is indicated by the brightness of the light. This is a handy device for checking simple circuits, but the common test light should not be used on the sensor portion of solid-state circuits. The current flow through the test light is high enough to damage the relatively fragile transistors and integrated circuits in some solid-state electronic equipment. These circuits can be checked using a high-impedance test light that draws very little current. These test lights use a light-emitting diode (LED), which has a very high resistance, and this allows it to draw very

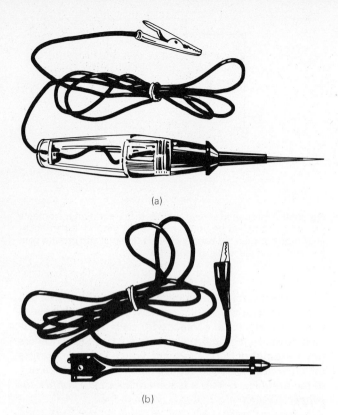

Fig. 6-32 (a) Common and (b) LED test lights are inexpensive and fast tools to determine if a circuit has voltage. The ground clip is connected to a good clean ground while the probe is touched to a wire or connector, and if the light comes on, there is voltage. Only the LED lights should be used on computer-controlled circuits.

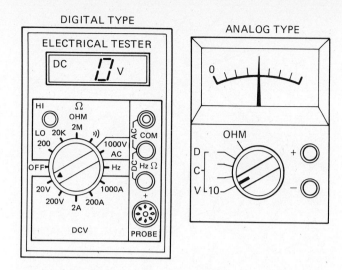

Fig. 6-33 Volt-ohmmeters are commonly available in either digital or analog form to determine the voltage in a circuit or resistance in a component.

little current. Some test lights are self-powered, using an internal battery. They are used to check for continuity (complete circuits) just like an ohmmeter.

Voltage and resistance can be accurately measured using a volt-ohmmeter. These meters are available in both analog and digital forms (Fig. 6-33). Analog meters use a needle that sweeps over a scale to give a reading. Digital meters display numbers that are the actual reading. Analog meters should not be used to check certain solid-state electronic units. Like the test light, they draw enough current from the circuit that it is possible to damage the electronic components. The internal resistance of the meter determines the current draw. If the internal resistance is 10 MΩ (10,000,000 Ω) or greater, the meter can be safely used on solid-state circuits. A digital volt-ohmmeter normally has over 10 MΩ of internal resistance. It is a good practice to use a digital meter or a high-impedance test light to check solid-state circuits.

When measuring voltage, the negative lead of the voltmeter is connected to a good, clean body ground and the positive lead is connected to various connection points in the insulated portion of the circuit. The reading on the meter is the voltage of the circuit at that point. It is important to select and set the meter to the correct voltage range before taking a reading on many meters. Many meters will be damaged if you try to measure a voltage higher than the meter setting. You should always select a range higher than the value you expect to read.

After measuring the voltage, the meter can be reset to a lower voltage scale if desired as long as the range of the lower scale exceeds the voltage just read (Fig. 6-34). Reading a voltage on the lowest range possible so the meter reads at the upper end of the scale usually produces a more accurate reading. The amount of resistance is measured by connecting both leads of an ohmmeter to the two connections of a component or to both ends of a wire. A reading indicates that the circuit is complete and the amount of resistance it has. An ohmmeter is self-powered and has its own internal battery. It causes a small amount of current flow through the circuit and measures the amount of flow to determine the resistance. *Never connect an ohmmeter to a circuit that contains voltage or is connected to a battery. The added voltage from the circuit will damage the meter.*

Before using an ohmmeter, the range of the ohmmeter is selected and set to the value above which you expect to read on the meter. Most meters have a range selector switch with ranges like ×1, ×1000 (1k), ×10,000 (10k), and so on. With the range selector set at ×1, a reading of 6 on the meter would be 6 Ω; at ×1000, this reading would be 6000 Ω (6k); and at ×10,000, this reading would be 60,000 Ω (60k). The reading on the meter should be multiplied by the range setting to determine the total amount of resistance. When using an analog meter, after selecting the range, the meter leads should be connected together and the meter read. With no resistance between the leads, it should read zero. If it does not read zero, most meters have a knob that is used to calibrate the meter to zero and do so at this time. The meter should read at the top of the scale, often marked as infinity when the leads are separated. A digital meter will read "OL" for out of limits. The leads of the meter are then connected to the two terminals of a component or to the ends of a wire (Fig. 6-35). A complete circuit with no resistance is indicated if the reading is zero (0). A large amount of resistance or a possible open circuit is indicated by a high reading. It is good practice to switch the meter to a higher scale,

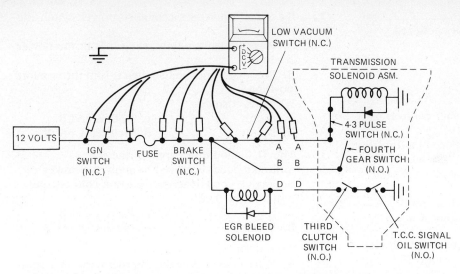

Fig. 6-34 A voltmeter is commonly used to locate an open circuit by connecting the negative lead to a good ground and moving the positive lead along the circuit. The open circuit will be after the last good reading.

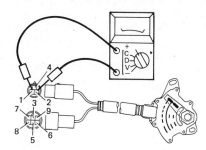

Fig 6-35 An ohmmeter is commonly used connected to the terminals of a component to see if the internal circuits are complete. This neutral start switch has six different circuits that should be complete as shown in the chart. For example, in park there should be a complete circuit between terminals 1 and 2 and another complete circuit between terminals 3, 4, 5, 6, 7, and 8. *(Courtesy of Toyota Motor Sales, U.S.A., Inc.)*

recheck the calibration, and remeasure the resistance when high readings occur.

It should be noted that a technician must be thoroughly familiar with the circuit, the components, and the test equipment before trying to diagnose a problem.

REVIEW QUESTIONS

The following questions are provided so you can check the facts you have just learned. Select the response that best completes each statement.

1. In most electronic-controlled transmissions, the shift valves are
   A. Controlled by an electric solenoid
   B. Replaced by an electric solenoid
   Which is correct?
   a. A only   c. Either A or B
   b. B only   d. Neither A nor B

2. Electrical pressure is measured in
   A. Amperes   C. Ohms
   B. Volts   D. Watts

3. The amount of current flowing through a circuit is measured in
   A. Amperes   C. Ohms
   B. Volts   D. Watts

4. The amount of resistance to electrical flow through a component is measured in
   A. Amperes   C. Ohms
   B. Volts   D. Watts

5. In comparing hydraulics to electronics,
   A. Hydraulic fluid flow is similar to voltage
   B. An orifice in a hydraulic circuit has the same effect as a capacitor in an electrical circuit
   Which is correct?
   a. A only   c. Both A and B
   b. B only   d. Neither A nor B

6. The controlling unit for an electronic-shifted transmission is
   A. The vehicle speed sensor   C. The ETCU
   B. The throttle position sensor   D. None of these

7. The _____ is an electromagnetic device used to control fluid pressure in a particular circuit.
   A. Neutral start switch   C. ETCU
   B. Solenoid   D. Vehicle speed sensor

179

8. A normally vented solenoid _____ allows fluid to flow when it is not energized.
   A. Will
   B. Will not
   C. Will but at a reduced flow
   D. Any of these

9. A transmission with electronic shift controls uses a
   A. Throttle position sensor in place of a throttle valve
   B. Vehicle speed sensor in place of a governor
   Which is correct?
   a. A only   c. Both A and B
   b. B only   d. Neither A nor B

10. If both solenoids fail in a transmission with electronic shift controls, the transmission will
    A. Be ruined if the car is driven
    B. Not have any forward gears
    C. Operate in reverse only
    D. Operate in reverse and one forward gear

11. The neutral start switch in a transmission with electronic shift controls is
    A. Basically the same as the ones in all other transmissions
    B. Operated by the manual shift linkage
    Which is correct?
    a. A only   c. Both A and B
    b. B only   d. Neither A nor B

12. If a transmission uses a comfort–power switch, the switch will
    A. Raise the shift points when it is in the power position
    B. Produce firmer shifts when it is in the comfort position
    Which is correct?
    a. A only   c. Both A and B
    b. B only   d. Neither A nor B

13. The TCC should not apply if the
    A. Brake pedal is moved to apply the brakes
    B. Engine coolant sensor senses a cold engine
    Which is correct?
    a. A only   c. Both A and B
    b. B only   d. Neither A nor B

14. Technically speaking, a short
    A. Can occur only if the internal wires in a component make contact
    B. Stops a component from operating by shutting off the current flow
    Which is correct?
    a. A only   c. Both A and B
    b. B only   d. Neither A nor B

15. A check to determine if there is voltage at a component can be made using a
    A. Voltmeter   B. Test light
    Which is correct?
    a. A only   c. Both A and B
    b. B only   d. Neither A nor B

TORQUE CONVERTERS

OBJECTIVES

After completing this chapter, you should:

- Be able to identify the various portions of a torque converter and explain their purpose.
- Have a basic understanding of the fluid flows that occur inside a converter.
- Realize the importance of torque converter clutches and how they differ in construction and operation from a standard converter.
- Be familiar with the variety of converters that have been used in the past and present.

7.1 INTRODUCTION

The torque converter is a type of fluid coupling that connects the engine's crankshaft to the transmission input shaft. It serves two major purposes as it transfers the engine torque to the transmission. At stops, it serves as an automatic clutch so the car can be stopped with the engine running and the transmission in gear. It also serves to multiply torque while the car is accelerating to improve acceleration and pulling power (Fig. 7-1). A fluid coupling can only perform the first function, to transfer power.

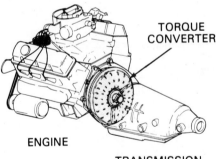

Fig. 7-1 The torque converter is mounted on the engine's crankshaft and is used to transfer power to the transmission. *(Copyrighted material reprinted with permission from Hydra-matic Div., GM Corp.)*

A driver uses the throttle to control the converter's output. At idle, there is very little power transfer—not enough to move the car. This is especially true if the brakes are applied. Opening the throttle speeds up the engine and causes the converter to begin transferring enough power to move the car. At this time the converter begins multiplying the torque so there is more torque to move the car. Since the power transfer is through fluid, it will be very smooth and free from the severe shocks sometimes found with standard transmissions and clutches (Fig. 7-2).

181

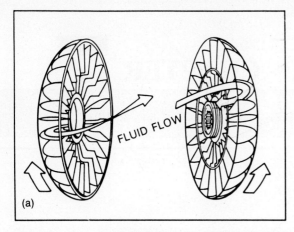

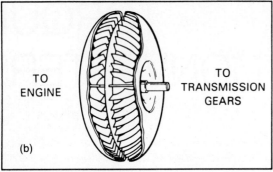

Fig. 7-2 The power flowing through a converter is transferred through the fluid flow from one set of fins to the other. *(Copyrighted material reprinted with permission from Hydra-matic Div., GM Corp.)*

7.2 CONSTRUCTION

Most torque converters have three major portions—the *pump* or *impeller*, the *turbine*, and the *stator*—and each is made up of a set of fins that act against the fluid (Fig. 7-3). In this text, we will use the term *impeller* instead of *pump* to reduce any confusion with the transmission pump.

The impeller is the converter's input. The fins that make up the pumping portion are attached to the rear, transmission end, of the converter housing or cover (Fig. 7-4). This assembly is bolted to the *flexplate*, which connects to the crankshaft. The flexplate and converter replace the flywheel used with standard transmissions. Whenever the engine runs, the flexplate and converter spin with the crankshaft (Fig. 7-5). The flexplate is flexible enough to allow the front of the converter to move forward or backward if the converter expands or contracts slightly from heat or pressure. The pilot (used to center the converter to the crankshaft) of the converter slides in the crankshaft recess at this time.

The turbine is the converter's output member. The center hub of the turbine is splined to the transmission input shaft. The turbine is positioned in the front, engine end, of the converter housing so the turbine fins face the impeller fins.

The stator is the reaction member of the torque converter. The stator assembly is about one-half the diameter of the impeller or turbine. Remember that for every action, there is an equal and opposite reaction. A converter cannot multiply torque without this reaction. The outer edge of the stator fins usually forms the inner edge of the three-piece fluid *guide ring* that is also part of the impeller and turbine fins (Fig. 7-6). The stator fins are mounted on a one-way clutch that is attached to the reaction shaft splines, which extend from the front of the transmission (Fig. 7-7). The one-way clutch allows the stator to rotate in a clockwise direction but blocks counterclockwise rotation (Fig. 7-8).

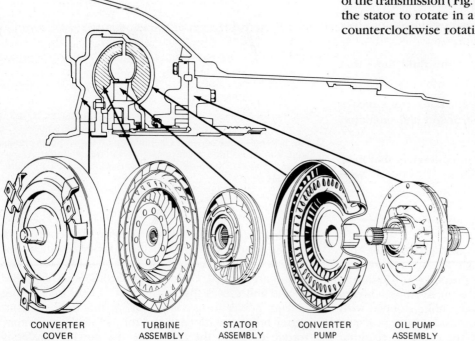

Fig. 7-3 The major parts of a converter are the converter cover, turbine assembly, stator assembly, and converter pump/impeller as shown here. *(Copyrighted material reprinted with permission from Hydra-matic Div., GM Corp.)*

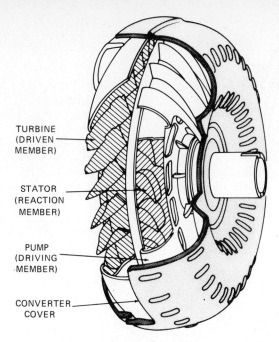

Fig. 7-4 A cutaway view of an assembled torque converter shows how the impeller fins are secured in the cover and how close the various parts are. *(Copyrighted material reprinted with permission from Hydra-matic Div., GM Corp.)*

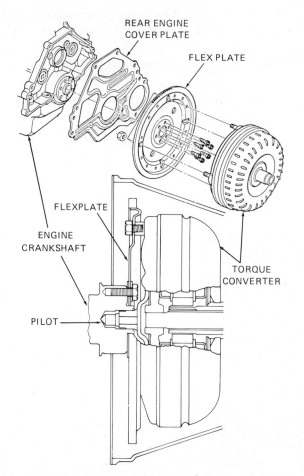

Fig. 7-5 The flexplate is used to connect the converter to the crankshaft. The converter is centered to the crankshaft as the snug fit of the converter pilot fits into the recess in the crankshaft. *(Courtesy of Ford Motor Company.)*

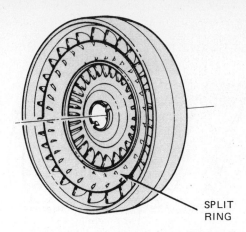

Fig. 7-6 The split guide ring directs the fluid into a smooth, turbulence-free flow. *(Courtesy of Ford Motor Company.)*

In the late 1940s and 1950s, some transmissions used four- and five-element torque converters. The fourth or fifth part or element was either a second turbine, stator, or impeller assembly. In a few cases, the five elements were three separate turbines along with one impeller and stator. The major reason for the additional elements was to obtain a torque converter with a wider range, a lower reduction ratio with the ability to change ratios inside the converter.

Some torque converters used with smaller engines are air cooled. They have a shroud with fins attached to the rear of the converter cover (Fig. 7-9). When the engine runs, air is pumped through the shroud to remove converter heat. Air-cooled converters do not use a cooler in the engine radiator.

7.3 OPERATION

A torque converter is a hydrodynamic unit. It transfers power through the dynamic motion of the fluid while most other hydraulic units transfer power through the static pressure of the fluid. When the engine runs, the converter impeller acts as a centrifugal pump. Fluid is thrown from the outer edge of the fins, and because of the curved shape of the converter cover, the fluid is thrown forward into the turbine. Because the impeller is turning in a clockwise direction, the fluid also rotates in a clockwise direction as it leaves the impeller fins (Fig. 7-10). The mechanical power entering the converter is transformed into the fluid as fluid velocity or speed.

The rotating fluid tries to turn the turbine in a clockwise direction, but if the turbine is stationary or turning at a speed substantially slower than the impeller, only part of the energy leaves the fluid to drive the turbine. Most of the fluid energy is lost as the fluid bounces back away from the turbine. The fluid flows toward the center of the turbine, being driven there by the force of the continuous flow of fluid from the impeller.

As energy leaves the fluid, the flow slows down, and this easily allows the fluid to return to the center of the impeller fins where the impeller can pick it up and keep it circulating. This flow is called a *vortex flow*,

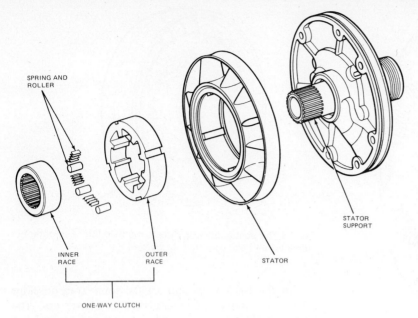

Fig. 7-7 The stator is connected to the stator support through a one-way clutch. The outer race of the clutch is secured to the stator while the inner race is splined to the stator support. *(Courtesy of Ford Motor Company.)*

that is, a continuous circulation of fluid (outward in the impeller and inward in the turbine) around the guide ring (Fig. 7-11). The guide ring directs the vortex flow so a smooth, turbulence-free flow occurs. The clockwise flow of fluid leaving the impeller is called *rotary flow*.

When the impeller is rotating substantially faster than the turbine, the fluid tends to bounce off the turbine fins and change the rotary flow to a counterclockwise direction. This flow still has quite a bit of energy left in it. It can be compared to a tennis ball thrown against a wall. The ball bounces back and travels in a different direction, but it has most of the energy of motion left in it (Fig. 7-12). A strong counterclockwise fluid flow would tend to work against impeller rotation.

7.3.1 Torque Multiplication

The stator's job is to redirect the fluid flow returning to the impeller to a clockwise direction and recover the energy remaining in the fluid. It does this by the curved shape of the stator fins. Fluid leaving the turbine in a counterclockwise direction tries to turn the stator in

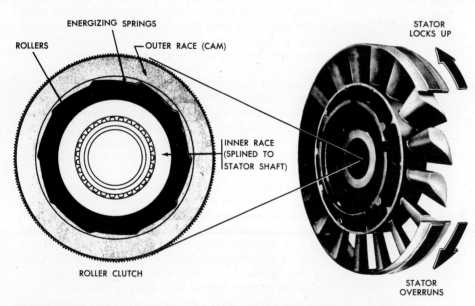

Fig. 7-8 A counterclockwise motion of the stator and outer race forces the rollers to the small area of the cam and locks the rollers. A clockwise motion of the stator moves the rollers to the wide area of the cam where they can roll freely. *(Copyrighted material reprinted with permission from Hydra-matic Div., GM Corp.)*

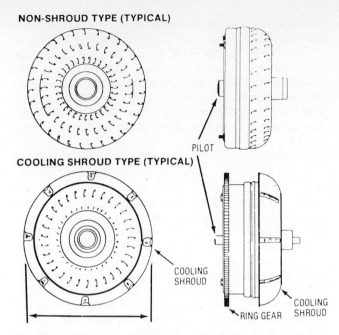

Fig. 7-9 A standard nonshrouded converter and an air-cooled one with a cooling shroud. *(Courtesy of RPM-Merit.)*

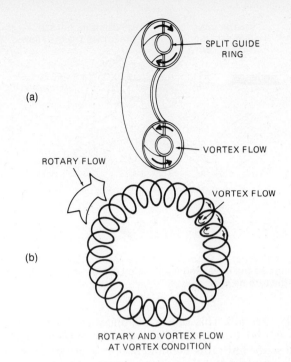

Fig. 7-11 (a) Fluid flow around the guide ring is called a vortex flow. (b) The fluid flow around the converter is called a rotary flow. *(Courtesy of Chrysler Corporation.)*

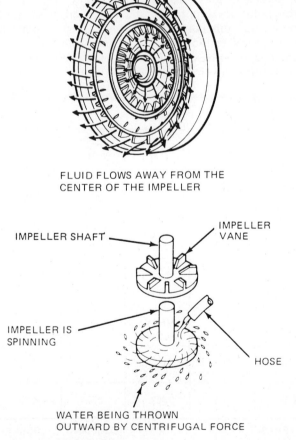

Fig. 7-10 When the engine runs, fluid in the impeller is thrown outward much like water directed onto a spinning impeller. *(Courtesy of Ford Motor Company.)*

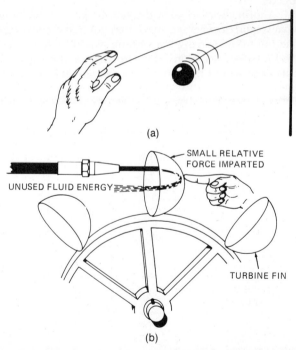

Fig. 7-12 (a) If a ball were thrown against a wall, the kinetic energy in the ball would cause it to bounce back from the wall. (b) The energy remaining in the fluid will cause it to bounce back from the turbine fins in a similar manner.

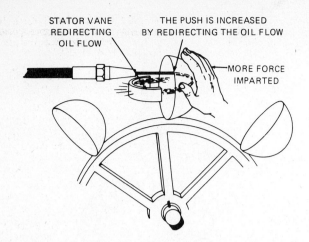

Fig. 7-13 A set of stator fins can be used to turn the fluid flow around so the remaining fluid energy can be sent back to push on the turbine again.

that direction. This causes the one-way clutch to lock up and hold the stator stationary. The smooth shape of the stator fins turns the fluid flow in a clockwise direction (Fig. 7-13). This flow leaving the stator is picked up by the impeller and speeded up for the next contact with the turbine. It is this action that returns the leftover energy in the fluid that produces the torque increase in a torque converter. Most passenger car converters increase torque by about 2:1 to 2.5:1 at stall speeds.

Another way to view the stator's job is to imagine what would occur if the stator was not in the converter. The fluid flow leaving the turbine would be in a counterclockwise direction, opposite that of the impeller. This flow would fight, not help, the impeller rotation, and this action would cause a power loss and slow down the impeller.

Torque multiplication occurs because of the stator's redirection of the fluid flow, and this occurs only when the impeller is rotating faster than the turbine. As the turbine speed increases, these flows change and the converter becomes more of a coupling—merely transferring power from the engine to the transmission (Fig. 7-14).

7.3.2 Coupling Phase

As the car starts moving, the turbine begins rotating and vice versa. As the vehicle speed increases, the turbine speed increases relative to the impeller, and as the turbine speed increases, changes occur in the fluid flow.

One change is that the fluid bounce-back from the turbine reduces, and this along with the rotation of the turbine causes the flow to leave the turbine in a clockwise direction. This moves the fluid pressure from the front to the back of the stator fins, and this in turn pushes the stator in a clockwise direction. Because of this action, the stator clutch releases; the stator freewheels along with the impeller and turbine; and there is a clean, undisturbed flow of fluid from the turbine back to the impeller (Fig. 7-15).

When the turbine reaches a speed of about 90 percent (nine-tenths of the impeller), coupling occurs. Centrifugal force acting on the fluid in the spinning turbine is great enough to stop the vortex flow (Fig 7-16). At this point, there is torque multiplication. It should be noted that this coupling speed is a relative point between the speeds of the impeller and turbine. Therefore, the coupling phase occurs at various speeds depending on throttle position and vehicle speed.

ENGINE SPEED (IMPELLER)	1800 RPM							
	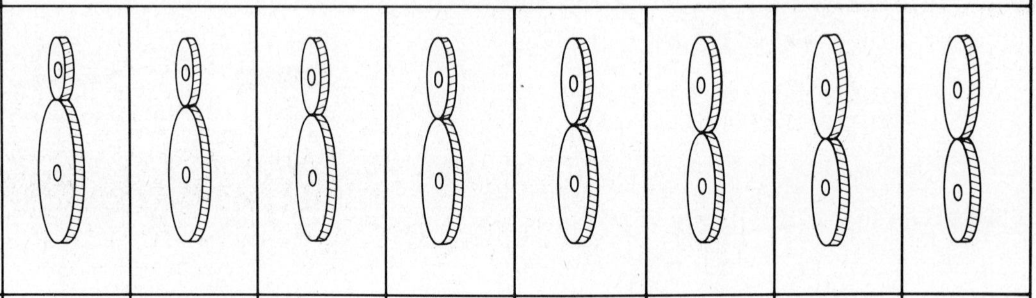							
TURBINE SHAFT SPEED	ZERO	175	350	600	950	1350	1500	18000
VORTEX FLOW	VERY HIGH	VERY HIGH	HIGH	MEDIUM	LOW	VERY LOW	ZERO	ZERO
TORQUE MULTIPLICATION	2.25:1	2.1:1	2:1	1.8:1	1.5:1	1.1:1	1.05:1	1:1
EFFICIENCY	0%	20%	40%	60%	80%	85%	90%	100%
	TORQUE MULTIPLICATION PHASE						COUPLING PHASE	LOCKUP

Fig. 7-14 Chart illustrating how fluid flow, torque multiplication, and the efficiency of a converter change as the turbine speed increases to that of the impeller.

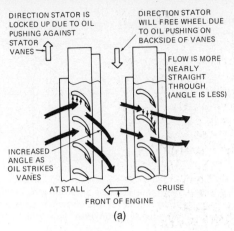

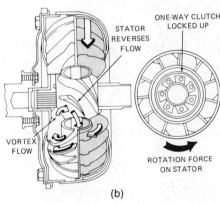

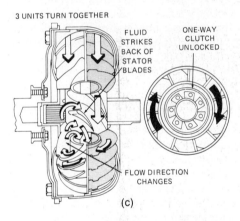

Fig. 7-15 (a) At stall, fluid flow from the turbine strikes the concave side of the stator fins while at cruise it strikes the other side. (b) At stall, the fluid force causes the stator clutch to lock while at cruise it causes it to release. *[(a) Courtesy of Chrysler Corporation; (b) Courtesy of Ford Motor Company.]*

A certain amount of slippage occurs during the coupling phase, and if power and load demands require it, the converter can return to the torque multiplication phase. In a nonlockup converter, the turbine almost never turns at the same speed as the engine and impeller, and this is commonly referred to as converter slippage.

A torque converter is a relatively simple and inexpensive device for transferring power, but it is not a completely efficient device. In fact, at stall speeds it is totally inefficient; of the power that enters it, no power leaves to the transmission. All the power entering the converter is lost to the fluid as heat. The converter's

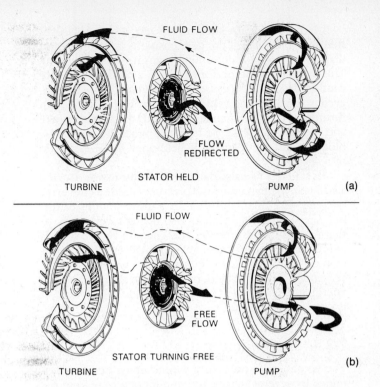

Fig. 7-16 (a) During the torque multiplication phase the fluid leaving the turbine is rather strong in a counterclockwise direction and causes the stator to lock. (b) During the coupling phase the fluid leaving the turbine rotates clockwise and flows freely through the rotating stator. *(Copyrighted material reprinted with permission from Hydra-matic Div., GM Corp.)*

efficiency steadily improves during torque multiplication and the coupling phases to about 90 to 95 percent.

7.3.3 Stall Speed

Stall condition is when the turbine is held stationary while the converter is spinning. This is done by shifting the transmission into gear and applying the brakes to hold the drive wheels stationary. Stall occurs to some degree each time a car starts moving, either forward or backward, and each time a car stops at a stop sign. *Stall speed* is the fastest rpm that an engine can reach while the turbine is held at stall (Fig. 7-17).

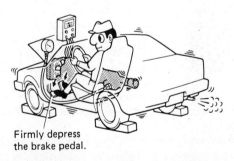

Fig. 7-17 A technician measures the stall speed while making a stall test to help locate any transmission problems. During this test, the wheels are blocked and the brakes are firmly applied while the gas pedal is depressed and a tachometer is read to determine the speed. A stall test should be carefully conducted within a time span of 5 seconds. *(Courtesy of Toyota Motor Sales, U.S.A., Inc.)*

187

As we will learn later, a *stall test*, which measures stall speed in each of the gear positions, is an important diagnostic test to determine transmission condition. This test should be performed with a certain degree of caution because it operates the car in a potentially dangerous situation; the vehicle is in gear with a wide-open throttle. It is recommended that both the parking brake and the service brake be firmly applied, the wheels be blocked, and the throttle be held open for a maximum of 5 seconds.

A stall test can also damage the transmission. During a stall test, the dynamic fluid pressures inside a converter become very high, and there is a great amount of turbulence. Fluid temperatures also become very high. At this time, all of the power the engine can produce is going into the converter, and no mechanical power is coming out. Because of the natural law of energy conservation—energy cannot be created or destroyed—the energy going into the converter must go somewhere; it is converted to heat.

7.3.4 Stall Factors

The actual stall speed of a torque converter is determined by several factors: the amount of torque from the engine, the diameter of the converter, the angle of the impeller fins, and the angle of the stator fins. For the time being, we will refer to a high stall speed as a loose converter and a low stall speed as a tight converter. It should be remembered that when a vehicle is standing still, the turbine is not rotating. When the vehicle is accelerated, the engine rpm rises fairly quickly to the torque capacity of the converter and then stabilizes to converter ratio speed above turbine rpm. As the turbine speeds up, the engine also speeds up and vice versa. A loose converter allows a higher rpm relative to the turbine rpm.

A large, strong engine has the ability to turn the impeller faster against a stalled turbine than a small or weak engine. A weaker engine is normally equipped with a looser converter because one that is too tight would not allow engine rpm to increase to the point of maximum power, and therefore, the vehicle would lose pulling power and performance. Too loose of a converter would cause the engine to operate at too high of a speed, and this would produce poor fuel economy, excessive noise, and again reduced performance. In production, the torque converter capacity—stall speed—is matched to the engine size and vehicle weight to produce the best vehicle performance and fuel economy. At the present time, one manufacturer uses eight different converters with one transmission model. This places an added burden on the technician to ensure that the correct replacement converter is used when one is replaced.

Diameter is one of the major factors in determining converter capacity or tightness. Passenger car converters vary in size from about 8 to 12 in. (20.3 to 30.5 cm) (Fig. 7-18). The rotary speed of the fluid in the impeller almost matches the speed of the converter's outer diameter—the circumference; the circumference of a circle is a product of the diameter times π (pi), which is equal to 3.1416. The circumference of an 8-in. converter

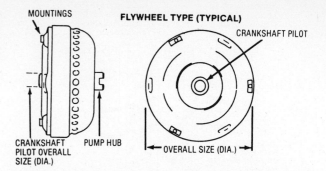

Fig. 7-18 The converter diameter helps determine the converter's stall speed. Along with the other factors shown, it is used to help select the correct replacement converter. *(Courtesy of RPM-Merit.)*

is 25.1 in. (63.7 cm), and the circumference of a 12-in. converter is 37.7 in. (95.8 cm), about 50 percent greater. From this, it should be fairly easy to see how a larger-diameter converter can put about 50 percent more speed and energy into the fluid. A 9-in. converter will have a stall speed that is about 30 percent faster than a 10-in. converter in the same vehicle. It should be noted, though, that the increased fluid speed and energy requires more power from the engine. The larger the converter, the tighter it becomes, with a higher torque capacity.

The impeller fins can be angled forward, straight, or backward relative to the direction of converter rotation (Fig. 7-19). A forward or positive angle produces greater fluid speed and therefore a tighter converter with a lower stall speed. A rearward or negative angle increases the stall speed. Changing the impeller fin angle is an easy method for a manufacturer to change converter torque capacity.

Another relatively simple method of changing converter capacity is the angle of the stator fins. The greater the curvature, the more the fluid will return to the direction of impeller rotation, and the stall speed and converter looseness will be increased. Straighter fins produce a tighter converter.

In the late 1960s, some THM 400's and THM 300's were produced with variable-pitch stator fins; these transmissions were called the Super 400 and the Super 300. A hydraulic piston was built into the stator assembly, and fluid to this position was controlled by an electric solenoid-controlled valve (Fig. 7-20). At very low vehicle speeds or high throttle openings, the stator fins are moved to a high angle for maximum stall speed, torque multiplication, and acceleration. Otherwise, the stator fins are kept at a low angle position for maximum fuel economy and quiet, smooth vehicle operation (Fig. 7-21). The switch to control stator fin position was mounted in the speedometer or throttle linkage depending on the particular car.

7.4 LOCKUP CONVERTERS

Since the late 1970s, most torque converters include an internal clutch that can be applied to eliminate the slippage that normally occurs during the coupling phase. Eliminating this slippage makes a significant improve-

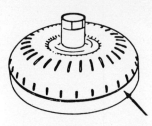 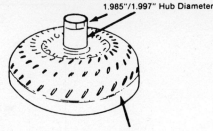 

Fig. 7-19 On some converters it is possible to see the impeller fin angle from the outside. Note the different fin angles and the stall ratings. *(Courtesy of RPM-Merit.)*

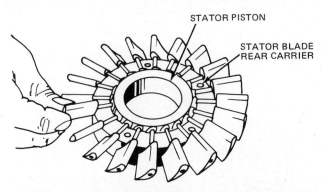

Fig. 7-20 Variable-pitch stator. The fin angle can be changed while the converter is operating. This changes the stall speed and the torque multiplication ratio. *(Courtesy of Chevrolet Div., GM Corp.)*

ment in fuel mileage. There are two methods used to apply the converter clutch: hydraulic or centrifugal force. When the converter clutch applies, the converter locks up, and vehicle operation is very similar to a car with a standard transmission and clutch.

The converter clutch is a large disc (about the same diameter as the turbine) that has friction material and a damper assembly attached to it (Fig. 7-22). The clutch disc is splined to the turbine so that it can drive the turbine mechanically when the friction material is forced against the torque converter cover. As the friction material makes contact, a mechanical link is formed that brings the turbine up to the same speed as the engine. The clutch releases when the hydraulic controls or mechanical action changes to switch the converter back to pure hydraulic operation.

A hydraulically applied clutch applies as the fluid flow through the converter is reversed by a valve, as described in Chapter 5. Fluid normally circulates from the front to the rear of a lockup converter. When the torque converter clutch control valve moves, the flow

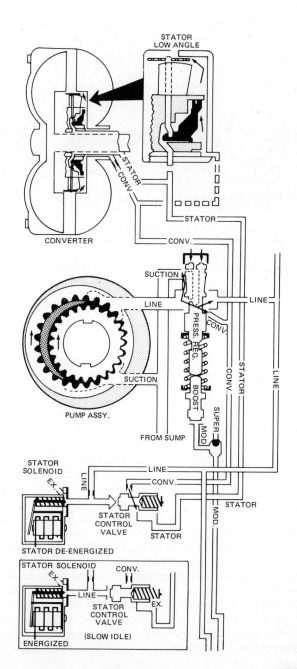

Fig. 7-21 A THM 400. The stator solenoid controls the fluid pressure at the stator piston. The solenoid is deenergized so there is no pressure at the piston and the stator is at low angle. Energizing the solenoid moves the stator control valve, which sends pressure to the stator piston and moves the stator fins to a high angle. *(Copyrighted material reprinted with permission from Hydra-matic Div., GM Corp.)*

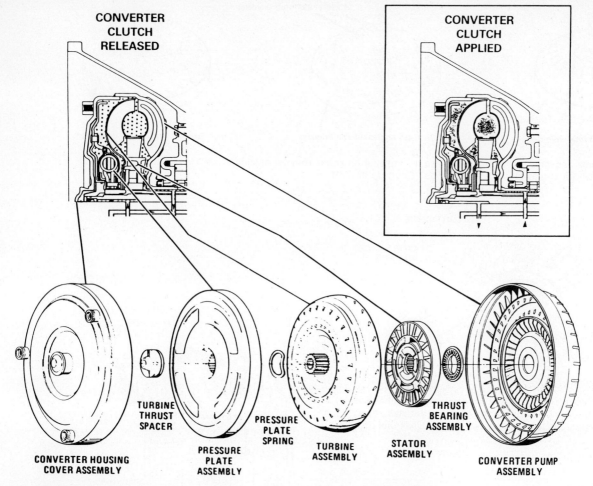

Fig. 7-22 Cutaway and exploded views of lockup converter. Note how the pressure plate assembly is splined onto the turbine. *(Copyrighted material reprinted with permission from Hydra-matic Div., GM Corp.)*

tries to reverse, and this forces the clutch disc and friction material against the front of the converter cover (Fig. 7-23). Application of the clutch then blocks the fluid flow through the converter, but with the clutch applied, the converter does not generate any heat. As the clutch is being applied, the fluid that must leave the area in front of the clutch disc must be forced out; the exiting fluid acts somewhat like an accumulator and softens the clutch application. The friction material is a large ring of paper friction material. It can be secured to the front of the clutch disc, to the inside of the converter cover, or left free between the two.

The friction material in a centrifugally applied clutch is attached to the outer edge of a group of shoes arranged around the outer diameter of the clutch disc (Fig. 7-24). Each shoe is spring loaded to move toward the center of the disc, which releases the clutch. As the turbine reaches a certain rpm, centrifugal force pushes the shoes and friction material outward, and this makes the contact to produce a mechanical transfer of torque from the converter cover to the clutch disc. Ford Motor Company calls this converter a centrifugal lockup converter (CLC) (Fig. 7-25).

Both types of clutch discs include a damper assembly that directs the power flow through a group of coil springs (Fig. 7-26). In some converters, the damper springs are grouped at the center; in others they are grouped around the outer edge. These springs are used to dampen torsional vibrations from the engine. All automotive engines produce torsional vibration at some operating speed. Torsional vibrations are small speed increases and slowdowns during portions of the crankshaft revolution, and these vibrations can produce gear noise in the transmission and drive train as well as a noticeable shaking in the car (Fig. 7-27).

7.4.1 Viscous Converter Clutch

Some cars are equipped with a viscous converter clutch. The operation is the same as the hydraulically applied converter clutch except that a viscous element is built into the clutch disc. The viscous element, much like that in a fan clutch, includes a rotor, a body, the clutch cover, and silicone fluid. The rotor and the body have a series of concentric, intermeshed ridges that provide two large

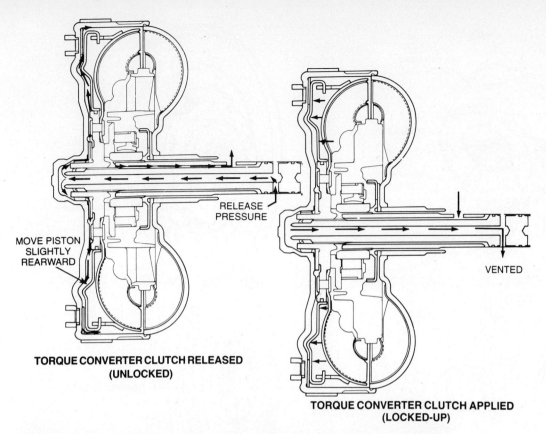

Fig. 7-23 When the converter clutch is unlocked, fluid flows down the turbine shaft, past the pressure plate assembly, through the converter, and out past the outside of the turbine shaft. The clutch locks when the flow is reversed, and fluid pressure forces the pressure plate assembly against the front of the converter. *(Courtesy of Chrysler Corporation.)*

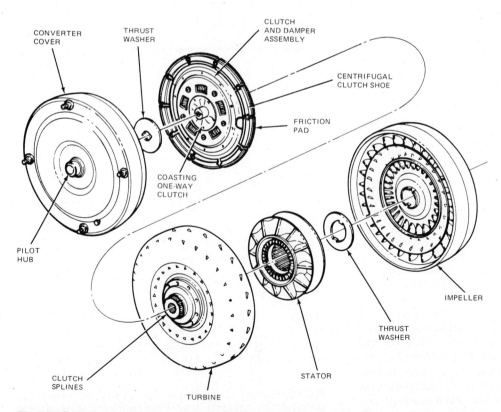

Fig. 7-24 Centrifugal lockup converter (CLC). Note the clutch shoes on the outer diameter of the clutch and the damper assembly. *(Courtesy of Ford Motor Company.)*

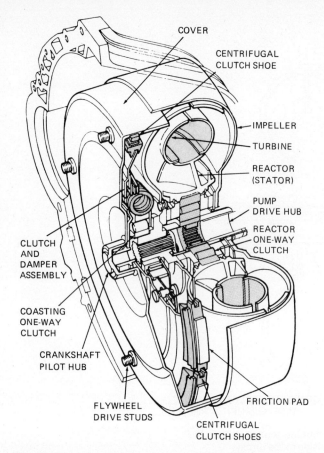

Fig. 7-25 When the turbine is rotating fast enough, the centrifugal clutch shoes move outward and rub on the inside of the converter cover to lock the clutch. *(Courtesy of Ford Motor Company.)*

surface areas (Fig. 7-28). The viscous drag of the thick silicone fluid transfers power from the body to the rotor through the viscous drag of the fluid. The viscous clutch transfers power in a very smooth manner with no engagement shock. There is a very minor amount of slippage in the viscous unit, about 40 rpm at 60 mph (96.5 km/h).

7.5 PLANETARY GEAR TORQUE CONVERTERS

Some of the torque converters used in the Ford ATX transaxle and some aftermarket (non-OEM) transmission component manufacturers have a converter that includes a planetary gearset. The Dual-Path transmission used in Buick Specials in the early 1960s also used a planetary gear converter with an internal multiple-disc clutch assembly. Ford refers to this converter as a split-torque converter because there are two ways that torque can pass through it. In the ATX split-torque converter, a sun gear is splined to the center of the turbine and to the input shaft; a ring gear is splined to the converter cover and damper assembly; and a planetary carrier is splined to the intermediate shaft that enters the transmission (Fig. 7-29). In second and third gear, the

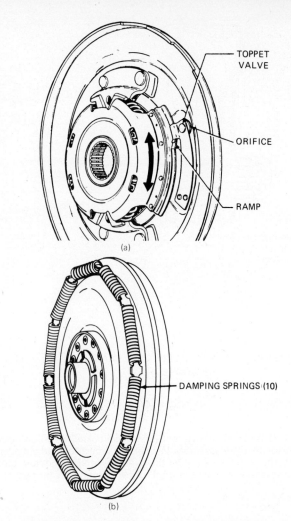

Fig. 7-26 (a) A clutch pressure plate showing the damper springs at the center and (b) one with the damping springs at the outer diameter. Note that the poppet valve, orifice, and ramp are only used with diesel engines. *[(a) Copyrighted material reprinted with permission from Hydra-matic Div., GM Corp. (b) Courtesy of Chrysler Corporation.]*

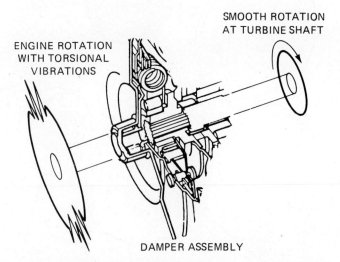

Fig. 7-27 In a four-cylinder engine, there are two power impulses per revolution, and these impulses cause the crankshaft to speed up momentarily. During this time the damper springs compress to absorb the speed fluctuation.

192

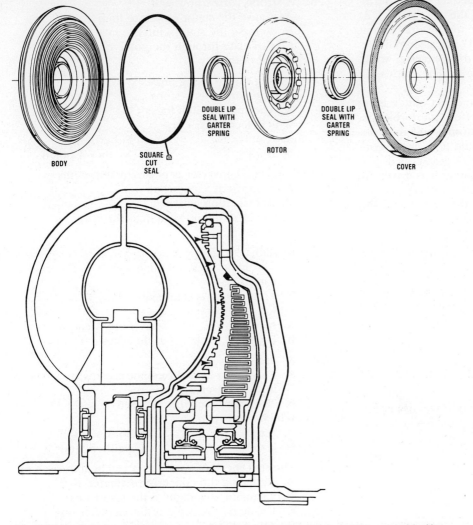

Fig. 7-28 Exploded and cutaway views of viscous converter clutch damper assembly. Note that the silicone fluid between the body and rotor is separate from the automatic transmission fluid in the converter. *(Copyrighted material reprinted with permission from Hydramatic Div., GM Corp.)*

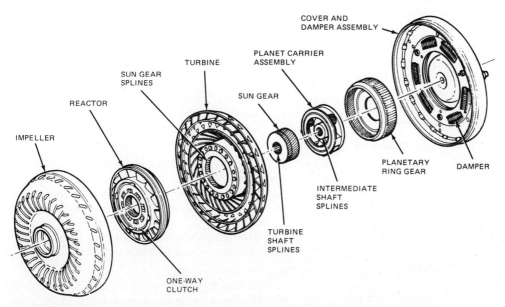

Fig. 7-29 An ATX converter with an internal planetary gearset. Note that the ring gear is splined into the damper; the sun gear is splined into the turbine; and the carrier is splined to the turbine shaft. *(Courtesy of Ford Motor Company.)*

193

intermediate clutch inside the transaxle locks the intermediate shaft to the ring gear in the main gearset. At this time, a major percentage of the torque (62 percent in second and 93 percent in third) is transferred mechanically through the gearset in the converter while the hydraulic power transfer to the turbine carries the remaining torque.

In second gear in the ATX, the direct clutch is released so no power is transferred from the turbine to the main gearset. However, the turbine and the converter ring gear is driven by the fluid flow in the converter. The ring gear in the converter gearset is driven by the converter cover while the sun gear is driven by the turbine. This normally causes the converter planetary gearset to lock up and drive the carrier and the intermediate shaft at a 1:1 ratio. Under load, the turbine can slow down, and this causes the planet gears to walk around the slower sun gear. This produces a reduction ratio in the converter gearset.

In third gear, both the intermediate clutch and the direct clutch in the main gearset are applied to place the main gearset in a 1:1 ratio. This in turn locks up the converter gearset, which stops any gear action in the converter.

REVIEW QUESTIONS

The following questions are provided so you can check the facts you have just learned. Select the response that best completes each statement.

1. A torque converter is used to
   A. Serve as a clutch
   B. Increase torque to the transmission
   Which is correct?
   a. A only   c. Both A and B
   b. B only   d. Neither A nor B

2. The input member of the converter is
   A. the turbine   C. The stator
   B. The impeller   D. All of these

3. The flexplate
   A. Connects the converter to the crankshaft
   B. Can bend sideways to align the converter with the transmission
   Which is correct?
   a. A only   c. Both A and B
   b. B only   d. Neither A nor B

4. Fluid motion inside a converter is controlled by the
   A. Guide ring
   B. Stator fins
   Which is correct?
   a. A only   c. Both A and B
   b. B only   d. Neither A nor B

5. The fluid flow
   A. From the impeller through the turbine and stator is called rotary flow
   B. Around the circumference of the converter is called vortex flow
   Which is correct?
   a. A only   c. Both A and B
   b. B only   d. Neither A nor B

6. A strong vortex flow will
   A. Cause the stator clutch to lock
   B. Release the stator clutch
   C. Circulate through the guide ring
   D. All of these

7. During the coupling phase
   A. The vortex flow stops
   B. The stator clutch locks
   Which is correct?
   a. A only   c. Either A or B
   b. B only   d. Neither A nor B

8. A torque converter is most efficient while
   A. Stopped at a stop sign
   B. Accelerating hard
   C. Accelerating under a light throttle
   D. Cruising

9. At stall,
   A. The impeller is spinning as fast as the engine can drive it
   B. The turbine is stationary
   Which is correct?
   a. A only   c. Both A and B
   b. B only   d. Neither A nor B

10. At stall
    A. The power from the engine is converted to heat in the converter
    B. The stator clutch is released
    Which is correct?
    a. A only   c. Both A and B
    b. B only   d. Neither A nor B

11. The stall speed of a converter can be increased by
    A. Reducing the converter diameter
    B. Changing the angle of the stator fins
    C. Changing the angle of the impeller fins
    D. All of these

12. A lockup converter includes a
    A. Lockup clutch plate and lining
    B. Damper used during lockup operation
    Which is correct?
    a. A only   c. Both A and B
    b. B only   d. Neither A nor B

13. In a CLC converter, lockup occurs when the
    A. Fluid flow in the converter is reversed
    B. Converter switch valve is moved by the lockup solenoid
    C. Turbine speed reaches a certain rpm
    D. Converter speed reaches a certain rpm

14. In most lockup converters, lockup occurs when the
    A. Fluid flow in the converter is reversed
    B. Converter switch valve is moved by the lockup solenoid
    C. Turbine speed reaches a certain rpm
    D. Converter speed reaches a certain rpm

15. In a lockup converter,
    A. The clutch plate is splined to the turbine shaft
    B. The clutch lining transfers power through a mechanical connection during lockup
    Which is correct?
    a. A only   c. Either A or B
    b. B only   d. Neither A nor B

CHAPTER 8

TRANSMISSION DESCRIPTION

OBJECTIVES

After completing this chapter, you should:

- Be familiar with the various transmission families that are used in domestic cars.
- Be able to identify which transmission is used in a particular car.
- Have a working knowledge of how a particular transmission operates.

8.1 INTRODUCTION

In the past 20 years, about 60 types of transmissions and transaxles have been used in domestic and imported cars. It is difficult to understand each one of these with the information provided in a book of this type. We will not try to describe each of them. We will concentrate only on those units installed in domestic cars.

The transmission technician's problem of having a working knowledge of so many transmissions is made easier by the fact that many transmissions operate just like other types. We can place the more recent domestic transmissions and transaxles into 16 categories as described in Chapter 3, and some of these categories are similar except for the use of a multiple-disc clutch instead of a band. Most transmissions used in import cars fit into these categories. It should be remembered that these transmissions and transaxles use planetary gearsets; they use multiple-disc clutches, bands, or one-way clutches to control the power flow; they apply and release the application devices by hydraulic pressure; and they use a torque converter.

In the following sections, we will briefly describe the major transmission and transaxle types used in domestic cars. The units will be described by major manufacturer and cross-referenced by the gear train types described earlier. The gear train categories are based on the type of gear train and the type of friction devices used.

Transmissions of a particular type also have many very small differences. In any single year, there may be as many as 50 versions of a particular transmission depending on the combinations of various parts. These differences might be the number of lined and unlined plates in each clutch pack, the ratio of the band levers, the size of the band servos, the programming of the valve body, the strength of the spring in an accumulator, the stall speed of the torque converter, the location of the mounting pad bolts, the shape of the torque converter housing, or the exact number of teeth on the gears in the gearsets. Each transmission must be matched to the weight of the car, the strength of the engine, emission and fuel mileage requirements, and sometimes where the car will be sold.

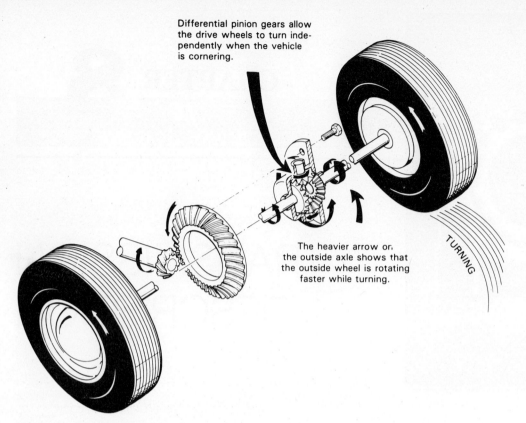

Fig. 8-1 The differential gears allow the outer drive wheel to turn faster when a car goes around a corner. *(Copyrighted material reprinted with permission from Hydra-matic Div., GM Corp.)*

A listing and brief description of the automatic transmissions used in domestic cars of the three major manufacturers is given in Appendices 2 to 7.

8.1.1 Transaxles

As mentioned earlier, a transaxle is a combination of a transmission, either standard or automatic, and a drive axle. The drive axle portion is usually a set of reduction gears and the differential. To refresh your memory, the differential is made up of two pairs of gears, the differential pinion gears mounted on the differential pinion shaft and the axle or side gears, which are attached to the axle shafts (Fig. 8-1). In a front-wheel drive (FWD) car, the axle shafts are often called drive shafts and occasionally called half-shafts. The purpose of the differential is to allow the drive wheels to be driven at different speeds while the car is going around a corner.

There are essentially three styles of final drive and differential combinations used in transaxles. Chrysler Corporation transaxles and the Ford ATX use helical gears and an intermediate/idler gear or shaft to transfer power from the transmission planetary gear assembly to the differential case (Fig. 8-2). The helical gears provide the necessary reduction ratio and are relatively efficient, quiet, and inexpensive.

General Motors transaxles and the Ford AXOD use a planetary gearset mounted next to the transmission gearset for the final drive reduction gears (Fig. 8-3). The output of the transmission is a sun gear that is the input

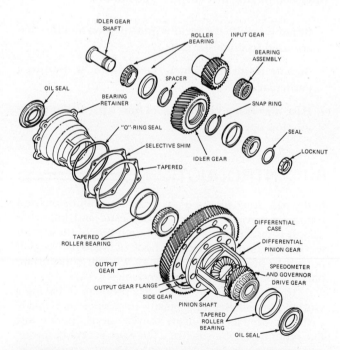

Fig. 8-2 Ford ATX uses a set of helical final drive gears to transfer power from the transmission gears to the differential and drive shafts. *(Courtesy of Ford Motor Company.)*

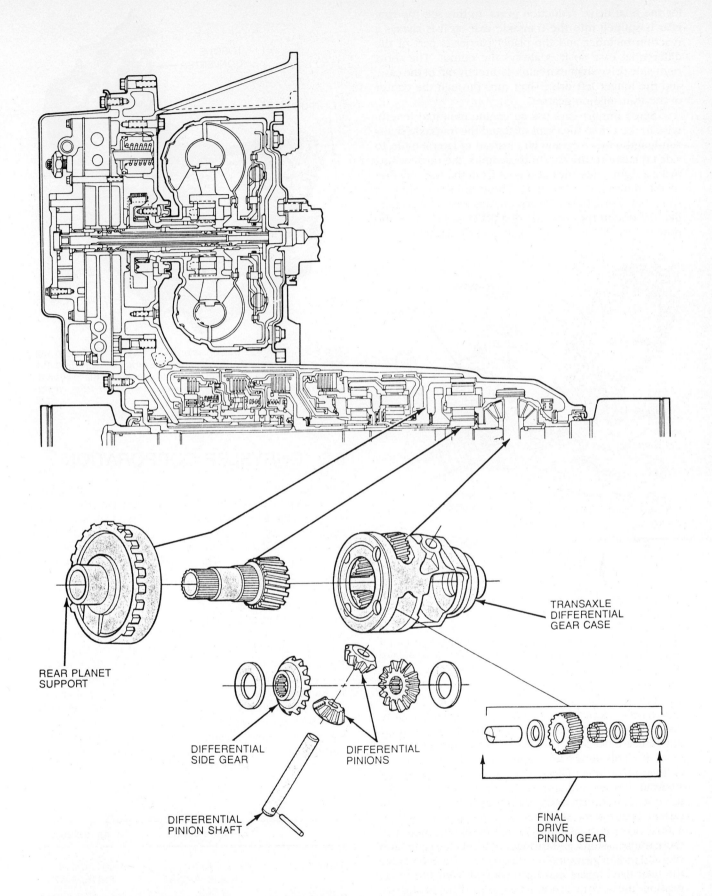

Fig. 8-3 Ford AXOD transaxle uses a planetary gearset for the final drive gears. Note that the differential case is a part of the carrier. *(Courtesy of Ford Motor Company.)*

for the final drive reduction gears. In this set, the ring gear is splined into the transaxle case so it is always a reaction member and the planet carrier is part of the differential case so it is always the output. The short right-side drive shaft exits almost directly out of the case, and the longer left drive shaft runs through the center of the transmission gearset.

Some import cars use an engine mounted lengthwise to the car so the crankshaft and the transaxle shafts run lengthwise (fore and aft) instead of lateral (side to side) relative to the car. In these units, the power must make a right-angle turn as it goes from the transmission to the differential (Fig. 8-4). These units use either a spiral bevel or a hypoid gearset in this area. The gearset, like the one in the rear axle of a RWD car, changes the power flow 90° and provides the gear reduction.

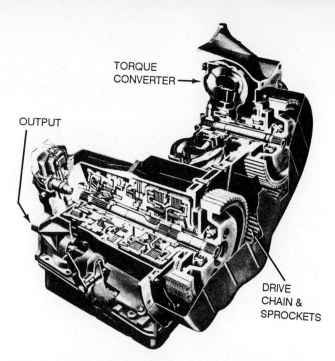

Fig. 8-5 A THM 425 transmission. The engine attaches to the torque converter with the front of the crankshaft toward the upper left, and the front-axle assembly attaches to the lower left. *(Copyrighted material reprinted with permission from Hydramatic Div., GM Corp.)*

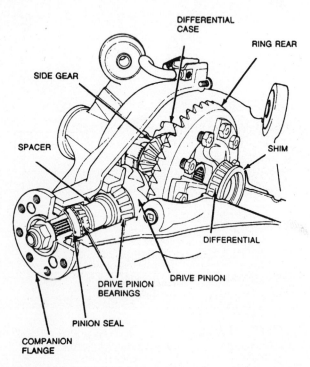

Fig. 8-4 In a rear-wheel drive (RWD) car, the final drive gears and differential are part of the rear-axle assembly. *(Courtesy of Ford Motor Company.)*

Another approach to providing FWD was used in the THM 325, THM 325-4L, and THM 425 transmissions in Cadillac Eldorados, Oldsmobile Toronados, and later Buick Rivieras until it was replaced by the THM 440-T4 in the mid-1980s. These transmissions are mounted alongside the engine and use a pair of sprockets and a drive chain to transfer power from the torque converter turbine shaft to the transmission input shaft (Fig. 8-5). A final drive gear assembly is bolted to the output of these transmissions, and this unit contains the reduction ring and pinion gears and the differential (Fig. 8-6). Since the final drive gears can be separated from the transmission, these units are not transaxles. This final drive gearset uses a hypoid gearset to turn the power flow 90°.

8.2 CHRYSLER CORPORATION

Chrysler Corporation produces three slightly different three-speed transmissions for passenger car use, called *Torqueflites*. The term *Loadflite* is used for transmissions intended for pickups, four-wheel drives (4WDs), and light-truck use. A four-speed Loadflite, the A-500, is also being produced. The major difference is the size and number of speeds: the two three-speed units are the larger A-727 and the smaller A-904 (Fig. 8-7). At one time, the A-727 was called the Torqueflite 8, and the A-904 was called the Torqueflite 6. The A-998 and A-999 Torqueflites are variations of the A-904. The operation and internal parts are almost identical except for size. If the A-904 and A-727 are placed side by side, it is easy to see the size difference, but normally, the simplest way to identify an A-727 from an A-904 is by the shape of the oil pan (Fig. 8-8). The four-speed A-500 is essentially an A-999 with overdrive gear train added to it (Fig. 8-9). The various Chrysler transmissions and transaxles and the categories (as described in Chapter 3) in which they are placed are as follows:

	Transmissions	Transaxles
Type 1	A-727, A-904, A-998, A-999	A-404, A-413, A-415, A-470
Type 5	A-500	—
Type 15	—	A-604

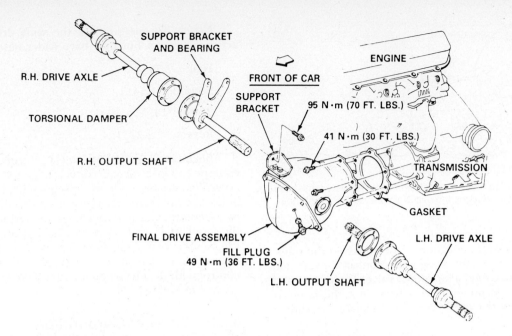

Fig. 8-6 A FWD front axle/final drive assembly of the type used with a THM 425, THM 325, or THM 325-4L transmission. *(Courtesy of Oldsmobile Div., GM Corp.)*

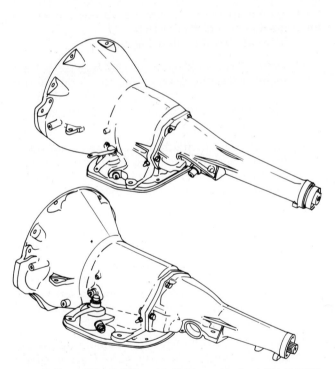

Fig. 8-7 Torqueflite A-727 and A-904 transmissions. The A-727 is the larger one; also note the more noticeable indentation around the area between the main case and the torque converter housing of the A-904.

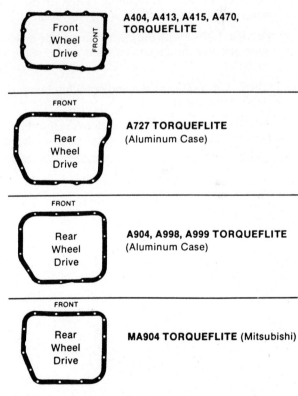

Fig. 8-8 The quickest way to identify an A-727 from an A-904 is the shape of the oil pan.

Each Torqueflite has a serial number located on the pad just above the oil pan flange on the left side of the transmission case (Fig. 8-10). This serial number provides positive identification of a transmission; it identifies the date of manufacture and the car and engine size in which it was used.

8.2.1 Chrysler Corporation Transmissions: Type 1

Both the A-727 and the A-904 are three-speed Simpson gear train transmissions and can be placed into our Type 1 category (Fig. 8-11). As indicated, the clutch termi-

199

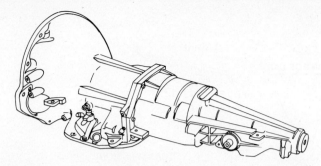

Fig. 8-9 Torqueflite A-500 transmission. Note the very large case extension, which contains the overdrive gearset for fourth gear.

nology refers to the location of the clutch in the transmission. All Torqueflites use a mechanical throttle and detent system called a kickdown valve. The terms used by Chrysler Corporation, which are not the same as those used by the majority of the industry, are shown in Fig. 8-12.

The gear ratios for early A-727 and A-904 transmissions are:

- First: 2.45:1 with the front clutch as the driving member and the overrunning clutch or the low and reverse band as the reaction member.
- Second: 1.45:1 with the front clutch as the driving member and the kickdown band as the reaction member.
- Third: 1:1 with the front and rear clutches as the driving members.
- Reverse: 2.22:1 with the rear clutch as the driving member and the low and reverse band as the reaction member.

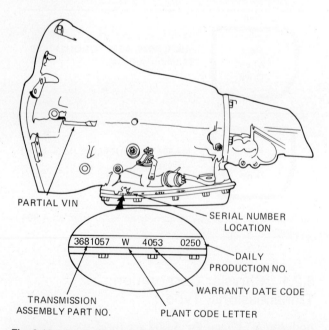

Fig. 8-10 A Torqueflite transmission can be identified by the serial number located on the oil pan flange pad on the left side. *(Courtesy of Chrysler Corporation.)*

Newer Torqueflites use the same driving and reaction members, but they have wider ratios:

First: 2.74:1.

Second: 1.54:1.

Third: :1.

Reverse: 2.22:1.

Beginning in 1978, most Torqueflites were equipped with a torque converter clutch with lockup controlled by the hydraulic programming in the valve body. Lockup converter transmissions can be identified by a smooth extension in front of the turbine splines on the input shaft, and the area where this extension fits can be seen by looking down into the converter. Recent transmissions have added electronic converter clutch control with the electronic control linked to the engine control module (ECM).

8.2.2 Chrysler Corporation Transmission: Type 15

The A-500 is a Simpson gear train four-speed unit. The gear train is a combination of a Type 1 (A-999) transmission plus an overdrive unit. The overdrive gearset uses two multiple-disc clutches plus a one-way clutch for control members (Fig. 8-13).

The A-500 has the following ratios:

- First: 2.74:1 with the front clutch as the driving member and the overrunning clutch or low and reverse band as the reaction member in the main gearset and the direct clutch as the driving member in the overdrive gearset.
- Second: 1.54:1 with the front clutch as the driving member and the kickdown band as the reaction member in the main gearset and the direct clutch as the driving member in the overdrive gearset.
- Third: 1:1 with the front and rear clutches as driving members in the main gearset and the direct clutch as the driving member in the overdrive gearset.
- Fourth: 0.69:1 with the front and rear clutches as the driving members in the main gearset and the overdrive clutch as the reaction member in the overdrive gearset.
- Reverse: 2.22:1 with the rear clutch as the driving member and the low and reverse band as the reaction member in the main gearset and the direct clutch as the driving member in the overdrive gearset.

The A-500 is unique in several ways: the three-speed section is a slightly modified A-999 transmission with the overdrive section built into the extension housing; the direct and overdrive clutches share the same piston; and the overdrive shift is controlled by the ECM (Fig. 8-14). The direct clutch is applied by a strong spring [800-pound (363.6-kg) force]; the same spring releases the overdrive clutch. When the overdrive apply piston strokes, it releases the direct clutch and then ap-

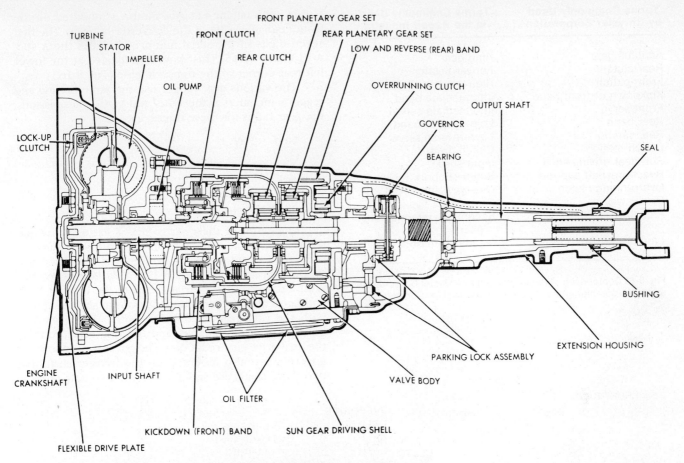

Fig. 8-11 Cutaway view of a Torqueflite A-904 transmission with clutch and band application for the various speeds. *(Courtesy of Chrysler Corporation.)*

plies the overdrive clutch. The overdrive one-way clutch transfers power from the sun gear to the ring gear during this period of time.

Fluid flow to the overdrive clutch piston is controlled by an electric solenoid. When the engine controller sees the right combination of engine temperature, engine speed, manifold vacuum, throttle position, and vehicle speed, it completes a ground circuit to activate the shift solenoid. Operation in fourth gear can be locked out by a switch on the instrument panel. Torque converter lockup is also controlled by a solenoid.

8.2.3 Chrysler Corporation Transaxles: Type 1

There are four Type 1 FWD transaxles produced by Chrysler, and they all use essentially the same gear train as the A-904. The basic difference between the four transaxles is the shape of the torque converter housing (Fig. 8-15). The A-404 is used with 1.7-L (104-CI) engines, the A-413 with 2.2-L (134-CI) engines, the A-415 with 1.4- or 1.6-L (85- or 98-CI) engines, and the A-470 with 2.6-L (159-CI) engines. The A-470 is easiest to identify

201

Terms Commonly Used by Chrysler Corporation	Terms Commonly Used in the Repair Industry
Annulus gear	Ring gear
Rear clutch	Forward clutch
Front clutch	High–reverse clutch
Kickdown or front band	Intermediate band
Front servo	Intermediate servo
Rear band	Low–reverse band
Rear servo	Low–reverse servo
Clutch retainer	Clutch drum
Sun gear driving shell	Input shell
Reaction shaft support	Stator support
Overrunning clutch	One-way clutch
Breakaway	First, or low, gear
Kickdown	Second, or intermediate, gear
Line pressure	Mainline pressure

Fig. 8-12 Comparison of the terms used by Chrysler Corporation and those commonly used by others in the transmission industry.

because the engine's starter motor is mounted on the front (radiator) side of the converter housing. The first seven digits of the serial number identifies the A-404, A-413, and A-415. This should be located at the lower left rear corner of the transaxle case (Fig. 8-16).

The various gear ranges use the same driving and reaction members as the A-727 and A-904 transmissions; the gear ratios for these transaxles are:

First: 2.69 : 1.

Second: 1.55 : 1.

Third: 1 : 1.

Reverse: 2.1 : 1.

Operation of these transaxles is easiest to understand if we think of them as shortened A-904 transmissions with an output shaft gear mounted where the governor is in an A-904 (Fig. 8-17). This gear is meshed with a transfer shaft gear attached to the transfer shaft, and a gear at the other end of the transfer shaft is meshed with the ring gear on the differential. The gear ratio between the transfer shaft and the ring gear produces the final drive gear reduction. The governor support and governor are also mounted on the transfer shaft. The two drive shafts are splined into the two side gears in

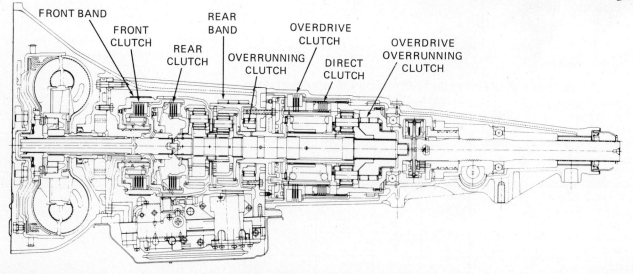

Elements in Use in Each Gear

	Clutches					Bands	
	Front	Rear	Overrunning	Direct	Overdrive	Front	Rear
Reverse	X			X			X
Neutral				X			
Drive							
First		X	X	X			
Second		X		X		X	
Third	X	X		X			
Fourth	X	X			X		

Fig. 8-13 Chrysler A-500 transmission. Note that the gear train to the left of the main case–extension housing juncture is similar to an A-904 and that the gears to the right of this juncture are the overdrive gearset. *(Courtesy of Chrysler Corporation.)*

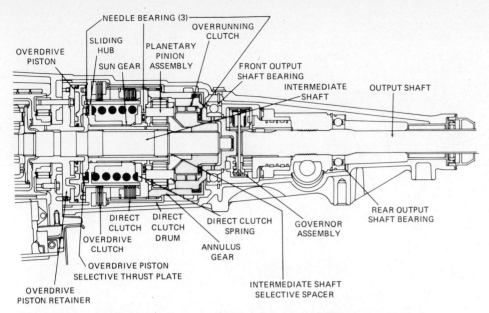

Fig. 8-14 Overdrive section of an A-500. Note that the overdrive piston will apply the overdrive clutch and release the direct clutch as it strokes to the right. *(Courtesy of Chrysler Corporation.)*

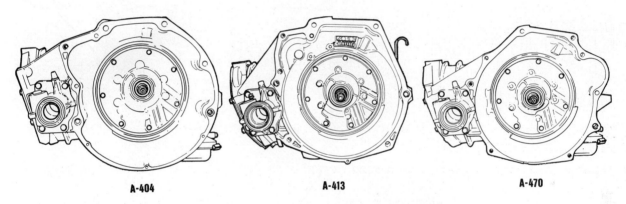

Fig. 8-15 A Chrysler three-speed transaxle is made in several different versions as determined by the bell/converter housing configuration. *(Courtesy of Chrysler Corporation.)*

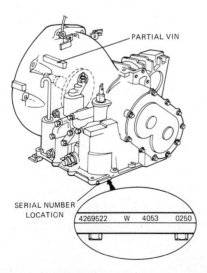

Fig. 8-16 A Chrysler three-speed transaxle can be identified by the serial number stamped onto a pad at the front of the left side of the oil pan flange. *(Courtesy of Chrysler Corporation.)*

the differential. Rotation of the output shaft causes the transfer shaft to rotate in the opposite direction, and this in turn causes the differential assembly to rotate in the same direction as the transaxle output shaft.

In these transaxles, the differential and the transmission portion share the same fluid and oil sump. They also share the same service procedures. Major service requirements of the transaxles include the output shaft bearings, transfer shaft and gears, and the differential assembly. These operations will be included in Chapters 11 and 12.

8.2.4 Chrysler Corporation Transaxle: Type 15

The A-604, which is also called the Ultradrive Electronic Automatic Transaxle (UEAT), is a unique four-speed transaxle that uses a gear train similar to the THM 700 (Fig. 8-18). This unit uses five multiple-disc clutches with no bands or one-way clutches. Three of the clutches are used as driving members, and two of them

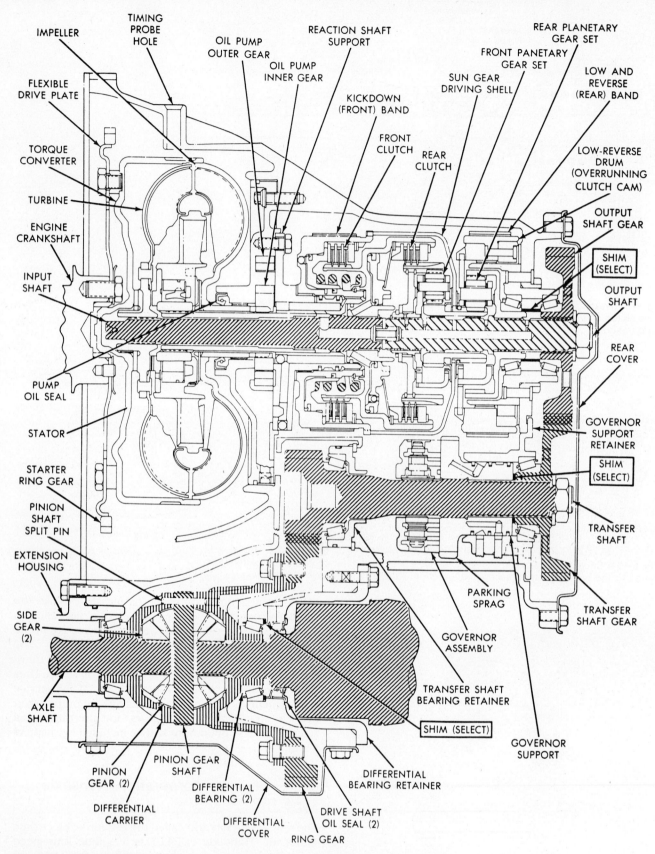

Fig. 8-17 Cutaway view of a Chrysler three-speed transaxle with a clutch and band application chart for the various speeds. *(Courtesy of Chrysler Corporation.)*

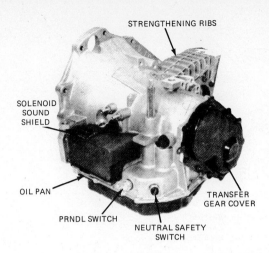

Fig. 8-18 Chrysler A-604 transaxle. This is a four-speed transaxle with electronic shift control. *(Courtesy of Chrysler Corporation.)*

are used as reaction members (Fig. 8-19). The use of electronic controls and no one-way clutches produces a driving feel more like that of a self-shifting standard transmission, especially on deceleration.

The gear ratios for the A-604 are:

- First: 2.84:1 with the underdrive clutch as the driving member and the low–reverse clutch as the reaction member.

- Second: 1.57:1 with the underdrive clutch as the driving member and the 2–4 clutch as the reaction member.

- Third: 1:1 with the underdrive and overdrive clutches as driving members.

- Fourth: 0.69:1 with the overdrive clutch as the driving member and the 2–4 clutch as the reaction member.

- Reverse: 2.21:1 with the reverse clutch as the driving member and the low–reverse clutch as the reaction member.

Besides using only multiple-disc clutches as control elements, the A-604 is unique in other ways. It is an electronically controlled transaxle that uses four solenoids along with the manual valve to apply and release the five gear train clutches and the lockup clutch in the torque converter. There are no shift valves. The electronic controller for the shift solenoids receives signals from a turbine speed sensor and an output shaft speed sensor in addition to the various engine sensors (Fig. 8-20). During a shift, the speed of these two units along with signals from the fluid pressure switches are monitored, and the electronic transmission control unit (ETCU) cycles the shift solenoid to adjust the clutch application rate for optimum shift quality. Thus, the shift quality is electronically programmed to changes in engine speed and load as well as to friction element torque. The solenoid assembly, the two-speed sensors, and the PRNDL (park/reverse/neutral/drive/low) and neutral safety switches are mounted on the outside of the transaxle case.

Like the other Chrysler Corporation transaxles, an intermediate transfer shaft and gears are used to connect the output shaft of the gear train to the ring gear and differential.

8.3 FORD MOTOR COMPANY

The Ford Motor Company has traditionally used the terms *Cruise-O-Matic* for its automatic transmissions. Recently it has used nine basic transmission types of which five can be placed into our Types 1 and 2 categories. One of them, the FMX, is a three-speed Ravigneaux gear train (Type 10), but the clutch and band application for the different gears is the same as the Type 1 units. The various Ford transmissions and the categories in which they are placed are shown in Table 8–1.

Ford Motor Company transmissions can be identified by the shape of the oil pan and whether the case is made from aluminum or cast iron or the torque converter housing is removable (Fig. 8-21). A much more exact identification can be made from the transmission code that is part of the car's identification tag (Fig. 8-22). There is also an identification ID tag secured to the outside of the transmission case by one of the servo cover bolts (Fig. 8-23).

8.3.1 Ford Motor Company Transmission: Type 1

The C3, C4, and C5 are three-speed Simpson gear train transmissions, and they all use a forward clutch, a high–reverse clutch, an intermediate band, and a low–reverse band (Fig. 8-24). All use an aluminum transmission case with a removable converter housing, a mechanical throttle linkage, and a vacuum modulator. Figure 8-25 gives some terminology differences between Ford and others in the industry. The C3 was introduced in 1974 for use in compact-size cars; the C4 was introduced in 1964 for small- and medium-size cars; and the C5 replaced the C4 in the early 1980s. All three transmissions operate in the same manner.

The gear ratios for the C3 are:

- First: 2.47:1 with the forward clutch as the driving member and the one-way clutch or low–reverse band as the reaction member.

- Second: 1.47:1 with the forward clutch as the driving member and the intermediate band as the reaction member.

- Third: 1:1 with the forward clutch and the high–reverse clutch as driving members.

- Reverse: 2.11:1 with the high–reverse clutch as the driving member and the low–reverse band as the reaction member.

The various gears for the C4 and C5 use the same driving and reaction members as the C3; the gear ratios are:

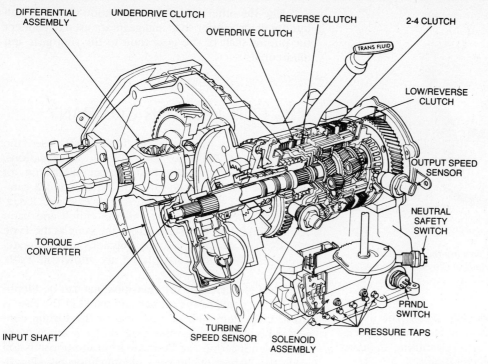

ELEMENTS IN USE SHIFT LEVER POSITION	CLUTCHES					
	Start Safety	Park Sprag	Underdrive	Overdrive	Reverse	2/4	Low/Reverse
P — PARK	X	X					X
R — REVERSE					X		X
N — NEUTRAL	X						X
OD — OVERDRIVE							
First			X				X
Second			X			X	
Direct			X	X			
Overdrive				X		X	
D — DRIVE*							
First			X				X
Second			X			X	
Direct			X	X			
L — LOW*							
First			X				X
Second			X			X	
Direct			X	X			

*Vehicle upshift and downshift speeds are increased when in these selector positions.

Fig. 8-19 Cutaway view of an A-604 transaxle with a clutch application chart for the various speeds. (Courtesy of Chrysler Corporation.)

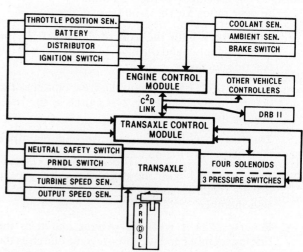

Fig. 8-20 The transaxle control module controls the four solenoids, which in turn control the shifts. The control module receives input signals from four transaxle switches, three engine sensors plus the battery, and other sensors acting through the ECM as well as any other vehicle controllers. (Courtesy of Chrysler Corporation.)

TABLE 8–1 Ford Transmission Categories

	Transmissions	Transaxles
Type 1	C3, C4, C5	—
Type 2	C6, Jatco	—
Type 6	A4LD	—
Type 10	FMX	—
Type 11	—	ATX
Type 12	AOD	—
Type 16	—	AXOD
	ZF-4HP-22	
	E4OD	

206

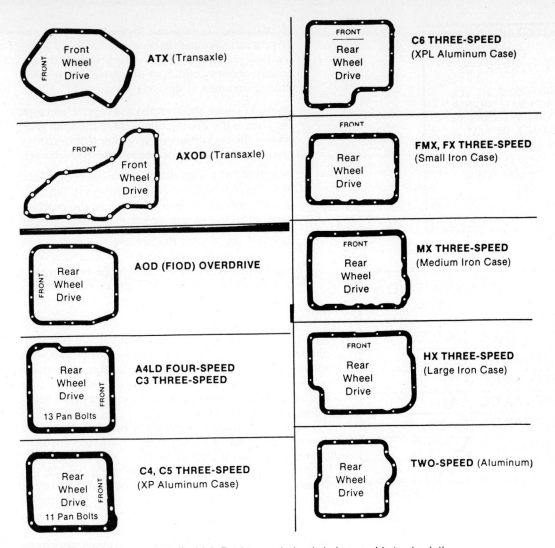

Fig. 8-21 The quickest way to tell which Ford transmission is being used is to check the shape of the oil pan.

First: 2.46:1.

Second: 1.46:1.

Third: 1:1.

Reverse: 2.19:1.

The C5 uses a CLC torque converter with lockup controlled by centrifugal force acting on the individual clutch shoes. A lockup converter can be identified by a flat, cylindrical area where the clutch shoes operate.

8.3.2 Ford Motor Company Transmission: Type 2

The C6 and Jatco transmissions use a multiple-disc clutch for the low–reverse reaction member instead of a band (Fig. 8-26). If we compare the Type 1 and Type 2 gear train schematics (Figs. 3-7 and 3-12), we see that they are identical except that the low–reverse band has been replaced by a low–reverse clutch.

The C6 is a heavy-duty transmission introduced in 1966 for use in full-size cars, pickups, and light trucks (Fig. 8-27). The Jatco transmission is manufactured by the Japanese Automatic Transmission Company and is designed for use in small-size cars, mostly of import manufacture. Like the C3, C4, and C5 transmissions, the C6 and Jatco use both a mechanical throttle linkage and a vacuum modulator. The Jatco has a removable torque converter housing while the C6, like most other large transmissions, has the converter housing cast as part of the transmission case.

The gear ratios for the C6 and Jatco transmissions are:

- First: 2.46:1 with the forward clutch as the driving member and the one-way clutch or low–reverse clutch as the reaction member.

- Second: 1.46:1 with the forward clutch as the driving member and the intermediate band as the reaction member.

- Third: 1:1 with the forward and high–reverse clutches as driving members.

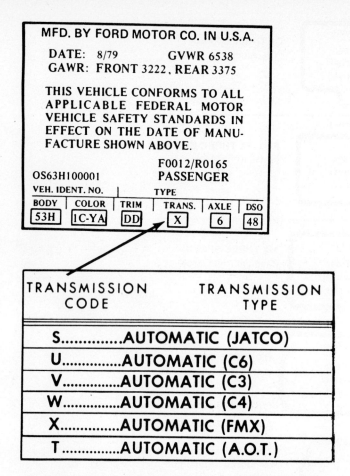

Fig. 8-22 (a) Vehicle identification label showing the transmission code X. (b) Transmission code X indicates FMX. The vehicle identification tag is on the left-side door post, and the key to the transmission code is in Ford service manuals. *(Courtesy of Ford Motor Company.)*

- Second: 1.47:1 with the forward clutch as the driving member and the intermediate band as the reaction member in the main gearset and the overdrive one-way clutch or overdrive clutch as the driving member in the overdrive set.

- Third: 1:1 with the forward and reverse and high clutches as the driving members in the main gearset and the overdrive one-way clutch or overdrive clutch as the driving member in the overdrive set.

- Fourth: 0.75:1 with the forward and reverse and high clutches as the driving members in the main gearset and the overdrive band as the reaction member in the overdrive set.

- Reverse: 2.1:1 with the reverse and high clutch as the driving member and the low and reverse band as the reaction member in the main gearset and the overdrive clutch as the driving member in the overdrive set.

The A4LD transmission uses an aluminum case with a removable torque converter housing. It uses a vacuum modulator and mechanical throttle linkage. This transmission uses a torque converter clutch with an electric solenoid so there is an electrical connector in the transmission case.

The Ford Motor Company has recently developed another four-speed overdrive transmission that is similar to the A4LD; this is the E4OD. The E4OD has been developed for pickups and light trucks to replace the C6. It uses an overdrive unit that is arranged at the front of the main gearset much like the A4LD. The major difference is the size of the gearset and that a Type 3 main gearset is used. The E4OD uses full electronic shift controls with two shift control solenoids, a converter clutch control solenoid, and a coast control solenoid that are operated by signals from the ECM.

- Reverse: 2.18:1 with the high–reverse clutch as the driving member and the low–reverse band as the reaction member.

8.3.3 Ford Motor Company Transmission: Type 6

The A4LD (Automatic Four-speed Light Duty) transmission is a Simpson gear train four-speed unit. The gear train is a combination of a Type 1 three-speed unit plus an overdrive unit that uses one multiple-disc clutch, one band, and a one-way clutch for control members (Fig. 8-28).

The gear ratios for the A4LD transmission are:

- First: 2.47:1 with the forward clutch as the driving member and the one-way clutch or low–reverse band as the reaction member in the main gearset and the overdrive one-way clutch or overdrive clutch as the driving member in the overdrive gearset.

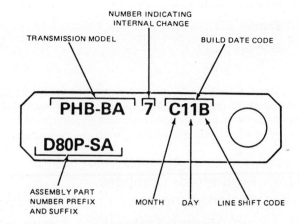

Fig. 8-23 A Ford transmission has an identification tag attached to one of the exterior bolts. The transmission model indicates the year, car line, and engine size for this transmission. *(Courtesy of Ford Motor Company.)*

208

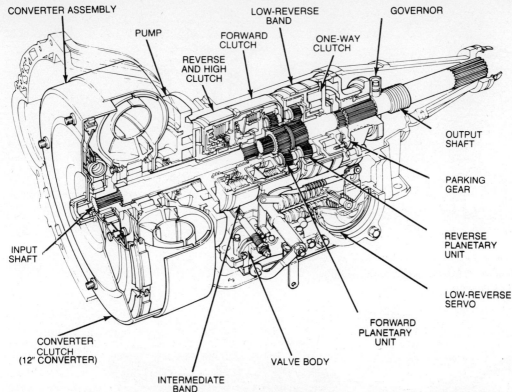

GEAR	RATIO	HOLDING MEMBERS	FORWARD PLANETARY GEAR SET			REVERSE PLANETARY GEAR SET		
			Driven	Held	Output	Driven	Held	Output
Manual Low 1	2.46	Forward Clutch Low-and-Reverse Band	Ring Gear	*Carrier	Sun Gear	Sun Gear	Carrier	Ring Gear
D Low	2.46	Forward Clutch One-way Clutch	Ring Gear	*Carrier	Sun Gear	Sun Gear	Carrier	Ring Gear
*The carrier is actually turning with the output shaft, but at a slower speed than the input.								
D Second or 2	1.46	Forward Clutch Intermediate Band	Ring Gear	Sun Gear	Carrier	EFFECTIVELY IN NEUTRAL		
High	1.00	Forward Clutch Reverse and High Clutch	Sun Gear Ring Gear	None	Carrier	TURNS AS A UNIT		
Reverse	2.19	Reverse and High Clutch Low-and-Reverse Band	EFFECTIVELY IN NEUTRAL			Sun Gear	Carrier	Ring Gear

Fig. 8-24 Cutaway view of a C5 transmission with a clutch and band application chart for the various speeds. The C3 and C4 are almost identical with the exception of the lockup torque converter. *(Courtesy of Ford Motor Company.)*

Terms Commonly Used by Ford Motor Company	Terms Commonly Used in the Repair Industry
Internal gear	Ring gear
Input shaft	Turbine shaft
Impeller	Torque converter pump
Reactor	Stator
Clutch cylinder	Clutch drum
1-way roller clutch	One-way clutch
Main control pressure	Mainline pressure
T.V. pressure	Throttle pressure

Fig. 8-25 Comparison of terms used by Ford Motor Company with those used by others in the transmission repair industry.

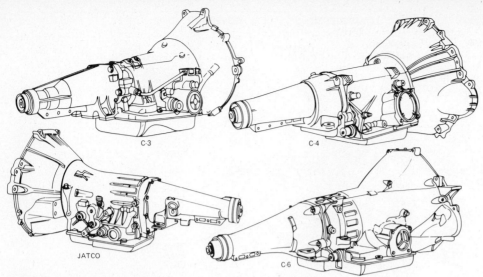

Fig. 8-26 External view of C3, C4, Jatco, and C6 transmissions. *(Courtesy of Ford Motor Company.)*

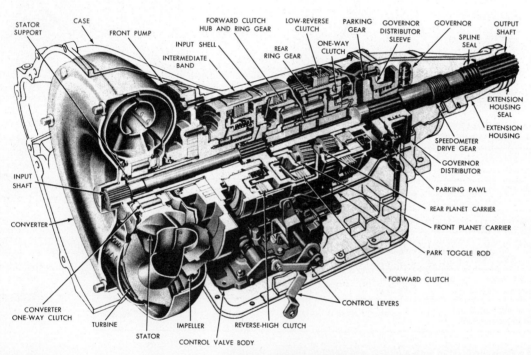

Clutch and Band Application Chart

Gear	Clutch				Intermediate Band
	Forward	Reverse–High	Low–Reverse	One-Way	
Reverse		X	X		
Neutral					
Drive					
First	X			X	
Second	X				X
Third	X	X			
Second					
First	X			X	
Second	X				X
Manual low	X		X	X	

Fig. 8-27 Cutaway view of a C6 transmission with a clutch and band application chart for the various speeds. *(Courtesy of Ford Motor Company.)*

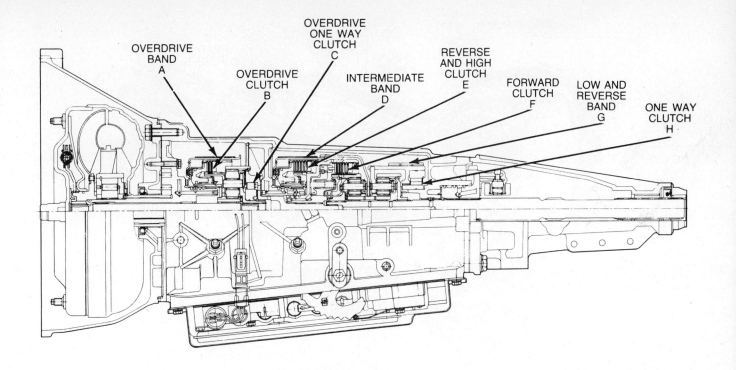

GEAR	OVER-DRIVE BAND A	OVER-DRIVE CLUTCH B	OVER-DRIVE ONE WAY CLUTCH C	INTERMEDIATE BAND D	REVERSE AND HIGH CLUTCH E	FORWARD CLUTCH F	LOW AND REVERSE BAND G	ONE WAY CLUTCH H	GEAR RATIO
1 — MANUAL FIRST GEAR (LOW)		APPLIED	HOLDING			APPLIED	APPLIED	HOLDING	2.47:1
2 — MANUAL SECOND GEAR		APPLIED	HOLDING	APPLIED		APPLIED			1.47:1
D — DRIVE AUTO. — 1ST. GEAR		APPLIED	HOLDING			APPLIED		HOLDING	2.47:1
ⓄD — O/D AUTO. — 1ST. GEAR			HOLDING			APPLIED		HOLDING	2.47:1
D — DRIVE AUTO. — 2ND. GEAR		APPLIED	HOLDING	APPLIED		APPLIED			1.47:1
ⓄD — O/D AUTO. — 2ND. GEAR			HOLDING	APPLIED		APPLIED			1.47:1
D — DRIVE AUTO. — 3RD. GEAR		APPLIED	HOLDING		APPLIED	APPLIED			1.0:1
ⓄD — O/D AUTO. — 3RD. GEAR			HOLDING		APPLIED	APPLIED			1.0:1
ⓄD — OVERDRIVE AUTOMATIC FOURTH GEAR	APPLIED				APPLIED	APPLIED			0.75:1
REVERSE		APPLIED	HOLDING		APPLIED		APPLIED		2.1:1

Fig. 8-28 Partially cutaway view of an A4LD transmission with a clutch and band application chart for the various speeds. *(Courtesy of Ford Motor Company.)*

8.3.4 Ford Motor Company Transmission: Type 10

The FMX transmission uses a gear train of the Ravigneaux design. The clutch and gear train arrangement is shown in Fig. 8-29. In the United States, the FMX is often referred to as a Borg-Warner gear train. This gear train is also used in the Fordomatic, Mercomatic, Flashomatic, and Ultramatic of the past and Borg-Warner Types 20 and 65 units of some import cars.

Besides the shape of the oil pan, it is fairly easy to identify an FMX because this transmission uses a cast-iron main case with an aluminum extension and torque converter housings. A vacuum modulator with a mechanical detent rod and lever are used on this unit. This is a medium-duty transmission.

The gear ratios for the FMX transmission (depending on the unit) are:

- First: 2.37:1, 2.4:1, or 2.46:1 with the forward clutch as the driving member and the one-way clutch or low–reverse band as the reaction member.

- Second: 1.46:1, 1.47:1, or 1.48:1 with the forward clutch as the driving member and the intermediate band as the reaction member.

- Third: 1:1 with the forward and high–reverse clutches as the driving members.

- Reverse: 2:1 with the high–reverse clutch as the driving member and the low–reverse band as the reaction member.

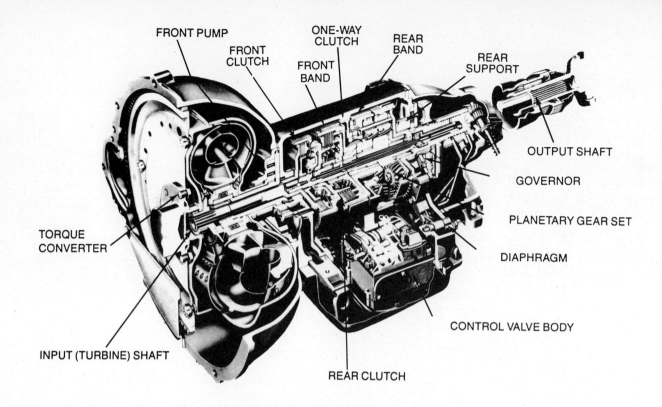

Gear	Clutches			Bands	
	Front	Rear	One-Way	Front	Rear
Reverse		X			X
Neutral					
Drive					
First	X		X		
Second	X			X	
Third	X	X			
Second					
First	X		X		
Second	X			X	
Manual low	X				X

Clutch and Band Application Chart

Fig. 8-29 Cutaway view of a FMX transmission with a clutch and band application chart for the various speeds. *(Courtesy of Ford Motor Company.)*

8.3.5 Ford Motor Company Transaxle: Type 11

The ATX (Automatic TransaXle) is a three-speed Ravigneaux gear train unit (Fig. 8-30). This transaxle uses a final drive input gear as the output of the transmission planetary gearset, and this gear meshes with an idler gear, which in turn drives the ring gear on the differential. The transmission and the final drive share the same fluid and oil sump. The governor is driven from the differential.

The gear ratios for the ATX are:

- First: 2.79:1 with the one-way clutch as the driving member and the band as the reaction member.
- Second: 1.61:1 with the intermediate clutch as the driving member and the band as the reaction member.
- Third: 1:1 with the intermediate and direct clutches as driving members.
- Reverse: 1.97:1 with the direct clutch as the driving member and the reverse clutch as the reaction member.

This transaxle is a little unusual in that the valve body is mounted under a cover at the top of the unit (Fig. 8-31). Also, the pump is mounted at the far end of the case, away from the engine, and the pump is driven by a small-diameter pump shaft from the front of the

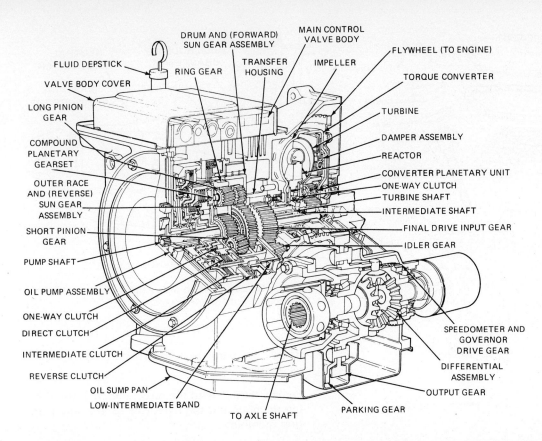

Range and Gear		Band	Direct Clutch	Intermediate Clutch	Reverse Clutch	One-Way Clutch
Park						Applied
Reverse			Applied		Applied	Applied
Neutral						Applied
D	1st	Applied				Applied
	2nd	Applied		Applied		
	3rd		Applied	Applied		
2	1st	Applied				Applied
	2nd	Applied		Applied		
1	1st	Applied	Applied			Applied

Fig. 8-30 Cutaway view of an ATX transaxle with a clutch and band application chart for the various speeds. *(Courtesy of Ford Motor Company.)*

torque converter. This unit uses a mechanical throttle linkage. The band is adjustable by changing the servo rod.

At this time, ATX transaxles used in the Taurus and Sable use a centrifugal lockup torque converter (CLC); the units used in the Escort and Lynx use a split-torque converter with an internal planetary gearset; and the units used in the Tempo and Topaz use a fluid link converter (FLC). In the CLC and FLC, both the turbine shaft and the intermediate shaft are driven by the turbine. In units using the split-torque converter, there are two input shafts, the turbine shaft connects to the direct clutch and the one-way clutch, and the intermediate shaft connects the planet carrier in the converter to the intermediate clutch (Fig. 8-32).

8.3.6 Ford Motor Company Transmission: Type 12

The Ford AOD (Automatic OverDrive) is a "one-of-a-kind" transmission. It is the only four-speed transmission based on the Ravigneaux gear train (Fig. 8-33).

The gear ratios for the AOD transmission are:

- First: 2.4:1 with the forward clutch as the driving member and the one-way clutch or low–reverse band as the reaction member.

- Second: 1.47:1 with the forward clutch as the driving member and the intermediate friction and roller clutches as the reaction member.

213

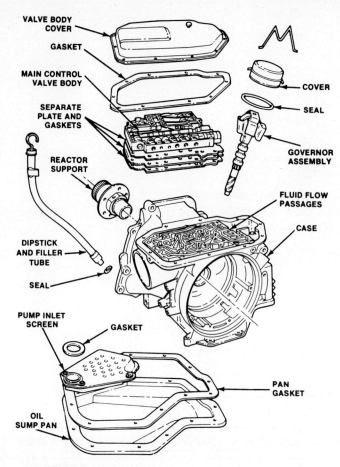

Fig. 8-31 Exploded view of an ATX. Note that the valve body cover is at the top and the oil sump pan is at the bottom. *(Courtesy of Ford Motor Company.)*

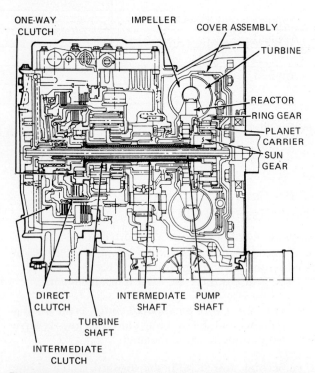

Fig. 8-32 An ATX using a split-torque converter. Note the pump shaft, intermediate shaft, and turbine shaft that enter the transaxle from the converter. *(Courtesy of Ford Motor Company.)*

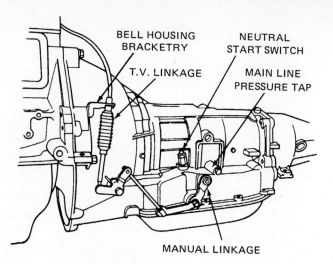

Fig. 8-33 An AOD transmission uses a mechanical throttle linkage. *(Courtesy of Ford Motor Company.)*

- Third: 1:1 with the forward and direct clutches as driving members.

- Fourth: 0.67:1 with the direct clutch as the driving member and the overdrive band as the reaction member.

- Reverse: 2:1 with the reverse clutch as the driving member and the low–reverse band as the reaction member.

The AOD uses a case with the converter housing cast as part of the case. It does not use a vacuum modulator and has only mechanical throttle linkage. It is also slightly unusual in that it uses two input shafts, one connected to the turbine in the converter and the other, the direct-drive shaft, connected to a damper assembly at the front of the converter. The direct clutch, connected to the direct-drive shaft, provides a purely mechanical input into the gear train in third and fourth gears (Fig. 8-34). This feature eliminates the need for a torque converter clutch.

8.3.7 Ford Motor Company Transaxle: Type 16

The AXOD (Automatic TransaXle OverDrive) is a four-speed overdrive unit that uses a gear train that is very similar to the General Motors THM 440-T4 transaxle (Fig. 8-35). The application devices in these two are different so they are classed as two slightly different types.

The AXOD, like the General Motors transaxles, uses a drive chain and pair of sprockets to transfer power from the torque converter turbine shaft to the planetary gearset. Also like the THM 125 and THM 440-T4, the valve body and pump are mounted at the left side of the transaxle under a separate cover. The final drive in the AXOD, like that in the THM 125 and THM 440-T4, is a planetary gearset that is in line with the transmission gear sets (Fig. 8-36).

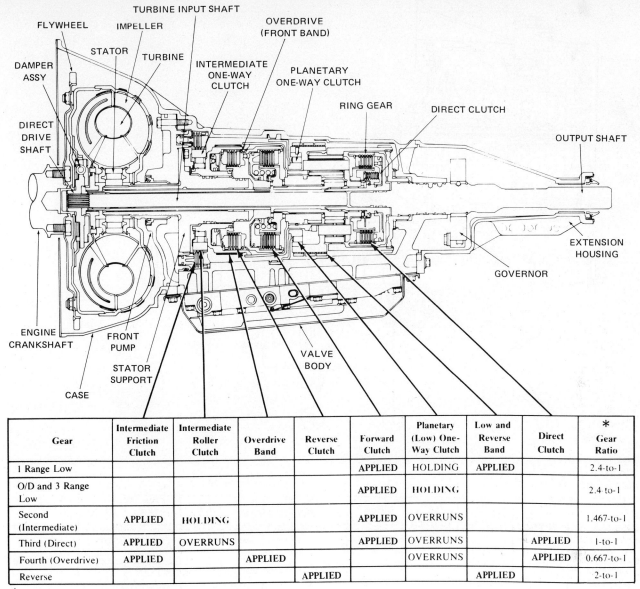

Gear	Intermediate Friction Clutch	Intermediate Roller Clutch	Overdrive Band	Reverse Clutch	Forward Clutch	Planetary (Low) One-Way Clutch	Low and Reverse Band	Direct Clutch	*Gear Ratio
1 Range Low					APPLIED	HOLDING	APPLIED		2.4-to-1
O/D and 3 Range Low					APPLIED	HOLDING			2.4-to-1
Second (Intermediate)	APPLIED	HOLDING			APPLIED	OVERRUNS			1.467-to-1
Third (Direct)	APPLIED	OVERRUNS			APPLIED	OVERRUNS		APPLIED	1-to-1
Fourth (Overdrive)	APPLIED		APPLIED			OVERRUNS		APPLIED	0.667-to-1
Reverse				APPLIED			APPLIED		2-to-1

*Not including torque converter reduction in 1st, Second and Reverse.

Fig. 8-34 Cutaway view of an AOD transmission with a clutch and band application chart for the various speeds. *(Courtesy of Ford Motor Company.)*

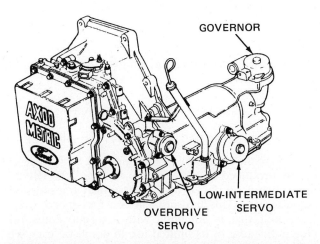

Fig. 8-35 An AXOD transaxle. *(Courtesy of Ford Motor Company.)*

The gear ratios for the AXOD are:

- First: 2.77:1 with the forward clutch driving and the low one-way clutch or the low–intermediate band as the reaction member.

- Second: 1.54:1 with the intermediate clutch as the driving member and the low–intermediate band as the reaction member.

- Third: 1:1 with the forward, intermediate, direct and direct one-way clutches as the driving members.

- Fourth: 0.69:1 with the intermediate clutch as the driving member and the overdrive band as the reaction member.

215

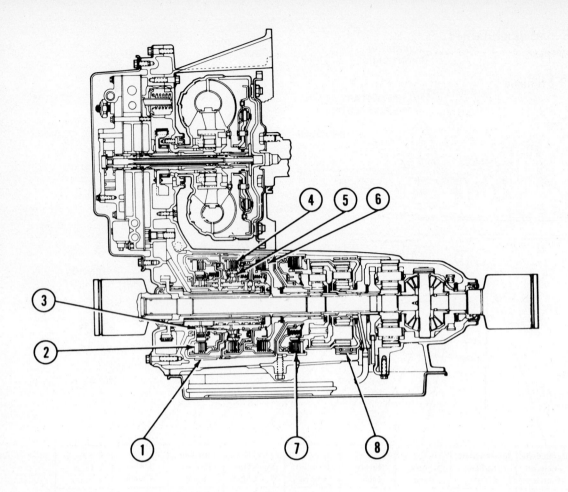

GEAR	OVER-DRIVE BAND ①	FOR-WARD CLUTCH ②	LOW ONE-WAY CLUTCH ③	DIRECT CLUTCH ④	DIRECT ONE-WAY CLUTCH ⑤	INTER-MEDIATE CLUTCH ⑥	REV CLUTCH ⑦	LOW INTER. BAND ⑧	RATIO
MANUAL LOW		APPLIED	HOLD	APPLIED				APPLIED	2.77:1
DRIVE 1st GEAR		APPLIED	HOLD					APPLIED	2.77:1
DRIVE 2nd GEAR		APPLIED	O/R			APPLIED		APPLIED	1.543:1
DRIVE 3rd GEAR		APPLIED		APPLIED	HOLD	APPLIED			1.000:1
DRIVE 4th GEAR	APPLIED			APPLIED	O/R	APPLIED			.694:1
REVERSE		APPLIED	HOLD				APPLIED		2.263:1

Fig. 8-36 Cutaway view of an AXOD transaxle with a clutch and band application chart for the various speeds. *(Courtesy of Ford Motor Company.)*

- Reverse: 2.26:1 with the forward clutch as the driving member and the reverse clutch as the reaction member.

The AXOD transmission uses both a vacuum modulator and a mechanical throttle linkage. The vacuum modulator is used to provide the correct shift feel while the mechanical linkage determines shift timing and downshifts. All AXODs use a lockup torque converter that is controlled electrically. A unique feature with some AXODs is an overtemperature switch mounted in the transaxle. When the transmission fluid rises above a certain point, this switch causes the converter clutch to lock. Locking the converter reduces the fluid friction, which in turn lowers the fluid temperature and reduces fluid breakdown.

8.3.8 Ford Motor Company: ZF-4HP-22

Some Lincoln automobiles are equipped with a diesel engine produced by BMW. These cars use a ZF-4HP-22 transmission produced by ZF, a German transmission manufacturer. This transmission is a four-speed, overdrive unit that is equipped with a torque converter clutch. Because of its limited use in domestic cars, we will not describe it any further.

8.4 GENERAL MOTORS CORPORATION

General Motors has traditionally used the terms *Hydra-matic*, *Turbo Hydra-matic*, or *THM* for its automatic transmissions even though some are not built at the Hydramatic plant in Ypsilanti, Michigan. Like other manufacturers, General Motors uses some slightly different terms for their transmission parts, as shown in Fig. 8-37.

Terms Commonly Used by General Motors Corporation	Terms Commonly Used in the Repair Industry
Internal gear	Ring gear
Direct clutch	High–reverse clutch
Clutch housing	Clutch drum
Input planetary gearset	Front planetary gearset
Reaction planetary gearset	Rear planetary gearset
Input drum	Input shell
Detent downshift	Full-throttle downshift

Fig. 8-37 Comparison of terms used by General Motors with those used by others in the transmission industry.

There are currently 10 different transmissions in use; there are 12 if we add the Powerglide and the THM 300, a similar unit. In addition to these transmissions, there are two transaxles. We can organize these units into our categories as shown in Table 8-2.

TABLE 8–2 Transmission/Transaxle Categories

	Transmissions	Transaxles
Type 2	THM 200, THM 250, THM 325	THM 125
Type 3	THM 400, THM 425	—
Type 4	THM 350	—
Type 7	THM 200-4R, THM 325-4L	—
Type 9	Powerglide, THM 300	—
Type 11	THM 180	—
Type 13	THM 700-R4	—
Type 15		THM 440-T4, 4T60-E

General Motors transmissions can also be identified by the shape of the oil pan (Fig. 8-38). Except for the THM 180, all are made with an aluminum case that has the torque converter housing cast as part of the case. The THM 180 has a removable converter housing. The THM 180, THM 200, THM 200-4R, and THM 700-R4

Fig. 8-38 The quickest way to determine which Hydra-matic transmission is used is to look at the shape of the oil pan.

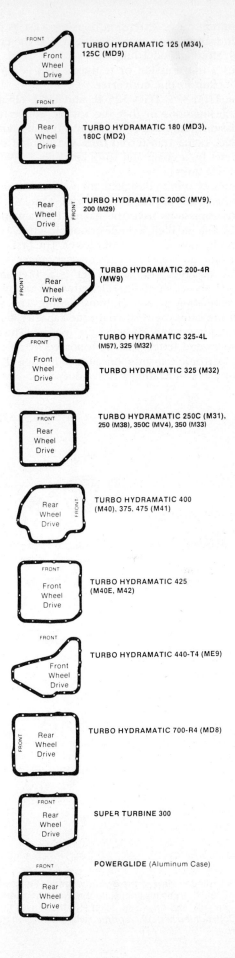

have nonremovable extension housings cast as part of the main transmission case. The letter C at the end of a General Motors model number (such as THM 200C) indicates a transmission or transaxle that is equipped with a lockup torque converter.

The THM 325, THM 325-4L, and THM 425 are unusual transmissions in that the torque converter and transmission pump are offset from the gear train with the power being transferred from the turbine shaft to the gearset by a chain and sprockets (Fig. 8-39). They are all FWD units.

A new system to designate the various transmission models has been developed by Hydra-matic. All Hydra-matic transmissions have the name "Hydra-matic" followed by a single digit, a single or double letter, a double digit, and in some cases another letter. The first number indicates the number of speeds or gear ranges—3, 4, 5, or V for variable. The first letter is a T (transverse mounted), an L [longitudinal (lengthwise) mounted], or an M, indicating a manual transmission. The second number indicates the relative torque capacity with a 10 being the lowest and 80 being the highest. This group of numbers and letters can be followed by an E, indicating electronic transmisison; an A, indicating all-wheel drive; or an H, indicating a heavy-duty unit.

8.4.1 General Motors Transmissions: Type 2

The THM 200, THM 250, and THM 325 are Simpson Gear train three-speed transmissions that use two driving clutches, an intermediate band, and a multiple-disc low–reverse clutch so they fit into the Type 2 category (Figs. 8-40, 8-41, and 8-42). The THM 200 and THM 325 use a cable-operated mechanical throttle linkage while the THM 250 uses both a vacuum modulator and a detent cable. The intermediate band is adjustable internally in the THM 200 and THM 325 by changing the servo rod, and the THM 250 has an external band adjustment at the right side of the case. A THM 200 can be identified by a nameplate attached to the right side of the extension housing; the THM 250 can be identified by a group of numbers and letters stamped on the servo cover or the governor cover; and the THM 325 can be identified by a nameplate on the left side of the torque converter housing (Fig. 8-43).

The gear ratios for the THM 200 and THM 325 are:

- First: 2.74:1 with the forward clutch as the driving member and the roller clutch or low–reverse clutch as the reaction member.

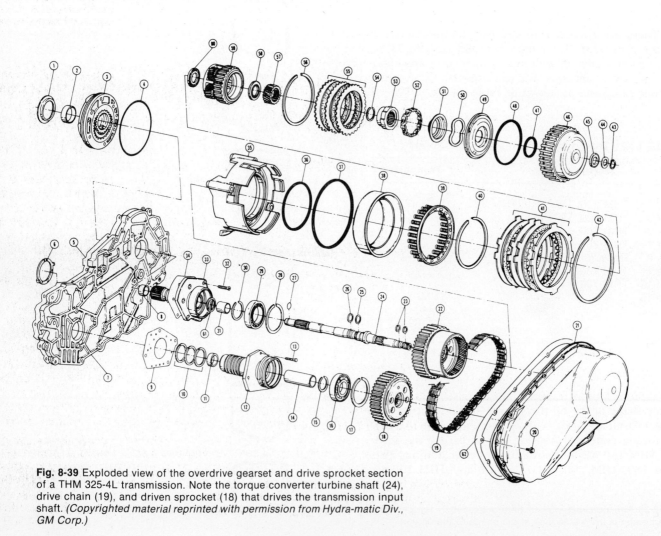

Fig. 8-39 Exploded view of the overdrive gearset and drive sprocket section of a THM 325-4L transmission. Note the torque converter turbine shaft (24), drive chain (19), and driven sprocket (18) that drives the transmission input shaft. *(Copyrighted material reprinted with permission from Hydra-matic Div., GM Corp.)*

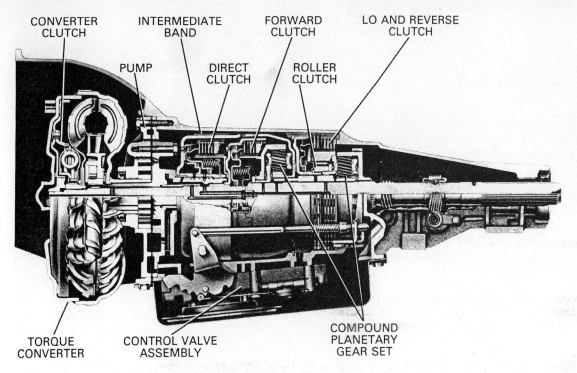

RANGE	GEAR	FORWARD CLUTCH	DIRECT CLUTCH	INTERMEDIATE BAND	ROLLER CLUTCH	LO-REVERSE CLUTCH
P-N						
D	1st	APPLIED			HOLDING	
	2nd	APPLIED		APPLIED		
	3rd	APPLIED	APPLIED			
2	1st	APPLIED			HOLDING	
	2nd	APPLIED		APPLIED		
1	1st	APPLIED			HOLDING	APPLIED
	2nd	APPLIED		APPLIED		
R	REVERSE		APPLIED			APPLIED

Fig. 8-40 Cutaway view of a THM 200C transmission with a clutch and band application chart for the various speeds. *(Copyrighted material reprinted with permission from Hydramatic Div., GM Corp.)*

- Second: 1.57:1 with the forward clutch as the driving member and the intermediate band as the reaction member.
- Third: 1:1 with the forward and direct clutches as driving members.
- Reverse: 2.07:1 with the direct clutch as the driving member and the low–reverse clutch as the reaction member.

The power flow for the THM 250 is the same, but the gear ratios are slightly different. They are:

First: 2.52:1.
Second: 1.52:1.
Third: 1:1.
Reverse: 1.93:1.

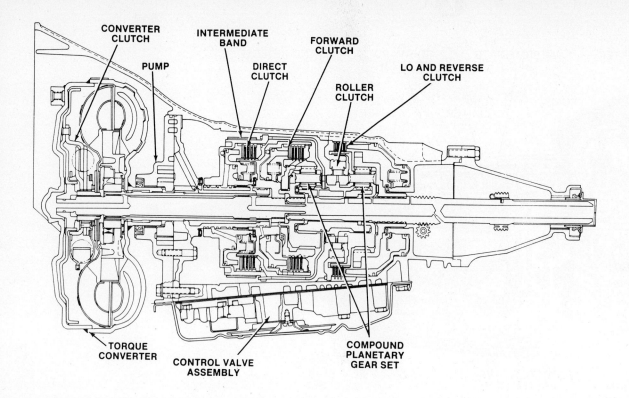

RANGE REFERENCE CHART

RANGE	GEAR	DIRECT CLUTCH	FORWARD CLUTCH	LO-REVERSE CLUTCH	INTER-MEDIATE BAND	LO-ROLLER CLUTCH
P-N						
D	1st		APPLIED			HOLDING
	2nd		APPLIED		APPLIED	
	3rd	APPLIED	APPLIED			
2	1st		APPLIED			HOLDING
	2nd		APPLIED		APPLIED	
1	1st		APPLIED	APPLIED		HOLDING
R	REVERSE	APPLIED		APPLIED		

Fig. 8-41 Cutaway view of a THM 250C transmission with a clutch and band application chart for the various speeds. *(Copyrighted material reprinted with permission from Hydramatic Div., GM Corp.)*

The later model THM 200C and the THM 250C are equipped with a lockup torque converter. The converter clutch operation is controlled by an electric solenoid at the valve body, and it has an electrical connector at the left side of the transmission case. A General Motors lockup converter can be identified by a flattened area where the converter clutch lining makes contact at the outer edge of the front of the converter cover.

As mentioned earlier, the THM 325 transmission uses a pair of sprockets and a drive chain to transfer power from the turbine shaft to the transmission input shaft. A silent chain, much like those used for engine timing chains, is used. This allows the power flow direction to be reversed toward the front of the car where the front wheels are driven. In a way, this also reverses the power flows in the transmission because the input end is now toward the rear of the car. Power flows through the gear train are the same as those through a THM 200.

8.4.2 General Motors Transaxle: Type 2

The THM 125 and THM 125C are Simpson gear train transaxles that, as far as the gear train is concerned, use the same components and power flow as the THM 200 with the final drive and differential parts added (Fig. 8-44). This transaxle is referred to as a Hydramatic 3T40 using the new designation system.

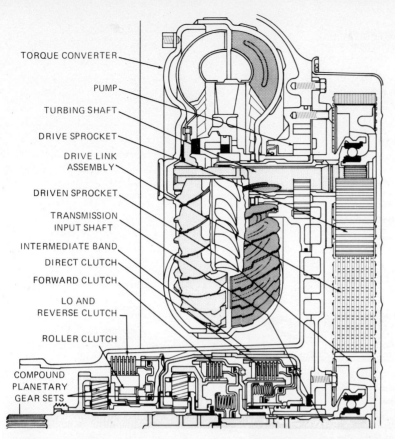

Fig. 8-42 Cutaway view of a THM 325 transmission. Clutch and band application is the same as that shown in Fig. 8-40. *(Copyrighted material reprinted with permission from Hydra-matic Div., GM Corp.)*

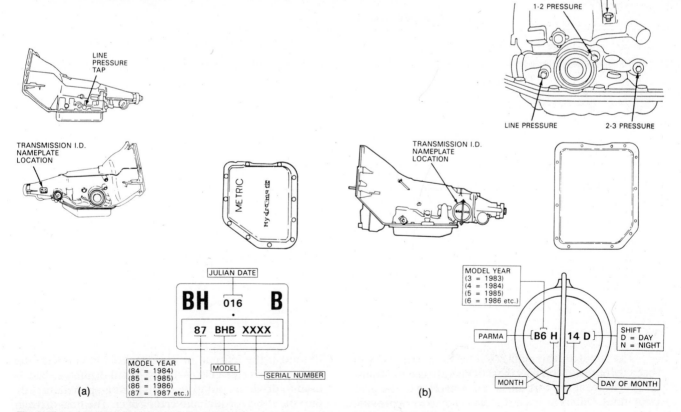

Fig. 8-43 (a) A THM 200C can be identified by a nameplate on the extension and (b) a THM 250C can be identified by the serial number on the governor cover. *(Copyrighted material reprinted with permission from Hydra-matic Div., GM Corp.)*

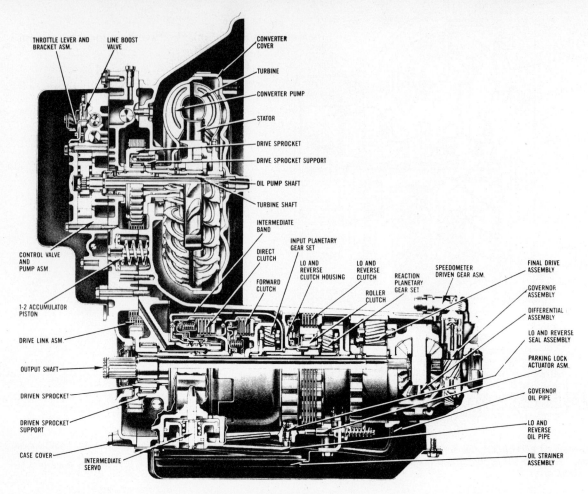

RANGE REFERENCE CHART

RANGE	GEAR	DIRECT CLUTCH	INTERMEDIATE BAND	FORWARD CLUTCH	ROLLER CLUTCH	LO-REVERSE CLUTCH
P - N						
D	1st			APPLIED	HOLDING	
	2nd		APPLIED	APPLIED		
	3rd	APPLIED		APPLIED		
2	1st			APPLIED	HOLDING	
	2nd		APPLIED	APPLIED		
1	1st			APPLIED	HOLDING	APPLIED
R	REVERSE	APPLIED				APPLIED

Fig. 8-44 Cutaway view of a THM 125 transaxle with a clutch and band application chart for the various speeds. *(Copyrighted material reprinted with permission from Hydra-matic Div., GM Corp.)*

The input in this unit resembles the THM 325 in that a drive chain and pair of sprockets are used to transfer power from the turbine in the converter to the gear train input. Different sized sprockets are used to provide a change in gear ratios. The major visual difference in these two is the addition of the final drive and differential and the location of the valve body and pump, which are located at the left end of the transaxle, in line with the torque converter (Fig. 8-45). A small-diameter shaft is used to drive the pump, and this shaft is splined directly into the front torque converter cover. The final drive in a THM 125 is a planetary gearset with the sun gear connected to the transmission output shaft and the ring gear splined to the case to lock it permanently in reaction,

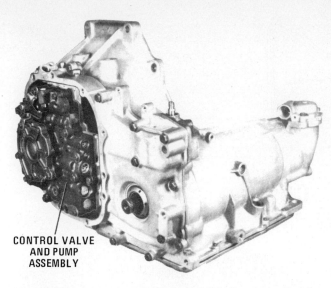

Fig. 8-45 The valve body cover has been removed from this THM 125 to show the control valve and pump assembly. *(Copyrighted material reprinted with permission from Hydra-matic Div., GM Corp.)*

the carrier becoming the differential case as the output member (Fig. 8-46). An axle drive shaft for the left front wheel extends from the differential and passes through the transmission gearset to the outside of the case. The inner CV (constant velocity) joint for the right-side drive shaft passes into the case and connects directly into the differential right-side gear.

The gear ratios for the THM 125 transaxle are:

- First: 2.84:1 with the forward clutch as the driving member and the roller clutch or low–reverse clutch as the reaction member.

- Second: 1.6:1 with the forward clutch as the driving member and the intermediate band as the reaction member.

- Third: 1:1 with the forward and direct clutches as driving members.

- Reverse: 2.07:1 with the direct clutch as the driving member and the low–reverse clutch as the reaction member.

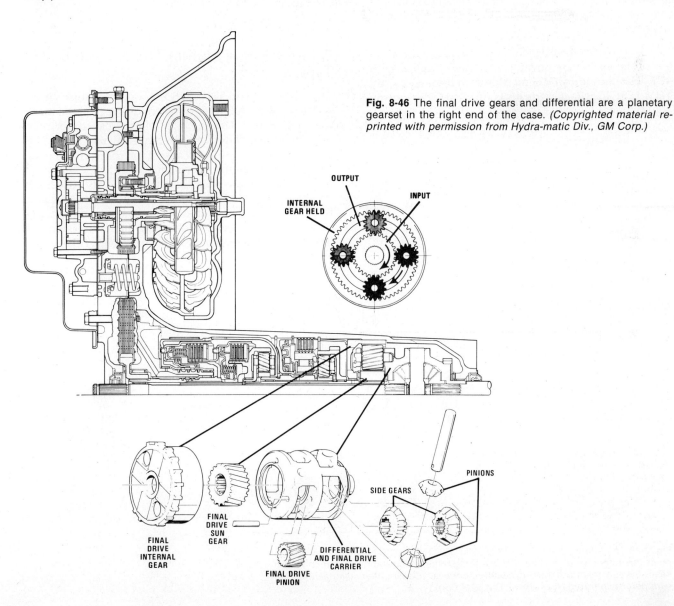

Fig. 8-46 The final drive gears and differential are a planetary gearset in the right end of the case. *(Copyrighted material reprinted with permission from Hydra-matic Div., GM Corp.)*

223

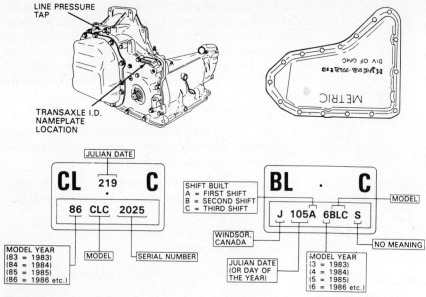

Fig. 8-47 A THM 125/125C transaxle can be identified by an identification nameplate in the location shown. There is a slight difference in the nameplate depending on the plant where the unit was built. *(Copyrighted material reprinted with permission from Hydra-matic Div., GM Corp.)*

A THM 125 can be identified by a nameplate at the top of the chain case cover, above the left axle (Fig. 8-47). Like the other General Motors transmissions, the THM 125C uses an electrically controlled lockup torque converter.

8.4.3 General Motors Transmission: Type 3

The THM 400 and THM 425 use a multiple-disc clutch and a one-way clutch for the second-gear reaction member (Fig. 8-48). The one-way clutch, called an intermediate roller clutch, can overrun in third gear. This simplifies the 2–3 shift because if the roller clutch overruns in third gear; the intermediate clutch does not need to be released. An intermediate band is also used, but this band is applied only in manual 2 to provide engine braking during deceleration. The bands in these transmissions are adjustable by changing the servo piston rod. The THM 400 is referred to as a Hydramatic 3L80 under

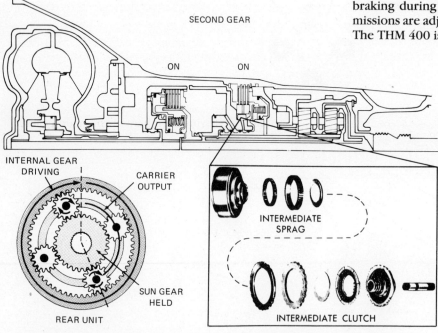

Fig. 8-48 The intermediate clutch in a THM 350, THM 400 (shown), and THM 425 holds the outer race of the intermediate sprag/roller clutch stationary, which in turn keeps the sun gear from rotating counterclockwise. *(Copyrighted material reprinted with permission from Hydra-matic Div., GM Corp.)*

the new system. A heavy-duty version of the THM 400 is called a THM 475 in the old system and a Hydramatic 3L80HD in the new system.

The gear ratios for the THM 400 and 425 are:

- First: 2.48:1 with the forward clutch as the driving member and the low roller clutch or rear band as the reaction member.

- Second: 1.48:1 with the forward clutch as the driving member and the intermediate clutch and intermediate roller/sprag or front band as the reaction member.

- Third: 1:1 with the forward and direct clutches as the driving members.

- Reverse: 2.08:1 with the direct clutch as the driving member and the rear band as the reaction member.

The THM 400 and THM 425 are heavy-duty transmissions with the gear train positioned end for end from all of the others (Fig. 8-49). By studying the internal parts, you can note that the sun gear is driven by a shaft instead of a shell and other end-for-end gear train differences. A THM 400 or THM 425 transmission can be identified by a nameplate located on the right side of the case (Fig. 8-50). These units use a vacuum modulator located at the right front of the case and an electric downshift solenoid at the valve body. All THM 400's and THM 425's have an electrical connector at the left side of the case; usually this is a single-terminal connector. Some THM 400's (1965 to 1967) use an electrically controlled variable-pitch stator in the torque converter. These units, called Super 400's, have a two-terminal electrical connector.

8.4.4 General Motors Transmission: Type 4

The THM 350 is smaller than the THM 400. It uses a low–reverse clutch while the THM 400 uses a low–reverse band (Fig. 8-51).

The gear ratios for the THM 350 are:

- First: 2.52:1 with the forward clutch as the driving member and the roller clutch or low–reverse clutch as the reaction member.

- Second: 1.52:1 with the forward clutch as the driving member and the intermediate clutch and intermediate roller clutch or intermediate band as the reaction member.

- Third: 1:1 with the forward and direct clutches as driving members.

- Reverse: 1.93:1 with the direct clutch as the driving member and the low–reverse clutch as the reaction member.

The THM 350 and the THM 250 are almost the same except the intermediate clutch is used in the THM 350 but not in the THM 250. The THM 250 uses a band. The THM 250 has a band adjusting screw on the right side of the case while a THM 350 does not. Many of their internal parts can be interchanged. Like the THM 250, the THM 350 has a vacuum modulator located at the right rear corner of the transmission and a throttle cable at the upper right front of the case. Also, like the THM 250, the exact type of THM 350 can be identified by markings on the servo cover or the governor cover. An electrically controlled lockup torque converter is used with later model THM 350C transmissions.

8.4.5 General Motors Transmission: Type 7

The gear train of a THM 200-4R is four-speed overdrive unit that combines a Simpson three-speed gear train with an overdrive gearset (Fig. 8-52). The THM 325-4L uses the same power flow as the THM 200-4R, but the parts are arranged differently. The overdrive unit of the THM 325-4L is in line with the torque converter and has the ability to cause the drive chain to travel at engine speed or at overdrive speed (Fig. 8-53). The THM 200-4R and THM 325-4L combine a Type 2 gearset with its overdrive gearset.

The gear ratios for the THM 200-4R and THM 325-4L are:

- First: 2.74:1 with the forward clutch as the driving member and the low roller clutch or low–reverse clutch as the reaction member in the main gearset and the overdrive roller clutch or overrun clutch as the driving member in the overdrive set.

- Second: 1.57:1 with the forward clutch as the driving member and the intermediate band as the reaction member in the main gearset and the overdrive roller clutch or overrun clutch as the driving member in the overdrive set.

- Third: 1:1 with the forward and direct clutches as driving members in the main gearset and the overdrive roller clutch or overrun clutch as the driving member in the overdrive set.

- Fourth: 0.67:1 with the forward and direct clutches as driving members in the main gearset and the overdrive roller clutch as the driving member and the fourth clutch as the reaction member in the overdrive set.

- Reverse: 2.07:1 with the direct clutch as the driving member and the low–reverse clutch as the reaction member in the main gearset and the overdrive roller clutch as the driving member in the overdrive set.

A THM 200-4R can be identified by a nameplate at the right rear of the extension housing, which like a THM 200 is nonremovable. A THM 325-4L can be identified by a nameplate at the left side of the torque converter housing (Fig. 8-54). Also, like the THM 200, neither unit uses a vacuum modulator; a throttle cable is used. All THM 200-4R's and THM 325-4L's use an electrically controlled lockup torque converter.

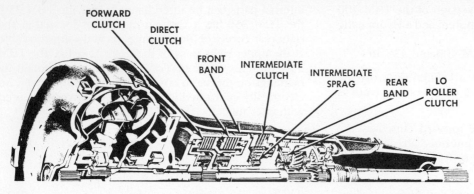

RANGE	GEAR	FORWARD CLUTCH	DIRECT CLUTCH	FRONT BAND	INT. CLUTCH	INT. SPRAG	LO ROLLER CLUTCH	REAR BAND
PARK—NEUT.		OFF	OFF	OFF	OFF	INEFFECTIVE	INEFFECTIVE	OFF
DRIVE	FIRST	ON	OFF	OFF	OFF	INEFFECTIVE	EFFECTIVE	OFF
	SECOND	ON	OFF	OFF	ON	EFFECTIVE	INEFFECTIVE	OFF
	THIRD	ON	ON	OFF	ON	INEFFECTIVE	INEFFECTIVE	OFF
INT.	FIRST	ON	OFF	OFF	OFF	INEFFECTIVE	EFFECTIVE	OFF
	SECOND	ON	OFF	ON	ON	EFFECTIVE	INEFFECTIVE	OFF
LO	FIRST	ON	OFF	OFF	OFF	INEFFECTIVE	EFFECTIVE	ON
	SECOND	ON	OFF	ON	ON	EFFECTIVE	INEFFECTIVE	OFF
REV.		OFF	ON	OFF	OFF	INEFFECTIVE	INEFFECTIVE	ON

(a)

(b)

Fig. 8-49 (a) Cutaway view of a THM 400 transmission with a clutch and band application chart for the various speeds and (b) cutaway view of a THM 425. *(Copyrighted material reprinted with permission from Hydra-matic Div., GM Corp.)*

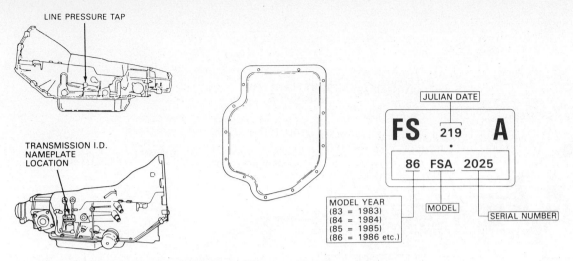

Fig. 8-50 A THM 400 transmission can be identified by a nameplate on the right side of the case. *(Copyrighted material reprinted with permission from Hydra-matic Div., GM Corp.)*

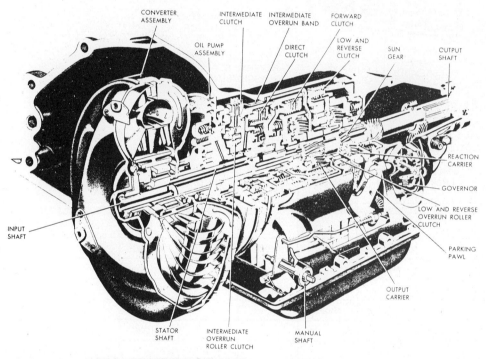

RANGE REFERENCE CHART

RANGE	GEAR	INTER-MEDIATE CLUTCH	DIRECT CLUTCH	FOR-WARD CLUTCH	LO-REVERSE CLUTCH	INTER-MEDIATE BAND	LO-ROLLER CLUTCH	INTER-MEDIATE ROLLER CLUTCH
P-N								
D	1st			APPLIED			HOLDING	HOLDING
D	2nd	APPLIED		APPLIED				HOLDING
D	3rd	APPLIED	APPLIED	APPLIED				
2	1st			APPLIED			HOLDING	HOLDING
2	2nd	APPLIED		APPLIED		APPLIED		HOLDING
1	1st			APPLIED	APPLIED		HOLDING	HOLDING
R	REVERSE		APPLIED		APPLIED			

Fig. 8-51 Cutaway view of a THM 350 transmission with a clutch and band application chart for the various speeds. *(Copyrighted material reprinted with permission from Hydra-matic Div., GM Corp.)*

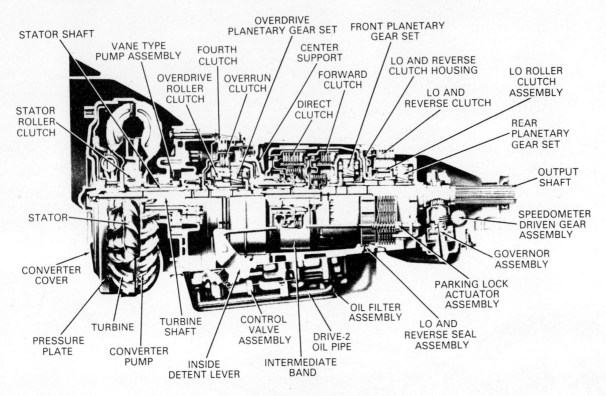

RANGE REFERENCE CHART

RANGE	GEAR	FOWARD CLUTCH	DIRECT CLUTCH	INTERMEDIATE BAND	FOURTH CLUTCH	OVERDRIVE ROLLER CLUTCH	OVERRUN CLUTCH	LO-ROLLER CLUTCH	LO-REVERSE CLUTCH
P-N						HOLDING			
⃞D	1st	APPLIED				HOLDING		HOLDING	
	2nd	APPLIED		APPLIED		HOLDING			
	3rd	APPLIED	APPLIED			HOLDING			
	4th	APPLIED	APPLIED		APPLIED				
D	1st	APPLIED					APPLIED	HOLDING	
	2nd	APPLIED		APPLIED			APPLIED		
	3rd	APPLIED	APPLIED				APPLIED		
2	1st	APPLIED					APPLIED	HOLDING	
	2nd	APPLIED		APPLIED			APPLIED		
1	1st	APPLIED					APPLIED		APPLIED
R	REVERSE		APPLIED			HOLDING			APPLIED

Fig. 8-52 Cutaway view of a THM 200-4R transmission with a clutch and band application chart for the various speeds. *(Copyrighted material reprinted with permission from Hydra-matic Div., GM Corp.)*

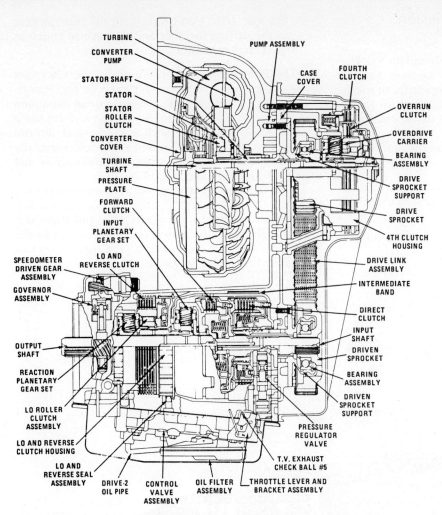

Fig. 8-53 Cutaway view of a THM 325-4L transmission. Clutch and band application is the same as for a THM 200-4R. *(Copyrighted material reprinted with permission from Hydra-matic Div., GM Corp.)*

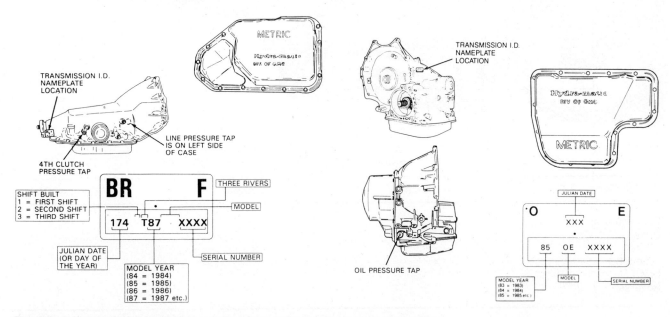

Fig. 8-54 A THM 200-4R identified by a nameplate at the right rear and a THM 325-4L identified by a nameplate on the left side. *(Copyrighted material reprinted with permission from Hydra-matic Div., GM Corp.)*

229

8.4.6 General Motors Transmission: Type 9

Both Type 9 transmissions (the Powerglide and the THM 300) have been out of production for some time. They are included here because of the past popularity of the Powerglide and the similarity of the THM 300 to the Powerglide (Fig. 8-55). Both units use a two-speed Ravigneaux gear train.

The gear ratios for a Powerglide transmission (depending on the model) are:

- Low: 1.76:1 or 1.82:1 with the rear sun gear as the driving member and the low band as the reaction member.
- High: 1:1 with the rear sun gear and the high clutch as driving members.
- Reverse: 1.76:1 or 1.82:1 with the rear sun gear as the driving member and the reverse clutch as the reaction member.

The Powerglide was only used by the Chevrolet division and was introduced in 1960. The first Powerglides used a cast-iron case; a change to an aluminum case was made in 1962 (Fig. 8-56). Powerglides use a vacuum modulator along with a mechanical throttle linkage. The modulator is located at the left rear of the case. The band is adjustable with an adjuster screw that extends through the left side of the case.

The THM 300 was a light-duty transmission used by Buick, Oldsmobile, and Pontiac divisions. It was similar in most aspects to the Powerglide, but there was little interchangeability between the two units. Like the THM Super 400, there were some THM Super 300's that used a variable-pitch stator in the torque converter.

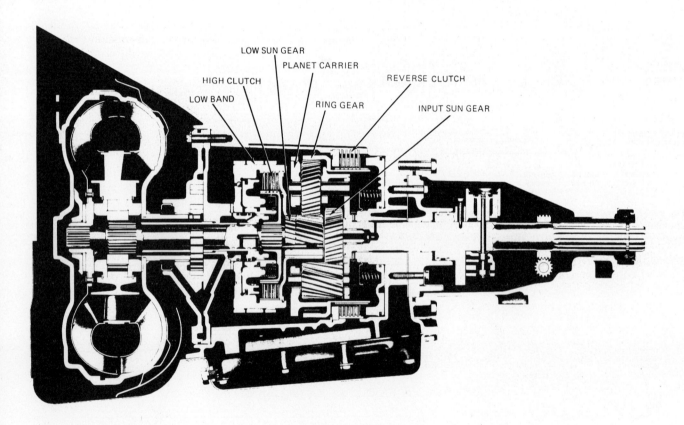

Range Reference Chart

Range Gear	High Clutch	Low Band	Reverse Clutch
Park–neutral			
Drive			
First		X	
Second	X		
Low		X	
Reverse			X

Fig. 8-55 Cutaway view of a Powerglide transmission with a clutch and band application chart for the various speeds. *(Courtesy of Chevrolet Div., GM Corp.)*

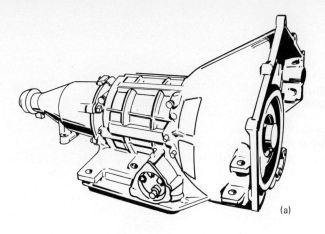

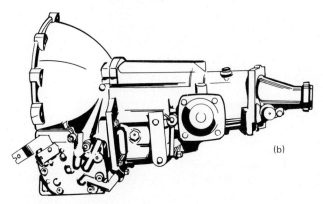

Fig. 8-56 (a) Aluminum case Powerglide and (b) an older cast-iron case Powerglide. *(Courtesy of Chevrolet Div., GM Corp.)*

These units had an electrical connector in the left side of the case.

8.4.7 General Motors: Type 11

The THM 180 is a three-speed Ravigneaux gear train transmission (Fig. 8-57). This transmission is referred to as a Hydramatic 3L30 using the new designation system. It is a light-duty unit that is used in the Chevrolet Chevette and smaller Chevrolet and General Motors pickups. It is manufactured in France.

The gear ratios for a THM 180 are:

- First: 2.4:1 with the sprag or third clutch as the driving member and the low band as the reaction member.

- Second: 1.48:1 with the second clutch as the driving member and the low band as the reaction member.

- Third: 1:1 with the second and third clutches as driving members.

- Reverse: 1.91:1 with the third clutch as the driving member and the reverse clutch as the reaction member.

A THM 180 can be identified by a nameplate at the left front of the transmission case (Fig. 8-58). This transmission has a removable torque converter housing. The low band is adjustable at the servo, which is inside the oil pan. Later model THM 180C transmissions are equipped with a lockup torque converter that, like other General Motors lockup converters, is controlled electrically and hydraulically.

A four-speed overdrive version of the THM 180/3L30 transmission has been developed. This new transmission is called the 4L30-E; it is a RWD unit with electronic shift controls. The gearbox section is essentially a 3L30 transmission with an overdrive unit, very much like the one used in a THM 200-4R, which is connected to the main gearset just like the THM 200-4R. The added gearset should improve fuel mileage, and the electronic controls should improve shift timing and feel.

8.4.8 General Motors Transmission: Type 13

The THM 700-R4 is another "one-of-a-kind" transmission. It is a four-speed overdrive transmission (Fig. 8-59). Using the new designation system, this transmission is called the Hydramatic 4L60.

The gear ratios for a THM 700-R4 are:

- First: 3.06:1 with the forward clutch and forward sprag or overrun clutch as driving members and the low roller clutch or the low–reverse clutch as the reaction member.

- Second: 1.63:1 with the forward clutch and forward sprag or overrun clutch as the driving member and the 2–4 band as the reaction member.

- Third: 1:1 with the forward clutch and forward sprag or overrun clutch and 3–4 clutch as driving members.

- Fourth: 0.7:1 with the 3–4 clutch as the driving member and the 2–4 band as the reaction member.

- Reverse: 2.3:1 with the reverse input clutch as the driving member and the low–reverse clutch as the reaction member.

A THM 700-R4 can be identified by a nameplate at the right rear of the case (Fig. 8-60). Like the THM 125, THM 200, THM 200-4R, and THM 325-4L, a vacuum modulator is not used; a cable is used for throttle input into the transmission. Also, like the THM 200-4R and THM 325-4L, all THM 700-R4's use an electrically controlled lockup torque converter.

8.4.9 General Motors Transaxle: Type 15

As mentioned earlier, the THM 440-T4 is a four-speed overdrive transmission that is similar to the Ford AXOD. The THM 440-T4 is an earlier design. The two gear trains are arranged slightly different so a different arrangement of friction apply devices is used. The THM 440-T4 is called a 4T60 using the new designation system (Fig. 8-61).

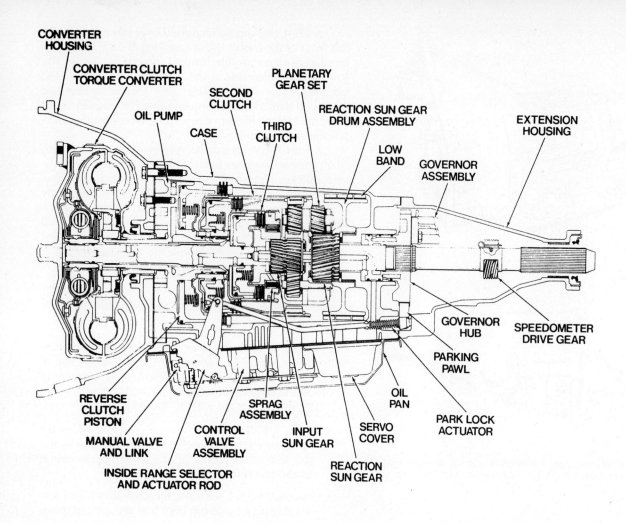

RANGE REFERENCE CHART

RANGE	GEAR	INPUT SPRAG	2ND CLUTCH	3RD CLUTCH	LOW BAND	REVERSE CLUTCH
P - N		HOLDING				
D	1st	HOLDING			APPLIED	
D	2nd		APPLIED		APPLIED	
D	3rd	HOLDING	APPLIED	APPLIED		
2	1st	HOLDING			APPLIED	
2	2nd		APPLIED		APPLIED	
1	1st	HOLDING		APPLIED	APPLIED	
R	REVERSE	HOLDING		APPLIED		APPLIED

Fig. 8-57 Cutaway view of a THM 180 transmission with a clutch and band application chart for the various speeds. *(Copyrighted material reprinted with permission from Hydramatic Div., GM Corp.)*

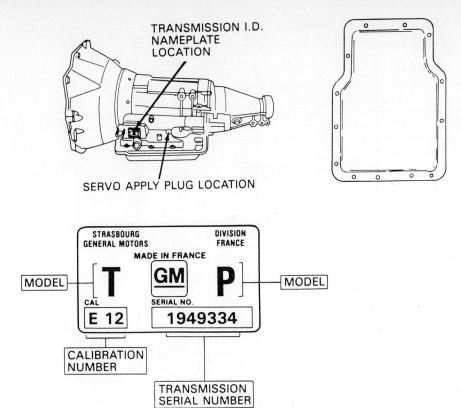

Fig. 8-58 A THM 180 identified by a nameplate on the left side of the case. *(Copyrighted material reprinted with permission from Hydra-matic Div., GM Corp.)*

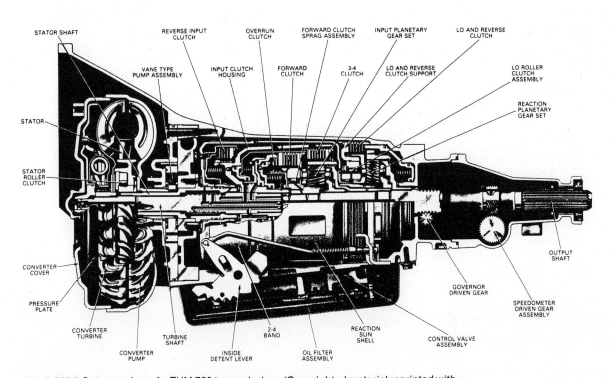

Fig. 8-59(a) Cutaway view of a THM 700 transmission. *(Copyrighted material reprinted with permission from Hydra-matic Div., GM Corp.)*

233

RANGE REFERENCE CHART

RANGE	GEAR	FORWARD CLUTCH	FORWARD SPRAG CL. ASSEMBLY	2-4 BAND	OVERRUN CLUTCH	3-4 CLUTCH	LO-ROLLER CLUTCH	LO-REV. CLUTCH	REVERSE INPUT CLUTCH
P-N									
D	1st	APPLIED	HOLDING				HOLDING		
D	2nd	APPLIED	HOLDING	APPLIED					
D	3rd	APPLIED	HOLDING			APPLIED			
D	4th	APPLIED		APPLIED		APPLIED			
D	1st	APPLIED	HOLDING		APPLIED		HOLDING		
D	2nd	APPLIED	HOLDING	APPLIED	APPLIED				
D	3rd	APPLIED	HOLDING		APPLIED	APPLIED			
2	1st	APPLIED	HOLDING		APPLIED		HOLDING		
2	2nd	APPLIED	HOLDING	APPLIED	APPLIED				
1	1st	APPLIED	HOLDING		APPLIED		HOLDING	APPLIED	
R	REVERSE							APPLIED	APPLIED

Fig. 8-59(b) Clutch and band application chart for various speeds of THM 700 transmission. *(Copyrighted material reprinted with permission from Hydra-matic Div., GM Corp.)*

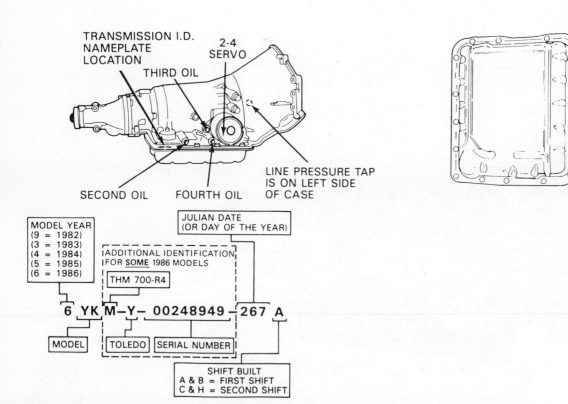

Fig. 8-60 A THM 700 identified by a nameplate on the right side of the case. *(Copyrighted material reprinted with permission from Hydra-matic Div., GM Corp.)*

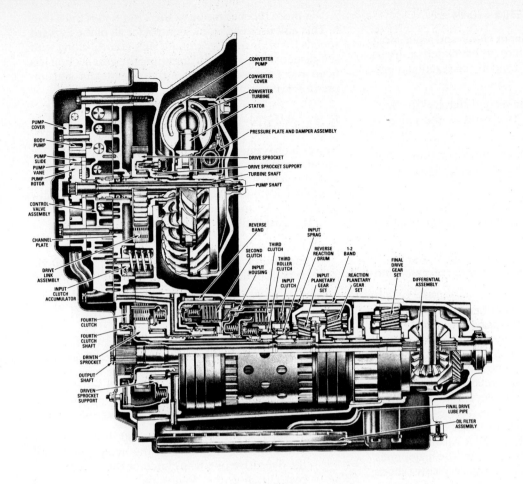

RANGE REFERENCE CHART

RANGE	GEAR	INPUT SPRAG	INPUT CLUTCH	1-2 BAND	2ND CLUTCH	3RD CLUTCH	3RD ROLLER CLUTCH	4TH CLUTCH	REVERSE BAND
P-N			*	*					
▢D	1st	HOLDING	APPLIED	APPLIED					
▢D	2nd	OVER-RUNNING	*	APPLIED	APPLIED				
▢D	3rd		OFF		APPLIED	APPLIED	HOLDING		
▢D	4th				APPLIED	*	OVER-RUNNING	APPLIED	
D	1st	HOLDING	APPLIED	APPLIED					
D	2nd	OVER-RUNNING	*	APPLIED	APPLIED				
D	3rd	HOLDING	APPLIED		APPLIED	APPLIED	HOLDING		
2	1st	HOLDING	APPLIED	APPLIED					
2	2nd	OVER-RUNNING	*	APPLIED	APPLIED				
1	1st	HOLDING	APPLIED	APPLIED			APPLIED	HOLDING	
R	REVERSE	REVERSE	HOLDING	APPLIED					APPLIED

*APPLIED BUT NOT EFFECTIVE

Fig. 8-61 Cutaway view of a THM 440-T4 transaxle with a clutch and band application chart for the various speeds. (Copyrighted material reprinted with permission from Hydra-matic Div., GM Corp.)

The gear ratios for the THM 440-T4 are:

- First: 2.92:1 with the input clutch and input sprag or third clutch and third roller clutch as input members and the 1–2 band as the reaction member.

- Second: 1.57:1 with the second clutch as the driving member and the 1–2 band as the reaction member.

- Third: 1:1 with the second, third, and third roller clutches or input clutch and input sprag as driving members.

- Fourth: 0.7:1 with the second clutch as the driving member and the fourth clutch as the reaction member.

- Reverse: 2.38:1 with the input clutch and input sprag as the driving member and the reverse band as the reaction member.

A THM 440-T4 can be identified by a nameplate attached to the left rear of the transaxle case (Fig. 8-62). This transaxle uses a vacuum modulator along with a throttle cable. The modulator is used to provide the correct shift feel relative to engine load while the cable produces the correct shift timing relative to throttle position. All THM 440-T4's use an electrically controlled lockup torque converter.

A version of the THM 440-T4 transaxle called the THM-F7 was introduced in the Cadillac Allante. A number of internal changes have been made. The major ones are that the THM F7 has a higher torque capacity (about one-third more), uses a viscous drive torque converter clutch, and features electronic shift controls in third and fourth gears. The electronic shift control uses 2–3 and 3–4 shift solenoids at the valve body that control fluid flows and move the shift valves (Fig. 8-63). The solenoids are connected to B^+ voltage and are grounded by the ECM to produce the shift.

A more recent version of the THM 440 is the 4T60-E. This unit uses electronic control for all of the forward speeds to produce more precise shift control along with self-diagnosis capability. In addition, the gear section has been reconfigured to reduce the complexity and improve serviceability.

8.5 AISIN-WARNER: TYPE 8

Newer Jeep wagons use an Aisin-Warner 4 (AW-4), four-speed overdrive transmission (Fig. 8-64). Both the transmission gear selection and the torque converter are electronically controlled using a transmission computer unit (TCU) for this purpose.

The gear ratios for an AW-4 transmission are:

- First: 2.8:1 with the forward clutch as the driving member and the one-way clutch or first and reverse brake as the reaction member in the main gearset and the overdrive one-way clutch or overdrive direct clutch as the driving member in the overdrive gearset.

- Second: 1.53:1 with the forward clutch as the driving member and the second brake and one-way clutch 2 or the second coast brake as the reaction member in the main gearset and the overdrive one-way clutch or overdrive direct clutch as the driving members in the overdrive gearset.

- Third: 1:1 with the forward and direct clutches as the driving members in the main gearset and the overdrive one-way clutch or the overdrive direct clutch as the driving members in the overdrive gearset.

- Fourth: 0.7:1 with the forward and direct clutches as the driving members in the main gearset and the overdrive one-way clutch as the driving member and the overdrive brake as the reaction member in the overdrive gearset.

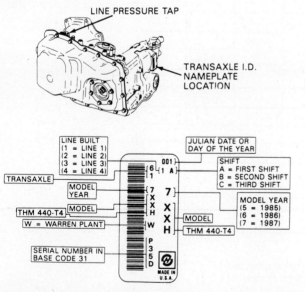
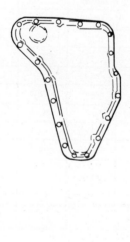

Fig. 8-62 A THM 440-T4 transaxle identified by a nameplate on the rear side of the case. *(Copyrighted material reprinted with permission from Hydra-matic Div., GM Corp.)*

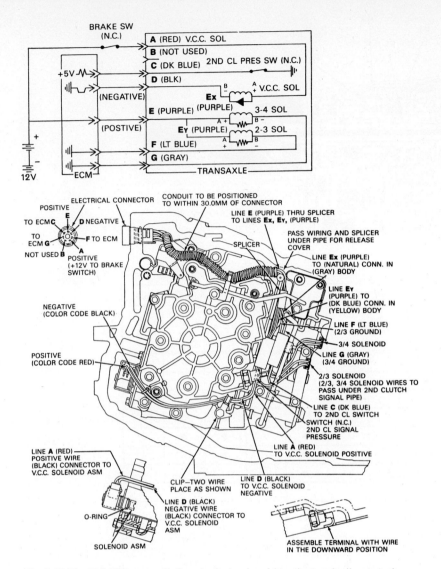

Fig. 8-63 The THM F7 uses a 2–3 and a 3–4 solenoid to electronically control the shift valves shown. The wiring is shown along with the valve body. *(Copyrighted material reprinted with permission from Hydra-matic Div., GM Corp.)*

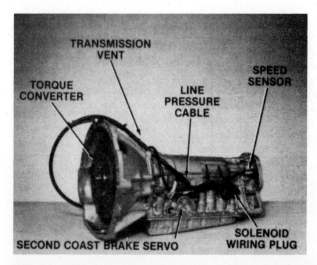
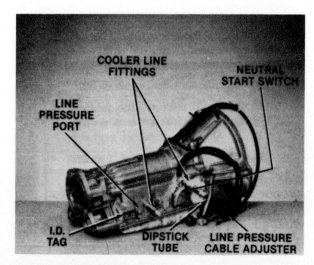

Fig. 8-64 An AW-4 transmission. Note the location of the various external components as well as the identification tag. *(Courtesy of Chrysler Corporation.)*

237

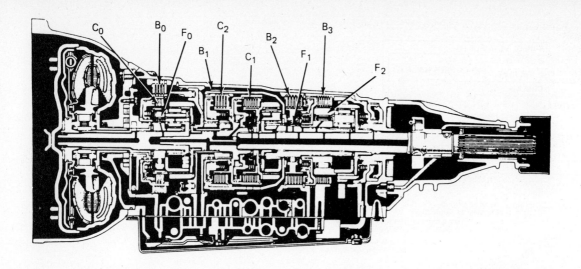

Apply Chart		Driving Clutches			Holding Clutches and Band				One-Way Clutches		
Range (i.e., Shift Lever Position)	Gear	C_0	C_1	C_2	B_0	B_1	B_2	B_3	F_0*	F_1	F_2
P	Park	●									
R	Reverse	●		●				●	●		
N	Neutral	●									
D	First	●	●						●		●
	Second	●	●				●		●	●	
	Third	●	●	●			●		●		
	OD		●	●	●		●				
3	First	●	●						●		●
	Second	●	●			●	●		●	●	
	Third	●	●	●			●		●		
1-2	First	●	●					●	●		●
	Second	●	●			●	●		●	●	

Fig. 8-65 Cutaway view of an AW-4 transmission with a clutch and band application chart for the various speeds. *Courtesy of Chrysler Corporation.)*

- Reverse: 2.39:1 with the direct clutch as the driving member and the first and reverse brake as the reaction member in the main gearset and the overdrive clutch as the driving member in the overdrive gearset.

The gear train of the AW-4 is a combination of a Simpson three-speed (Type 4) and an overdrive gear unit. Like the THM 200-4R, the AW-4 uses a driving multiple-disc clutch, a driving one-way clutch, and a holding multiple-disc clutch in the overdrive unit (Fig. 8-65). Since this transmission is electronically controlled, it has no modulator or throttle valve linkage.

8.6 IMPORT TRANSMISSIONS

For the most part, the transmission and/or transaxles used in import cars are very similar to those described earlier. All units use planetary gears (with the exception of the Honda transaxles described in Chapter 1), multidisc clutches, one-way clutches, hydraulic control systems, and torque converters. Many use one or more bands also. There is a rather large variety of units used, and to save space, they will not be described in detail. A detailed description of their operation is normally available in the manufacturer's service manual and in transmission service manuals.

In some cases, the vehicle manufacturer will design and build its own transmissions, much like our domestic manufacturers do. In other cases, the vehicle manufacturer will use units built by another manufacturer such as Aisin-Warner, Borg-Warner, Jatco, or ZF.

A listing of the transmission and/or transaxle usage for 1987 passenger cars with a brief description is given in Table 8-3.

TABLE 8-3 Transmission/Transaxle Usage, 1987 Passenger Cars

Car Manufacturer	Car Model	Transmission Model	1	2	3	4	5	6
Acura	Integra	CA	X	—	4	—		
	Legend	G4	X	—	4	—	—	X
Audi	Coupe	087	X	—	3	2		
	4000S, CS	089	X	—	3	2		
	Quattro	089	X	—	3	2		
	5000S, CS	087	X	—	3	2		
BMW	All	ZF 4HP 22/H	—	X	4	D	—	X
		ZF 4HP 22/EH	—	X	4	D	X	X
Chrysler Corporation	Colt	Mitsu.-KM 171	X	—	3	10/s	—	M
	Colt Vista	Mitsu.-KM 172	X	—	3	10/s	—	M
	Conquest	Mitsu.-JM 600	—	X	4	7/s	—	X
Ford Motor Co.	Merkur XR4Ti	C 3	—	X	3	1		
General Motors	Spectrum	KF 100	X	—	3	2		
	Sprint	Suzuki 3-speed	X	—	3	2	X	
Honda	Accord	F4	X	—	4	—		
	Civic, CRX	CA	X	—	4	—		
	Prelude (FI)	F4	X	—	4	—		
	Prelude (carb)	AS	X	—	4	—	—	X
Hyundai	Excel	Mitsu.-KM 170-2AP	X	—	3	10	—	X
		Mitsu.-KM 170-3AP2	X	—	3	10	—	X
Isuzu	I-Mark	KF 100	X	—	3	2		
	Impulse	AW03-70	X	—	4	8		
		AW03-72L	X	—	4	8	—	X
Jaguar	XJ6	BW-66	—	X	3	10		
	XJS	THM 400	—	X	3	3		
Mazda	323	F3A	X	—	3	2		
	626	F3A	X	—	3	2		
	RX7	Jatco L4N71B	—	X	4	8	—	X
Mercedes-Benz	190	W4A020	—	X	4	D		
	300	W4A040	—	X	4	D		
	420	W4A040	—	X	4	D		
	560	W4A040	—	X	4	D		
Mitsubishi	Cordia	KM 172	X	—	3	10	—	M
	Galant	KM 175	X	—	3	10	—	M
	Mirage	KM 171	X	—	3	10		
	Starion	JM 600	—	X	4	7/s	—	X
	Tredia	KM 172	X	—	3	10		
Nissan	Maxima	Jatco RL4F02A	X	—	4	14/s		
	Pulsar NX	Jatco RL3F01A	X	—	3	2		
	Sentra	Jatco RL3F01A	X	—	3	2		
	Stanza	Jatco RL3F01A	X	—	3	2		
	Stanza SW	Jatco RL4F02A	X	—	4	14/s		
	200 SX	Jatco L4N71B	—	X	4	7	—	X
	300 ZX	Jatco E4N71B	—	X	4	7	X	X
Peugeot	505	ZF 4HP 22	—	X	4	D		
Porsche	928S	A 28.02	X	—	4	D		
	944	087	X	—	3	2		
SAAB	900	BW model 37	X	—	3	10		
Suburu	GL	2WD-Gumna (C)	X	—	3	10/s		
		4WD-Gumna (F)	X	—	3	10/s		
Toyota	Camry	A140E	X	—	4	8/s	X	X
		A140L	X	—	4	8/s	—	X
	Celica	A140E	X	—	4	8/s	X	X
		A140L	X	—	4	8/s	—	X
	Corolla FWD	A131L	X	—	3	4/s	—	X
		A240L	X	—	4	8/s	—	X
	Corolla RWD	A42DL	—	X	4	D	—	X
	Cressida	A43DE	—	X	4	D	X	X
	Supra	A340E	—	X	4	D	X	X
	Tercel	2WD-A55	X	—	3	4/s		
		4WD-A55F	X	—	3	4/s		
Volkswagen	Quantum	087 or 089	X	—	3	2		
	All others	010	X	—	3	2		
Volvo	240	Aisin-Warner AW71	—	X	4	8	—	X
	740/760 Turbo (gas)	Aisin-Warner AW71	—	X	4	8	—	X
	740 GL, GLE	ZF 4HP 22	—	X	4	D	—	X
	740 GLE Turbo (diesel)	ZF 4HP 22	—	X	4	D	—	X
	760 GLE	Aisin-Warner AW71	—	X	4	D	—	X

Note: column 1, transaxle; column 2, transmission; column 3, number of speeds; column 4, gear train type (some of these are slightly different); column 5, electronic shifts; column 6, lockup converter. Abbreviations: D, different; s, similar; M, may use it; Mitsu., Mitsubishi. These transmissions/transaxles may have variations and will possibly be different for different model years.

REVIEW QUESTIONS

These following questions are provided so you can check the facts you have just learned. Select the response that best completes each statement.

1. A. Transaxles operate just like a transmission of the same type.
   B. Transaxles are a combination of a transmission, final drive reduction gears, and a differential.
   Which is correct?
   a. A only
   b. B only
   c. Both A and B
   d. Neither A nor B

2. Which of the following units is not a transaxle?
   A. AXOD
   B. THM 125
   C. THM 325
   D. A-404

3. The final drive reduction gears used in a General Motors transaxle are
   A. A planetary gearset
   B. Of the helical type
   Which is correct?
   a. A only
   b. B only
   c. Both A and B
   d. Neither A nor B

4. When discussing Chrysler transmissions,
   A. All passenger car transmissions are called Torqueflites
   B. The A-904 is a larger transmission than the A-727
   Which is correct?
   a. A only
   b. B only
   c. Both A and B
   d. Neither A nor B

5. When a Chrysler transmission is shifted into manual 1,
   A. The front clutch applies
   B. The rear band applies
   Which is correct?
   a. A only
   b. B only
   c. Both A and B
   d. Neither A nor B

6. A Chrysler A-400 series transaxle uses a gear train that is very similar to the
   A. A-727
   B. A-904
   Which is correct?
   a. A only
   b. B only
   c. Both A and B
   d. Neither A nor B

7. The Chrysler A-604 transaxle is a little unusual in that it has not
   A. Bands
   B. Multiple-disc clutches
   Which is correct?
   a. A only
   b. B only
   c. Both A and B
   d. Neither A nor B

8. When discussing Ford transmissions,
   A. The C4 is exactly the same as a C6 except that it is smaller
   B. The C5 is essentially a C6 with a lockup converter
   Which is correct?
   a. A only
   b. B only
   c. Both A and B
   d. Neither A nor B

9. While in second gear, a C6 has
   A. The forward clutch applied
   B. The intermediate band applied
   Which is correct?
   a. A only
   b. B only
   c. Both A and B
   d. Neither A nor B

10. While in reverse in a C4, the gearset
    A. Input member is the front ring gear
    B. Reaction member is the rear carrier
    Which is correct?
    a. A only
    b. B only
    c. Both A and B
    d. Neither A nor B

11. In second gear in an A4LD transmission,
    A. The overdrive gearset has the sun gear locked to the ring gear through a multiple-disc clutch
    B. The sun gear in the main gearset is held by the intermediate band
    Which is correct?
    a. A only
    b. B only
    c. Both A and B
    d. Neither A nor B

12. Torque converter lockup in an AOD transmission occurs when
    A. Centrifugal forces move the clutch shoes outward against the converter cover
    B. The lockup solenoid energizes to move the converter switch valve
    Which is correct?
    a. A only
    b. B only
    c. Both A and B
    d. Neither A nor B

13. When discussing Ford transaxles,
    A. The AXOD is an ATX with an overdrive section added
    B. The ATX shifts into first gear as the low–reverse band is applied
    Which is correct?
    a. A only
    b. B only
    c. Both A and B
    d. Neither A nor B

14. General Motors transmissions that use the Simpson gear train include
    A. The THM 180
    B. The THM 200
    C. The THM 700
    D. All of these

15. In manual 2 in a THM 350,
    A. The forward clutch and intermediate band is applied
    B. The intermediate clutch is applied so the intermediate roller clutch is effective
    Which is correct?
    a. A only
    b. B only
    c. Both A and B
    d. Neither A nor B

16. The THM 250 and THM 350 are similar transmissions except that the THM 250
    A. Is a two-speed transmission
    B. Does not have an intermediate clutch and intermediate roller clutch
    Which is correct?
    a. A only
    b. B only
    c. Both A and B
    d. Neither A nor B

17. The THM 325, THM 325-4L, and THM 425
    A. Are FWD transaxles
    B. Use a chain and sprocket drive to connect the turbine shaft to the torque converter
    Which is correct?
    a. A only
    b. B only
    c. Both A and B
    d. Neither A nor B

18. The driving members used in a THM 200-4R in fourth gear include
    A. The forward clutch
    B. The direct clutch
    C. The fourth clutch
    D. Both A and B

19. The THM 700-R4 uses a gearset that has
    A. Two planetary gearsets
    B. Carriers in one gearset connected to the ring gears of the other gearset
    Which is correct?
    a. A only
    b. B only
    c. Both A and B
    d. Neither A nor B

20. The THM 440-T4 transaxle uses
    A. A throttle valve operated by throttle cable to help provide the correct shift timing
    B. A vacuum modulator to provide the correct shift feel
    Which is correct?
    a. A only
    b. B only
    c. Both A and B
    d. Neither A nor B

CHAPTER 9

GENERAL TRANSMISSION SERVICE AND MAINTENANCE

OBJECTIVES

After completing this chapter, you should:

- Be able to correctly check the fluid level in an automatic transmission.
- Be able to determine if the fluid needs to be changed.
- Be able to change the fluid in an automatic transmission.
- Be able to check and adjust the manual shift linkage.
- Be able to check and adjust the throttle linkage.
- Be able to adjust a band.

Maintaining the correct fluid level and changing it when it becomes old is the primary transmission maintenance operation. Other things that should be checked occasionally, and corrected if found bad, include:

- Manual Linkage: To ensure that the manual valve and the internal park mechanism are positioned properly relative to the gear selector.
- Throttle Linkage: To ensure that the throttle or detent valve provides the correct fluid pressures relative to the throttle position.
- Band Adjustment: To ensure the correct band clearance. Band adjustments are found only on older transmissions.

9.1 INTRODUCTION

Many automatic transmissions and transaxles operate properly for many miles, well over 50,000, while almost totally neglected. Others fail because of any number of reasons. Many that fail could probably have operated for a longer period of time if they had been maintained. Several surveys of transmission shops have produced responses that over 80 percent of transmission failures were the result of neglected fluid checks or changes.

9.2 FLUID CHECKS

The operator of a vehicle should check the fluid level in an automatic transmission periodically; a good time is at every engine oil change. If the level is low, fluid of the correct type should be added. If the fluid is bad, it should be changed.

Most transmission dipsticks are marked with two fluid levels, one for cold and another for hot temperatures. The most obvious markings are for the hot level,

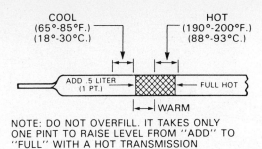

Fig. 9-1 Normal fluid level for a hot transmission should be somewhere in the hatched area of the transmission dipstick; note how the fluid level changes on a cool or warm transmission. *(Copyrighted material reprinted with permission from Hydra-matic Div., GM Corp.)*

which is the normal temperature operating range, about 150 to 170°F (66 to 77°C); some manufacturers use a higher operating range of 180 to 200°F (82 to 93°C). It usually takes 1 pint (0.5 L) to change the fluid level from low to full (Fig. 9-1). The cold operating range is indicated by two dimples or holes for the low and full points. Room temperature of about 65 to 85°F (18 to 29°C) is considered cold for transmission fluid. Many technicians use a rule of thumb that a dipstick from the hot fluid is too hot to hold in your fingers. In some transaxles, a thermostatic valve is used to raise the fluid level in the upper valve body pan as the transaxle warms up. These units have a lower hot level and a higher cold level (Fig. 9-2).

It is a good idea to never operate a transmission with the fluid level too high (an overfill) or too low (an underfill). An underfill is sometimes marked "Do not drive," and this is about the same level as the low, cold level. An underfill can allow air to enter the filter and pump intake, which can cause mushy operation, lack of engagement, or slipping.

An overfill can bring the fluid level up to the point where it contacts the spinning gearsets. This in turn causes foaming of the fluid and a loss of the foamy fluid out of the vent or filler pipe. There have been cases of car fires caused by the fluid spilling out of the filler pipe and onto a hot exhaust manifold. Overfilling can also cause slippage and mushy operation because of the air in the foamy fluid.

To check transmission fluid, you should:

1. Block the wheels, front and back; apply the parking brake securely; place the gear selector in park or neutral as required by the manufacturer; and start the engine.

2. With the service brakes firmly applied, move the gear selector to each of the operating ranges. Leave the selector in each position long enough for each gear to become completely engaged.

3. Return the selector level to park or neutral, depending on the transmission. Leave the engine running at idle speed. The exact checking procedure is often written on the dipstick.

4. Clean any dirt from the dipstick cap, and remove the dipstick (Fig. 9-3).

5. Wipe the dipstick clean, and return it completely back into the filler pipe making sure that it is seated.

6. Pull the dipstick out again, and read the fluid level. Carefully grip the end of the dipstick between your fingers to get an indication of the fluid temperature.

   a. If it feels cold, use the COLD marks.

   b. If it feels warm and you can hold onto it, the correct fluid level will be between the two sets of marks.

   c. If it is too hot to hold onto without burning yourself, use the HOT marks.

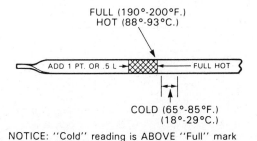

Fig. 9-2 Unusual transaxle in that it has a higher cold fluid level than a hot level. It is a good idea to refer to the vehicle owner's manual or a service manual if you are not sure of the correct checking procedure. *(Copyrighted material reprinted with permission from Hydra-matic Div., GM Corp.)*

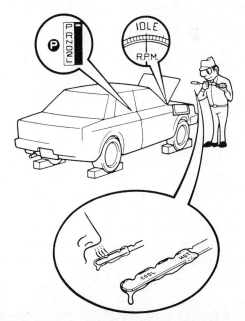

Fig. 9-3 When checking the fluid level, the gear selector should be positioned as required by the manufacturer and the engine should be at idle speed. Note the fluid level and compare it with the markings on the dipstick. It is also a good practice to note any unusual odors of the fluid. *(Courtesy of Toyota Motor Sales, U.S.A., Inc.)*

HELPFUL HINT: If the fluid is fairly new, it appears almost transparent and can be very difficult to read. You can rub the dipstick with ordinary carbon paper to darken the fluid and make it easier to see.

7. Replace the dipstick completely into the filler tube.

NOTE If the vehicle has just been operated under heavy loads or at high speeds, it is a good idea to let it sit for about 30 minutes to cool down before checking the fluid level.

NOTE Occasionally fluid suspended in the filler pipe causes a false high or normal reading. If there is any sign of fluid above the correct range, repeat Steps 5 and 6 several times. Also, it is a good practice to read both sides of the dipstick and believe the lower level.

9.2.1 Fluid Condition

Fluid condition should always be checked while checking the fluid level. A transmission technician will normally smell the fluid and check the color for anything unusual. We normally expect the fluid to be a bright reddish color with a smell that is similar to new fluid. It should be noted that some newer Dexron II fluids will normally darken and take on a definite odor after a few hundred miles; do not confuse this normal occurrence with the other signs of fluid breakdown.

If the fluid has foreign particles in it or has a dirty appearance, it is a good practice to wipe the dipstick clean using a clean white paper towel or facial tissue. The wetness of clean fluid will slowly spread through the paper leaving an even stain. Dirty fluid will leave specks or deposits of the very fine solid material while the fluid spreads through the paper (Fig. 9-4).

Indications of old, bad fluid are:

- Brown or black color, which indicates dirt or burned friction material.
- Pink or milky color, which indicates a coolant leak at the oil cooler in the radiator.
- A definite burned odor, which indicates slippage or overheating.
- A varnishlike odor, which indicates fluid oxidation and breakdown. This is often accompanied by a gold-brown varnish coating on the dipstick.
- Foam, which might indicate a leak in the pump intake system.

Dirty or contaminated fluid should be changed, and while this is done, the filter and the inside of the pan and transmission case should be checked for contamination, varnish, or other signs of possible transmission problems.

9.3 FLUID CHANGES

Most manufacturers recommend fluid changes every 100,000 miles (160,000 km) under normal driving conditions. This is usually accompanied with a recommendation that the change interval be shortened as low as 15,000 miles (24,000 km) under severe driving conditions. These are described as:

- Frequent trailer pulling.
- Heavy city traffic, especially in areas of relatively high temperature, above 90°F (32°C).
- Very hilly or mountainous conditions.
- Commercial use such as taxi or delivery service.
- Police or ambulance usage.

The key to transmission fluid life is heat or how hot the fluid is during vehicle operation. If the temperature can be kept below 175°F (79°C), the life should be 100,000 miles. At higher temperatures, oxidation causes fluid breakdown at a rate of one-half life for every increase of 20°F (11°C). A chart of transmission fluid life is shown in Table 9-1.

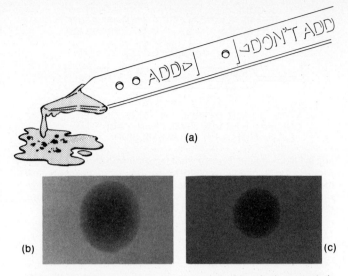

Fig. 9-4 (a) Fluid condition can be checked by placing a sample on clean, white, absorbent paper. (b) Clean fluid will spread out and leave only a wet stain. (c) Dirty fluid will leave deposits. (Courtesy of Ford Motor Company.)

TABLE 9-1 Transmission Fluid Life

Fluid Life (miles)	Temperature (°F)
100,000	175
50,000	195
25,000	215
12,500	235
6,250	255
3,125	275
1,560	295
780	315
390	335
195	355
97	375
48	395

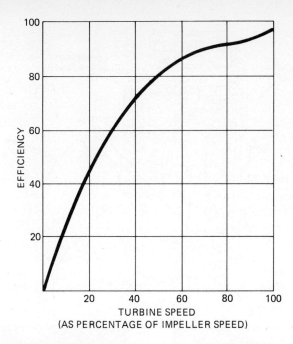

Fig. 9-5 The efficiency of a torque converter is very low during those occasions where there is much internal slippage. At this time much heat is being generated in the fluid inside the converter.

Besides affecting fluid life, high temperatures also cause hardening and early failure of the rubber seals.

You can relate much of the fluid temperature to the torque converter and its operating efficiency. Remember that a converter is zero percent efficient at stall (Fig. 9-5). Power is coming in, but no power is going out, so all of the mechanical energy that is going in is becoming heat in the fluid. Heat is generated whenever the converter is slipping and multiplying torque. A transmission also generates heat because of friction at the bushings and bearings and the action of the gears and in the pressurized fluid in the hydraulic system, but this is fairly small compared to the heat generated by the converter. It can be safely said that the fluid life is shortest in vehicle operation that has a lot of torque converter usage.

The fluid should be changed when it starts to break down, which is best indicated by the fluid appearance and smell. It is wise to change the fluid early, before transmission damage occurs. Many technicians fear that dirty fluid may cause the valves to start sticking, which in turn can cause sluggish shifts and slippage and thus more fluid heat, breakdown, contamination, and damage.

Most transmissions do not have a drain plug in the oil pan. The fluid is drained by allowing it to spill over the side of the pan as it is removed. The pan is first lowered at an angle to allow the fluid to spill and then removed completely (Fig. 9-6).

To change fluid in a transmission, you should:

1. Raise and securely support the vehicle. This is normally done on a hoist to allow complete access to the pan bolts.

2. Select the best direction for fluid to spill from the pan, place a large drain pan in this area, and remove all but two of the pan bolts. The remaining two bolts will serve as the "hinge" for lowering the pan.

3. Loosen the two remaining bolts about two turns (Fig. 9-7).

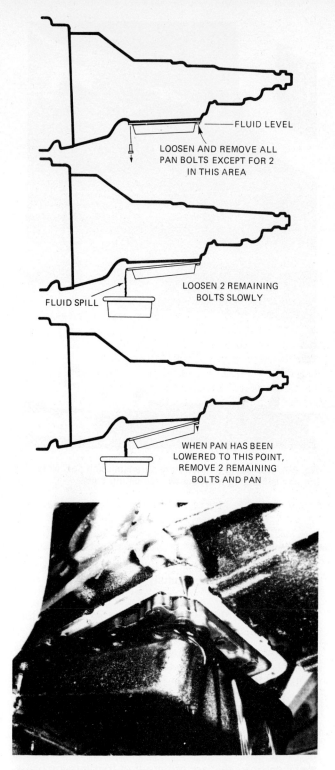

Fig. 9-6 Since transmissions do not have drain plugs, the fluid must be spilled in a controlled fashion by lowering the pan at an angle. To do this, all bolts but two on the left side are removed. As the last two bolts are loosened, the pan will lower, and fluid will spill over the right edge.

Fig. 9-7 Fluid being drained from a transmission pan. Note how the pan is hanging from the two remaining bolts.

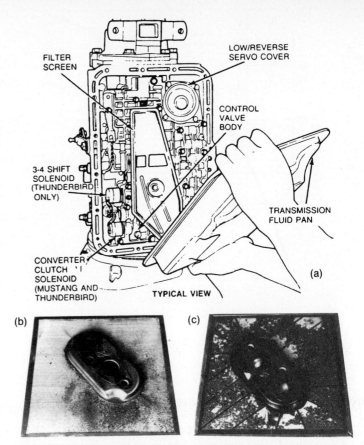

Fig. 9-8 (a) The filter is normally exposed when the pan is removed. It is usually an easy operation to replace the filter at this time. (b) A dirty filter and (c) a badly clogged and torn filter. *(Courtesy of Ford Motor Company.)*

4. If the oil pan does not come loose and start dropping on its own, carefully pry the pan loose from the transmission.

5. Loosen the two bolts further so a continuous and controlled spill occurs.

6. When the pan lowers to an angle of about 30° to 45°, support it by hand; remove the remaining two bolts and the pan; and finish draining the pan.

7. Remove the filter, which is usually attached to the valve body (Fig. 9-8).

8. Inspect the pan, filter, and transmission for debris and varnish buildup (Fig. 9-9). A small amount of metal particles that are the result of transmission break-in are considered normal.

9. Install a new filter using a new gasket or O-ring, and tighten the mounting bolts to the correct torque. If an O-ring is used, it should be lubricated with petroleum jelly or automatic transmission fluid (ATF).

10. Clean the oil pan and dry it with compressed air. Check the flanges for dimpling or bends where the pan bolts are. If the flange surface is bent upward, the dimples should be flattened using a hammer and block of wood, vise, or anvil (Fig. 9-10).

CHECK FLUID AND OIL PAN

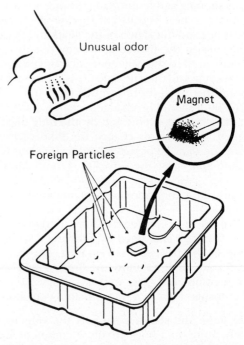

Fig. 9-9 After the pan has been removed, it should be inspected for foreign particles. A small amount of material with some iron attracted to the magnet is normal. *(Courtesy of Toyota Motor Sales, U.S.A., Inc.)*

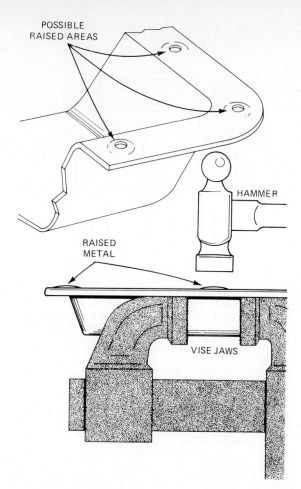

Fig. 9-10 The flange of the pan should be checked for distortion caused by overtightening the bolts. This can be corrected by tapping down the raised areas as shown.

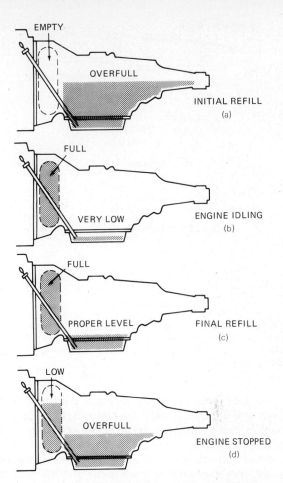

Fig. 9-11 (a) If a transmission and converter are drained, the proper amount of fluid will overfill the transmission. Enough fluid should be added to fill, (b) the engine should be started, and (c) more fluid should be added as needed for the proper fill. (d) This amount can show an overfull unit if it is checked with the engine stopped. *(Courtesy of Ford Motor Company.)*

11. Install a new gasket on the pan, and install the pan on the transmission. The bolts should be tightened in a back-and-forth, across-the-pan sequence and to the correct torque. Overtightening can cause a future leak by compressing the gasket too tightly and bending the flange.

12. Lower the vehicle, and add the proper amount of fluid.

13. Start the engine, check the fluid level as described earlier, and add additional fluid to correct the level if necessary.

It should be noted that this procedure only changes the fluid that is in the pan, and this is about a third to half that in the transmission (Fig. 9-11). The remaining fluid stays in the torque converter, clutch and band servos, accumulators, and fluid passages. It is possible to drain the converter by drilling a hole in it and installing a plug, but most technicians do not because of the time involved and the possible damage that can occur in the converter. How to drill a converter is described in Chapter 13. For very dirty transmissions, it is best to change the fluid in the pan as just described; operate the transmission long enough to thoroughly mix the new and old fluid; and then change the fluid again.

In cases of very dirty fluid, it is good practice to check for a plugged oil cooler. Debris can plug up the cooler, and this causes severe gear train and bushing wear because of the reduction of cooler and lubrication oil flow. The procedure used to check fluid flow through a cooler will be described in Chapter 11.

9.4 MANUAL LINKAGE CHECKS

Every automatic transmission has an adjustable portion of the linkage to allow the manual valve to be positioned correctly relative to the gear selector. It is wise to check this adjustment periodically because a misadjusted linkage can cause the manual valve to leak oil pressure into the wrong passage (Fig. 9-12). Imagine allowing a small amount of line pressure to leak into the drive passages while in neutral. This might cause the clutch to partially apply and produce only a slight creep condition, but even worse, this slippage will produce overheating and burnout. Another potential fear is for the transmission to shift into park without the gear selector entering the

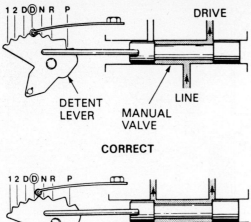

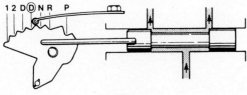

INCORRECT

Fig. 9-12 A misadjusted manual shift linkage will place the manual valve in the wrong position so fluid can leak into the wrong passage. *(Copyrighted material reprinted with permission from Hydra-matic Div., GM Corp.)*

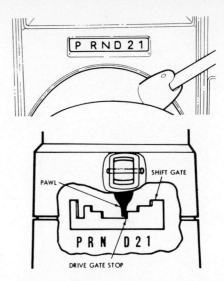

Fig. 9-13 The gate stops prevent the shift selector lever from being moved unless the pawl is ungated by moving the lever rearward or toward the steering wheel. You should be able to feel these gates as you move the selector lever across the shift quadrant. *(Courtesy of Ford Motor Company.)*

locking gate. This allows the selector level to slip out of park and the transmission to shift into reverse gear—a highly dangerous condition. This occurrence has caused fatalities.

Since the detents act on the transmission's internal linkage, they normally stay correctly aligned with the valve position. To check for correct linkage adjustment, you should:

1. Firmly set the parking brake and leave the engine shut off.

2. Move the selector lever lock or the selector lever to the unlocked or ungated position to allow moving the selector lever freely through the different ranges (Fig. 9-13).

3. Move the selector level through the ranges, and observe the range pointer as you feel the internal detents. You should feel the detent engage as the pointer aligns for the gear position (Fig. 9-14).

4. Move the selector lever to park, and release the lever/lock. The parking pawl should freely engage to lock the transmission, and the lever lock should freely enter the gate and be locked.

5. Test to ensure that the starter operates in park and neutral but not in other gear positions.

It should be noted that in some transmissions, the neutral start switch is activated by the manual lever quadrant inside the transmission so it will always be positioned correctly relative to the manual valve (Fig. 9-15). Improper starter operation on these cars indicates a misadjusted shift linkage (Fig. 9-16). Other transmissions mount the neutral start switch outside the transmission—on the outside of the case or at the shift lever—and these switches have a separate adjustment (Fig. 9-17).

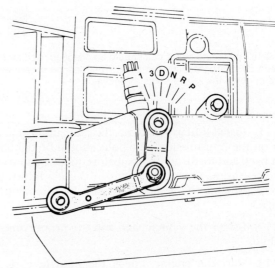

TRANSMISSION LEVER POSITIONS

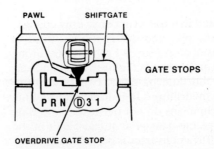

GATE STOPS

Fig. 9-14 As the shift selector lever is moved across the quadrant, you should feel the detents inside the transmission at the same time that the indicator shows a gear position. *(Courtesy of Ford Motor Company.)*

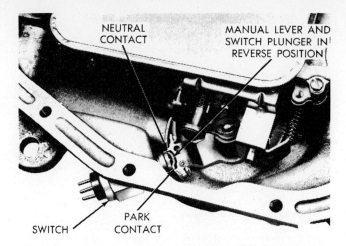

Fig. 9-15 This neutral safety switch is mounted in the transmission case and is activated by the same cam that positions the manual valve. It always stays in adjustment with the manual valve so it can be used to check the adjustment of the linkage between the transmission lever and the shift lever. *(Courtesy of Chrysler Corporation.)*

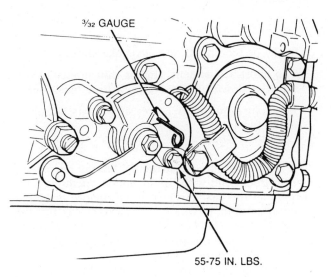

Fig. 9-16 Neutral safety switch mounted to the outside of the transmission. When the transmission is in neutral a $\frac{3}{32}$-in. gauge should enter completely into the switch if it is adjusted correctly. If the gauge does not enter, the mounting bolts are loosened so the switch can be readjusted. *(Courtesy of Ford Motor Company.)*

Also, in some cars, the range pointer is operated by a line operated by a lever on the shift column (Fig. 9-18). On these cars, it is possible for the pointer to be out of adjustment relative to the linkage and manual valve. Adjustment of the pointer position is usually a simple matter of changing the pointer or connecting link position.

While checking the adjustment, it is good practice to also watch for any binding or hard movement, which indicates a shift cable that is dirty, rusty, or beginning to fail. Also check for excessive lever slide or vertical movement or sloppy motion, which indicates failure of the selector lever.

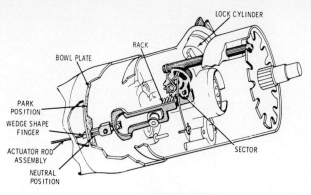

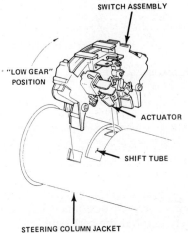

INSTALLATION PROCEDURE

1. Place gear selector in neutral.
2. Align acutator on switch with hole in shift tube.
3. Position rearward portion of switch (connector side) to fit into cutout in lower jacket.
4. Push down on front of switch: The two tangs on housing back will snap into place in rectangular holes in jacket.
5. Adjust switch by moving gear selector to park. The main housing and the housing back should ratchet, providing proper switch adjustment.

READJUSTMENT PROCEDURE:

1. With switch installed, move the housing all the way toward low.
2. Repeat step 5.

Fig. 9-17 Neutral safety switch mounted in the steering column. The correct adjustment procedure is shown. *(Courtesy of Oldsmobile Div., GM Corp.)*

9.4.1 Manual Linkage Adjustment

If the starter engagements occur in the wrong position or the transmission detents do not align correctly relative to the gear range pointer, the manual linkage should be adjusted. This operation will vary between car models, and it is wise to check a service manual for the exact procedure for a particular car (Fig. 9-19). The description that follows is general. To adjust manual shift linkage, you should:

249

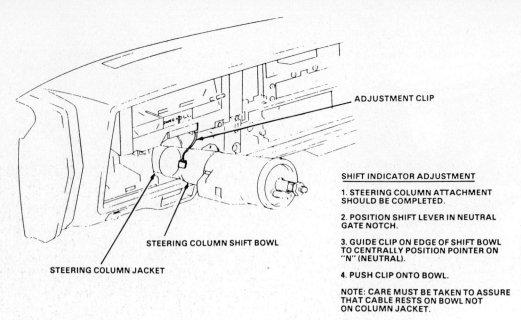

Fig. 9-18 The adjustment clip connects the shift lever to the range pointer at the gear indicator. The adjustment procedure is shown. *(Courtesy of Oldsmobile Div., GM Corp.)*

1. Shift the gear selector into park making sure the selector lever is correctly positioned and the lever pawl has completely entered the park gate.

2. Raise and securely support the car on a hoist or jack stand.

3. Locate the manual linkage and the adjustment in the linkage. The adjustable portion is a slot in the shift rod or cable, a swivel clamp, or a long threaded portion using a bracket and two nuts (Fig. 9-20). Loosen the clamping nut or bolt.

4. Move the transmission lever to park, making sure that the detent and the parking pawl are completely engaged. Test this by trying to rotate the drive shaft(s); it should be securely locked. Adjust the linkage to the proper length.

5. Tighten the clamping nut or bolt, making sure that the transmission lever or the shift rod do not move out of adjustment.

6. Repeat the linkage check described earlier to ensure correct adjustment.

9.5 THROTTLE LINKAGE CHECKS

In most transmissions, a throttle rod or cable connects the engine's throttle linkage to the transmission's throttle valve. This linkage is commonly called TV linkage. The TV linkage should produce a rise in TV or line pressure that is matched to the throttle (Fig. 9-21). This pressure controls shift timing—at what speeds the upshifts and downshifts occur—and usually shift feel also. A sticky, binding, or improperly adjusted TV linkage can cause the following problems:

Sticky or Binding: Erratic, irregular shifts.

Too Short Adjustment: Higher shift points; overly sensitive downshifts; excessively firm, harsh shifts.

Too Long Adjustment: Lower shift points; late or no downshifts; soft, mushy shifts.

It should be noted that the TV linkage movement begins at idle speed with the engine at operating temperatures so the fast idle cam is off. Before making a linkage adjustment, it is good practice to ensure that the idle speed is correct (Fig. 9-22). Also, the throttle should reach its wide-open throttle (WOT) position before the transmission's throttle valve bottoms.

TV linkage checks are fairly easy, and they are normally made with the engine off. In some cases, the linkage binds only with the engine running, so if there are problems that point to this, some of the checks should be repeated with the engine running. To check TV linkage, you should:

With the engine off:

1. Check to ensure that the throttle is closed against the idle-speed adjusting screw; check the cable or rod for any slack or free movement. In most cases, there should be none.

2. Roll the throttle to the WOT position and, while doing this, check the cable or rod for free or sloppy movement, binding, or sticking (Fig. 9-23). The TV linkage should move to the WOT position and release freely and smoothly.

3. With the throttle at WOT, try moving the TV linkage farther. It should be able to move a very small amount before contacting its stop.

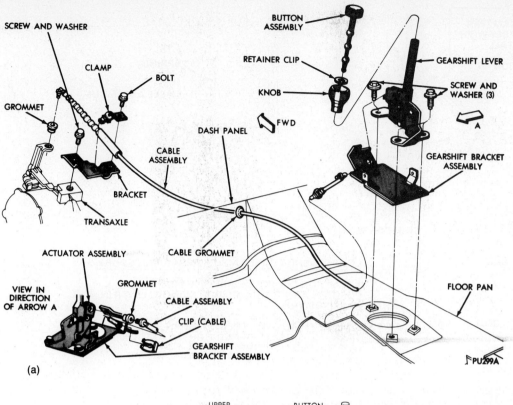
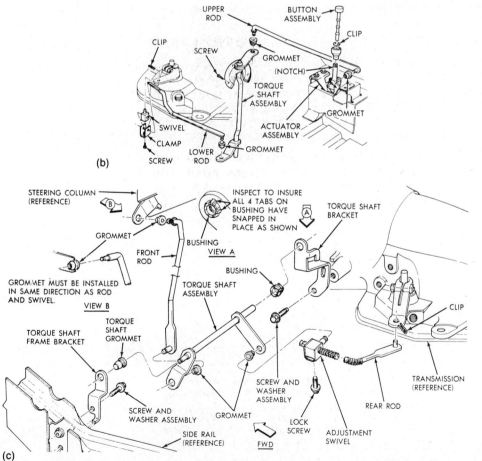

Fig. 9-19 (a) Floor-mounted shift lever with a cable shift linkage, (b) floor-mounted shift lever with a rod linkage, and (c) column-mounted shift lever with a rod linkage. Note the adjustment swivel on the two rod linkages; this cable shift is adjusted at the transaxle bracket and clamp. *(Courtesy of Chrysler Corporation.)*

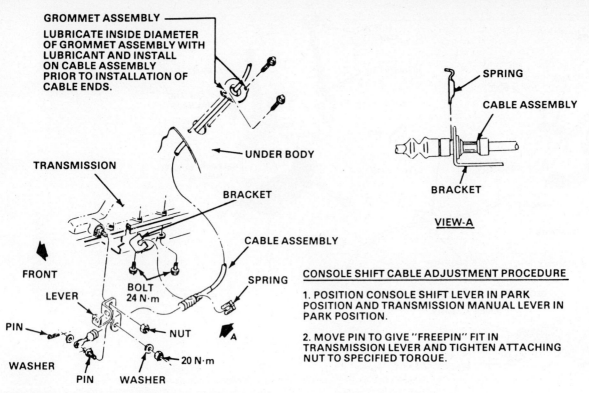

Fig. 9-20 Manual shift cable adjusted using the procedure shown. *(Copyrighted material reprinted with permission from Hydra-matic Div., GM Corp.)*

With the engine running:

1. Firmly set the parking brake, and shift the gear selector to neutral.

2. Pull on the end of the TV cable or move the TV rod so it moves through its travel and then release it. During the linkage movement, check for free or sloppy movement, sticking, or binding. There should be none.

NOTE A more exact check for proper TV operation can be made using a pressure gauge installed in the transmission (Fig. 9-24). This procedure is described in the next chapter.

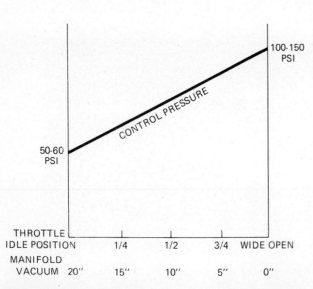

Fig. 9-21 As the throttle is opened, throttle pressure and mainline control pressure increases as shown.

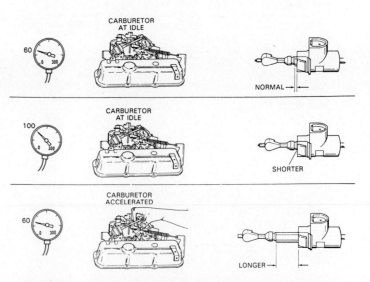

Fig. 9-22 A faulty throttle linkage adjustment causes incorrect transmission control pressures; they will be either too high (center) or too low (bottom). *(Copyrighted material reprinted with permission from Hydra-matic Div., GM Corp.)*

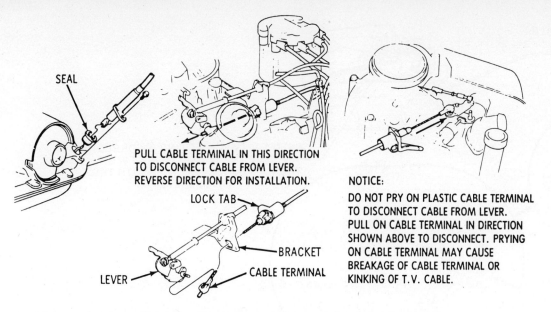

Fig. 9-23 Throttle cable attached to the transmission and throttle linkage as shown. Note how the cable and terminal are moved to disconnect them from the linkage. *(Copyrighted material reprinted with permission from Hydra-matic Div., GM Corp.)*

In many cases, it is possible to drop the pan and note the position of the TV plunger relative to throttle position (Fig. 9-25). It should move directly with the throttle. Improper TV operation should be corrected by readjusting, rerouting, repairing, or replacing the linkage.

9.5.1 Throttle Linkage Adjustments

Like the manual shift linkage, all TV linkage has an adjustment built into it. There is little standardization between manufacturers so it is necessary to consult the service manual for that particular car when making an adjustment. The descriptions that follow are examples of some of the methods that you can expect. All adjustments are made with the engine off and the throttle at normal idle speed.

"Sliding swivel" adjustment:

1. Loosen the swivel lock screw; the TV rod should slide in the swivel (Fig. 9-26).

2. Move the TV linkage forward against its internal stop, and retighten the lock screw.

"Snap lock" adjustment:

1. Lift the snap lock to the unlocked position; the cable housing should be free to slide in the assembly (Fig. 9-27).

2. Rotate the throttle to WOT, and push the snap lock inward to lock.

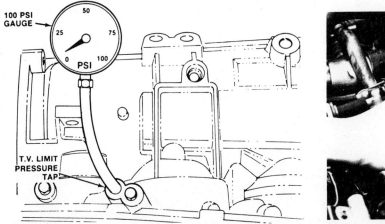
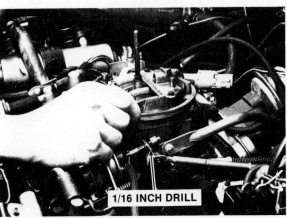

Fig. 9-24 It is recommended that the throttle linkage adjustment on this transmission be checked using a pressure gauge. At idle speed with a $\frac{1}{16}$-in. gauge, there should be less than 5 psi. If the gauge increases to $\frac{5}{16}$ in., there should be at least 22 psi. *(Courtesy of Ford Motor Company.)*

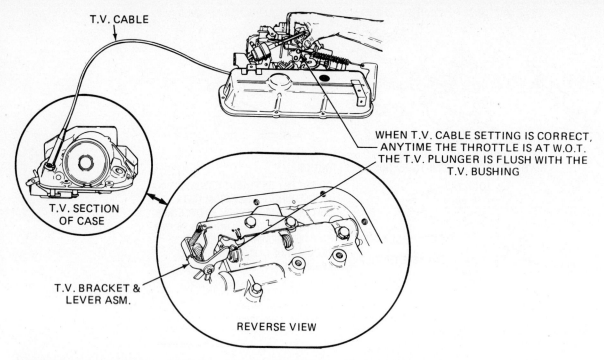

Fig. 9-25 When the throttle is wide open, the throttle valve plunger should be as shown *(Copyrighted material reprinted with permission from Hydra-matic Div., GM Corp.)*

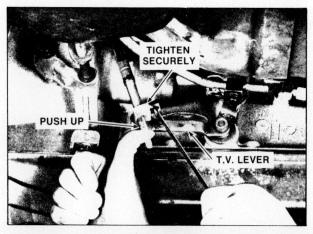

Fig. 9-26 This throttle linkage is adjusted by loosening the clamping bolt, moving the lever to its correct position, and then retightening the clamp. *(Courtesy of Ford Motor Company.)*

"Self-adjusting" adjustment:

1. Depress the readjusting tab, and move the slider inward against the base of the adjuster housing and release the tab (Fig. 9-28).

2. Rotate the throttle to WOT. As this occurs, the cable slider should move at least three "clicks" or about $\frac{1}{8}$ in.

"Self-adjusting at transaxle" adjustment:

1. Depress the adjuster button, and extend the adjuster to its longest position by pulling on the cable housing.

2. Using the proper special tool, rotate the throttle to the WOT position using a torque of 50 in.-lb (Fig. 9-29). The adjuster should automatically change to the proper length.

MANUAL TYPE T.V. CABLE

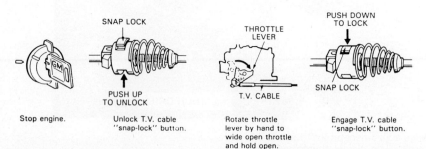

Fig. 9-27 A throttle cable using a "snap lock" adjuster is adjusted using the procedure shown. *(Copyrighted material reprinted with permission from Hydra-matic Div., GM Corp.)*

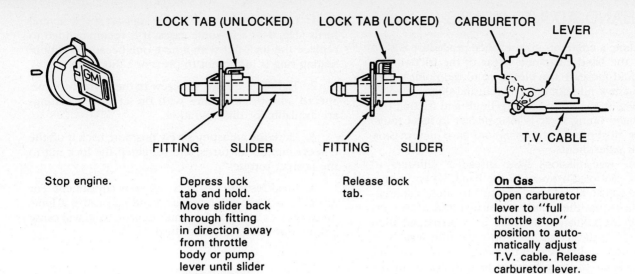

Fig. 9-28 "Self-adjusting" TV cable used on transmissions is adjusted using the procedure shown. *(Copyrighted material reprinted with permission from Hydra-matic Div., GM Corp.)*

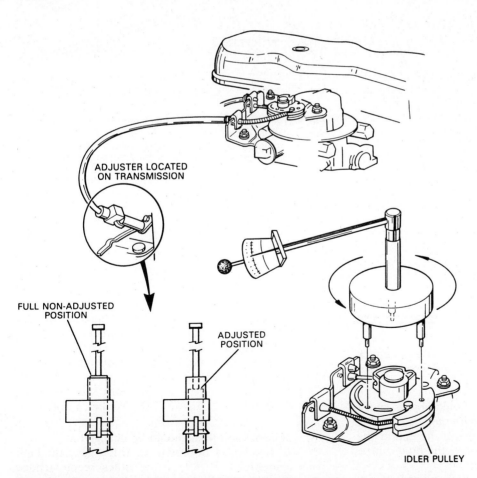

Fig. 9-29 "Self-adjusting" TV cable used with transaxles adjusted using the procedure shown. *(Copyrighted material reprinted with permission from Hydra-matic Div., GM Corp.)*

255

9.6 BAND ADJUSTMENTS

At one time, a common maintenance procedure was to readjust the band clearance. Wear of the friction material could increase the clearance to the point where engagement would not be complete and slippage would occur (Fig. 9-30). With modern fluids and friction materials, band lining wear is insignificant in most transmissions. Most modern transmissions have no provision for in-car adjustments.

Some transmissions have threaded adjusters to allow an easy readjustment of the band. When using a threaded adjuster, it is good practice to count the number of turns that the adjuster is turned inward. For example, if the adjuster is turned in 3 turns and then backed off $2\frac{1}{2}$ turns, you have an indication that there was not much lining wear. With some band adjustments, it is necessary to drop the pan to gain access to the adjuster.

It is a good practice to check the band adjustments during a fluid change or if there is a shift problem related to a band. A service manual should be checked to determine the exact adjustment procedure for each particular car. The description that follows is very general. To readjust a band, you should:

1. Loosen the lock nut on the adjuster screw several turns (Fig. 9-31). In some cases, it is recommended to replace the lock nut with a new one because a rubber sealing ring is in the nut to prevent a fluid leak.

2. Tighten the adjuster screw to the correct torque. Special adjuster wrenches with preset torque settings are available for this operation.

3. Mark the adjusting screw position, back it off the correct number of turns, and retighten the lock nut to the correct torque.

4. Road test the car to check your adjustment. For example, a loose intermediate band can cause a long, drawn-out 1–2 shift, a tight intermediate band will cause a drag in reverse, first, and third.

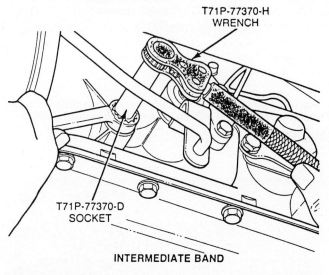

Fig. 9-31 The first step to adjust this band is to hold the adjuster screw stationary using a special socket and wrench while the lock nut is loosened with a box-end wrench. Step 2 is to turn the adjuster screw inward using the special wrench until it breaks away, indicating the correct tightness. Step 3 is to back the adjuster screw off the proper number of turns and tighten the lock nut to the correct torque. *(Courtesy of Ford Motor Company.)*

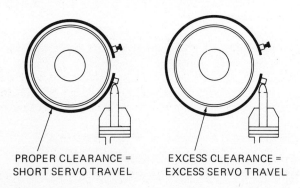

Fig. 9-30 When a band is new, the piston travels only a short distance to apply it. As the lining wears, the servo piston must travel further. When it is worn severely, the piston will not travel far enough to tighten the band.

REVIEW QUESTIONS

The following questions are provided so you can check the facts you have just learned. Select the response that best completes each statement.

1. Technician A says that you should always check the transmission fluid level with the engine idling in park. Technician B says that if the end of the dipstick is too hot to hold onto, the fluid can be considered hot. Who is right?
   A. A only
   B. B only
   C. Both A and B
   D. Neither A nor B

2. Technician A says that on most dipsticks, the distance from the bottom of the hatched area to the top is equal to 1 pint of fluid. Technician B says that too much fluid in a transmission can cause bad shifts. Who is right?
   A. A only
   B. B only
   C. Both A and B
   D. Neither A nor B

3. Transmission fluid should be changed if
   A. It has black or brown coloration
   B. It has a definite burned smell
   C. It has gone the limit of miles recommended by the manufacturer
   D. Any of these

4. Technician A says that too high a fluid level can cause foamy fluid to spill out of the vents or filler pipe. Technician B says that too low a fluid level can cause this same problem. Who is right?
   A. A only
   B. B only
   C. Both A and B
   D. Neither A nor B

5. Technician A says that all ATF is a bright medium red color. Technician B says that all ATF has the same oily smell. Who is right?
   A. A only
   B. B only
   C. Both A and B
   D. Neither A nor B

6. A pink transmission fluid color indicates
   A. Normal operation
   B. Too high a level and air in the fluid
   C. Water in the fluid
   D. None of these

7. Technician A says that transmission fluid life is dependent on the transmission's operating temperature. Technician B says that driving conditions that increase fluid temperature like trailer towing or delivery type operations shorten fluid life. Who is right?
   A. A only
   B. B only
   C. Both A and B
   D. Neither A nor B

8. Technician A says that most transmissions can be drained by removing the drain plug. Technician B says that you should also change the filter when the fluid is changed. Who is right?
   A. A only
   B. B only
   C. Both A and B
   D. Neither A nor B

9. Technician A says that you normally have to start the engine as you refill a transmission with fluid. Technician B says that a torque converter can be refilled with the engine off. Who is right?
   A. A only
   B. B only
   C. Both A and B
   D. Neither A nor B

10. Technician A says that if an engine cranks while the gear position indicator is at DRIVE, the manual shift linkage is out of adjustment. Technician B says that this problem can be caused by a misadjusted neutral safety switch. Who is right?
    A. A only
    B. B only
    C. Both A and B
    D. Neither A nor B

11. You can quickly check the adjustment of the manual shift linkage by
    A. Trying the operation of the starter in the various gear positions
    B. Feeling for the transmission internal detents as you move the gear selector from park to low
    C. Feeling for complete engagement of the lever pawl into the park gate
    D. All of these

12. Technician A says that a misadjusted manual linkage can cause vehicle creep in neutral. Technician B says that a gear selector can slip out of park if the linkage is not adjusted correctly. Who is right?
    A. A only
    B. B only
    C. Both A and B
    D. Neither A nor B

13. Technician A says that most manual linkages can be easily adjusted by repositioning the adjustable end of the rod or cable. Technician B says that the selector lever should be positioned in reverse while adjusting the linkage. Who is right?
    A. A only
    B. B only
    C. Both A and B
    D. Neither A nor B

14. If the throttle linkage is too short, the transmission will probably
    A. Shift early
    B. Have soft, mushy shifts
    C. Have very late shifts
    D. Do both A and B

15. Technician A says that sticky throttle linkage can cause erratic shifts. Technician B says that passing gear engagement is affected by the throttle linkage adjustment. Who is right?
    A. A only
    B. B only
    C. Both A and B
    D. Neither A nor B

16. Technician A says that the bands in most transmissions can be adjusted using an adjuster screw that extends out at the side of the transmission case. Technician B says that if the intermediate band (three-speed transmission) is too tight, there will be a drag in reverse. Who is right?
    A. A only
    B. B only
    C. Both A and B
    D. Neither A nor B

CHAPTER 10

PROBLEM SOLVING AND DIAGNOSIS

OBJECTIVES

After completing this chapter, you should:

- Understand what checks can be used to determine the causes of the various possible transmission problems.
- Be able to properly perform these different checks.
- From the results of the checks that you perform, be able to determine the cause of a transmission problem and recommend the proper repair.

10.1 INTRODUCTION

Most transmission service operations begin as problems that need to be solved. An automatic transmission is a rather complicated device, and there are many possibilities for complete or partial failure. When experienced technicians are given a problem, based on past experiences, they can often go directly to the solution, or if not absolutely sure of the cause, they will usually perform those tests necessary to locate the specific cause of the problem.

A sound knowledge of how an automatic transmission operates helps greatly when diagnosing troubles. An example of this is a transmission that does not move in drive but starts in manual 1 and then, if shifted back into drive, operates normally. This problem is probably caused by a faulty one-way clutch. This reaction member is only used in drive 1, and the car won't move in drive without it. Another example is a transmission that moves in drive but does not upshift. Knowledge of this transmission tells us that the intermediate band is not applying, but why isn't it? In this case, there are several possible faults, for example, no governor pressure, a modulator or throttle valve fault, or a stuck 1–2 shift valve.

The following is a series of tests and checks that can be used to help determine the exact cause of problems:

Fluid Check Checks that the fluid is at the correct level and in good condition.

Modulator Checks Checks the operation of the modulator and vacuum throttle valve system.

Road Test Tests the actual operation of the transmission and can be used to confirm the original complaint; allows checking shift timing and quality under different throttle openings, torque converter clutch operation, unusual noises, and vibrations.

Torque Converter Clutch Tests Tests made to determine if the torque converter clutch is operating properly and, if not, what is wrong.

Hydraulic Pressure Test Checks the hydraulic system pressures, which give a good indication of the condition of the pump, pressure control valves, various seals, and so on.

Electrical System Checks Checks to ensure that the electronic controls are operating correctly.

Stall Test Loads the clutches and bands to check for slippage; also checks the torque converter stator clutch.

Oil Pan Debris Check An inspection of the oil pan to determine if an abnormal amount of debris is present and, if so, the nature of the debris.

Air Test Tests made to determine if the seals in the hydraulic circuits are operating correctly.

Leak Checks Checks that help locate a fluid leak.

Noise and Vibration Checks Checks to help locate the cause of a noise or vibration.

Depending on the nature of the problem, these checks are made in the most logical order for that particular problem.

10.2 PROBLEM-SOLVING PROCEDURES

One of the difficulties in describing a diagnostic procedure is caused by the varied types of problems that are encountered. There is no single set procedure. The process of finding and curing a leak or noise problem is different from the procedure to locate a faulty upshift. With experience, a technician makes the checks or tests of the various things that can cause a particular problem. It is recommended that a procedure such as that shown in Fig. 10-1 be followed.

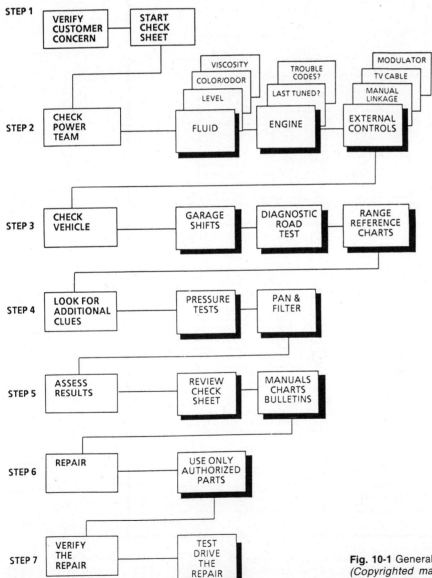

Fig. 10-1 General automatic transmission diagnosis procedure. (Copyrighted material reprinted with permission from Hydramatic Div., GM Corp.)

Problem solving should always begin with identifying the problem. This usually starts with the person who brings the problem to you. It is important to realize that that person feels that something is wrong and may not have much of an understanding of automatic transmissions or automotive drive trains. In many cases, it is good practice to go for a road test with the customer so he or she can more easily explain the problem to you. At this time, it often helps to take notes. Some manufacturers recommend that you follow a check sheet like that shown in Fig. 10-2. This gives you a more professional appearance and helps keep you from skipping important checks.

Ford Automatic Transmission Diagnosis Guide
Ford Customer Service Division

General: This form must be completely filled in throughout the steps required to diagnose the condition covering transmission malfunction complaints (e.g., erratic shifting, slippage during shifts, failure to shift, harsh and delayed shifts, noise, etc.). It is not necessary to complete this form on complaints involving external leaks.

Transmission Model _____ Transmission Date Code/or Serial No. _____
R.O. No. _____ Axle Ratio _____ Tire Size _____

DIAGNOSIS PROCEDURE

Following steps will provide complete data necessary to perform an accurate diagnosis of transmission difficulties.

1. Check transmission fluid level ☐ Room Temp. ☐ Operating Temp. ☐ OK ☐ Overfilled ☐ Low
2. Engine (CID) and Calibration Number _____
   Idle RPM in Drive _____
   Specification _____ As Received _____ Set To _____
   Check EGR System (if so equipped)
   Valve Operation ☐ OK ☐ Other (Explain) _____
   Restriction ☐ OK ☐ Other (Explain) _____
3. Check downshift and manual linkage ☐ OK ☐ Other (Explain) _____
4. Drive the car in each range, and through all shifts, including forced downshifts, observing any irregularities of transmission performance.

Throttle Opening	Range	Shift	Shift Points (MPH) Record Actual	Record Spec.
Minimum (Above 12" Vacuum)	D	1-2		
	D	2-3		
	D	3-1		
	1	2-1		

Throttle Opening	Range	Shift	Shift Points (MPH) Record Actual	Record Spec.
To Detent (Torque Demand)	D	1-2		
	D	2-3		
	D	3-2		
Thru Detent (WOT)	D	1-2		
	D	2-3		
	D	3-2		
	D	2-1 or 3-1		

5. Control Pressure Test _____ AND _____ Stall Speed Data
   Transmission fluid must be normal operating temperatures. DO NOT hold throttle open over five seconds during tests.
   CAUTION: Release throttle immediately if slippage is indicated.
   After each stall test move selector lever to neutral with engine running at 1000 RPM to cool the transmission.

Engine RPM	Manifold Vacuum In-Hg	Throttle	Range	PSI Record Actual	Record Spec.
Idle	Above 12	Closed	P		
			N		
			D		
			2		
			1		
			R		
As Required	10 ①	As Required	D, 2, 1		
As Required	Below 3	Wide Open	D		
			2		
			1		
			R		

Above Specified Engine RPM
1. Transmission slippage
2. Clutches or bands not holding
Below Specified Engine RPM
1. Poor engine performance, such as need for tune-up
2. Converter one way clutch slipping or improperly installed

Specified Engine RPM	Record Actual Engine RPM

① On units equipped with a dual area diaphragm, the front port of diaphragm must be vented to atmosphere (hose disconnected and plugged) during this check only.

After the tests, you should know the following items:
- CONTROL PRESSURE – Does the transmission have the CORRECT CONTROL PRESSURE? ☐ Yes ☐ No
- CONTROL VALVES – Beyond the manual valve are all the CONTROL VALVES FUNCTIONING? ☐ Yes ☐ No
- HYDRAULIC CIRCUITS – If the first two items check out good, then check the transmission's internal hydraulic circuits that are beyond the VALVE BODY. These circuits must be checked during transmission disassembly.

6. TORQUE CONVERTER AND OIL COOLER (where applicable)
   - Was torque converter flushed with a mechanical cleaner? ☐ Yes ☐ No
   - Was oil cooler flushed with a mechanical cleaner? ☐ Yes ☐ No
7. The problem was diagnosed to be: _____

8. If it was necessary to disassembly the transmission, record the actual problem found: _____

Mark Corp Nov 72 — Litho in U.S.A.

Fig. 10-2 Diagnosis guide recommended by Ford Motor Company provides a series of systematic checks and a place to record the test results in a professional manner. *(Courtesy of Ford Motor Company.)*

10.3 DIAGNOSTIC PROCEDURE

All transmission problem solving should begin with a check of the fluid level and condition. Many problems can be caused by low or high fluid level, and the condition of the fluid will often give you a clue as to the cause of the problem. While checking the level, you should also check the condition as described in Chapter 9. Many technicians make it a practice also to check the manual shift linkage and the throttle linkage at this time. The linkage checks don't take very long, and there can be several problems if the linkages are not correct. It is also good practice to ensure that they are correct so the transmission will work correctly when you are done with your checks.

The next check to make varies with the nature of the problem, and there can be over a dozen different problem areas. The normally encountered problems and the checking procedure is shown in Chart 10–1.

Before proceeding very far in the diagnostic procedure, don't forget that the engine must be in a good state of tune. A poor running engine can cause problems that may appear as faults in the transmission.

10.4 MODULATOR CHECKS

A vacuum modulator is usually used to control shift feel. It uses the manifold vacuum signal from the engine to adjust TV, line, accumulator, or shift valve pressure. Modulator checks should include diaphragm leakage, vacuum supply, sleeve/stem alignment, spring pressure, and on-car operation. The on-car operation will be described in a later section along with the other hydraulic pressure checks.

10.4.1 Modulator Diaphragm Leakage

There are several simple ways to check that the modulator has a sound diaphragm. The easiest is to remove the vacuum hose from the modulator and inspect the inside of the hose for automatic transmission fluid

DIAGNOSTIC CHECKS

A. Fluid level and condition checks
B. Manual linkage checks
C. Throttle linkage checks
D. Check engine idle speed
E. Modulator tests
F. Road test
G. Torque converter clutch checks
H. Band adjustment
I. Pressure test
J. Stall test
K. Oil pan debris check
L. Air test
M. Oil leak checks
N. Vibration checks

Problems	Check
1. Noise	A, F, H, K
2. Vibration	A, E, N
3. Smell	A, F, M
4. Leaks	A, M
5. Blows oil out vent or filter	A, I, L, M
6. No forward or reverse gears	A, B, F, I, L
7. Slips in any gear	A, B, F, H, I, J, K, L
8. Slow initial engagement	A, B, C, E, I
9. Harsh initial engagement	A, C, D, E
10. No upshifts or downshifts	A, C, E, F, H, I, L
11. Harsh, jerky shifts	A, C, E, F, H, I
12. Engine flare during shift	A, C, E, F, H, I
13. Jerky action while driving	A, F
14. Converter clutch does not engage	A, F, G, I
15. Stalls at stops	A, F, G
16. Creeps at stops	A, D, F, K
17. Creeps in neutral	A, B, F
18. Early, late, or irregular shifts	A, C, E, F, I
19. Skips intermediate gear	A, B, F, H
20. Drags or locks up on shifts	A, B, F
21. No power—poor acceleration	A, F, J

CHART 10-1 Diagnostic Check Procedure

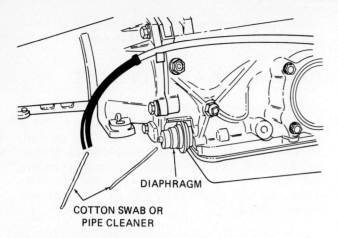

Fig. 10-3 A cotton swab or pipe cleaner should show no sign of ATF it it is used to wipe the inside of the modulator port or vacuum hose. *(Courtesy of Ford Motor Company.)*

Fig. 10-5 Push the modulator plunger/rod inward and cover the vacuum port with your finger. As you release the pressure on the rod, you should feel a vacuum at the vacuum port.

(ATF). Some technicians wipe the inside of the hose with a cotton-tipped swab (Fig. 10-3). Any red ATF easily shows up on the white cotton. Any sign of ATF in the hose indicates a faulty modulator diaphragm.

Another way to check for diaphragm leakage is to connect a hand vacuum pump to the modulator port and operate the pump to about 20 in. Hg (inches of mercury) (Fig. 10-4). The vacuum should hold and not leak down for at least 30 seconds. This check can be made with the modulator off the car or mounted in the transmission. If this test is made with the modulator off the car, you should be able to see the sleeve/plunger move smoothly inward as the vacuum is applied and back out again as the vacuum is released.

A third way of checking the diaphragm is simpler but not as accurate. Push the sleeve/plunger all the way inward; it should move smoothly inward against a spring pressure. With the sleeve all the way inward, place your finger over the vacuum port and release the plunger/sleeve. You should be able to feel a definite vacuum at the port (Fig. 10-5). A modulator with a leaky diaphragm must be replaced.

10.4.2 Modulator Vacuum Supply

A modulator must have a good vacuum signal in order to operate correctly. Many technicians make this check first when they suspect a modulator problem. You can check this by installing a tee and a vacuum gauge at the modulator port (Fig. 10-6). Now start the engine, and let it idle in park (parking brake applied). With the engine at operating temperature, the vacuum reading at the modulator should be about 15 to 20 in. Hg or greater. This reading is affected by altitude and engine condition. Higher altitudes produce lower readings; 10 to 15 in. Hg is normal for an elevation of 5000 ft. Worn or out-of-tune engines also produce weaker vacuum signals. While checking the vacuum, the throttle should be goosed or opened and closed quickly while watching the vacuum gauge. The gauge needle should quickly drop as the throttle is opened and then raise as it is closed.

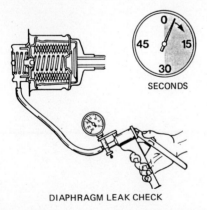

Fig. 10-4 As the hand vacuum pump is operated, the modulator plunger should move inward. The modulator should hold 15 to 20 in. Hg vacuum for at least 30 seconds. *(Copyrighted material reprinted with permission from Hydra-matic Div., GM Corp.)*

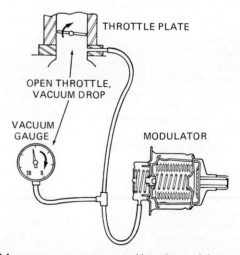

Fig. 10-6 A vacuum gauge connected into the modulator vacuum line using a tee fitting to check source vacuum. A good strong signal that varies rapidly with throttle position should be available. When the throttle is opened, the vacuum gauge should show an immediate drop. When the throttle is closed, the reading should go back to 15 to 20 in. Hg. *(Copyrighted material reprinted with permission from Hydra-matic Div., GM Corp.)*

Some technicians disconnect the vacuum line from the tee fitting while the engine is running to check for restricted lines. The vacuum leak created should cause the engine to stall or at least pick up speed and run rough. A small change in the engine speed or sluggish movement of the vacuum needle as the throttle is goosed indicates a restricted or kinked vacuum supply line.

10.4.3 Modulator Stem Alignment

The body of the modulator must be aligned with the stem or else the plunger will probably bind. The modulator is in a fairly well protected location, but if a transmission is removed or replaced, there is a possibility of bumping the modulator and bending the stem. The simplest way to check for a bent modulator is to roll it across a bench top and watch for a wobble motion of the stem, which indicates a problem (Fig. 10-7). You can expect it to roll in a circle, and units with side-mounted vacuum ports will bump as the port contacts the bench top. A bent modulator will produce an easily seen up and down wobble.

An alternate check is to place the modulator so only the stem is contacting the bench top and, again, roll the modulator (Fig. 10-8). If the body wobbles up and down, the stem is bent, and the modulator should be replaced.

10.4.4 Modulator Spring Pressure

Modulator spring pressure is also called *modulator weight* by some technicians because it is measured by "weighing" the modulator. Place the modulator with the stem end down on a 0-to-20-lb scale that is graduated in pounds (Fig. 10-9). On some modulators, an extension made from a ⅜-in. rod, a small bolt, or an old modulator valve must be used.

Next, simply push downward on the modulator as you watch the extension and the scale. At about 10 to 15 lb of pressure, you should be able to see the extension start to move inward into the modulator as the spring compresses, and as you push harder, the extension will move further inward. At some point if you keep

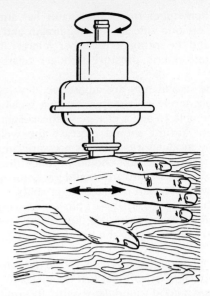

Fig. 10-8 Another way of checking for a bent stem is to place the modulator as shown and roll it back and forth. A good modulator will not wobble. *(Copyrighted material reprinted with permission from Hydra-matic Div., GM Corp.)*

pushing, the spring will be fully compressed, and the inward movement of the extension will stop. This is the maximum spring pressure, and the first motion you saw was at the minimum spring pressure.

A modulator with a higher spring pressure produces higher hydraulic pressures at a given vacuum than one with a weaker spring. This in turn produces firmer shifts. A heavier vehicle with a stronger engine generally needs a stronger, heavier modulator while a lighter vehicle or one with a weaker engine needs a softer, lighter one. Some manufacturers use different sizes of modulators, and the strength of the modulator is affected by its di-

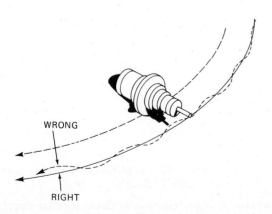

Fig. 10-7 If a modulator is rolled across a bench top, the stem section of a good modulator should not wobble up and down. A wobble motion indicates a bent stem that will probably bind or stick.

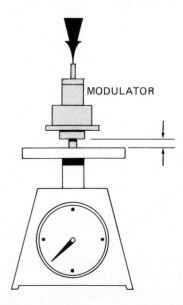

Fig. 10-9 Weighing a modulator to check the strength of its spring is done by pushing the modulator downward until the stem starts to contract into the modulator sleeve. At this point, check the reading on the scale.

ameter. Remember that a modulator has atmospheric pressure on one side of the diaphragm and manifold vacuum and the spring on the other. A larger diaphragm makes atmospheric pressure stronger relative to the spring.

Some modulators are adjustable. Try turning the adjustment screw of one after you have weighed it. You should be able to see a change in the weight. Turning the screw inward should increase the weight. This causes the modulator-controlled pressures in the transmission to increase and should produce firmer shifts. Turning the screw outward should lower the modulator weight, which should produce softer shifts.

With modulators that use a pin between the vacuum diaphragm unit and the valve, modulator pressure can be adjusted by changing the length of the pin. A longer pin will increase pressures.

A similar modulator check is called the *load check* or *bellows comparison check*. It compares the strength of a "known-good" modulator with the one you are checking. A known-good modulator is one that you know is operating correctly, and it must be of the same type and part number as the one you are checking. This check requires a special gauge that can be either purchased or made. The gauge is placed in the stem ends of the two modulators; then the modulators are pushed slowly toward each other (Fig. 10-10). If the modulators compress at the same rate while they come together, the one you are checking is good. If the modulator being checked compresses easier so the center of the gauge disappears while it is $\frac{1}{16}$ in. or more from the other modulator, the questionable modulator is bad.

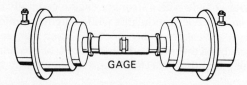

Fig. 10-10 Modulator load check that requires a special gauge and a known good modulator. The two modulators are pushed toward each other, and the gauge is read to determine if the one being tested is faulty. *(Copyrighted material reprinted with permission from Hydra-matic Div., GM Corp.)*

10.5 ROAD TEST

A road test is used to check the general overall condition of the transmission and the specific parts of its operation. Depending on the problem, a technician will install one or more of the following: a hydraulic pressure gauge to measure line, governor, shift circuit, or cooler line pressure; a tachometer to measure engine speed; a vacuum gauge to indicate engine load; a test light or voltmeter to monitor electrical circuits; or an electronic scanner tool to monitor engine or transmission operation (Fig. 10-11).

The following are some of the things that are normally checked during a road test:

- Quality of the garage shifts (neutral–drive and neutral–reverse).
- Time lag for the garage shifts.
- Quality of each upshift and downshift.

Fig. 10-11 Electronic scanner tool, a hand-held diagnostic computer. It plugs into the diagnostic link that is built into the car's electrical wiring and is programmed for the particular car being checked. A master cartridge (bottom) is plugged into the tool to program it to check transmission operation. *(Courtesy of Kent-Moore.)*

- Any busyness (hunting) between gear ranges.
- Operation of the torque converter clutch.
- Timing of each upshift and downshift.
- Slipping in any gear range.
- Binding in any gear range.
- Noise or vibration in any gear range.

The terms used to describe abnormal shifts include the following:

Bump A sudden, harsh application of a clutch or band.

Chuggle A bucking or jerking condition similar to the sensation of clutch chatter or acceleration in too high of a gear with a standard transmission (could be engine related).

Delayed The operation occurs sometime after normally expected. This is also called "late."

Double Bump Two sudden, harsh applications of a clutch or band. This is also called "double feel."

Early The operation occurs before normally expected. An early shift results in a laboring engine, poor acceleration, and sometimes chuggle.

End Bump A shift feel that becomes noticeably firmer as it is completed. This is also called "end feel" or "slip bump."

Firm A quick, easily felt shift that is not harsh or rough.

Flare A rapid increase in engine speed usually caused by slippage.

Harsh An unpleasantly firm band or clutch application. This is also called "rough."

Hunting A repeated up and down shifting sequence that produces noticeable repeated changes in engine rpm.

Initial Feel The feel at the beginning of a shift.

Shudder A more severe form of chuggle.

Slipping A noticeable loss of power transfer that results in an increase in engine rpm.

Soft A very slow shift that is barely noticeable.

Surge An engine condition that is a very mild form of chuggle.

Tie-Up A very noticeable drag that causes the engine to slow down and labor.

A technician will often use a check sheet such as the one shown in Fig. 10-12 to make sure that no important points are skipped and to keep a record of the findings.

Normally the technician will drive the car in a normal manner and listen and feel for each upshift or downshift. It is a good practice to install a tachometer (if there is not one in the instrument panel) to provide a more positive indication of an upshift or downshift. A tachometer is necessary when checking for torque converter clutch application and release. A scanner tool can record the speed, timing, and other factors involved with the shifts for information retrieval at the end of the road test (Fig. 10-13).

During the road test, light-, medium-, and full-throttle upshifts are made so the different operating conditions can be checked. The throttle positions you will use are as follows:

```
TECH 1 HYDRAMATIC TRANSMISSION CARTRIDGE
1988 J CAR        2.8L PFI LB6           VIN: W
                  4 SPD AUTO
                  ___/___/___
       VIN:                              ODO:
                   SHIFT POINT TEST
                   LIST OF SHIFT POINTS
 TIME   RPM   SPD   TPS   MAP   GEAR   SHTIME

  8.35  1989   18    25    17            0.30
 14.55  2409   27    25    17    3       0.30
 19.60  2659   37    25    17    34      0.35
 22.65  2819   38    25    17    34      0.15
```

Fig. 10-13 This scanner tool printout is a recording of a small portion of the road test. It shows us the time, engine rpm, vehicle speed, throttle position, manifold air pressure, gear, and time it took to complete a shift.

Minimum The least throttle opening that produces acceleration.

Light About one-fourth open.

Medium About one-half open.

Heavy About three-fourths open.

Wide-Open Throttle (WOT): Full downward travel of the pedal.

Road Test Checklist

Stall Test

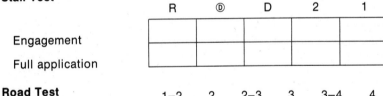

Engagement

Full application

Road Test Upshifts

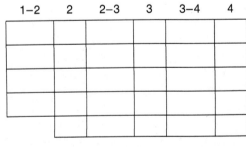

Light throttle

Medium throttle

Wide-open throttle

Manual

Torque converter clutch

Downshifts

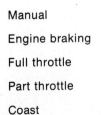
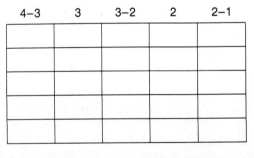

Manual

Engine braking

Full throttle

Part throttle

Coast

D = delay
H = harsh
S = slip

E = early
L = late
H = harsh
M = miss
N = noisy
S = slip
So = soft

Fig. 10-12 A road test sheet such as this allows you to quickly record your observations so you can refer back to them or show them to the car owner. The code letter (from the right side) is simply placed in the proper square if an improper operation should occur.

Detent WOT, which forces a downshift.

Zero A complete release of the throttle, which results in coasting.

Engine Braking A manual downshift with zero throttle to produce a condition where engine compression slows the car.

Many manufacturers publish shift points for their various vehicle combinations; these are the speeds at which upshifts and downshifts should occur at the different throttle openings. It is easy to check the shift points. Accelerate the car using different throttle openings, and watch the tachometer or listen for the engine speed change that indicates a shift. Note and record the speed for comparison with the specifications. It should be noted that shift points are related to tire diameter and drive axle ratio. A change in either changes shift points. If no specifications are available, the approximate shift point will be as shown in Table 10-1.

Observe all speed limits and traffic regulations during a road test to ensure your safety and that of those around you. A road test is not grounds for violation of any traffic law or regulation.

While on a road test it is good practice to make sure that park works and will hold the car stationary on a grade in both forward and reverse. Simply select a fairly steep grade, stop the car on that grade, shift into park, and carefully release the brake. The car should only roll a small amount to completely engage park and then stop securely. Next you should turn the car around and repeat this test in the other direction.

10.6 TORQUE CONVERTER CLUTCH TESTS

Most torque converter clutch problems fall in the areas of failure to apply, failure to release, hunting or busyness, early application, or harsh application. These problems are fairly easy to confirm on a road test using a tachometer to note the engine speed change during clutch application.

In the electronically controlled units, a light tap or very light application of the brake pedal should cause the converter clutch to release for a few seconds, and this should produce a noticeable increase in engine speed of 50 rpm or more. This test cannot be made on a hoist because of the very small change in rpm that occurs in a nonloaded condition. It is recommended that this test be made with the transmission in its highest gear and then repeated at the same speed in the next highest gear. A greater rpm drop should occur in the higher gear because of the greater load on the converter, which produces more slippage and therefore a greater rpm change during lockup.

When troubleshooting busy or jerky converter clutch application, you should remember that these problems can be caused by rough or weak engine operation. When the converter clutch applies, especially at low vehicle speeds, the engine speed can drop down below the power band, and any engine malfunction becomes very evident.

Because there are different styles of converter clutch operation, there are different procedures for troubleshooting each type. The description that follows will cover the most common, electronically controlled operation. There are essentially two major checks: hydraulic pressure and electrical control circuit.

10.6.1 Converter Clutch Hydraulic Checks

Most torque converter clutches apply when the oil flow through the converter is reversed. If a pressure gauge is installed in the oil cooler line, a slight drop (or in a few transmissions a slight increase) in oil pressure can be observed during clutch apply.

To make a converter clutch pressure test, you should:

1. Install a 0-to-100-psi (0-to-689-kPa) oil pressure gauge in the transmission-to-cooler feed line at the radiator (Fig. 10-14). You should use a tee fitting for this so that the cooler flow will not be blocked.

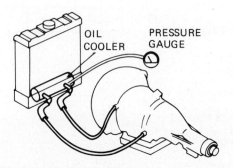

Fig. 10-14 Converter clutch hydraulic operation can be checked by installing a pressure gauge in the transmission to the cooler line as shown. A change in pressure will be seen when the clutch applies. Note that the gauge can be connected at either end of this line.

TABLE 10-1 Shift Points

Closed Throttle	Speed	Part Throttle	Speed	WOT	Speed
1-2	5-10	1-2	15-30	1-2	35-45
2-3	15-25	2-3	25-45	2-3	55-65
3-4	30-45	3-4	40-55	3-4	Above 55-65
4-3	30-40	4-3	35-45	4-3	Above 55-65
3-2	10-15	3-2	30-40	4-3	55-65
2-1	5-10	2-1	10-20	2-1	25-40

NOTE See the note concerning pressure gauge selection in Section 10.7.

2. Raise the car's drive wheels, and securely support it on a hoist or jack stands. Alternatively, run the hose from the pressure gauge through the driver side window, and close the hood as far as possible.

3. Operate the car in drive range until the 2–3 shift and maintain a speed of 55 mph (88 km/h) as you watch the pressure gauge. A drop of 5 to 10 psi (35 to 70 kPa) indicates operation of the torque converter clutch control valve.

NOTE If you miss the pressure change, tap the brake pedal; this should cause the converter clutch to release and then reapply. The gauge pressure should fluctuate as this occurs.

> *SAFETY NOTE:*
> When running a vehicle with the drive wheels raised, never exceed 60 mph (96 km/h) on the speedometer. The differential can allow one tire to overspeed to 120 mph (193 km/h), and it can explode from the centrifugal force.

- *If the pressure changed and the converter clutch applied:* This is normal operation.
- *If the pressure changed and the converter clutch did not apply:* There is a problem inside the converter or wear or a cut O-ring on the input shaft (Fig. 10-15).
- *If the pressure did not change:* There is a problem in the converter clutch valve, control solenoid, or solenoid control circuit.

Fig. 10-15 The badly worn end of a turbine shaft causes converter clutch pressure loss and improper operation. A cut or damaged O-ring would have a similar effect.

10.6.2 Torque Converter Electrical Checks

All electrical checks should begin with an identification of the circuit. A variety of circuits are used depending on the manufacturer and the engine–vehicle combination. It is impossible to test a circuit if you don't know where the switches are or where B^+ power or ground should be. Wiring diagrams are available that show the relationship of the various components (Fig. 10-16).

In some systems, the control solenoid has a ground connection inside the transmission (Fig. 10-17). This can be through a gear range switch or a governor switch that closes and completes the ground circuit in a particular gear or at a certain speed. In most modern systems, the control solenoid always receives B^+ voltage (except when the brake is applied) and is grounded at the electronic control unit to cause the solenoid to operate (Fig. 10-18).

Begin your electrical checks to determine if B^+ power is provided to the transmission electrical connector when it should. To perform this test, you should:

1. Locate the connector; unplug it from the transmission; and determine which terminal should be power in or B^+.

2. Connect the probe of a test light or the positive lead (+) of a voltmeter to this terminal, and connect the ground lead of the test light or the negative lead (−) of the voltmeter to the transmission case (Fig. 10-19).

3. Raise the drive wheels off the ground, and securely support the car on a hoist or jack stand.

4. Start the engine; shift into drive; and use very light throttle to operate the transmission at minimum speed in third gear. The test light should come on or the voltmeter should indicate battery voltage.

5. If the light comes on or B^+ is indicated, tap the brake pedal. The light should go off and in a few seconds come back on.

- *If the light comes on and then goes off as the brake pedal is tapped:* This is normal operation.
- *If the light does not come on:* There is a problem in the electrical harness; a fuse is burned out; or there is an open switch in the circuit.
- *If the light comes on but does not respond to the brake pedal:* There is a faulty brake switch or it needs adjustment. It should be noted that the brake switch in some cars is located in the circuit leaving the transmission, and this is normal operation for these units.

If the power circuit is good, you should continue this test by seeing if the solenoid is switched at the proper time. To perform this check, you should:

1. Carefully attach a small wire into the power in or B^+ terminal of the connector or scrape a small amount of insulation from this wire so you can connect the test light to it.

2. Attach one lead of the test light to the jumper or bared wire and the other lead to ground at the transmission case (Fig. 10-20).

267

DIESEL 2 WHEEL DRIVE
NON-E.C.M. CONTROLLED

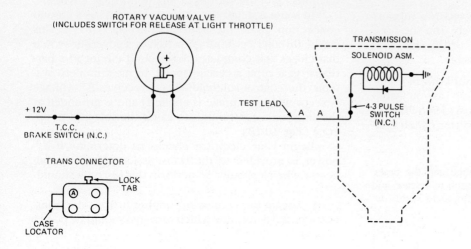

LT. TK. GAS — 2 OR 4 WHEEL DRIVE
NON-E.C.M. CONTROLLED

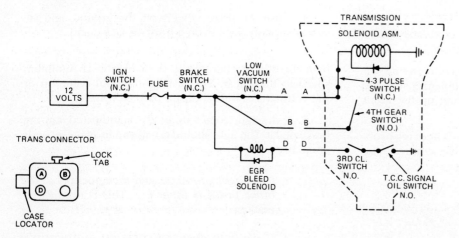

700-R4 CORVETTE C.C.C. OR E.F.I.
E.C.M. CONTROLLED

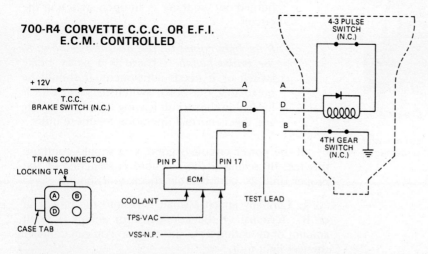

Fig. 10-16 Three General Motors converter clutch control electrical circuits; it is important to use the right circuit when checking a particular car. Note that terminal A is battery positive, B^+. *(Copyrighted material reprinted with permission from Hydra-matic Div., GM Corp.)*

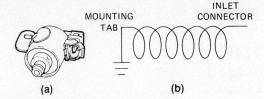

Fig. 10-17 Solenoid using a single terminal; the ground circuit is completed through the mounting bolt. A schematic drawing would be the case connector, solenoid coil, and ground connection.

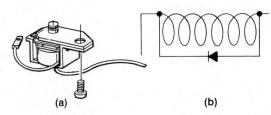

Fig. 10-18 Solenoid using two wire connections; the ground is external. A schematic drawing would be the two wire connections, coil, and a diode used with many computer controls.

3. With the drive wheels still in the air and the vehicle securely supported, start the engine; place the gear selector into drive; and allow the transmission to reach about 30 mph (48 km/h) using a light throttle. The test light should be on.

4. Increase the speed to about 50 to 55 mph (80 to 88 km/h). The test light should dim when the circuit applies the converter clutch.

- *If the light dims and the converter clutch applies:* This is normal operation.
- *If the light dims and the converter clutch does not apply:* There is a faulty converter clutch or a stuck converter clutch control valve.
- *If the light stays bright:* There is a faulty solenoid, internal transmission switch, or control module; at this point, a circuit diagram is absolutely necessary.

There are several checks that can be made on a solenoid and its diode, and these are described in a later section.

10.7 HYDRAULIC SYSTEM PRESSURE TESTS

The operation of an automatic transmission is dependent on the hydraulic system, and problems in the hydraulic system produce problems in transmission operation. A technician performs pressure checks as a part of problem diagnosis so the simplest repair procedure can be made. Some of these hydraulic system problems can be cured with the transmission in the car. For example, it would be a shame to remove and replace (R&R) a transaxle if the problem was caused by a loose valve body or faulty governor.

All transmissions have a pressure test port; some have more than one. If there is only one port, it will be line pressure. The additional ports provide the apply or release pressure of a particular servo, governor, or throttle valve pressure (Fig. 10-21). Most automatic transmission service manuals include illustrations to identify these ports.

When measuring fluid pressures, you should always use a pressure gauge that has the capability of reading higher pressures than you expect to read. A pressure gauge will be damaged if subjected to higher pressures than its rating. In other words, a 0-to-100-psi (0-to-690-kPa) pressure gauge does not have the capability of reading pressures over 100 psi, and it will be ruined if you try. One or more pressure gauges are often combined into a transmission test unit along with a vacuum gauge and a tachometer. This provides a convenient tool because the pressure specifications for a transmission are usually given for certain speeds, and modulator pressure checks are often specified for particular amounts of engine vacuum.

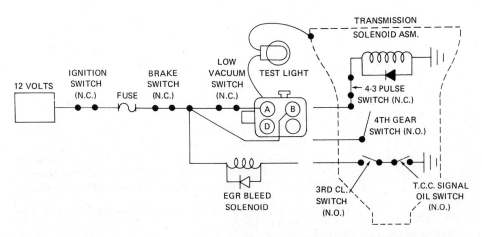

Fig. 10-19 Test light connected to determine if B^+ is at the transmission. The light should light. *(Copyrighted material reprinted with permission from Hydra-matic Div., GM Corp.)*

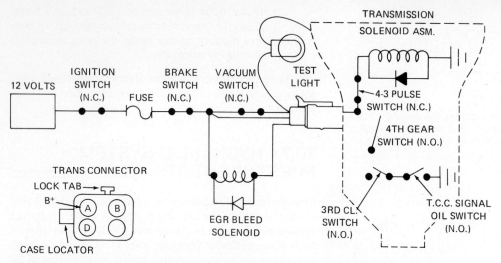

Fig. 10-20 Test light connected as shown should become dimmer when the ECM completes the ground path to apply the solenoid. *(Copyrighted material reprinted with permission from Hydra-matic Div., GM Corp.)*

To test the pressures in a transmission, you should:

1. Raise and securely support the car on a hoist or jack stand.

2. Locate the pressure ports; remove the plugs; and connect the gauge(s) to the port (Fig. 10-22). The port will usually use $\frac{1}{8}$ in., National Pipe Thread (NPT).

3. If the tester is equipped with a vacuum gauge and the transmission uses a vacuum modulator, remove the vacuum hose from the modulator, and connect a tee to the vacuum gauge, vacuum supply line, and modulator.

4. If the tester is equipped with a tachometer, connect it to the engine's ignition system following the instructions for the tester.

5. Route the various lines and wires from the connections to the driver's side window. Be sure to keep them away from the hot exhaust system and rotating parts like the drive shaft.

6. Place the gear selector in park; securely apply the brakes; start the engine; and note the readings on the various gauges. Some transmissions will not read line pressure in park; others will. The vacuum gauge should read about 15 to 20 in. Hg, and the tachometer should be reading idle speed.

NOTE: If a pressure gauge is reading the maximum amount, stop the engine immediately, and use a pressure gauge with a greater scale.

7. Run the engine at idle speed; shift the gear selector through each of the different gear ranges; and record the pressure readings. On vacuum modulator transmissions, there should be at least 15 in. Hg of vacuum.

8. Make sure the brakes are securely applied; increase the engine speed to 1000 rpm; and repeat Step 7.

9. On mechanical throttle valve linkage transmissions, disconnect the linkage from the throttle and pull it to the WOT position.

On vacuum modulator transmissions, disconnect the vacuum line and plug the engine end for normal engine operation.

Next repeat Step 7.

10.7.1 Interpreting Pressure Readings

A technician compares the various pressure readings and checks them against the specifications given in service manuals to determine if the system is operating correctly. If no specifications are available, the approximate pressures in most transmissions will be as follows:

Neutral, park, and drive at idle—50 to 60 psi (345 to 414 kPa).

Neutral, park, and drive with wide open throttle linkage or 0 in. Hg to modulator—75 to 125 psi.

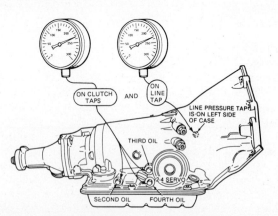

Fig. 10-21 A THM 700 with four pressure ports to be used for making oil pressure checks. *(Copyrighted material reprinted with permission from Hydra-matic Div., GM Corp.)*

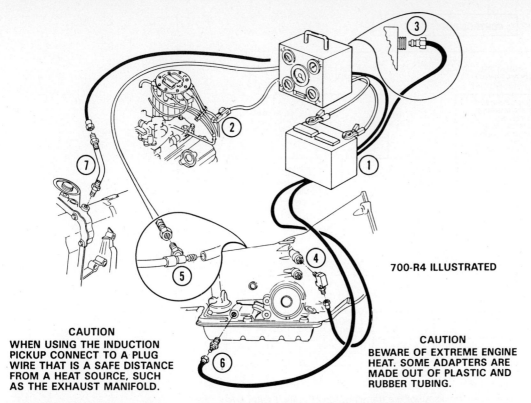

Fig. 10-22 A multifunction test unit connected to the battery (1) for power, a spark plug wire (2) to indicate engine speed, to the cooler outlet (4), to the vacuum hose (5), and to transmission pressure ports (6, 7). Note that special adapters are used at 4, 5, 6, and 7. *(Copyrighted material reprinted with permission from Hydra-matic Div., GM Corp.)*

Manual 1 and 2 (in some transmissions)—100 to 125 psi.

Reverse—150 to 250 psi.

It should be noted that any transmission that develops normal pressure in reverse is sure to have a good pump and pressure control circuit.

Remember that the basic source for these pressures is the pump, intake filter, and pressure regulator and that the path the oil follows is usually through the filter, valve body, transmission case, pump assembly, and back through the transmission case to the valve body. In park and neutral, the throttle valve, torque converter, and cooler are open to pressure but the flow to the rest of the transmission is shut off at the manual valve. High or low pressures in neutral are usually caused by a problem in the circuits or parts just mentioned (Fig. 10-23).

Making pressure checks at high and low vacuum to the modulator or normal and full throttle valve linkage movement allow us to look at the throttle valve and modulator circuits. If the line pressure increases in Step 9 (neutral range), the throttle valve/modulator valve is working (Fig. 10-24).

Moving the gear selector to the different ranges allows us to check the clutch and band servos and the passages leading to them. For example, normal pressure throughout the test except for low pressure in drive 3 and reverse indicates there is a bad seal in the high–reverse clutch circuit, and this pressure loss probably causes slippage in high and reverse. It should be noted that you can't check all of the servos with the drive wheels stationary. Some technicians take the pressure gauge on a road test or raise the drive wheels as just described to allow upshifts to second, third, and fourth gear.

When a transmission has test ports for individual circuits as well as line, the condition of that circuit can be easily determined by comparing its pressure with line pressure. For example if line pressure is 75 psi (517 kPa) and third gear pressure is also 75 psi, you know that the third gear circuit, all of the piston seals, and sealing rings are doing a good job and nothing is wrong with them. If the third gear circuit is more than 10 psi (69 kPa) lower than line pressure, you know that there is a problem in the third gear circuit that must be corrected.

10.7.2 Fluid Flow Diagrams

Every manufacturer provides fluid flow diagrams that show the fluid passages and valves in a transmission (Fig. 10-25). These are used by a technician to help locate the cause of a problem such as low fluid pressure or no upshift. The diagrams are used to trace the fluid flow through a circuit in the same manner as you would use a street map to follow a series of roads from one point to another.

Some students have difficulty the first time they try to interpret a fluid diagram. After a little practice, though, they find it is not hard, and the diagram becomes a useful tool. Here are some tips that make it easier to read a diagram.

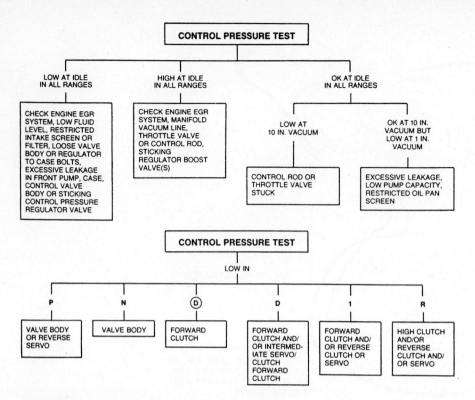

Fig. 10-23 A complete control pressure checks all of the transmission circuits. Faults are determined by the gear selector position and the gear that is engaged when the pressure is lower than normal. *(Courtesy of Ford Motor Company.)*

- When first looking at a diagram, don't try to read it all at once. Look for the circuit you need, and concentrate on it.
- Most manufacturers provide fluid diagrams that illustrate the fluid flows for the different gears. If you need to identify a gear position, look at the position of the manual valve in the diagram.
- Many fluid diagrams are color coded, and a legend normally identifies the various colors.

- A circuit is easiest to locate by starting from the servo or clutch and following it through the valves toward the pump.
- Fluid passages often have a different name as they pass through a valve.
- The passages in and out of a valve are laid out in the same position as the real passages at the valve.
- The valves are normally shown in their actual position for the gear status of the diagram. For example, in drive 1, the shift valves should be in a downshift position.
- Some of the symbols (such as for check valves or exhaust ports) vary between manufacturers. These are shown in Fig. 10-26.

10.7.3 Checking Governor Pressure

When a transmission won't upshift and stays in first gear or starts out in third gear, there is the question of whether the problem is caused by a stuck governor or shift valve. Some transmissions provide a governor test port that makes checking fairly easy. Simply connect a pressure gauge to this port; raise the drive wheels; run the car in drive; and watch for a governor pressure (Fig. 10-27). A steadily increasing pressure relative to wheel speed indicates a good governor and circuit.

Many transmissions do not have a governor pressure port, but many have a governor cutback of line pressure as the car is moving at 10 to 20 mph. This cutback can be seen on a pressure gauge connected to

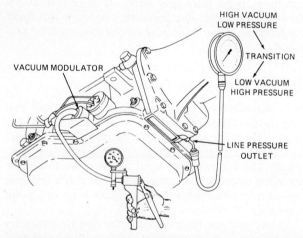

Fig. 10-24 A properly acting modulator and throttle pressure circuit produces line pressures that are inversely proportional to modulator vacuum. *(Copyrighted material reprinted with permission from Hydra-matic Div., GM Corp.)*

DRIVE RANGE-SECOND GEAR

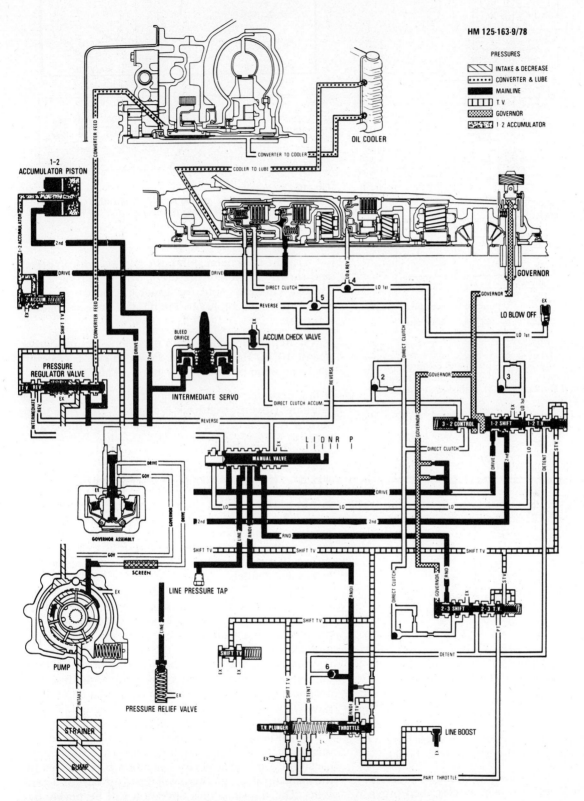

Fig. 10-25 A fluid diagram shows the fluid paths through the different circuits in a transmission. On the diagram, note that it is in drive 2 so the 1–2 shift valve is upshifted and the 2–3 shift valve is downshifted. Also note the legend (with color coding) showing the pressures in some of the circuits; how realistically a manual valve is shown, and how an exhaust, orifice, and check ball are shown. *(Copyrighted material reprinted with permission from Hydra-matic Div., GM Corp.)*

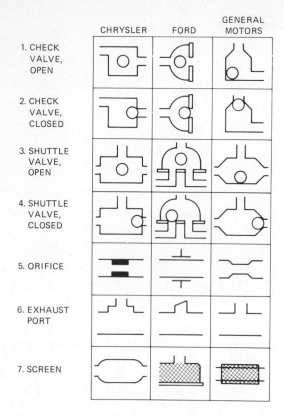

Fig. 10-26 Symbols used by Chrysler, Ford, and General Motors are for an open check ball (1), closed check ball (2), unseated shuttle valve (3), seated shuttle valve (4), orifice (5), exhaust port (6), and filter screen (7).

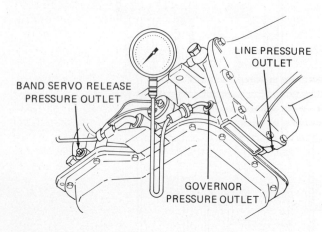

Fig. 10-27 Checking governor pressure is easy if there is a test port. Gauge pressure should steadily increase from zero as the drive shaft increases speed from a stop.

the line pressure port (Fig. 10-28). To check for governor operation on these transmissions, you should:

1. Raise and support the car, and connect a pressure gauge into the line pressure port as described previously.

2. Disconnect and plug the engine vacuum line.

3. Connect a vacuum pump to the modulator, and maintain a constant 2 in. Hg of vacuum during the test.

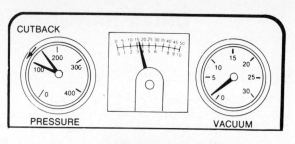

Fig. 10-28 Governor pressure is indicated if line pressure reduces/cuts back as the vehicle slowly accelerates from 10 to 20 mph. *(Courtesy of Ford Motor Company.)*

4. Start the engine; place the gear selector in drive; and slowly accelerate the drive wheels. During acceleration, carefully note the gauge pressures. At some point above 10 mph (16 km/h), the pressure should decrease.

5. Repeat this check with the gear selector in drive 2 and 10 in. Hg of vacuum. Pressure cutback should occur at some point above 5 mph (8 km/h).

If line pressure decreases as indicated, the governor is working. Governor cutback is also indicated by a pressure increase as the car coasts down through these same speeds.

10.8 ELECTRONIC CONTROL SYSTEM CHECKS

Electronic shift controls require that the technician learn additional diagnostic and test procedures, and these will vary somewhat with the different transmission makes and models.

A common problem when diagnosing electronic transmission troubles is determining if the cause is faulty electronic controls, hydraulic system faults, or something in the mechanical system (clutches, bands, etc.). A quick way to start isolating the problem area is to road test the car so you are familiar with the problem. Conduct a second road test after you remove the fuse that supplies power to the transmission control unit. Without power, the electronic transmission control unit (ETCU) and the electronic controls will shut down, causing the transmission operation to revert to a "limp-in" mode. In this mode, forward operation is limited to a single gear, usually second. Second-gear operation is controlled by the manual valve, and there will be no upshifts or downshifts. The transmission's operation will be purely hydraulic and mechanical so it will start in a higher gear than first, probably second or fourth, and have no automatic upshifts. If the second road test shows the same problems as the first or no additional problems, you can be sure the problem is in the electronic controls.

Some electronically controlled transmissions are designed to store and display a problem code if a problem occurs. The problem code is displayed on a special tester or scan tool, and the code number indicates the electrical component or circuit that is not functioning correctly.

Without a problem code, the technician must check the circuits, one at a time, using an ohmmeter, voltmeter, test light, or scanner. With a problem code, an ohmmeter or voltmeter is normally used to locate the exact cause of the trouble. Checking a circuit requires a wiring diagram as well as a good understanding for that particular circuit. Understanding the circuit often leads the technician close to the source of the problem. For example, if an electronically controlled transmission operates normally but has no third or fourth gear, one particular shift control solenoid is probably not allowing one of the clutches to operate.

When checking electrical circuits, it is wise to remember that the most probable faults are wire connectors or parts that move to operate. These include all wire connectors, switches, speed sensors, and solenoids. When checking wire connectors, look for loose, dirty, or bent and misaligned prongs (Fig. 10-29). The least probable cause of failure will be the ETCU. It is fairly well protected, and since there are no moving parts, there is nothing to wear out. However, the ETCU can be damaged by the technician if he or she causes excessive current to flow through it or a high-voltage spike to occur in one of its circuits. Testing a circuit using the wrong test light or voltmeter is discussed in the next section, and this can cause excessive current flow. The most common causes for a high-voltage spike are a solenoid with an open diode, disconnecting the ETCU while the ignition switch is on, and touching the prongs of the ETCU with your fingers. Moving about the car can cause the technician to build up a static electric charge with a voltage high enough to damage the ETCU. Some simple rules to follow while testing and working on electronic-controlled transmissions are:

- Make sure the ignition is off before disconnecting or connecting the ETCU.
- Never touch the ETCU prongs without first grounding yourself to the car's ground.
- Use the proper test equipment.

A general description of some test procedures follows to give you an idea of the process. However, you should always follow the procedure described in a service manual for a particular transmission.

10.8.1 Testing Computer-Controlled Circuits

Current flow is critical when testing circuits controlled by an ETCU, engine control module (ECM), transmission control module (TCM), or other types of computer control units. Excessive current can damage the unit. A common test light draws several amperes of current when connected to a circuit. A common analog (the meter uses a needle that sweeps across a scale) voltmeter also draws enough current to damage a computer circuit. Test lights using light-emitting diodes (LEDs) and digital voltmeters have rather high internal electrical resistance. They will not allow much current to pass through and will not excessively load and damage a circuit or cause changes that might affect the readings.

The wire of a test light is normally connected to ground, and the probe is touched to wire connections. If the light goes on, we know that voltage is available at that point in the circuit (Fig. 10-30). Test lights are relatively inexpensive and quite durable and trouble free.

The negative ($-$) wire of a voltmeter is normally connected to ground at clean, unpainted metal, and the positive ($+$) lead is touched to a wire connection. A voltage reading indicates the amount of voltage that is available in the circuit at that point (Fig. 10-31). When using a voltmeter, you should remember always to select a higher voltage range than you expect to measure and then reset to a lower range if possible. An analog meter is most accurate at the upper end of the scale. Also, don't use ac ranges when measuring direct current or vice versa because the readings will not be accurate. Volt-

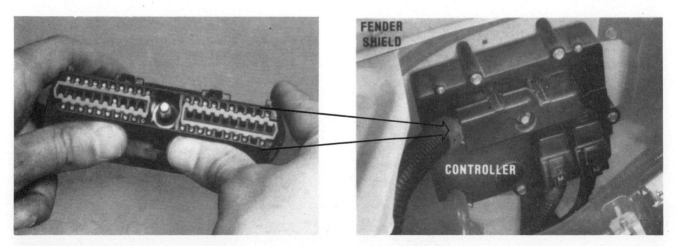

Fig. 10-29 This 60-way/pin connector makes the circuit contacts between the transaxle controller and the car's wiring. Each of the pins must make a good connection. *(Courtesy of Chrysler Corporation.)*

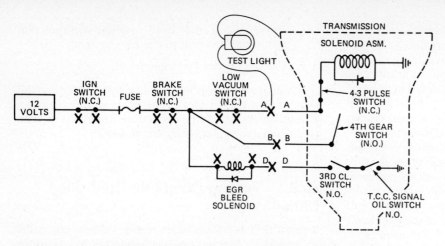

Fig. 10-30 The ground clip of the test light is connected to a good ground such as the transmission case and the probe is touched to various points in the circuit. In this example, the light should light at the point shown and all the locations marked with an X. *(Copyrighted material reprinted with permission from Hydra-matic Div., GM Corp.)*

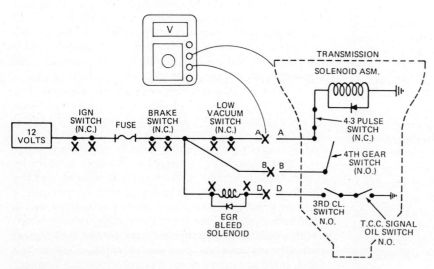

Fig. 10-31 A voltmeter is used by connecting the negative (−) lead to a good ground and the positive (+) lead to the circuit. A voltage reading, almost the same as the battery, should occur at the point shown and at the points marked by an X. *(Copyrighted material reprinted with permission from Hydra-matic Div., GM Corp.)*

meters are more expensive and fairly delicate when compared to a test light, and they should be treated with care.

An ohmmeter has its own battery supply. The two leads are connected to the connections of a component to allow measuring the resistance of that component (Fig. 10-32). The units of resistance are measured in ohms (Ω). An ohmmeter is used to see if a circuit is complete and to measure the resistance in the circuit. After measuring the resistance through a component, we often move one of the leads to the body or case to check for an unwanted short to ground. Too little resistance in a component indicates a short circuit, but don't forget that the meter might be reading an unwanted parallel path from another part of the circuit. Too much resistance indicates a faulty or poor connection. Like a voltmeter, an ohmmeter is fairly expensive and fairly delicate; ohmmeters are often combined into the same unit with a voltmeter. The combination meter is commonly called a volt-ohmmeter in the automotive field and a multimeter in the electrical field. An ohmmeter should never be connected to a circuit that is live— it has its own voltage. The power from the circuit can damage the meter.

Special scanners like the Tech I shown in Fig. 10-12 or the DRB II shown in Fig. 6-20 are available to test particular circuits. These units can be used only on specific circuits for which they have been designed or for which special adapters are available. These units are fairly expensive but can save considerable time when locating an electrical fault.

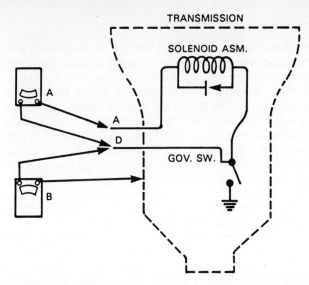

Fig. 10-32 An ohmmeter is used by connecting the two leads to the two ends of a circuit or component. When checking diodes, the meter leads are connected and then reversed and connected again. Because of the diode at the solenoid, meter A should show a circuit with one connection but not when reversed. Meter B should show no circuit until the vehicle speed increases enough to close the governor switch. *(Copyrighted material reprinted with permission from Hydra-matic Div., GM Corp.)*

10.8.2 Solenoid Checks

Electrically, a solenoid is a coil of thin wire that should have a certain amount of resistance. The amount of resistance can be easily checked using an ohmmeter. Simply connect one ohmmeter lead to each of the solenoid electrical connectors; read the amount of resistance; and compare the reading to the specifications (Fig. 10-33). Follow this by moving one of the leads to the solenoid body or base to check for a ground circuit. A two-wire solenoid should not have a circuit to the ground. If the first reading is within specifications and the second one infinite or OL (out of limits), the solenoid should operate.

Some solenoids include a diode to eliminate any voltage spikes that might occur as the solenoid is de-energized. As the solenoid is tested, the diode is also checked to see if it allows a flow in one direction but not the other. You can check the diode and the solenoid coil by connecting an ohmmeter to each of the connections. This can be done with the solenoid in the transmission or on the bench. With the solenoid out of the transmission and on the bench, you should connect the positive ohmmeter lead and solenoid leads together and the negative leads together. With the solenoid in the transmission, the positive terminal is at the B^+ terminal of the case connector and the negative terminal is at the transmission case or one of the other terminals of the case connector. In some transmission circuits, there is a pressure switch between the solenoid negative terminal and the case connector.

With the ohmmeter connected to the solenoid as just described, the reading should be 20 to 40 Ω depending on the temperature. If the reading is less than 20 Ω, the coil or diode is shorted. If the reading is greater than 40 Ω or infinite, there is an open or broken circuit. In either case, the solenoid is faulty.

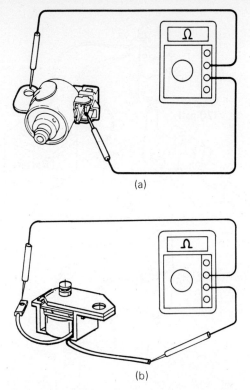

Fig. 10-33 (a) A single-terminal solenoid being checked with an ohmmeter; this particular solenoid should have a reading of 11 to 15 Ω. A lower reading indicates a short; a higher reading indicates excessive resistance. (b) A two-terminal solenoid is being checked. The amount of resistance allowed might be different; you should check the specifications. Also, if a diode is used, you should reverse the meter leads to check the diode. *(Courtesy of Chrysler Corporation.)*

Next reverse the meter connections so they are connected positive to negative and negative to positive. The meter reading should now be lower than it was before, usually about 2 to 15 Ω. If the reading does not change, the diode is open, and the solenoid is faulty.

A solenoid can also be checked by connecting it to a 12-V battery, but you must be careful if doing this. The diode will be destroyed if you connect the solenoid backward. As the solenoid is connected to the power source, you should be able to feel and hear it "click" as it changes position (Fig. 10-34). Some solenoids are normally closed; you should not be able to blow through it while it is not activated.

The mechanical operation of a solenoid also should be checked. Since solenoids are basically electromagnets operating in an area that might have some metal debris, they tend to attract metal particles, which can cause sticking or binding of the solenoid plunger or blocking of the fluid passage. While connected to the power source, you should be able to blow through a normally closed solenoid. The passage should close so you can't blow through it when it is not energized. Other solenoids are normally open; they will test just the opposite. You should be able to blow through it when not energized, but not when it is energized. A tester is available that provides a power source and uses air pressure to completely test solenoid and switch operation (Fig. 10-35).

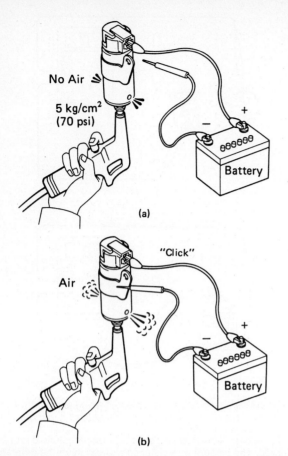

Fig. 10-34 (a) You should not be able to blow through this solenoid if it is not activated. (b) If a solenoid is connected to a 12-V battery, it should make a "click"; be careful to connect the leads with the correct polarity on solenoids that use an internal diode. After it is connected to a battery, you should be able to blow through it. (Courtesy of Toyota Motor Sales, U.S.A., Inc.)

10.8.3 Switch Checks

A control switch is usually checked by connecting it to an ohmmeter and then operating the switch. The meter leads are connected to the two terminals of the switch. If there is only one terminal, one meter lead is connected to it, and the other lead is connected to the switch body. Some switches are normally open, and the reading should be a very high value—infinite or OL. Some switches are normally closed, and the reading should be a very low value—zero. When the switch is operated, the reading should change to the opposite value. A pressure switch can usually be operated using the tester shown in Fig. 10-35 or air pressure and a rubber-tipped air gun (Fig. 10-36).

A switch can also be checked on the car using a voltmeter by connecting the negative lead to a good ground or the switch body and the positive lead to the B^+ wire entering the switch. Voltage must be available in order to test the switch. Next move the positive lead to the second switch terminal, and operate the switch. As the switch is operated, the output voltage should change from zero to the same as the input voltage or vice versa.

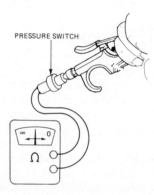

Fig. 10-36 An ohmmeter is used to check a pressure switch. It should change from an open circuit to one with very little resistance when pressure is applied or removed.

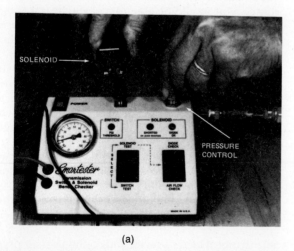

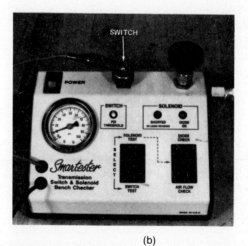

Fig. 10-35 A special tester is used to check (a) a solenoid and (b) a pressure switch. The solenoid should open and allow air to flow through it when it is activated. The switch should either open or close as the pressure is changed.

The neutral start switch used in some electronic transmissions has several circuits and terminals. These switches are checked using an ohmmeter and a guide to determine when a circuit should be complete between two particular terminals. For example, using the switch connections shown in Fig. 10-37, there should be no resistance between terminals *B* and *C* in park and neutral and infinite resistance between these two terminals in all other gear positions. There should be no connection to the other terminals in any gear position. The other terminals are checked in a similar manner.

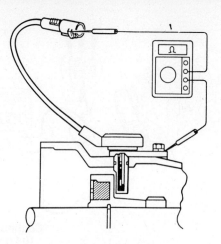

Fig. 10-38 An ohmmeter connected between ground and the connector terminal of a vehicle speed sensor. The meter needle should deflect from one end of the scale to the other as the transmission output shaft is rotated. *(Courtesy of Chrysler Corporation.)*

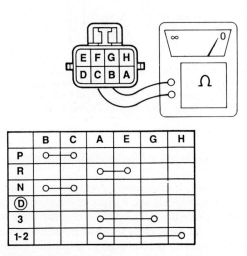

Fig. 10-37 An eight-prong connector for a neutral start switch should have continuity, as shown. For example, in park an ohmmeter should show very little resistance if the leads are connected to terminals B and C but high resistance to the other six terminals. *(Courtesy of Chrysler Corporation.)*

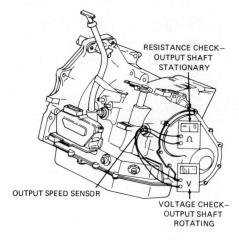

Fig. 10-39 An ohmmeter connected to the two terminals of the output shaft speed sensor should show the correct resistance. A voltmeter connected in the same manner should show a fluctuating voltage as the output shaft is rotated. *(Courtesy of Chrysler Corporation.)*

10.8.4 Speed Sensor Checks

Two styles of speed sensors are used: one style is a switch that opens and closes as a transmission part next to it rotates, and the other is a unit that generates an alternating voltage signal as the transmission part rotates.

The switch type normally has a single wire connector and is tested using an ohmmeter (Fig. 10-38). Connect one lead of the ohmmeter to a good ground and the other ohmmeter lead to the switch's connector. Next rotate the transmission output shaft while watching the ohmmeter. The reading should change from a very low reading (zero to a few ohms) to a very high reading (infinite or OL) at each output shaft revolution.

The voltage-generating unit in a speed sensor normally uses a two-wire connector and is checked using both an ohmmeter and a voltmeter. First connect the two ohmmeter leads to the two sensor terminals (Fig. 10-39). There should be a complete circuit through the unit, and the amount of resistance should fall within the range given in the specifications for that unit. Excessive or infinite resistance indicates a high resistance or open circuit; too low of a reading indicates a short circuit. Next move one of the ohmmeter leads to the sensor body. The reading should be infinite or OL. A low resistance indicates a grounded sensor. The next check is to connect the two leads of a voltmeter to the two sensor connectors, and rotate the transmission output shaft. As the shaft rotates, the voltmeter should show a fluctuating voltage reading.

10.8.5 Throttle Position Sensor Checks

A throttle position sensor (TPS) is normally checked using an ohmmeter or voltmeter. Connect the two ohmmeter leads to two of the TPS connectors and rotate the throttle (Fig. 10-40). As the throttle is opened and closed, the ohmmeter reading should change from a low to a high reading and vice versa.

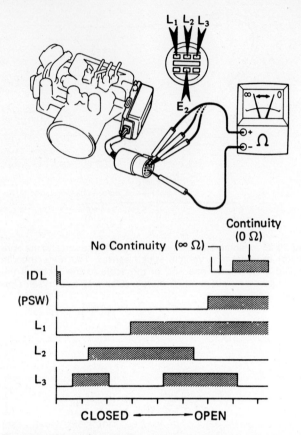

Fig. 10-40 Ohmmeter used to check a throttle position sensor. One lead is connected to the E2 terminal and the other is moved as indicated. As the throttle is opened, the needle should deflect between no resistance and high resistance as indicated. (*Courtesy of Toyota Motor Sales, U.S.A., Inc.*)

When using a voltmeter, leave the TPS connector connected; connect the negative lead to a good ground, and use the positive lead to probe the input voltage to the connector (Fig. 10-41). It should be the specified voltage indicated in the service manual. Next move the positive voltmeter lead to the output connector, and measure the voltage as you open and close the throttle. The output voltage should change as the throttle is opened and closed.

10.9 STALL TESTS

Stall tests are used to check the strength of the various apply devices. A band and clutch application chart for that particular transmission is normally used when making a stall test to verify which clutch or band is applied in each gear position (Fig. 10-42).

Caution should be exercised when performing a stall test for several reasons; personal safety, the safety of those around you, and the chance of possible damage to the car and transmission. You will be operating the car with the transmission in gear and the throttle wide open. There is a possibility that the car can get away from you and run into things or people. All four wheels must be on the ground, the brakes must be in good operating condition, and the parking and service brakes must be firmly applied while performing a stall test.

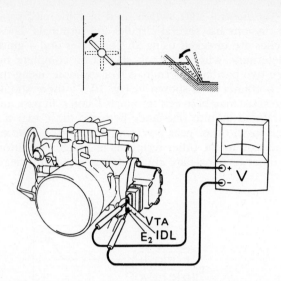

Fig. 10-41 Voltmeter used to check a throttle position sensor. As the throttle is opened, the voltage should match the manufacturer's specifications. (*Courtesy of Toyota Motor Sales, U.S.A., Inc.*)

Many technicians place blocks at the front and rear tires also.

Stalling the transmission creates rather high pressure, much heat inside the torque converter, and severe loads to the drive train. The period that the transmission is at stall should be kept as short as possible, never over 5 to 10 seconds. It is also a good practice to shift into neutral between the different checks and run at fast idle for 30 seconds or so to cause a change of fluid inside the converter.

A stall check also places a heavy load on the engine's mounts. It is good practice to have the hood open and keep an eye on the engine while doing these checks. If the engine appears to lift an excessive amount, double check the condition of the mounts (Fig. 10-43). Faulty mounts usually cause a heavy clunk or knock if you close the throttle and allow the engine to drop on the mounts during a stall check.

Most manufacturers publish stall speed specifications for their engine–transmission combinations. This allows for an easy comparison between your readings and the specifications. If no specifications are available, you can compare the readings from the different gear ranges. Remember that the brakes of the car hold the transmission output shaft stationary, and the correct operation of the clutches and bands should hold the transmission and the torque converter turbine stationary. Failure of these parts allows the turbine shaft speed and therefore the engine speed to increase. Too high of a stall speed indicates slippage inside the transmission. To perform a stall test, you should:

1. Connect a tachometer to the engine following the directions for the tachometer.

2. Position the vehicle with all four wheels securely on the ground. It is a good practice to place blocks in front and back of the drive wheels (Fig. 10-44).

3. Start the engine, and note the tachometer reading to become familiar with the instrument's readout. It is

ELEMENTS IN USE AT EACH POSITION OF THE SELECTOR LEVER

LEVER POSITION	START SAFETY	PARKING SPRAG	CLUTCHES			BANDS	
			FRONT	REAR	OVER-RUNNING	(KICKDOWN) FRONT	(LOW-REV.) REAR
P—PARK	X	X					
R—REVERSE			X				X
N—NEUTRAL	X						
D—DRIVE							
First				X	X		
Second				X		X	
Direct			X	X			
2—SECOND							
First				X	X		
Second				X		X	
1—LOW (First)				X			X

Fig. 10-42 Band and clutch chart indicates which control elements are used at each position of the gear selector lever. *(Courtesy of Chrysler Corporation.)*

recommended that you mark the upper and lower stall speed limits on the face using a grease pencil or non-permanent felt marker (Fig. 10-45).

SAFETY NOTE:
Direct any bystanders away from the front or back of the vehicle.

4. Apply the brakes firmly; move the gear selector to reverse; move the throttle to wide open; and watch the tachometer. The speed should increase to somewhere between 1500 and 3000 rpm. As soon as the speed stops increasing or goes higher than 3500 rpm, quickly note the reading, and close the throttle. Record the speed.

NOTE The throttle should not be held open longer than 5 to 10 seconds.

5. Shift to neutral, and run the engine at fast idle for 30 to 60 seconds to cool the converter.

6. Repeat Steps 4 and 5 with the gear selector in drive and low.

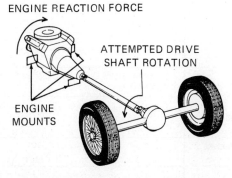

Fig. 10-43 During a stall test, torque reaction tries to lift the left side of the engine. Bad engine mounts allow the engine to lift upward.

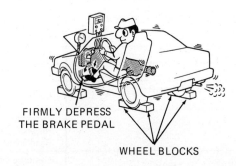

Fig. 10-44 The wheels should be blocked and the parking and service brake firmly applied during a stall test. A tachometer is used to measure engine speed, and sometimes a pressure gauge is used to measure system pressure. *(Courtesy of Toyota Motor Sales, U.S.A., Inc.)*

☐ H — STALL TEST

Range	Specified Engine RPM	Record Actual Engine RPM
D		
2		
1		
R		

Results _____

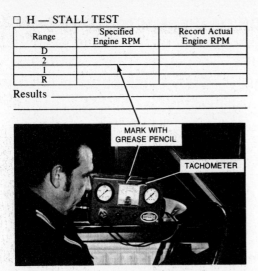

Fig. 10-45 Before starting the actual stall test, it is good practice to mark the tachometer with the specified speeds. You should not keep the throttle depressed for more than 5 seconds so you must be able to read the tachometer quickly. *(Courtesy of Ford Motor Company.)*

10.9.1 Interpreting Stall Test Readings

- *If the three stall speeds were all within the specification range:* The apply devices for the three gear ranges are all sound and in good shape. The apply devices for intermediate gear(s) cannot be applied with the vehicle at rest so they cannot be stall tested.

- *If the stall speeds were all equal but low:* Either the engine is weak or out of tune or the stator one-way clutch is slipping. Checking the vehicle performance on a road test should let you know which is at fault.

- *If the stall speed is high in one or two of the gear ranges:* One or more of the apply devices is slipping. Consult a band or clutch application chart for that transmission to determine which ones. For example, let's say we are checking a Type 2 transmission (Chapter 2) and the stall speed is too high only in drive 1. The chart indicates that the forward clutch and the one-way clutch are the apply devices. The forward clutch must be good because it is also applied in low, and the stall speed is OK in low. Therefore, the high stall speed in drive 1 is caused by a faulty one-way clutch. If the stall speed was too high in both drive 1 and low, the forward clutch would probably be the problem (Fig. 10-46).

10.10 OIL PAN DEBRIS CHECK

When we decide that something is probably wrong with the transmission, the next step is to drain the oil and remove the pan for inspection. The debris that is in the pan can give a good indication of what is occurring in the transmission.

A small amount of debris with a blackish oil film is normal. We attribute the metal to the small amount of wear that occurs during break-in and normal operation.

An excess of loose, black material is burned lining material from a slipping band or clutch. It usually has a burned smell.

A heavy golden brown coating is from badly oxidized, old fluid. The lower part of the case and valve body also has this varnish coating. It usually has a strong varnishlike odor.

An excess of metal is from the gearset, thrust washers, or bushings. Steel is usually from the gears, needle bearings, or a spring or spring retainer. Aluminum is from the case, a carrier, or a clutch piston. Brass or bronze is from a bushing or thrust washer.

Any plastic debris (broken or melted) is from a thrust washer or a clutch spring retainer.

Selector Position	Stall Speeds High	Stall Speeds Low
Ⓓ Overdrive, D and 1	Overdrive One-Way Clutch, Rear One-Way Clutch	—
D, 2 and 1	Overdrive Clutch, Forward Clutch	—
Ⓓ Overdrive	Forward Clutch	—
Ⓓ Overdrive, D, 2, 1 and R	General Problems Pressure Test	Converter Stator One-Way Clutch or Engine Performance
R Only	Overdrive Clutch, Overdrive One-Way Clutch, Reverse and High Clutch and Low and Reverse Band/Servo	—
2 Only	Overdrive One-Way Clutch and Intermediate Band/Servo	—
1 Only	Low and Reverse Band/Servo	—

Fig. 10-46 Chart indicates the problem area if the stall speed is too high in one or more of the gear selector positions. *(Courtesy of Ford Motor Company.)*

10.11 AIR TESTS

Air tests are a diagnostic tool and are also used as a final check during transmission assembly. Air tests are used to tell if a clutch or band servo operates, if a governor can operate, and if the passages are tight. A tight hydraulic circuit holds air pressure quite well.

Air tests often are made with the valve body removed. Most transmission circuits end at the valve body so there is a passage that leads from the valve body to each hydraulic component. Most manufacturers identify these passages for us (Fig. 10-47).

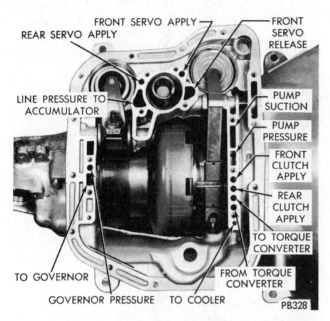

Fig. 10-47 Passages in a Torqueflite transmission are identified so the different components can be checked using air pressure. They were exposed when the valve body was removed. *(Courtesy of Chrysler Corporation.)*

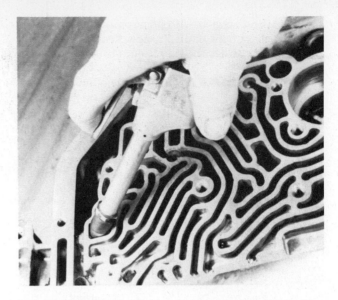

Fig. 10-48 A rubber-tipped air gun is normally used when making air tests. Pushing it against the port opening is usually all that is required to make an airtight seal. It must be removed from the opening to allow the air to escape and release the component.

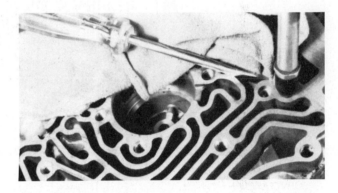

Fig. 10-49 When the test port is a long slot, part of the slot can be sealed off by forcing a shop cloth into the slot.

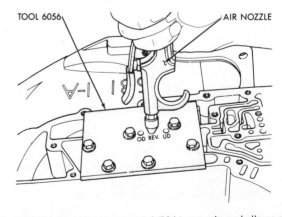

Fig. 10-50 Test plate fits onto an A-604 transaxle and allows easy air tests of the overdrive, reverse, and underdrive clutches. After fastening the test plate in position, an air nozzle is used to air check the clutches. *(Courtesy of Chrysler Corporation.)*

A rubber-tipped air gun is normally used when making air tests. It is merely pushed against the end of the passage to make a seal (Fig. 10-48). If the passage ends in a long slot, a shop cloth can be forced into part of the opening to close it down to a manageable size (Fig. 10-49). Some technicians use an air gun that is modified with a piece of tubing to reach into hard-to-get-to locations. Special test plates are also available to provide air pressure access to difficult passages (Fig. 10-50). Pressures of about 50 psi (345 kPa) should be used while air checking. Very high pressures can cause normal seepage to appear like a major leak. Remember that air passes through openings easier than ATF.

To make an air test, you should:

1. Remove the valve body and identify the necessary fluid passages.

2. Force air pressure into the passages one at a time. As you air test the different components, you should get the following results:

a. Band servos—an operation of the band with a very small amount of air leakage. Removal of the air gun from the passage should result in band release (Fig. 10-51).

b. Clutches—a "kachunk" type of noise indicating clutch apply with a very small amount of leakage. Removal of the air gun should result in another noise indicating clutch release. Good clutch seals should hold air pressure for about 5 seconds or more if the air nozzle is kept in place.

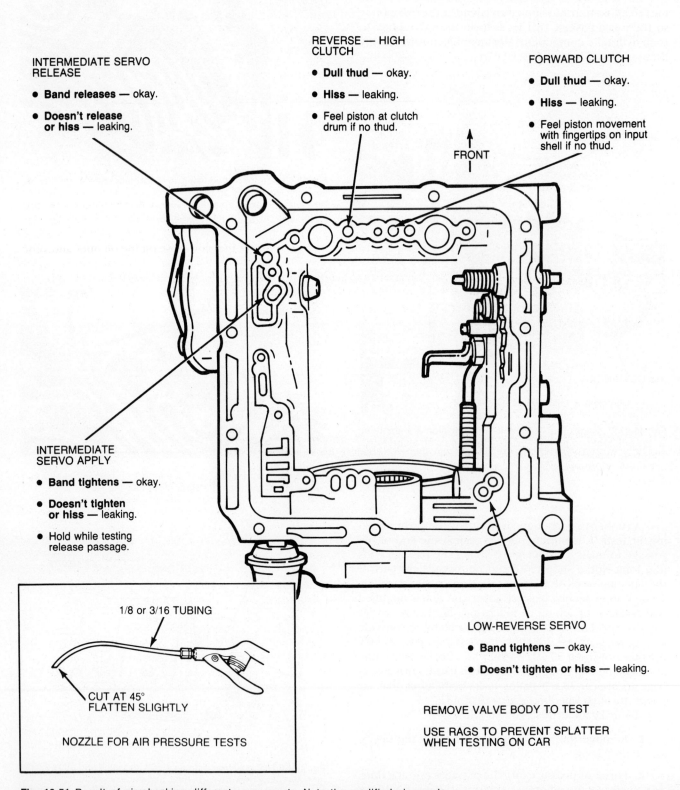

Fig. 10-51 Result of air checking different components. Note the modified air nozzle. *(Courtesy of Ford Motor Company.)*

NOTE Some clutches have two piston areas—an inner and an outer. When air is blown into one chamber, it can normally leak into the other and pass out through the other port. You can cover this second port with your finger while testing.

c. Governor—will vary between different makes. There are two governor ports to check: one is a feed from the drive circuit and the other a return from the governor. Check both of them. In some cases, you will get a "click" or "snap" and then a very small leak, a "click" and an open exhaust leak, or a rather heavy "buzz" sound depending on the type of governor, the position of the governor shaft, and whether you blow air into the inlet or outlet passage. The response should change if you plug the second passage with your finger.

It is also possible to air check the whole transmission or an individual circuit by blowing air into the pressure test ports. If checking the whole transmission, insert the rubber tip of the air gun, set to about 150 psi (1,034 kPa), into the "line" test port, and pull the trigger (Fig. 10-52). Depending on the location of the test port in the hydraulic circuit, this will pressurize the line circuit, and you can control the flow using the manual valve. In some cases with the transmission on the bench, the pump will rotate, but it can be held stationary by installing the torque converter. With some transmissions, the governor weights can be blocked outward, and the shift valves will operate to an upshift position. This allows pressure checking of each of the gear positions, which you can control by selecting drive, manual 1, manual 2, or manual 3 using the manual valve. If the transmission has pressure test ports for the different circuits, each of these circuits can be checked by pressurizing each port. Note that this pressure will try to exhaust at the shift valve, but you can usually block this flow with your finger. In this case, you might have to drop the air pressure to about 40 psi (275 kPa). While performing these checks, you will be listening for escaping air to indicate leaks, just like checking air at the valve body ports.

10.12 OIL LEAK TESTS

These tests are made to locate the cause of oil leaks, a common complaint. Most leaks originate at the joints or connections between the various parts of the transmission; between the cooler line connections; or at the front-torque converter seal, rear extension housing/universal joint seal, or with transaxles, the two CV (constant velocity) joint seals (Fig. 10-53).

The cause of a leak can often be found by raising the car on a hoist and making a visual inspection. You can usually follow the red ATF upward and often forward to the cause (Fig. 10-54).

A leak coming from inside the torque converter housing usually indicates a bad front seal, but this can also be caused by a leaking torque converter or faulty pump gasket or seal (Fig. 10-55). In either case, it is necessary to remove the transmission for a closer inspection and repair.

When there is so much oil that you cannot locate the source of the leak, it is necessary to steam clean the transmission and car underbody. Direct the steam wand so as to clean all around the transmission. Next drive the car briefly so the wind and heat dries things off. Also operate the transmission in all gear ranges at high throttle openings to move the high-pressure oil so the source of leak can be seen. Then raise the car on a hoist, and using a good, strong light, carefully check the transmission area for the red ATF. A fluorescent dye is available that can be added to the transmission fluid. Using a black light, this dye allows a leak to show up better when inspecting the transmission. The black light causes the dye to glow (Fig. 10-56).

If you have located a leak in a general area but can't find the specific location, thoroughly clean the area using the steam and then dry it. Next coat the area using an aerosol-applied foot powder. Most foot powders dry quickly and leave a white, powder coating. Any leak under this coating will show as an obvious red stain (Fig. 10-57).

If the exact location does not show up using the methods just described, it is possible to pressurize the transmission case with air and then locate the escaping air. This requires that you close off the oil filler and vent openings if the transmission is in the car plus the converter, extension housing, and cooler line openings if the unit is out of the car. Caps and plugs can be fabricated for this, as shown in Fig. 10-58.

10.13 VIBRATION CHECKS

Occasionally a transmission repair or replacement results in a vibration or a vehicle vibration diagnosis leads to the possibility of the transmission as the cause. Most transmission vibrations are caused by an unbalanced torque converter, an off-center torque converter, or a faulty output shaft–universal joint connection.

Torque converter problems cause an engine-speed-related problem. Output shaft problems are vehicle speed related. These often are accompanied by drive line clunk. Drive line clunk usually occurs in the garage shifts, neutral–drive or neutral–reverse. They can often be confirmed by looseness as you shake the universal joint side to side relative to the transmission. The cure for this is usually a replacement of the rear extension housing bushing and seal or universal joint.

Engine-speed-related vibrations occur at a particular engine speed range; these vibrations change when the transmission shifts gears. The causes are several, for example, belt-driven accessories like the fan, alternator, or air-conditioning compressor or internal engine unbalance. Belt-driven problems can be identified by running the engine with the belt removed; if the vibration is gone, you have located the vibration source. The only way to identify a torque converter problem is careful inspection with the torque converter cover removed. Look for a wobble (called runout) of the converter during engine operation. If there is no wobble, the next thing to do is to remove the torque converter and either check its balance or replace it.

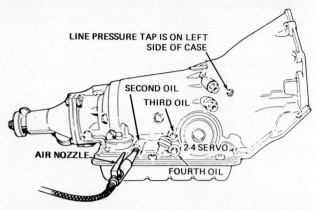

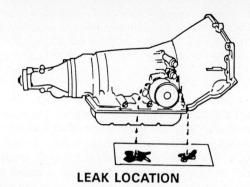

Fig. 10-52 Air can be blown into the pressure test port when checking for leaks in the hydraulic circuits. The whole circuit can be checked by blowing into the line port/tap, or individual circuits can be checked as shown here. Note that the pan should be removed so you can locate any escaping air.

Fig. 10-54 Normally the fluid from a leak runs downward. The air passing under a car also tends to blow it toward the rear. *(Copyrighted material reprinted with permission from Hydra-matic Div., GM Corp.)*

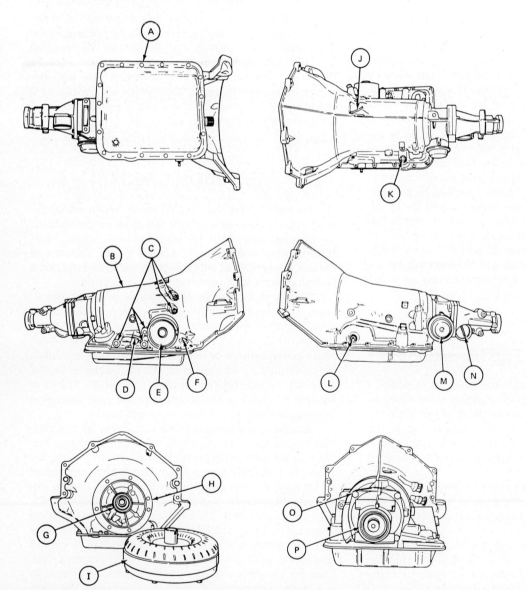

A OIL PAN
B CASE
C COOLER CONNECTORS & PLUGS
D T.V. CABLE SEAL
E SERVO COVER
F OIL FILL TUBE SEAL

G OIL PUMP SEAL ASSEMBLY
H OIL PUMP TO CASE SEAL
I CONVERTER
J VENT
K ELECTRICAL CONNECTOR SEAL

L MANUAL SHAFT SEAL
M GOVERNOR COVER
N SPEEDO SEAL
O EXTENSION TO CASE SEAL
P EXTENSION OIL SEAL ASSEMBLY

Fig. 10-53 Possible location of fluid leaks from a THM 700 transmission. *(Copyrighted material reprinted with permission from Hydra-matic Div., GM Corp.)*

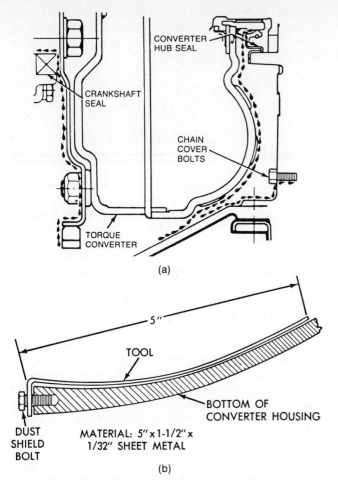

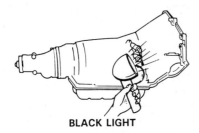

BLACK LIGHT

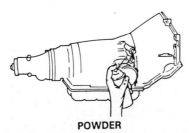

POWDER

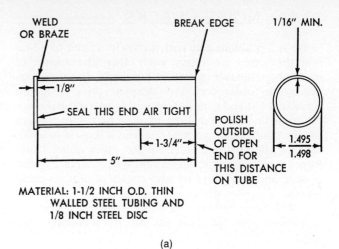

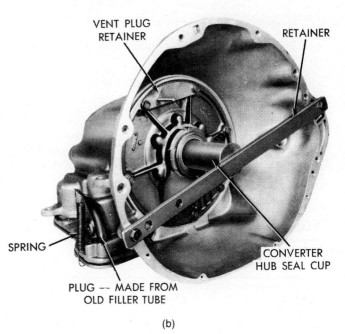

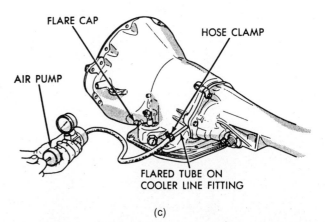

Fig. 10-55 (a) Fluid leaking from the converter housing might come from any of the locations shown. (b) Tool used to determine if the fluid is coming from the back of the converter housing. *[(a) Courtesy of Ford Motor Company. (b) Courtesy of Chrysler Corporation.]*

Fig. 10-56 If a fluorescent dye is added to the transmission fluid, a small leak shows up quite readily using a black light. *(Copyrighted material repeated with permission from Hydra-matic Div., GM Corp.)*

Fig. 10-57 If the transmission is clean and dry, an aerosol foot powder can be sprayed onto the transmission. After operating the transmission, any leaks will be obvious because of the red stain on the white powder. *(Copyrighted material reprinted with permission from Hydra-matic Div., GM Corp.)*

Fig. 10-58 If the openings into the transmission are closed, air can be pumped into the transmission through the cooler return line. Soapy water can now be sprayed onto the transmission, and bubbles expose the location of the leak. The converter hub seal in (a) can be shop made. *(Courtesy of Chrysler Corporation.)*

287

10.14 NOISE CHECKS

Noise is a problem area that is closely related to vibration; they often accompany each other. The sources of most objectionable noises in automatic transmissions can be the torque converter, flexplate, drive chain and sprockets (in some units), planetary gearset, band application, and transaxle final drive. Noise complaints are usually confirmed on a road test, and it is good practice to have the customer identify the exact noise. Don't forget that the engine, suspension system, tires, exhaust system, and drive line are also causes of objectionable noises.

When troubleshooting noise problems, it helps to determine how the noise fits into the following categories:

Speed Relation: Engine or vehicle speed related.

Gear Range: Related to which gears.

Pitch/Frequency: Low (rumble), medium, or high (squeal).

Load Sensitive: Heavy throttle, light throttle, or coast.

The problem areas of transmission noises usually fit into these categories as shown in Table 10-2.

The next step is disassembly and careful inspection of the suspected component. The exact cause might be a rough bearing, a loose drive chain, a cracked flexplate, or any number of defects. Some manufacturers recommend the use of a diagnostic guide to locate the source of noise problems (Fig. 10-59).

10.15 TROUBLESHOOTING CHARTS

Another important diagnostic tool is the troubleshooting chart or diagnostic guide (Fig. 10-60). These are published for each transmission and normally found in service manuals. They are also available as handy diagnosis wheels. They normally list the more common problems encountered with each make and model of transmission. For each problem, there is a list of the most probable causes and what should be done to repair or correct the problem. The causes are normally listed in descending order starting with the most common or easiest to check.

As mentioned earlier, it is a good practice to determine what is wrong with a transmission before you disassemble it for repairs. This allows you to make the necessary checks during disassembly to locate the exact cause of the original problem and helps ensure that you cure all problems as you make your repairs.

TABLE 10–2 Transmission Noise Categories

	Speed Relation	Gear Range	Pitch	Load Sensitive
Torque converter	Engine	All	High, whine	Yes
Flexplate	Engine	All	Low, growl	Yes
Pump	Engine	All and park and neutral	High, whine	
Drive chain	Engine	All	Low to high, growl to whine	Yes
Planetary gears	Engine	All but direct	Medium to high, whine	Yes
Band application	—	When used	Medium to high, screech	Yes
Final drive	Vehicle	All	Low to medium, hum	Yes

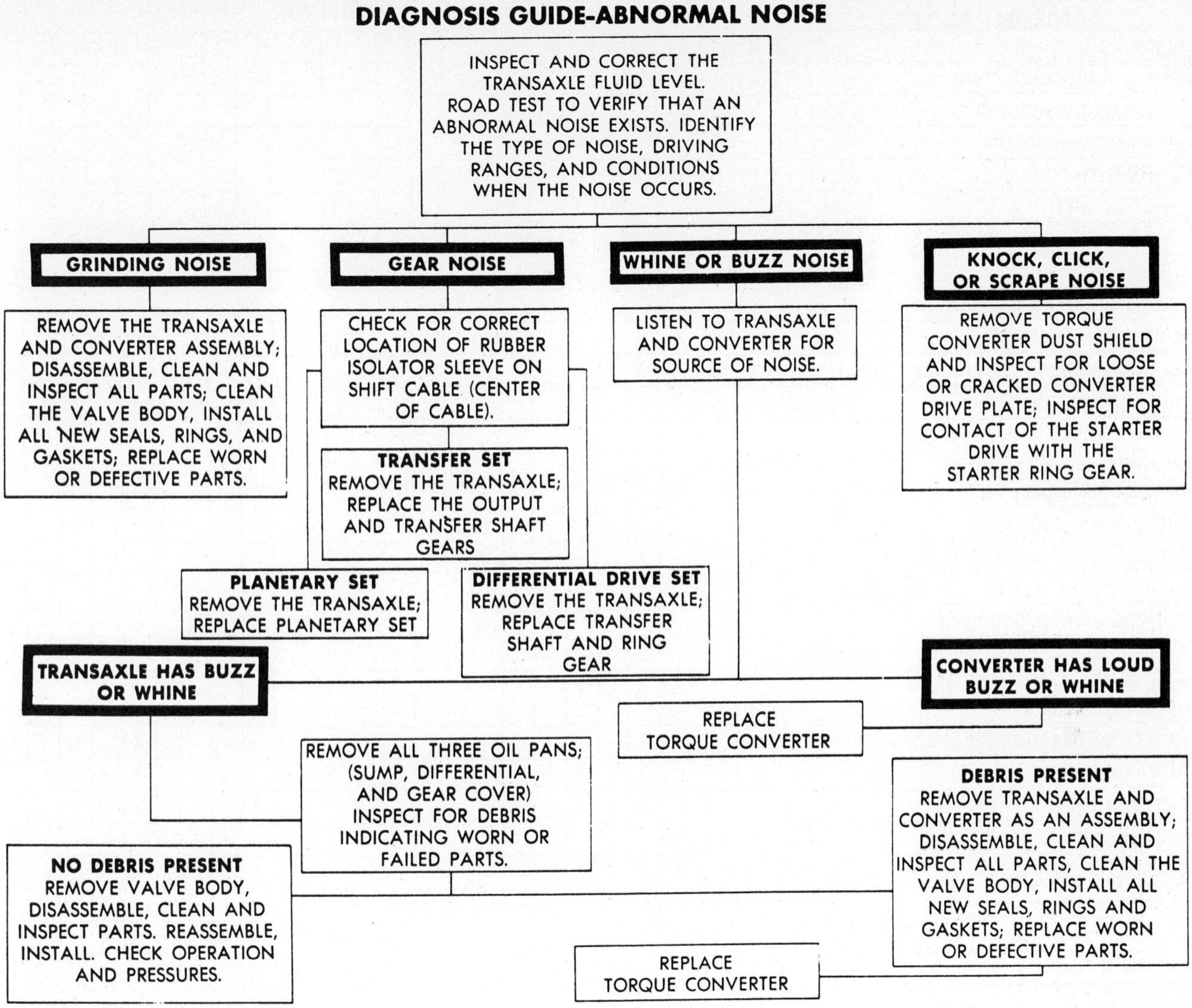

DIAGNOSIS GUIDE-ABNORMAL NOISE

```
INSPECT AND CORRECT THE TRANSAXLE FLUID LEVEL.
ROAD TEST TO VERIFY THAT AN ABNORMAL NOISE EXISTS.
IDENTIFY THE TYPE OF NOISE, DRIVING RANGES, AND
CONDITIONS WHEN THE NOISE OCCURS.
```

GRINDING NOISE
REMOVE THE TRANSAXLE AND CONVERTER ASSEMBLY; DISASSEMBLE, CLEAN AND INSPECT ALL PARTS; CLEAN THE VALVE BODY, INSTALL ALL NEW SEALS, RINGS, AND GASKETS; REPLACE WORN OR DEFECTIVE PARTS.

GEAR NOISE
CHECK FOR CORRECT LOCATION OF RUBBER ISOLATOR SLEEVE ON SHIFT CABLE (CENTER OF CABLE).

TRANSFER SET
REMOVE THE TRANSAXLE; REPLACE THE OUTPUT AND TRANSFER SHAFT GEARS

PLANETARY SET
REMOVE THE TRANSAXLE; REPLACE PLANETARY SET

DIFFERENTIAL DRIVE SET
REMOVE THE TRANSAXLE; REPLACE TRANSFER SHAFT AND RING GEAR

WHINE OR BUZZ NOISE
LISTEN TO TRANSAXLE AND CONVERTER FOR SOURCE OF NOISE.

KNOCK, CLICK, OR SCRAPE NOISE
REMOVE TORQUE CONVERTER DUST SHIELD AND INSPECT FOR LOOSE OR CRACKED CONVERTER DRIVE PLATE; INSPECT FOR CONTACT OF THE STARTER DRIVE WITH THE STARTER RING GEAR.

TRANSAXLE HAS BUZZ OR WHINE

REMOVE ALL THREE OIL PANS; (SUMP, DIFFERENTIAL, AND GEAR COVER) INSPECT FOR DEBRIS INDICATING WORN OR FAILED PARTS.

REPLACE TORQUE CONVERTER

CONVERTER HAS LOUD BUZZ OR WHINE

NO DEBRIS PRESENT
REMOVE VALVE BODY, DISASSEMBLE, CLEAN AND INSPECT PARTS. REASSEMBLE, INSTALL. CHECK OPERATION AND PRESSURES.

REPLACE TORQUE CONVERTER

DEBRIS PRESENT
REMOVE TRANSAXLE AND CONVERTER AS AN ASSEMBLY; DISASSEMBLE, CLEAN AND INSPECT ALL PARTS, CLEAN THE VALVE BODY, INSTALL ALL NEW SEALS, RINGS AND GASKETS; REPLACE WORN OR DEFECTIVE PARTS.

Fig. 10-59 Diagnosis guide designed so technicians can locate the cause of abnormal noises in an A-604 transaxle. *(Courtesy of Chrysler Corporation.)*

DIAGNOSIS CHART "B"

POSSIBLE CAUSE	Harsh engagement from neutral to D	Harsh engagement from neutral to R	Delayed engagement from neutral to D	Delayed engagement from neutral to R	Poor shift quality	Shifts erratic	Drives in neutral	Drags or locks	Grating, scraping, growling noise	Knocking noise	Buzzing noise	Buzzing noise during shifts only	Hard to fill, oil blows out filler tube	Transaxle overheats	Harsh upshift	No upshift into overdrive	No lockup	Harsh downshifts	High shift efforts	Harsh lockup shift
Engine Performance	X	X			X										X		X			
Worn or faulty clutch(es)	X	X	X	X	X	X	X								X	X	X			
—Underdrive clutch	X		X		X	X	X										X			
—Overdrive clutch					X	X	X								X	X				
—Reverse clutch		X		X	X	X														
—2/4 clutch					X		X								X		X			
—Low/reverse clutch	X	X			X		X										X			
Clutch(es) dragging					X															
Insufficient clutch plate clearance					X								X							
Damaged clutch seals			X	X													X			
Worn or damaged accumulator seal ring(s)	X	X	X	X													X			
Faulty cooling system													X							
Engine coolant temp. too low																X	X			
Incorrect gearshift control linkage adjustment			X	X		X	X						X							
Shift linkage damaged																		X		
Chipped or damaged gear teeth								X	X											
Planetary gear sets broken or seized								X	X											
Bearings worn or damaged								X	X											
Driveshaft(s) bushing(s) worn or damaged								X												
Worn or broken reaction shaft support seal rings			X	X	X	X											X			
Worn or damaged input shaft seal rings			X	X												X				
Valve body malfunction or leakage	X	X	X	X	X	X	X			X							X	X	X	
Hydraulic pressures too low			X	X	X	X							X	X			X			
Hydraulic pressures too high	X	X											X				X			
Faulty oil pump			X	X	X								X				X			
Oil filter clogged			X	X	X	X						X								
Low fluid level			X	X	X	X				X				X			X	X		
High fluid level													X	X						
Aerated fluid			X	X	X	X				X			X	X			X	X		
Engine idle speed too low			X	X																
Engine idle speed too high	X	X												X			X			
Normal solenoid operation											X									
Solenoid sound cover loose											X									
Sticking lockup piston																				X

Fig. 10-60 Diagnostic guide designed so technicians can locate the cause of improper operation of an A-604 transaxle. It lists the most probable causes for a series of problems. (Courtesy of Chrysler Corporation.)

REVIEW QUESTIONS

The following questions are provided so you can check the facts you have just learned. Select the response that best completes each statement.

1. Technician A says that you should begin by checking the fluid when you are diagnosing a transmission problem. Technician B says that proper use of a pressure gauge will disclose any transmission problem. Who is right?
   A. A only
   B. B only
   C. Both A and B
   D. Neither A nor B

2. Technician A says that ATF in a modulator vacuum hose is a sign of a bad modulator. Technician B says that the engine should run differently if you pull the vacuum hose off the modulator. Who is right?
   A. A only
   B. B only
   C. Both A and B
   D. Neither A nor B

3. When making a road test, you should pay particular attention to
   A. The quality of the "garage shifts"
   B. The timing of the upshifts and downshifts
   C. Any unusual noises that might occur
   D. All of these

4. Technician A says that a minimum-throttle 1–2 upshift should be smooth and soft and occur at about 20 mph. Technician B says that a harsh light-throttle upshift could be caused by a sticky modulator. Who is right?
   A. A only
   B. B only
   C. Both A and B
   D. Neither A nor B

5. Technician A says that both a tachometer and an oil pressure gauge are necessary to diagnose torque converter clutch problems. Technician B says that a tap of the brake pedal causes the converter clutch to release. Who is right?
   A. A only
   B. B only
   C. Both A and B
   D. Neither A nor B

6. If a transmission has a single pressure test port, it will be for
   A. Governor pressure
   B. Line pressure
   C. Modulator pressure
   D. None of these

7. Technician A says that low fluid pressure in all forward ranges is caused by a faulty pump or filter. Technician B says that fluid pressures should increase when the gear selector is moved to reverse. Who is right?
   A. A only
   B. B only
   C. Both A and B
   D. Neither A nor B

8. The most probable cause of a problem with electronic-controlled transmission is a bad
   A. Electronic control unit
   B. Vehicle speed sensor
   C. Throttle position sensor
   D. Electrical connection

9. When checking electronic transmission controls, you should not use an analog voltmeter or ordinary test light because
   A. They are obsolete
   B. They are slow and clumsy
   C. They draw too much current and can change damage the circuit
   D. All of these

10. Technician A says that you can ruin an electronic control unit by touching its terminals with your finger. Technician B says that you can ruin a solenoid by connecting it to a battery backward. Who is right?
    A. A only
    B. B only
    C. Both A and B
    D. Neither A nor B

11. When testing a pressure switch, it should show
    A. No circuit without pressure at the switch
    B. A complete circuit with pressure at the switch
    C. A complete circuit without pressure at the switch
    D. All of these depending on the switch

12. When making a stall test, you should
    A. Connect a tachometer to the engine
    B. Completely apply the parking and service brakes
    C. Make sure there is no one in front or behind the car
    D. Do all of these

13. The stall speed is high in drive 1 but normal in manual 1 and reverse. Technician A says that this indicates a faulty forward clutch. Technician B says that this is caused by a bad one-way clutch. Who is right?
    A. A only
    B. B only
    C. Both A and B
    D. Neither A nor B

14. A Simpson gear train transmission has a slipping second gear but the fluid pressures are normal. Technician A says that this problem could be caused by a loose intermediate-band adjustment. Technician B says that the band must be replaced on some transmissions. Who is right?
    A. A only
    B. B only
    C. Both A and B
    D. Neither A nor B

15. A transmission has no high gear or reverse, but all other gears are normal. Technician A says you should drop the pan and valve body and air check the high–reverse clutch. Technician B says that you should make a complete oil pressure check first. Who is right?
    A. A only
    B. B only
    C. Both A and B
    D. Neither A nor B

16. When checking the pan contents,
    A. A small amount of metal particles is considered normal
    B. A golden brown coating is a sign of a needed fluid change
    C. Much metal or melted plastic shows a need for a transmission overhaul
    D. All of these are true

17. Technician A says that you should be able to hear a clutch apply and release when you apply air pressure to the passage leading to it. Technician B says that an excessive air leak is an indication of bad seals. Who is right?
    A. A only
    B. B only
    C. Both A and B
    D. Neither A nor B

18. Technician A says that you can use foot powder to help trace the location of a transmission leak. Technician B says that the actual location of a leak is usually slightly forward of where the fluid is found. Who is right?
    A. A only
    B. B only
    C. Both A and B
    D. Neither A nor B

19. A car vibrates rather violently just before a full-throttle 1–2 or 2–3 shift. Technician A says that this can be caused by a faulty extension housing bushing or universal joint. Technician B says that this can be caused by an engine problem or unbalanced torque converter. Who is right?
    A. A only
    B. B only
    C. Both A and B
    D. Neither A nor B

20. When diagnosing transmission noise problems, it is important to note
    A. When the noise occurs
    B. The frequency or pitch of the noise
    C. If the noise changes with the load on the car
    D. All of these

CHAPTER 11

IN-CAR TRANSMISSION REPAIR

OBJECTIVES

After completing this chapter, you should:

- **Understand which repair operations can be performed without removing the transmission from the car.**
- **Be able to determine the correct procedure for performing the in-car repair operations and make those repairs in an approved manner.**
- **Be able to repair faulty threads in an aluminum casting.**

11.1 INTRODUCTION

A technician should repair a problem in a thorough manner, but one that is also as quick, and efficient as possible. Many transmission problems require "bench overhaul"; the unit must be removed from the car, repaired, and then replaced. Whenever possible, transmission repairs are made with the unit in the car to save the time needed to remove and replace (R&R); R&R time for a front-wheel drive (FWD) transaxle is about 4 to 10 hours depending on the car.

In-car service operations include those already described in Chapter 9 plus certain repairs that vary somewhat between transmission makes and models. This normally depends on the access that there is to the component. In most cases, the following are considered in-car repair operations:

Manual shift linkage and seal replacement and adjustment.

TV linkage or cable and seal replacement and adjustment.

Vacuum modulator replacement.

Cooler line and fitting replacement or repair.

Neutral start switch replacement.

Valve body removal, repair, and replacement.

Servo piston and cover seal replacement (when accessible).

Accumulator piston and cover seal replacement (when accessible).

293

Speedometer gear and seal replacement.

Governor removal, repair, and replacement.

Extension housing seal, bushing, or gasket replacement [rear-wheel drive (RWD)].

Drive shaft seal replacement (FWD).

Portions of these operations such as valve body or governor repair will be described in Chapter 12 along with other transmission repair procedures.

It should be remembered that the exact procedure for the service operations that follow vary between transmission makes and models. A service manual should be consulted. The operations described in this chapter are normally found under the heading "Service (in vehicle)" or "On-Car Service" in the service manual. Here they are described in as general a manner as possible.

11.2 MANUAL SHIFT LINKAGE AND SEAL

The manual shift linkage attaches to a lever on the outside of the transmission case. In many cars, a cable and housing are used, and in most modern transmissions the cable connects to a shift lever using a metal clip or plastic grommet (Fig. 11-1). The cable housing is secured to a bracket on the transmission case. The other end of the cable and housing is connected to the gear selector in a similar manner. The removal of a cable is usually a matter of disconnecting each end and sliding the cable and housing out through any handy body opening. Cable replacement is the reverse of the removal procedure and should be followed by an adjustment to ensure correct manual valve positioning.

A lip seal is normally used to prevent fluid leaks where the manual selector shaft passes through the transmission case (Fig. 11-2). This seal is easier to R&R if the manual selector shaft is removed, but that would require valve body removal in most cases. It is possible to remove this seal using either a slide hammer and self-tapping screw or a sharp chisel (Fig. 11-3). Be careful that you don't damage the seal bore. The new seal can be driven into the bore using a correctly sized socket as a driving tool (Fig. 11-4).

11.3 LIP SEAL REPLACEMENT

A standard metal-backed lip seal has to seal against two different surfaces: a dynamic seal with the moveable shaft at the inner bore and a static seal where it fits into its bore (Fig. 11-5). The static seal is made when the slightly oversize seal backing is driven into the bore. This is often made more positive by a coating on the seal case or by applying a sealant on this area before installing the seal (Fig. 11-6). A leak occurs if the bore for the seal is scratched or damaged or if the seal case is bent or distorted. A driver that fits completely against the back of the seal should always be used when installing a seal (Fig. 11-7). This ensures that the seal is driven in straight and that it will not be dented.

The seal lip is a flexible, elastic compound. It must remain flexible and resilient in order to maintain a fluid-tight seal against the moving shaft. The sharp lip of the seal is easily damaged by cuts or tears. When installing a seal over a shaft, it is good practice to protect this lip with a seal protector, especially if there are rough or sharp edges on the shaft. A seal protector is often just a piece of slick paper wrapped around the shaft (Fig. 11-8). The lip of the seal should always be lubricated to prevent wear. Either petroleum jelly or automatic transmission fluid (ATF) is often used for this.

11.4 TV LINKAGE AND SEAL

In most modern transmissions, the throttle valve (TV) cable is a cable and housing, and its replacement is essentially the same as a shift cable (Fig. 11-9). In these units, the TV cable is attached to a lever on a shaft that extends into the transmission, just like the shift lever and shaft. Also like the shift lever, a lip seal fits over the shaft and into the transmission (Fig. 11-10).

In many modern General Motors units, the TV cable extends into the transmission. The transmission end of the cable is disconnected by unbolting the mounting bracket and disconnecting the cable (Fig. 11-11). The seal can be easily replaced while the bracket and cable are loose.

11.5 VACUUM MODULATOR REMOVAL AND REPLACEMENT

Most modern vacuum modulators are held in the transmission by a bolt and retaining clamp. Clamp removal allows the modulator to be pulled out of the transmission case. In some transmissions, a control rod or pin is used with the modulator to connect the modulator stem to the valve (Fig. 11-12). The control rod and modulator valve can usually be removed once the modulator is out.

In some older transmissions, the modulators are threaded into the case. To R&R the modulator on these units usually requires a special modulator wrench, a relatively short, thin wrench with a jaw opening of $\frac{7}{8}$ or 1 in. (Fig. 11-13).

To R&R a vacuum modulator, you should:

1. Disconnect the vacuum hose and remove the retaining bolt and clamp.

2. Pull the modulator out of the transmission case.

3. If necessary, remove the control rod, modulator valve, or both. Perform the necessary service on these and replace them.

4. Install a new seal on the modulator stem, wet the seal with ATF, and push the modulator into the case.

5. Replace the retaining clamp and bolt and tighten the bolt to the correct torque (Fig. 11-14).

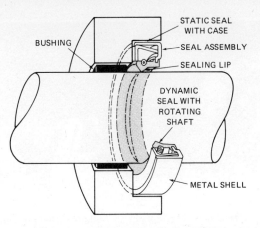

Fig. 11-5 Cutaway view of a seal. Note that the seal must make a static seal with the housing bore and a dynamic seal with the shaft.

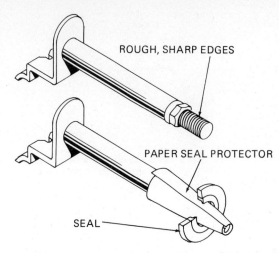

Fig. 11-8 When a seal is installed, it is good practice to protect the sealing lip from contact with any rough or sharp areas of the shaft. A piece of slick paper such as a playing card can be used.

Fig. 11-6 The seal at the top has a sealant around its metal backing. The seal on the bottom has a coating of shop-applied sealant.

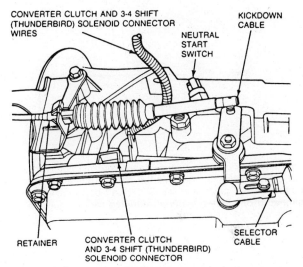

Fig. 11-9 This kickdown/throttle cable is attached to the retainer bracket and lever at the transmission. Removal of these and the other end connections at the throttle linkage allows replacement of the cable. *(Courtesy of Ford Motor Company.)*

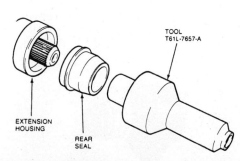

Fig. 11-7 Tool T61L-7657-A is a seal driver; note how the seal fits the driver so it can be driven straight into the bore with no damage. *(Courtesy of Ford Motor Company.)*

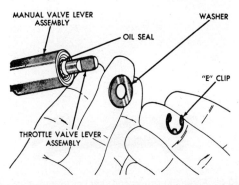

Fig. 11-10 Removal of the exterior throttle lever, E-clip, and washer allows access to the small seal at the throttle lever shaft. *(Courtesy of Chrysler Corporation.)*

297

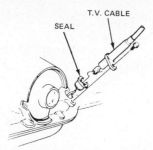

Fig. 11-11 The throttle cable is sealed to the transmission using a rubber seal. *(Copyrighted material reprinted with permission from Hydramatic Div., GM Corp.)*

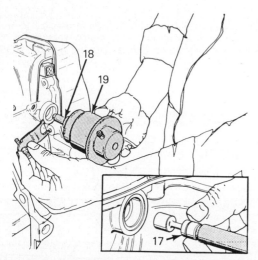

Fig. 11-12 Modulator is secured to the transmission by the bolt and clamp in the technician's left hand. After removal of the modulator, the modulator valve (17) can be removed using a magnet. *(Copyrighted material reprinted with permission from Hydra-matic Div., GM Corp.)*

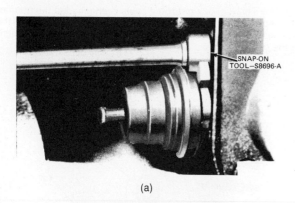

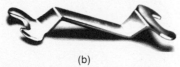

Fig. 11-13 (a) A special crows-foot socket (S8696-A) is used to remove or install a modulator. (b) Special wrenches available to install modulator. *[(a) Courtesy of Ford Motor Company; (b) Courtesy of Snap-On Tools.]*

Fig. 11-14 When a modulator is replaced, the retaining bolt should be tightened to the correct torque.

11.6 ALUMINUM THREAD REPAIR

Transmission cases and extension housings are made from relatively soft, cast aluminum. Bolt threads begin to wear after a couple of uses, and these threads easily pull out if the bolt is overtightened or if too short of a bolt is used (Fig. 11-15). The most commonly used thread repair method is to install a special steel coil thread insert; one brand is called a Heli-Coil. This thread insert provides a stronger and much longer lasting thread. Another recently developed repair method is to form new threads using an epoxy compound. This method is not recommended for threads that will be in contact with ATF.

Thread damage is often caused by the use of a bolt or cap screw that is too long or too short. Too long of a bolt will bottom in the hole before it is tight, and too short of a bolt will grip fewer threads than needed. Normally a bolt will enter the threaded hole a distance that is about $1\frac{1}{2}$ to 2 times the bolt diameter (Fig. 11-16). In other words, a $\frac{1}{4}$-in. (6.35-mm) bolt will have about $2 \times \frac{1}{4}$, or $\frac{1}{2}$, in. (12.7 mm) of thread contact.

To install a thread insert, you should:

1. Drill out the worn threads, if necessary, using a drill of the correct size for the special tap (Fig. 11-17).

2. Use the special tap to cut new threads slightly longer than the length of the insert.

3. Install the thread insert onto the installing tool and thread the insert completely into the hole. The end of the last thread should enter into the hole.

4. Break off the installing tang or resize the insert as required by the manufacturer (Fig. 11-18).

Fig. 11-15 This bolt was too short, and when it was tightened, it pulled the threads out of the aluminum case.

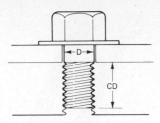

Fig. 11-16 The thread contact distance (CD) of a bolt should be about twice the diameter of the threads (D) to prevent thread stripping.

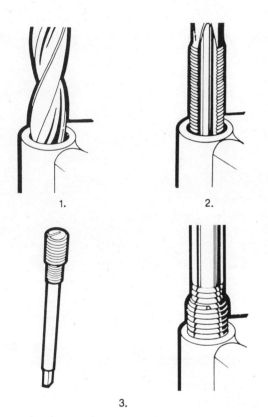

Fig. 11-17 Step 1 in repairing damaged threads is to drill the hole to the proper size for the special tap. Step 2 in repairing a thread is to use the special tap to cut new threads. Step 3 in repairing thread is (a) to thread the insert onto the installing mandrel so (b) the tang of the insert has engaged the mandrel. *(Courtesy of Oldsmobile Div., GM Corp.)*

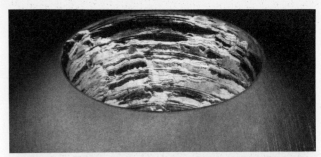

Fig. 11-18 The damaged threads (top) have been repaired (bottom) by installing a thread insert. They are better than new because of the hardness of the insert. *(Courtesy of Helicoil.®)*

11.7 COOLER LINE AND FITTING REPAIR

A crushed or leaking cooler line must be repaired or replaced. A leaking fitting is often fixed by merely tightening it. A cooler line is steel tubing that usually ends at a tube nut that is threaded into an adapter that in turn is threaded into the transmission case or radiator cooler (Fig. 11-19). The adapter usually uses tapered NPT (National Pipe Threads) threads to form a seal in the case. Some adapters use straight, machine threads, but these require a gasket or a seal.

The best repair for a damaged steel cooler line is replacement, but the replacement line has to be the correct length and have the correct bends. Never use copper line for replacement because it has a tendency to fatigue, crack, and break. Steel cooler lines have been successfully repaired by cutting out the damaged portion, flaring the cut ends, and using a coupler to rejoin the lines (Fig. 11-20). Compression fittings can also be used to join tubing. If a portion of the line has been cut away, an additional piece of tubing can be added. This

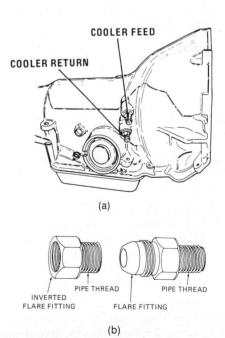

Fig. 11-19 (a) The two cooler fittings threaded into the case adapt the cooler line to the case. (b) These fittings usually have a NPT thread on one end where they thread into the case and a flare fitting at the other end to match the cooler line. *[(a) Copyrighted material reprinted with permission from Hydra-matic Div., GM Corp.]*

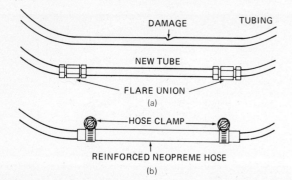

Fig. 11-20 A damaged cooler tubing can be repaired by cutting out the damaged section and replacing it with a section of (a) steel or (b) rubber. (a) Requires that the cut ends of the tubing be flared so a flare union can be used to join them to the new line section. (b) A rubber hose clamped over the cut ends.

tubing is available in various sizes and lengths at many automotive parts houses. If you make your own tubing flare, don't forget that steel tubing must have a double flare. Another repair method is to cut out the damaged portion and replace it with a rubber hose that is clamped onto the tubing. With this repair method, reinforced fuel hose of the correct size is used; it must be clamped securely onto the tubing; and it must be located so it will not be cut, burned, or otherwise damaged.

When removing and replacing a cooler line, two flare nut or tube nut wrenches should be used (Fig. 11-21). If a single wrench is used, the tube nut and the adapter often turn together, and this causes twisting and damage to the tubing. If the tube nut is tight, a common open end wrench will tend to round off the corners on the tube nut; a flare nut wrench has a much better grip.

11.8 OIL COOLER CHECK

If the fluid in a transmission is dirty or there has been a failure of the torque converter clutch, the oil cooler should be checked to ensure it is not plugged and restricting flow. To check this, you should:

1. Remove the transmission dipstick and place a funnel into the filler tube.

2. Remove the cooler return line from the transmission case and slide a tight-fitting hose over the end of the cooler line. Run the other end of this hose into the funnel.

3. Set the parking brake, start the engine, place the gear selector in neutral, and run it at a speed of about 1000 rpm while you observe the cooler oil flow from the end of the hose into the funnel (Fig. 11-22). After the air bubbles clear up, a substantial, solid flow of oil (about 1 quart per 20 seconds) should occur.

4. If the flow appears too small, disconnect the cooler feed line (out from the transmission) at the cooler and transfer the hose from the return line to this line (Fig. 11-23). Repeat Step 3. If the oil flow is substantially greater, the cooler is clogged.

A clogged cooler should be flushed (this procedure is described in Chapter 13).

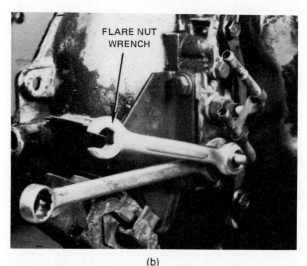

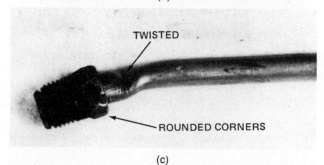

Fig. 11-21 (a) Two flare nut wrenches should be used to loosen a tube nut. (b) Use of a standard open-end wrench can result in a rounded-off tube nut. (c) Use of a single wrench can result in a twisted line. [(a) Courtesy of Snap-On Tools.]

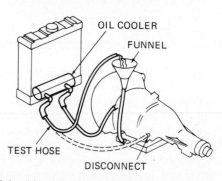

Fig. 11-22 A rubber hose has been attached to the cooler return line, a funnel has been placed in the transmission filler, and the hose has been placed into the funnel. When the engine is started, a substantial flow of fluid should be observed from the return hose.

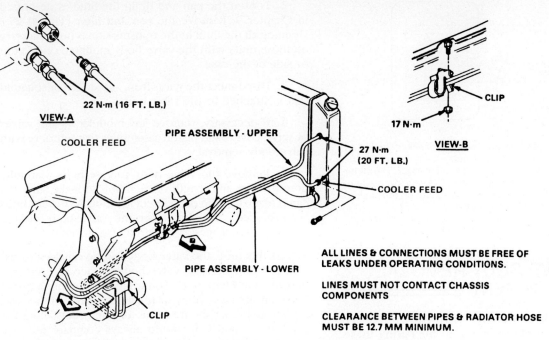

Fig. 11-23 If the cooler flow test shows a poor fluid flow, transfer the hose to the cooler feed line at any of the points indicated and repeat the cooler flow test.

11.9 NEUTRAL START SWITCH SERVICE

The neutral start switch allows the starter to operate in park and neutral and turns on the backup lights in reverse. Some transmissions mount this switch in the transmission case so a lever or cam (usually the "rooster comb" of the manual shift) operates a plunger on the switch. The commonly encountered neutral start switch problems are fluid leaks or a faulty switch.

Fluid leaks at the base can often be cured by replacing the gasket or O-ring. Leaks through the switch require switch replacement. Switch replacement is usually performed by simply disconnecting the wire connection; unscrewing the switch from the transmission case; installing the new switch and sealing ring; tightening the switch to the correct torque; and reconnecting the wires (Fig. 11-24). In some cases, it is necessary to use a special socket when removing and replacing the switch.

The electrical functions of the switch can be checked using an ohmmeter. With the wires unplugged, connect the two ohmmeter leads to the two terminals for each function (Fig. 11-25). In the starter portion, there should be very little resistance (about 0 to 1 Ω) in park and neutral and a very high resistance (infinite) in the other gear positions. With the backup light function, there should be a low resistance in reverse and a high resistance in the other gear positions. Except for single-terminal switches, there should be no continuity (an open circuit) between each terminal and the body of the switch.

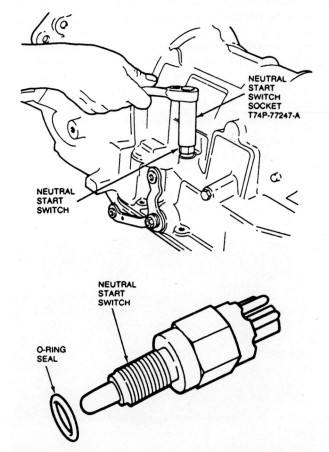

Fig. 11-24 Case-mounted neutral start switch is removed and replaced using a special deep socket. A new O-ring should be used when replacing this style of switch. *(Courtesy of Ford Motor Company.)*

301

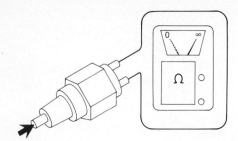

Fig. 11-25 A neutral start switch can be checked using an ohmmeter. Connect one of the ohmmeter leads to one of the start switch terminals and push inward on the plunger. The meter reading should go from a very high reading to a very low reading. Don't forget that these switches also serve to turn on the backup lights. *(Courtesy of Ford Motor Company.)*

11.10 VALVE BODY REMOVAL AND REPLACEMENT

There are many possibilities for valve body problems, and the valve body can be removed for curing these problems. This operation is rather messy due to the fluid in the transmission and the fact that valve body removal opens many fluid passages so they can drain. Depending on the transmission, you may find check balls, one or more filter screens, a servo piston and spring, or an accumulator assembly above the valve body. Be prepared for these so they are not misplaced or damaged (Fig. 11-26).

According to some technicians, it is important that a valve body be removed from a transmission at room temperature, especially with aluminum valve bodies. Removal of a hot valve body can produce warpage, which might cause sticky valves.

To remove a valve body, you should:

1. Raise and support the vehicle on a hoist so you have access to the transmission pan.

2. Loosen the pan and drain the fluid as described in Chapter 9. Remove the pan and filter (Fig. 11-27). Draining all the fluid in the transmission is not necessary on those units with the valve body mounted on the top or side of the case.

3. Disconnect the wires from any switch or solenoid that is attached to the valve body.

4. If necessary, remove any modulators and valves or servo covers and piston assemblies that interfere with valve body removal.

5. Remove the valve body to case bolts. Carefully note the size and length of each bolt to make sure you can replace them in the correct location. As the last bolts are removed, carefully lower the valve body disconnecting any linkages as required (Fig. 11-28).

At this time, the valve body can be repaired or exchanged as necessary. Valve body service and repair is described in Chapter 12. Some valve bodies use a gasket between the valve body or separator plate and the case. A new gasket(s) should be used when reinstalling the valve body, and you should always compare the new and old gaskets to ensure the correct replacement. Often several gaskets are almost but not quite the same; they will not work (Fig. 11-29).

To replace a valve body, you should:

1. If used, position a new gasket or any check balls on top of the valve body or separator plate. The check balls can be held in place with petroleum jelly.

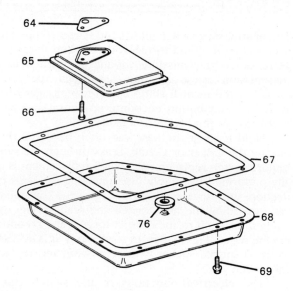

ILL NO.	DESCRIPTION
64	GASKET, TRANSMISSION OIL PUMP SCREEN
65	SCREEN ASSEMBLY, TRANSMISSION OIL
66	BOLT, HEX HEAD 6.3 X 1 X 55
67	GASKET, TRANSMISSION OIL
68	PAN, TRANSMISSION OIL
69	SCREW & CONICAL WASHER ASM., HEX HEAD
76	MAGNET

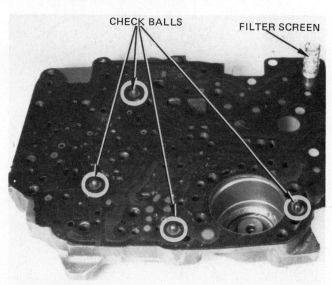

Fig. 11-26 As the valve body is removed on some transmissions, you should be ready to catch some check balls or accumulator springs.

Fig. 11-27 The oil pan and filter are removed before removing the valve body. *(Copyrighted material reprinted with permission from Hydra-matic Div., GM Corp.)*

Fig. 11-28 As the valve body is removed from a Torqueflite transmission, the output shaft might have to be rotated so you can pull the park control rod (arrow) past the park pawl.

Fig. 11-29 The old valve body gasket should be saved to compare with the new gasket to ensure that you have the right one.

2. Carefully reposition the valve body, making sure to correctly position the manual and TV linkage as necessary. It is recommended that two guide pins or bolts be threaded into the case to hold the transfer plate and gasket(s) in proper alignment. Loosely install a few bolts to hold it in position (Fig. 11-30). Remember that the bolts should thread into the case a distance that is about twice the bolt diameter.

3. Reinstall the valve body bolts making sure the correct lengths and sizes are in the proper position (Fig. 11-31).

4. Tighten the bolts to the correct torque. Normally, you should start at the centermost bolts and work to the outer bolts using a spiral pattern (Fig. 11-32).

5. Replace the filter and pan and refill the transmission with fluid as described in Chapter 9.

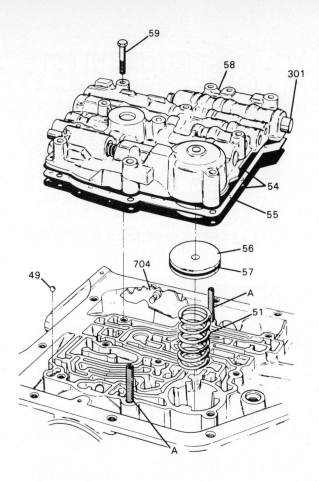

ILL NO.	DESCRIPTION
A	GUIDE PINS
49	BALL, CHECK (5)
51	SPRING, 1-2 ACCUMULATOR PISTON
54	GASKET, VALVE BODY/SPACER PLATE
55	PLATE, VALVE BODY SPACER
56	PISTON, 1-2 ACCUMULATOR
57	SEAL, ACCUMULATOR PISTON
58	VALVE, CONTROL
59	BOLT, HEX HEAD 6.3 X 1 X 45
301	MANUAL VALVE
704	DETENT LEVER

Fig. 11-30 Two guide pins are being used to ensure proper alignment of the gaskets and spacer plate. Note that the 1–2 accumulator piston (56) and spring (51) will be installed along with the valve body and that the tang on the detent lever (704) must be aligned with the manual valve (301). *(Copyrighted material reprinted with permission from Hydra-matic Div., GM Corp.)*

11.11 SERVO AND ACCUMULATOR PISTON, COVER, AND SEAL SERVICE

A faulty servo piston or cover seal can cause a loss in servo apply pressure and band slippage or a bad cover seal can cause a fluid leak. A faulty accumulator piston or cover seal can cause a poor shift quality or a bad cover seal can cause a fluid leak. The service procedures for servos and accumulators are similar so they will be

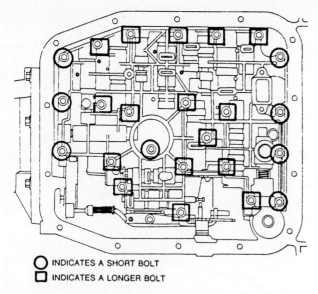

Fig. 11-31 This AOD valve body is retained by bolts of two lengths. The holes for the shorter bolts are marked with a circle, and those for the longer bolts are marked with a square. *(Courtesy of Ford Motor Company.)*

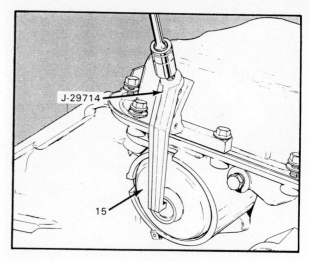

Fig. 11-33 Tool J-29714 is used to hold the THM 700 2-4 servo cover (15) inward and the spring compressed while the retaining ring is removed or replaced. *(Copyrighted material reprinted with permission from Hydra-matic Div., GM Corp.)*

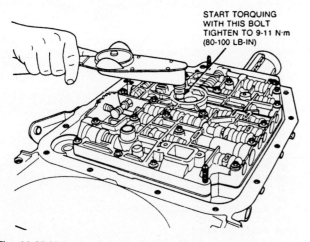

Fig. 11-32 Valve body bolts should always be tightened to the correct torque using a pattern that starts in the center and works outward. *(Courtesy of Ford Motor Company.)*

described together. If the servo or accumulator cover is accessible, it can be removed and replaced and the piston can be serviced on the car.

A service manual should be checked before disassembling the servo or accumulator. In some cases, the valve body serves as the cover so servo or accumulator service becomes an easy operation once the valve body is removed (like the 1-2 accumulator piston in Fig. 11-30). In others, a separate cover is above the valve body so cover removal is done after valve body removal. Some servos are rather complex and have two or more pistons, up to four or five fluid passages, and fluid control orifices or check balls inside them. In some cases, a strong return spring is under the piston. Depending on the transmission, a special tool is sometimes required to keep the spring compressed so the cover can be removed (Fig. 11-33).

The servo or accumulator cover seal is usually either a gasket or an O-ring; it is fairly easy to R&R once the cover is removed. The servo or accumulator piston seal is usually either a metal or rubber ring. Seal ring checks are described in Chapter 12.

11.12 EXTENSION HOUSING SEAL AND BUSHING SERVICE

The seal and bushing at the rear of the extension housing can be removed and replaced as on-car service operations. In some cases the entire extension housing is removed for bearing, seal, governor, speedometer gear, or park mechanism service.

A faulty seal can be easily removed using a special puller that fits over the output shaft and into the seal. It can also be removed using a slide hammer and self-tapping screw or sharp chisel and hammer, but be careful not to damage the seal bore. A special puller and installer tool is usually required for on-car bushing replacement.

To remove and replace an extension housing seal, you should:

1. Raise and support the car on a hoist so you have access to the transmission and drive shaft.

2. Place alignment marks for assembly on the rear universal joint and rear-axle pinion flange and disconnect the drive shaft from the rear axle (Fig. 11-34). Do not let the drive shaft hang from the transmission; either remove it completely or support it until removal. Check the drive shaft slip yoke for wear at the seal and bushing surfaces (Fig. 11-35).

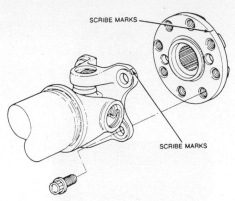

Fig. 11-34 It is good practice to scribe a pair of index marks before removing the drive-shaft-to-rear-axle connections so they can be replaced in the same position. *(Courtesy of Ford Motor Company.)*

Fig. 11-36 A transmission stopoff tool is used to stop off the fluid leak from the extension housing when the drive shaft is removed. *(Courtesy of K-D Tools.)*

NOTE ATF will begin leaking out of the drive shaft opening. You should either raise the rear of the car enough to stop the flow or place a container under it to catch the flow. A stopoff tool or old U-joint slip yoke can be used to temporarily stop the fluid (Fig. 11-36).

3. Pull or pry out the rear seal using a suitable tool (Fig. 11-37).

4. If desired, pull out the rear bushing using a suitable tool (Fig. 11-38).

5. If the bushing is removed, use a suitable tool to drive the new bushing completely into place (Fig. 11-39).

6. Use a correctly sized driver to drive the new seal into place (Fig. 11-40). If the replacement seal does not have an outer sealant coating, a film of sealant should be spread around the outer surface of the seal case.

HELPFUL HINT: When driving a seal into place, it is possible for the seal's garter spring to pop out of position. Some technicians fill the seal cavity with petroleum jelly to help hold the spring in position (Fig. 11-41).

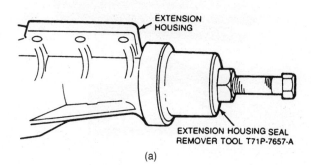

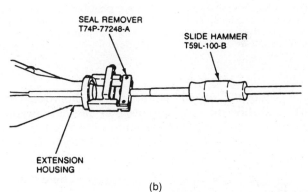

Fig. 11-37 Several styles of pullers are available that attach to the seal so the seal can be pulled out (a) by tightening the puller screw or (b) using the slide hammer. *(Courtesy of Ford Motor Company.)*

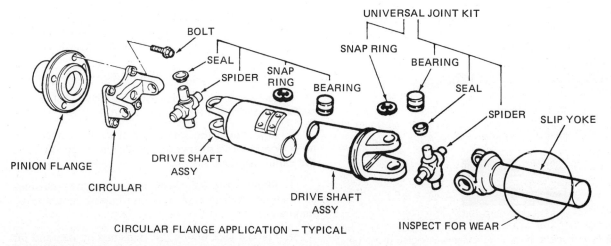

Fig. 11-35 As a drive shaft is removed, you should check the slip yoke for wear in the extension housing bushing and seal area. *(Courtesy of Ford Motor Company.)*

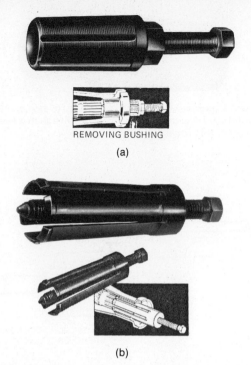

Fig. 11-38 Several styles of pullers are available that attach to the bushing so it can be removed by tightening the puller screw. Note that the transmission output shaft is in place. *(Courtesy of OTC Tool & Equipment Div. of SPX Corp.)*

7. Wet the seal lip and drive shaft slip yoke with ATF and slide the yoke into place in the transmission; align your marks; connect the universal joint to the rear-axle flange; and tighten the retaining bolts to the correct torque.

8. Check and correct the transmission fluid if necessary.

11.13 EXTENSION HOUSING REMOVAL

The extension housing is removable on most RWD transmissions, and depending on the transmission, this allows replacement of the extension housing gasket or service

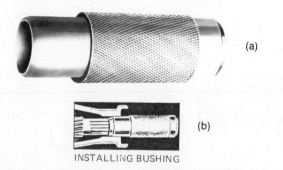

Fig. 11-39 Tool made to hold a particular bushing so it can be easily driven into place using a hammer. *(Courtesy of OTC Tool & Equipment Div. of SPX Corp.)*

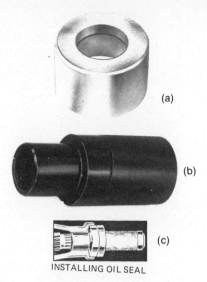

Fig. 11-40 Collar that fits over the driving tool shown in Fig. 11-39 to allow the seal to be driven into place using a hammer. *(Courtesy of OTC Tool & Equipment Div. of SPX Corp.)*

Fig. 11-41 The cavity of a seal has been filled with petroleum jelly to hold the garter spring in position during installation. The petroleum jelly will melt and mix with the ATF when the transmission gets hot.

of the seal, bushing, governor, speedometer gear, or park mechanism. In the case of the Chrysler A-500, extension housing removal allows service of the overdrive gear train.

To remove an extension housing, you should:

1. Raise the car and remove the drive shaft as described in Section 11.12, Steps 1 and 2.

2. Disconnect the speedometer cable. If necessary for access, remove the parking brake equalizer and move the cables aside.

3. Remove the bolts securing the engine–transmission mount to the transmission support/center cross member.

4. Place a jack under the transmission pan and lift the transmission just enough to clear the cross member (Fig. 11-42).

NOTE A transmission jack should be used so the transmission pan is not damaged. If a conventional jack is used, blocks of wood should be placed in a position (under the stronger areas) so the pan will not be dented. Also check the transmission mount at this time. A faulty mount should be replaced when the transmission support/cross member is reinstalled.

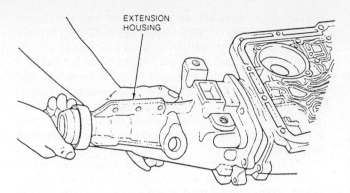

Fig. 11-43 After the retaining bolts have been removed, it is usually possible to slide the extension housing off, over the output shaft. *(Courtesy of Ford Motor Company.)*

Fig. 11-42 A transmission jack has been positioned so it will support the transmission while the rear mount and cross member is removed.

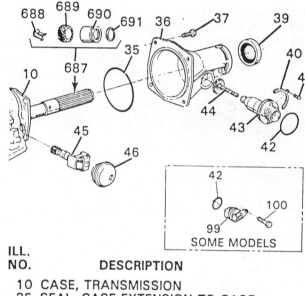

ILL.
NO.   DESCRIPTION
10   CASE, TRANSMISSION
35   SEAL, CASE EXTENSION TO CASE
36   EXTENSION, CASE

Fig. 11-44 A new O-ring (35) or gasket should be installed when the extension housing is replaced. *(Copyrighted material reprinted with permission from Hydra-matic Div., GM Corp.)*

5. Remove the cross member.

6. Place a drip pan under the transmission and lower the transmission just enough to gain access to the upper extension housing bolts.

7. Remove the retaining bolts, the extension housing, and gasket (Fig. 11-43). On some transmissions, it is necessary to expand a bearing retainer during removal.

Extension housing replacement is essentially a reversal of the procedure used to remove it. To replace an extension housing, you should:

1. Clean off all traces of old gasket and sealant from the back of the transmission and front of the extension housing.

2. Place the new gasket in position; place the extension in position; install the bolts; and tighten them to the correct torque (Fig. 11-44). If necessary, petroleum jelly can be used to stick the gasket in position on the case.

NOTE In some cases, it is recommended that a sealant be placed on the gasket or a sealant or thread lock on the bolt threads.

3. Raise the transmission; place the cross member in position; install the bolts; and tighten them to the correct torque.

4. Lower the transmission onto the cross member as you install the attaching bolts and tighten the bolts to the correct torque. Remove the jack.

5. Reconnect the speedometer cable and the parking brake mechanism if necessary.

6. Complete the reassembly by following Steps 7 and 8 of Section 11.12.

11.13.1 Speedometer Gear Replacement

Speedometer gear problems show up as a nonoperating speedometer. The cause can be a faulty cable, driven gear, or drive gear. On some transmissions, the speedometer drive gear is a set of teeth cut into the output shaft. An example of this is shown in Fig. 11-43. On these transmissions, a rough drive gear can sometimes be cured using a small three-corner file. Scale on the gear teeth usually can be removed by wire brushing. The other choice is to dismantle the transmission for output shaft replacement. On other transmissions, this drive gear is slid over the output shaft, and it can be easily removed and replaced once the extension housing has been removed.

The speedometer drive gear in some transmissions is held in place by a special clip. To remove the gear, simply push inward on the clip and slide the gear over the clip and down the shaft (Fig. 11-45). On some transmissions, the gear is driven by a steel ball that is halfway in the output shaft and gear. To remove the gear, remove the retaining ring and slide the gear off the shaft and drive ball.

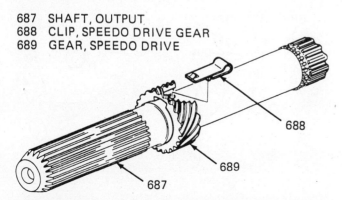

687 SHAFT, OUTPUT
688 CLIP, SPEEDO DRIVE GEAR
689 GEAR, SPEEDO DRIVE

Fig. 11-45 To remove the speedometer gear (689), push the tang of the clip (688) downward and slide the gear off. *(Copyrighted material reprinted with permission from Hydra-matic Div., GM Corp.)*

11.13.2 Park Gear and Pawl Service

On some transmissions, the park pawl and spring are mounted in the extension housing, and the park gear is mounted on the output shaft (Fig. 11-46). On these units, a certain amount of service such as inspection or replacement of the parking pawl, return spring, or park gear can be done. The operating rod coming from the transmission case can also be checked for correct movement.

11.14 GOVERNOR SERVICE

If the governor is shaft driven, removal of the extension housing allows access to the governor for inspection or disassembly and cleaning (Fig. 11-47). General Motors transmissions and transaxles and Ford transaxles use gear-driven governors. These units have a separate cover that retains the governor in the transmissions (Fig. 11-48). Removal of this cover allows the governor to be removed.

Normal on-car service includes disassembly of the valve and weights to cure sticking, replacement of the assembly, or replacement of the drive gear.

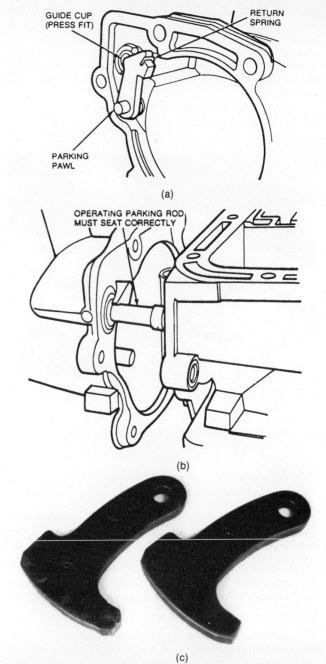

Fig. 11-46 (a) The park lever and return spring are mounted in an A4LD extension housing. (b) When the extension housing is replaced, be sure to position the operating rod correctly. (c) A worn-out park lever is shown at left. *(Courtesy of Ford Motor Company.)*

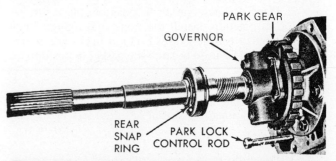

Fig. 11-47 Torqueflite mounts its shaft drive governor on the governor support and parking gear. *(Courtesy of Chrysler Corporation.)*

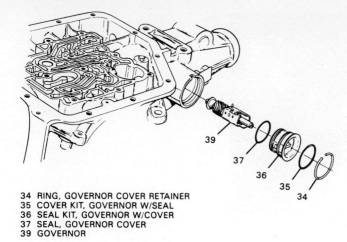

Fig. 11-48 Gear driven-governor (39) is held in the transmission by its cover (36). *(Copyrighted material reprinted with permission from Hydra-matic Div., GM Corp.)*

34 RING, GOVERNOR COVER RETAINER
35 COVER KIT, GOVERNOR W/SEAL
36 SEAL KIT, GOVERNOR W/COVER
37 SEAL, GOVERNOR COVER
39 GOVERNOR

3. Remove the governor. In some cases, it will slide out as the cover is removed; in others, turning the output shaft will move the governor outward.

NOTE   You should inspect the governor-driven gear and the governor bore in the transmission to make sure it is in good condition. A faulty governor gear will stop governor operation (Fig. 11-50). Reaming and installing a bushing in a worn bore will be described in Chapter 12.

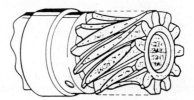

Fig. 11-50 Applecored governor gear can cause erratic or no governor operation. It is usually the result of rough output shaft governor drive gear teeth or a governor that is trying to seize in the bore. *(Copyrighted material reprinted with permission from Hydra-matic Div., GM Corp.)*

To remove a gear-driven governor, you should:

1. Raise and support the car on a hoist.

2. Remove the governor cover attaching bolts or retaining ring and the cover (Fig. 11-49). In some cases, it is necessary to lower the transmission as described in Section 11.12, Steps 1 to 6.

To replace a gear-driven governor, you should:

1. Slide the governor into position in its bore. Some governors use a seal ring at one end; make sure this seal is in good condition before installing.

2. Install a new gasket or O-ring for the governor cover and place the cover in position. In cases where the cover slides into the transmission and O-ring, wet the O-ring with ATF and carefully tap the cover into place using a brass drift or suitable driver (Fig. 11-51).

3. Install the retaining ring or bolts. If bolts are used, tighten them to the correct torque.

4. Check and adjust the fluid level if necessary.

11.14.1 Governor Disassembly

A shaft-driven governor usually consists of the governor body, valve(s), weight(s), and support, and sometimes a spring or shaft/pin is used to connect the weight to the valve (Fig. 11-52). Service of these units is disassembly, cleaning, inspection to ensure the valve(s) and weight(s) move freely, and reassembly. It should be noted that some units use a gasket between the governor body and support. Also, on units where the body fits completely around the output shaft, it is good practice to tighten the body-to-support bolts after the two have been placed onto the output shaft (Fig. 11-53).

A gear-driven governor can also be disassembled for service or replacement of the drive gear (Fig. 11-54). Correct movement of the valve can be checked by holding the governor with the gear end upward and squeezing and releasing the weights as shown in Fig. 11-55.

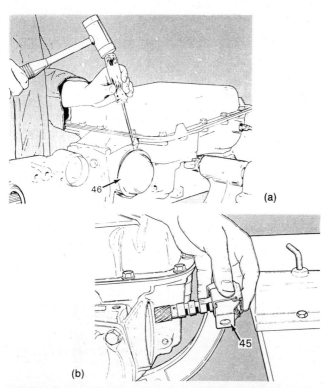

Fig. 11-49 (a) After the cover has been removed, (b) the governor can be slid out of its bore. *(Copyrighted material reprinted with permission from Hydra-matic Div., GM Corp.)*

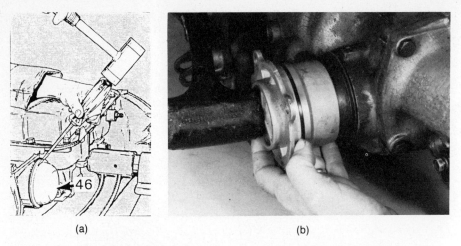

Fig. 11-51 (a) When the governor is replaced, the cover can be driven into place using a screwdriver; (b) an old C6 servo cover is the right size to use as a driver tool. [(a) Copyrighted material reprinted with permission from Hydra-matic Div., GM Corp.)]

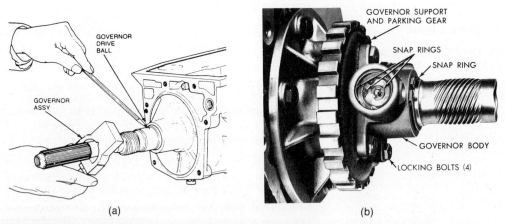

Fig. 11-52 (a) Governor is retained by a retaining ring and is driven by a ball in the output shaft. (b) Governor uses a valve shaft/pin that passes through the output shaft to connect the valve to the weights. [(a) Courtesy of Ford Motor Company; (b) Courtesy of Chrysler Corporation.)]

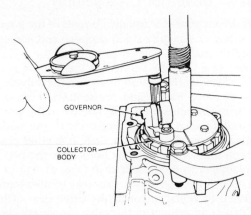

Fig. 11-53 It is good practice to tighten the governor mounting bolts to the correct torque after the governor and support are in position on the output shaft. (Courtesy of Ford Motor Company.)

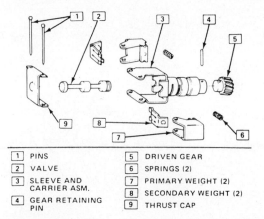

Fig. 11-54 Typical General Motors early style, gear drive governor that has been disassembled for cleaning or repair. (Copyrighted material reprinted with permission from Hydra-matic Div., GM Corp.)

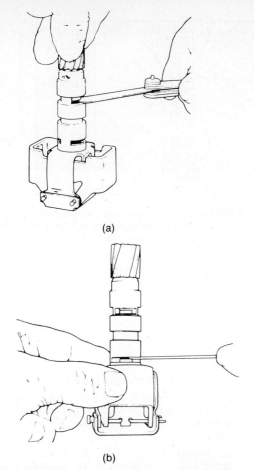

Fig. 11-55 (a) With the governor weights fully extended, there should be a minimum opening of 0.020 in. at the entry port of the General Motors governor shown. (b) With the weights compressed, there should be a minimum opening of 0.020 in. at the exhaust port. *(Copyrighted material reprinted with permission from Hydra-matic Div., GM Corp.)*

To disassemble a gear-driven governor, you should:

1. Cut the retaining pins and remove the thrust cap, weights, springs, and valve (Fig. 11-56).

2. Support the governor and drive out the gear retaining pin (Fig. 11-57).

3. Support the governor using two $\frac{7}{64}$-in. (2.7-mm) plates positioned in the exhaust slots and press the gear off of the sleeve.

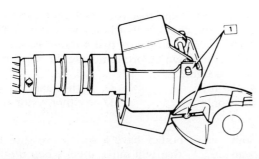

Fig. 11-56 The first step in disassembling a governor is to cut the pins using a pair of cutting pliers. *(Copyrighted material reprinted with permission from Hydra-matic Div., GM Corp.)*

Fig. 11-57 To replace the governor gear, it is necessary to drive out the retaining pin; note the shop-made governor support fixture. With the retaining pin out, the gear can be driven off.

To assemble this governor, you should:

1. Support the governor as just described and press the new gear completely into position (Fig. 11-58).

2. Drill a new hole, $\frac{1}{8}$ in. (3.17 mm) in diameter, through the new gear and the sleeve (Fig. 11-59). This new hole should be at a right, 90°, angle to the original hole.

3. Support the governor and install a new retaining pin. Stake both ends of the retaining pin hole to prevent the pin from coming out.

4. Wash the assembly thoroughly in solvent and air dry.

5. Install the valve, weights, springs, and thrust cap; position new retaining pins in place; and crimp both ends of the pins to keep them in place.

Newer-style gear-driven governors that use check balls in place of a spool valve should be carefully inspected for missing check balls, damaged springs, a blocked oil passage, binding weights, or a damaged seal (Fig. 11-60).

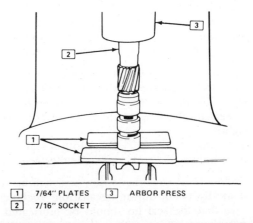

| 1 | 7/64″ PLATES | 3 | ARBOR PRESS |
| 2 | 7/16″ SOCKET | | |

Fig. 11-58 The new gear must be pressed onto the governor; note how the governor is supported during this operation. *(Copyrighted material reprinted with permission from Hydra-matic Div., GM Corp.)*

311

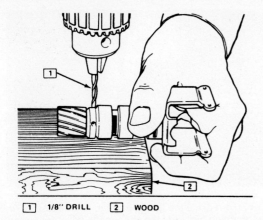

Fig. 11-59 A new hole must be drilled through the governor and new gear so the retaining pin can be installed. *(Copyrighted material reprinted with permission from Hydra-matic Div., GM Corp.)*

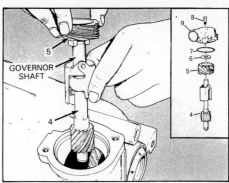 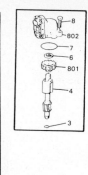

Fig. 11-60 Late-model General Motors gear drive governor. Note that there is very little to go wrong with this unit. *(Copyrighted material reprinted with permission from Hydra-matic Div., GM Corp.)*

REVIEW QUESTIONS

The following questions are provided so you can check the facts you have just learned. Select the response that best completes each statement.

1. Technician A says that a faulty governor can be repaired with the transmission in the car. Technician B agrees but says that gear-driven governors used on General Motors cars require removal of the extension housing. Who is right?
   A. A only
   B. B only
   C. Both A and B
   D. Neither A nor B

2. Of the following operations, which cannot usually be done with the transmission in the car?
   A. Valve body removal and replacement
   B. Manual shift linkage seal replacement
   C. Front torque converter seal replacement
   D. Vacuum modulator replacement

3. Technician A says that a seal can be pried out of a bore using a sharp chisel. Technician B says that smaller metal-backed seals can often be driven into place using a standard socket. Who is right?
   A. A only
   B. B only
   C. Both A and B
   D. Neither A nor B

4. Technician A says that a metal-backed seal has to seal against two surfaces, the housing bore and shaft. Technician B says that a sealing lip is used to seal the backing to the housing bore. Who is right?
   A. A only
   B. B only
   C. Both A and B
   D. Neither A nor B

5. Technician A says that a new seal will be ruined if the lip is cut during installation. Technician B says that the seal lip should be lubricated with petroleum jelly or ATF before installing. Who is right?
   A. A only
   B. B only
   C. Both A and B
   D. Neither A nor B

6. Technician A says that all throttle/kickdown linkages into the case use a metal-backed lip seal. Technician B says that all the cable or rods connect to a lever on the outside of the case. Who is right?
   A. A only
   B. B only
   C. Both A and B
   D. Neither A nor B

7. Technician A says that stripped threads in an aluminium case can be repaired by installing a thread insert. Technician B says that a $\frac{3}{8}$-in. bolt should have at least $\frac{5}{16}$ in. of thread contact. Who is right?
   A. A only
   B. B only
   C. Both A and B
   D. Neither A nor B

8. Technician A says that the best repair for a kinked, leaky cooler line is to cut out the damaged section and replace it with copper tubing. Technician B says that you should always use two wrenches to loosen a tube nut. Who is right?
   A. A only
   B. B only
   C. Both A and B
   D. Neither A nor B

9. Technician A says that you need to be careful not to lose the check balls when you remove a valve body. Technician B says that a transmission must be upside down before removing the valve body. Who is right?
   A. A only
   B. B only
   C. Both A and B
   D. Neither A nor B

10. Some valve bodies have a spring pressure from a/an _____ that will move them when the bolts are loosened.
    A. Accumulator
    B. Shift valves
    C. Servo
    D. Pump relief valve

11. Technician A says that you should be careful to fit the bolts with the correct length into the proper holes as a valve body is replaced. Technician B says that these bolts should always be tightened to the correct torque. Who is right?
    A. A only
    B. B only
    C. Both A and B
    D. Neither A nor B

12. Technician A says that all servo covers are held in place by a group of three or four bolts. Technician B says that many servos contain a spring that must be kept compressed while the cover is removed. Who is right?
    A. A only
    B. B only
    C. Both A and B
    D. Neither A nor B

13. Which of the following is not a reason for removing an extension housing?
    A. Speedometer gear replacement
    B. Governor repair
    C. Park pawl or spring replacement
    D. Extension housing to drive shaft seal replacement

14. Technician A says that you need to lift the transmission slightly in order to remove the rear mount and extension housing. Technician B says that all transmissions use a paper gasket between the extension housing and case. Who is right?
    A. A only
    B. B only
    C. Both A and B
    D. Neither A nor B

15. Technician A says that a faulty governor gear can keep a transmission from upshifting. Technician B says that it is possible to disassemble a governor so it can be thoroughly cleaned and repaired. Who is right?
    A. A only
    B. B only
    C. both A and B
    D. Neither A nor B

CHAPTER 12

TRANSMISSION OVERHAUL

OBJECTIVES

After completing this chapter, you should:

- Be able to remove or replace the automatic transmission or transaxle of a car.
- Understand the various repair procedures needed to overhaul an automatic transmission.
- Be able to determine if a used transmission part is usable.
- Be able to overhaul an automatic transmission.

12.1 INTRODUCTION

A badly worn or broken transmission is removed from the vehicle for repair or an overhaul. The repairs can be specific operations to cure a particular problem such as a noisy gearset or worn-out band. An overhaul usually implies a "rebuild," and this is generally considered to include:

Transmission teardown or disassembly.

Replacement of all rubber and paper gaskets and seals.

Replacement of all friction materials.

Replacement of all worn bushings.

Cleaning and thorough inspection of the planetary gears.

Cleaning and thorough inspection of the valve body and all other valves.

Cleaning and thorough inspection of the torque converter.

Reassembly with a check of all necessary end plays.

Some state bureaus regulate what must be done to rebuild or overhaul a transmission.

The parts needed to overhaul a transmission are available from different sources. The vehicle manufacturer can usually supply all of the "hard" and "soft" parts needed. Major components in a transmission such as the front pump, a clutch drum, or a gearset are called hard parts. Soft parts are those that are normally replaced during an overhaul such as the gaskets, seals, and friction material. Soft parts can be purchased from several aftermarket sources (a supplier other than the vehicle manufacturer). These parts are usually available as individual components or as part of a kit, and there are several types of kits. A kit is more convenient and often less expensive than buying individual parts (Fig. 12-1).

Some of the more popular kits are as follows:

314

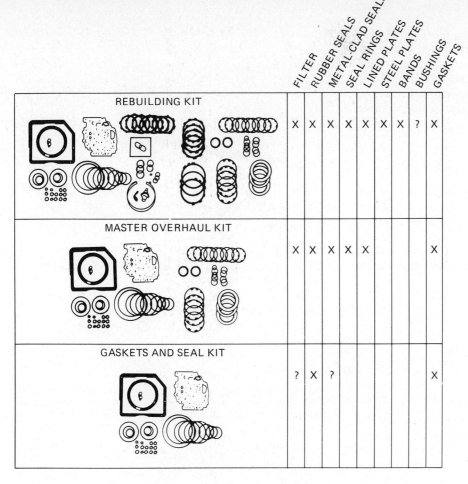

Fig. 12-1 Most soft parts needed to rebuild a transmission are available in kits. The contents of each type of kit varies with the supplier.

Gasket and Seal Kit All paper and rubber components.

Master Kit All paper, rubber, and metal seals and lined and unlined clutch plates.

Overhaul Kit All parts of a master kit plus bushings.

The contents of a kit will vary somewhat between suppliers.

The overhaul procedure varies somewhat between transmissions and transaxles and between makes and models of these. A service manual should always be followed as you overhaul a particular transmission. This makes the job easier and quicker and ensures that you don't skip important steps. The quickest yet completely thorough repair procedure has been developed by experts for a particular transmission. Service manuals are available from car manufacturers, publishers who serve the automobile repair field (such as Chilton's, Mitchell, and Motors), and automatic transmission trade organizations such as the Automatic Transmission Rebuilder's Association (ATRA) and the Automatic Transmission Service Group (ATSG). In general, the procedure normally followed to overhaul a transmission is:

1. Removal.

2. Precleanup.

3. Measurement of input shaft end play.

4. Disassembly along with any additional end play checks.

5. Subassembly repair—pump, valve body, each clutch assembly, each gearset, and the differential in transaxles.

6. Case inspection.

7. Replacement of worn bushings.

8. Reassembly, which includes air checks and setting of the various end plays.

9. Rebuilding or inspection and cleaning of the torque converter.

Most newer transmissions have gone through a series of improvements after their original introduction. These changes have been made to improve the units' durability and reliability. It is usually good practice to make these changes on ("update") the unit on which you are working if possible.

As you are learning to become a transmission service technician, don't forget that the technician who rebuilds a transmission is the final quality control of the unit. There is no valid excuse for building a faulty transmission. As each new or used part is assembled, it should be thoroughly checked to ensure that it will operate correctly.

12.2 TRANSMISSION REMOVAL

As previously mentioned, transmission removal varies somewhat between car makes and models, especially with transaxles. The descriptions given here are very general and only serve to give you an idea of what is involved in these operations. A service manual should be consulted to determine the exact procedure for a particular car.

To remove a transmission, you should:

1. Disconnect the battery ground cable. On some cars it will be necessary to remove the dipstick or disconnect the engine end of the TV linkage.

2. Raise and support the car on a hoist so you have access to the transmission and drive shaft. On some cars, it is necessary to remove portions of the exhaust system.

3. Remove the torque converter cover, mark the converter and flywheel/flexplate/drive plate for reassembly, and remove the converter-to-flexplate bolts (Fig. 12-2). There are usually three or four bolts so you need to rotate the engine to get to all of them. The engine can be rotated using a flywheel turner or a socket and handle on the vibration damper bolt (at the front of the crankshaft) (Fig. 12-3). On some cars, it is necessary to remove the starter.

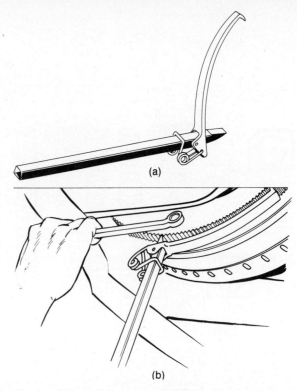

Fig. 12-3 (a) Flywheel turner being used to (b) hold the flexplate while the converter bolt is loosened and to rotate the flexplate to the next bolt. *[(a) Courtesy of Snap-On Tools.]*

4. Remove the transmission pan, and drain the fluid as described in Chapter 9. Be sure to check the pan for debris that indicates internal problems. Replace the pan and install three or four bolts to hold it in place.

5. Remove the drive shaft as described in Chapter 11.

6. Disconnect the manual shift linkage, speedometer cable, oil cooler lines, and any electrical connectors attached to the transmission.

7. Support the transmission using a transmission jack, and remove the center cross member/transmission support as described in Chapter 11 (Fig. 12-4).

8. Lower the transmission just enough to gain access to the upper transmission-to-engine bolts, and remove these bolts (Fig. 12-5). On some cars, it is necessary to install an engine support fixture (Fig. 12-6).

9. Pull the transmission back far enough to clear the converter housing alignment dowels, converter pilot, or converter studs, and lower the transmission out of the car (Fig. 12-7).

The engine always needs to be supported when removing a transaxle. Special fixtures are available that provide a safe, secure support with no damage to the car. These units are usually placed on the fender flanges and radiator support (Fig. 12-8). Some shops have made support fixtures using a section of strong pipe or square tubing, a piece of threaded rod, and a pair of thick phone

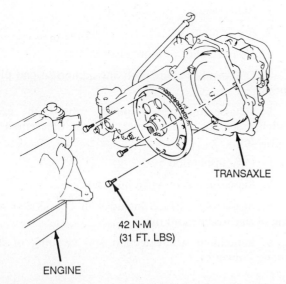

Fig. 12-2 The torque converter is disconnected from the flexplate by removing three or four nuts or bolts. This is usually done before the transmission-to-engine bolts are removed. *(Copyrighted material reprinted with permission from Hydra-matic Div., GM Corp.)*

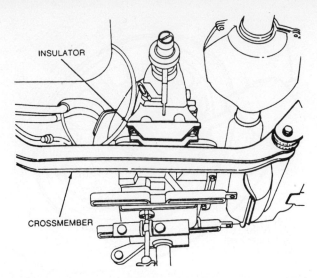

Fig. 12-4 A transmission jack is positioned under the transmission pan so it can lift the transmission to allow removal of the insulator/mount and cross member. *(Courtesy of Ford Motor Company.)*

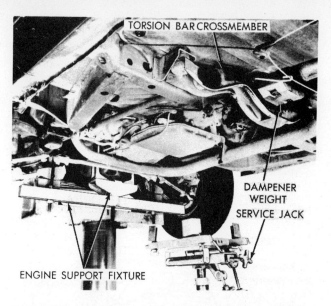

Fig. 12-6 In some cases, it is necessary to install an engine support fixture to hold the engine in place while the transmission is removed. *(Courtesy of Chrysler Corporation.)*

books or catalogs placed on the fenders. Another transaxle removal problem is caused by the drive shafts. There are several removal methods and all require a certain amount of front suspension disassembly as described in the appropriate service manual. In some cases only one of the drive shafts needs to be removed. The other can be disconnected from the transaxle as the transaxle is removed from the engine.

To remove a transaxle, you should:

1. Disconnect the battery ground cable and any accessible oil cooler lines, shift linkage, TV linkage, speedometer cable, and electrical connectors from under the hood.

2. Install an engine support fixture, and adjust it to begin supporting the engine's weight (Fig. 12-9).

3. Remove any accessible transaxle-to-engine bolts.

4. Loosen the front hub nuts. This usually requires that the front wheels remain on the ground, the parking brake be applied, and the transmission be shifted into park (Fig. 12-10).

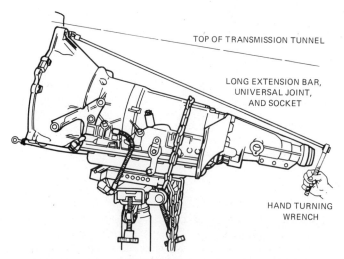

Fig. 12-5 Many transmissions can be lowered enough so one can remove the transmission-to-engine bolt(s) working over the top of the transmission.

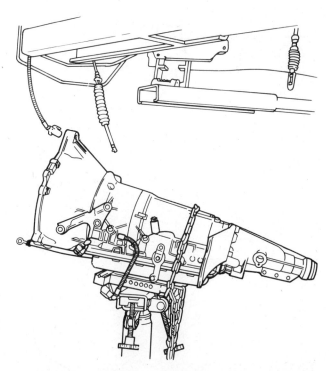

Fig. 12-7 After everything has been disconnected, most transmissions can be moved to the rear and downward for removal. A transmission jack is a very helpful tool for this operation. *(Courtesy of Ford Motor Company.)*

317

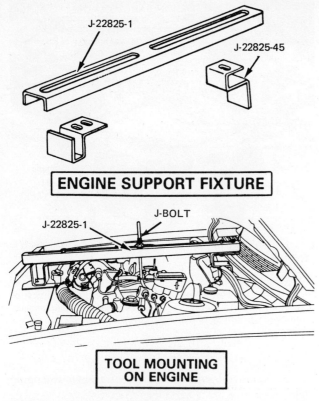

Fig. 12-8 Engine support fixture used to hold the engine while the transaxle is removed. The J-bolt is hooked into the engine support bracket. *(Copyrighted material reprinted with permission from Hydra-matic Div., GM Corp.)*

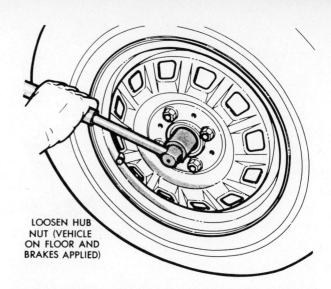

Fig. 12-10 Before a front-wheel drive (FWD) shaft can be removed, the front hub nuts should be removed. It is usually necessary to have the tire on the floor and the brakes applied in order to loosen the hub nuts. *(Courtesy of Chrysler Corporation.)*

5. Raise and support the car so you have access to the transaxle and front suspension.

6. Remove the oil pan, drain the fluid, and replace the pan using a few loosely tightened bolts.

7. Remove the drive shaft(s) (Fig. 12-11).

NOTE On some transaxles, a special tool is necessary to hold the differential gears positioned correctly if both drive shafts are removed.

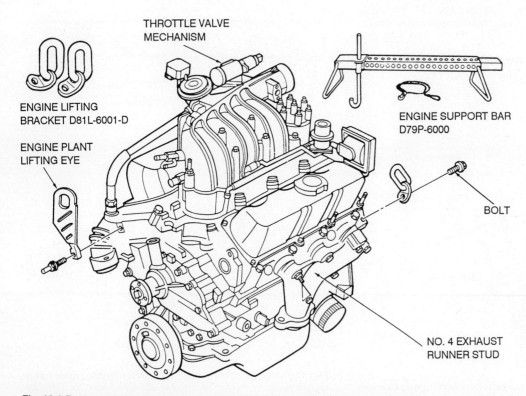

Fig. 12-9 Engine should be supported at two points to prevent tilting as the transaxle is removed. *(Courtesy of Ford Motor Company.)*

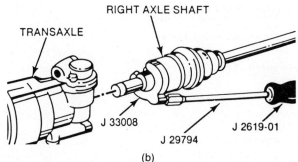

Fig. 12-11 (a) Special tool hooks onto the inner CV (constant velocity) joint (b) so a slide hammer can be used to remove the drive shaft/axle shaft from the transaxle. *(Courtesy of Kent-Moore.)*

8. Remove the torque converter cover/dust shield, mark the converter and flexplate, and remove the converter-to-flexplate bolts (Fig. 12-12).

9. Disconnect any remaining oil cooler lines, shift linkage, speedometer cable, or electrical connectors. On some cars, it is necessary to remove the starter.

10. Place a suitable transmission jack under the transaxle, and adjust it to support the transaxle (Fig. 12-13).

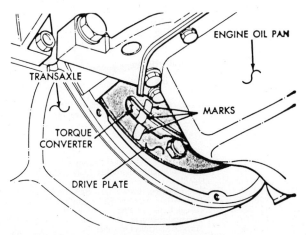

Fig. 12-12 The converter should be disconnected by removing the drive plate/flexplate-to-converter bolts. *(Courtesy of Chrysler Corporation.)*

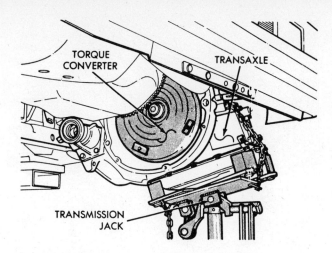

Fig. 12-13 A transmission jack has been positioned under the oil pan and is being used to support the transaxle as it is pulled away from the engine and lowered downward. *(Courtesy of Chrysler Corporation.)*

11. Remove any transaxle support mounts. On General Motors cars, it is necessary to remove the engine cradle (Fig. 12-14).

12. Finish removing the transaxle-to-engine bolts.

13. Pull the transaxle away from the engine far enough to clear the engine-to-transaxle alignment dowels or torque converter pilot, and lower the transaxle out of the car.

12.3 TRANSMISSION DISASSEMBLY

As previously mentioned, the disassembly procedure of a transmission varies between makes and models. In many cases, special tooling or procedures are required (Fig. 12-15). A service manual should be followed. The descriptions given here cover the commonly used procedures and equipment to give you an idea of what is involved.

12.3.1 Predisassembly Cleanup

Cleanliness is a must during transmission overhaul, and it begins before teardown. Most shops steam clean the outside of the transmission as soon as it is removed from the car (Fig. 12-16). This removes all of the exterior oil, dirt, and other debris from the work area. An alternate cleanup method is to use solvent or an engine degreaser and a parts cleaning brush. After cleanup, the transmission should be rinsed off with a clean water spray.

12.3.2 Torque Converter Removal

The converter is simply slid out from the front of the transmission, but be ready to support the weight. Converters are heavy. Also be ready for a fluid spill from the converter and transmission. The converter should be

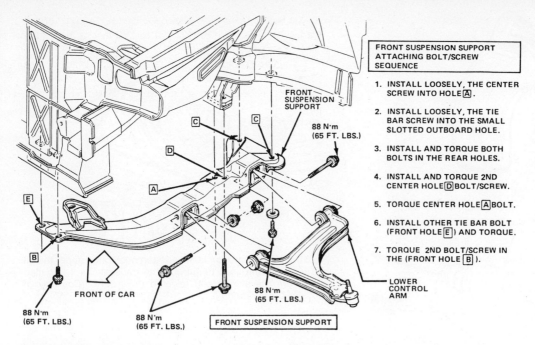

Fig. 12-14 The front suspension support and part of the front suspension must be removed to allow transaxle removal. *(Courtesy of Chevrolet Div., GM Corp.)*

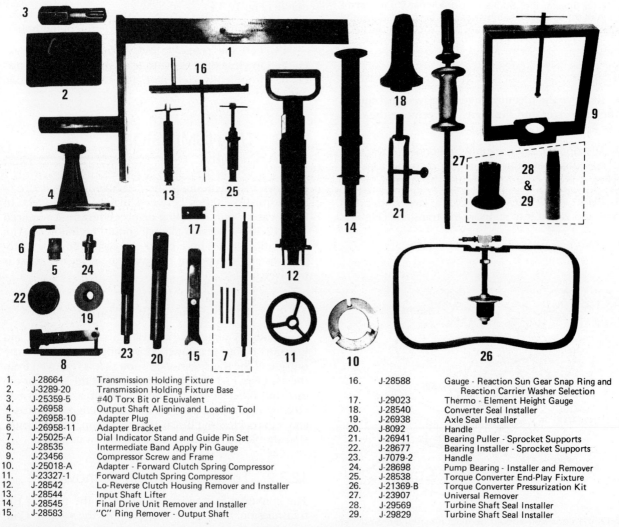

1.	J-28664	Transmission Holding Fixture	16.	J-28588	Gauge - Reaction Sun Gear Snap Ring and Reaction Carrier Washer Selection
2.	J-3289-20	Transmission Holding Fixture Base	17.	J-29023	Thermo - Element Height Gauge
3.	J-25359-5	#40 Torx Bit or Equivalent	18.	J-28540	Converter Seal Installer
4.	J-26958	Output Shaft Aligning and Loading Tool	19.	J-26938	Axle Seal Installer
5.	J-26958-10	Adapter Plug	20.	J-8092	Handle
6.	J-26958-11	Adapter Bracket	21.	J-26941	Bearing Puller - Sprocket Supports
7.	J-25025-A	Dial Indicator Stand and Guide Pin Set	22.	J-28677	Bearing Installer - Sprocket Supports
8.	J-28535	Intermediate Band Apply Pin Gauge	23.	J-7079-2	Handle
9.	J-23456	Compressor Screw and Frame	24.	J-28698	Pump Bearing - Installer and Remover
10.	J-25018-A	Adapter - Forward Clutch Spring Compressor	25.	J-28538	Torque Converter End-Play Fixture
11.	J-23327-1	Forward Clutch Spring Compressor	26.	J-21369-B	Torque Converter Pressurization Kit
12.	J-28542	Lo-Reverse Clutch Housing Remover and Installer	27.	J-23907	Universal Remover
13.	J-28544	Input Shaft Lifter	28.	J-29569	Turbine Shaft Seal Installer
14.	J-28545	Final Drive Unit Remover and Installer	29.	J-29829	Turbine Shaft Seal Installer
15.	J-28583	"C" Ring Remover - Output Shaft			

Fig. 12-15 Special tools are required to completely overhaul a THM 125 transaxle. *(Copyrighted material reprinted with permission from Hydra-matic Div., GM Corp.)*

Fig. 12-16 Transmission being steam cleaned prior to teardown. A thorough exterior cleanup makes it easier to locate and remove bolts and retaining rings along with keeping the work area cleaner.

either sent out for rebuilding or set aside for cleanup and inspection later. Torque converter service is described in Chapter 13.

12.3.3 Disassembly Fixtures

Automatic transmission disassembly is fairly messy. Used fluid drains out as each hydraulic component is removed or disassembled. Many shops do their teardown on a bench that has a steel top designed to catch the fluid and drain it into a catch pan. During disassembly, the transmission is simply placed on the bench and torn down. It is usually placed upside down and rolled over as needed. Some shops use transmission-holding fixtures (Fig. 12-17). These allow the unit to be easily rotated to the best working position. When using a holding fixture, a drain pan should be placed under the transmission to catch the dripping fluid.

During reassembly, clutch installation is easier if the transmission is turned with the main opening upward. Some shops use a simple fixture that resembles a three-legged stool; other shops simply use an old transmission case in a rear-side-up position (Fig. 12-18).

(a)

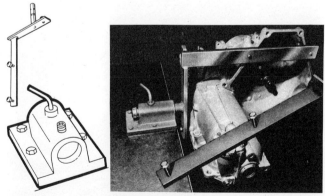

Fig. 12-17 Transmission-holding fixture and base. This attaches to a transmission so it can be hung from the service bench. Note that it allows the transmission to be easily rotated to different positions during disassembly and reassembly. *(Courtesy of Kent-Moore.)*

(b)

Fig. 12-18 (a) The output shaft of the transmission being serviced has been placed into the back end of an old transmission case. (b) The other transmission's output shaft has been placed into a shop-made stool. Caution should be exercised in using either method. Because of the narrow base, the transmissions can tip over.

321

12.3.4 Preliminary Disassembly

Normally the first teardown step is the removal of the oil pan, filter, and valve body. Many technicians use the oil pan to store the retaining rings, screws, bolts, and other small parts removed during disassembly. The normal procedure is to remove the pan, inspect the debris (if it hasn't been done already), wash the pan in solvent, and air dry it. Next the filter and gasket are removed and set aside for comparison with the new filter to be installed. The valve body is then removed and set aside for later cleaning and inspection (Fig. 12-19). Don't forget to check for different lengths and sizes of bolts, to remove any screens, and to be ready to catch any check balls as the valve body is removed. Also it is good practice to save the valve body gasket (if used) to compare with the new one to be installed.

On some transmissions, the valve body covers an accumulator or servo piston(s). Note the position of the piston(s) and spring (if used), and remove them at this time. In other transmissions, a separate cover is used for the accumulator or servo. Two styles of retaining rings are used: one is like a standard snap ring; the other looks like a round wire. A handy tool for removing the first style is a scribe, commonly called a *seal pick* (Fig. 12-20). The second style of ring often has a punch hole that allows a pin punch to be used to start ring removal (Fig. 12-21). Be careful during servo cover removal; some servos use a strong piston spring and require a special tool to hold the spring compressed during retainer ring removal and to allow the spring to be safely extended (Fig. 12-22).

On transmissions with gear-driven governors, remove the governor cover and governor at this time.

Special Notes on Retaining Rings. A retaining ring is normally used to prevent a part from moving sideways. For example, it can hold a gear in position on a shaft or a servo cover in a bore (Fig. 12-23). There

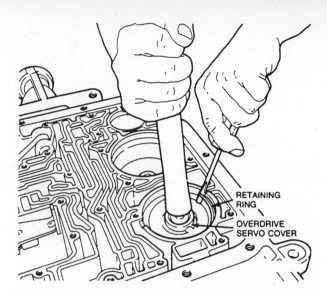

Fig. 12-20 Some retaining rings can be pried out of their groove and the bore using a small screwdriver or seal pick. Note the wooden dowel being used to keep the overdrive servo spring compressed at this time. *(Courtesy of Ford Motor Company.)*

is a variety of retaining rings; some examples are snap rings, tru-arc rings, and E-clips (Fig. 12-24). All types come in different sizes as required to fit the shaft or bore diameter.

The two major classifications of retaining rings are external and internal. External rings fit over a shaft and need to be expanded for removal or installation. Internal rings fit into a bore and are contracted or compressed for removal or installation. Many retaining rings can be removed by prying them with a tool such as a scribe, seal pick, or screwdriver. Special snap ring pliers are used to remove and install the harder to remove rings, and the correct type and size of snap ring pliers must be used (Fig. 12-25).

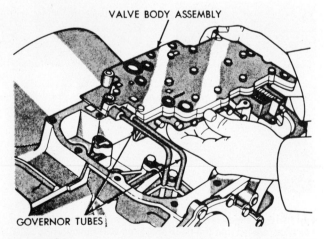

Fig. 12-19 After the pan has been removed, the valve body can be removed. Note that on some valve bodies, tubes are used to connect fluid passages to the transmission case. *(Courtesy of Chrysler Corporation.)*

Fig. 12-21 A punch passing through a hole in the case is being used to push the retaining ring out of its groove so a seal pick or small screwdriver can be used to pry it out.

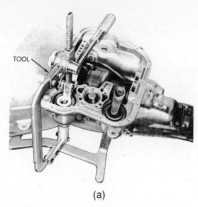

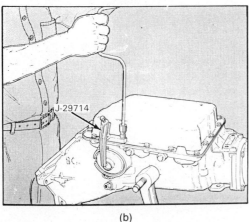

Fig. 12-22 (a) A modified valve spring compressor and (b) special compressing tool J-29714 are used to hold the servo spring compressed while the retaining ring is removed. Then they will be slowly released to allow the servo spring to extend in a controlled manner. [(a) Courtesy of Chrysler Corporation. (b) Copyrighted material reprinted with permission from Hydra-matic Div., GM Corp.)]

12.3.5 End Play Check

It is standard practice to measure the input shaft end play before removing the pump and disassembling the transmission further. End play is the amount of free movement that most rotating objects have in a direction that is parallel to the shaft (Fig. 12-26). If there is no end play, there will be drag; too much end play allows sloppy motion. If the end play is good, the internal thrust washers are probably in fairly good shape. If the end play is excessive, there is internal wear, which must be corrected during the rebuild.

End play is normally measured using a dial indicator.

To measure input shaft end play, you should:

1. Place the transmission in a vertical position with the input shaft upward. Some transmissions require that a special fixture be used to hold the output shaft in the correct position during end play checks (Fig. 12-27).

2. Clamp a dial indicator onto the input shaft and position the body so the measuring stylus is against the transmission case or front pump (Fig. 12-28). Or clamp a dial indicator to the transmission case and position the measuring stylus at the end of the input shaft.

NOTE The measuring stylus must be exactly parallel to the input shaft, and the indicator body should be adjusted to load the stylus about one full needle rotation inward (Fig. 12-29).

3. Pull the input shaft upward slightly and then inward as far as it will go. Now adjust the face of the indicator so that it reads zero.

4. On many transmissions, pull upward on the input shaft, and read the amount of travel on the dial indicator (Fig. 12-30). This is the amount of end play.

NOTE On some transmissions, the input shaft is not attached and can be pulled right out. On these units, end play is checked by prying upward on the gear train (Fig. 12-31).

5. Repeat Steps 3 and 4 until you get consistent, reliable readings. Then make three more measurements and, if there is a slight difference, average them.

6. Record your reading, locate the end play specification in a service manual, and compare your reading to the specification.

Some transmissions require additional end play checks to locate wear in specific areas or to determine if the correct selective thrust washers or spacers are used. For example, any transmission with a center support will have a certain amount of end play on each side of the stationary center support. Selective washers or snap rings are produced in groups of exact sizes that vary slightly in thicknesses (Fig. 12-32). This allows the technician to measure the amount of play and select the correct size to produce the proper amount of end play or clearance.

12.3.6 Other Predisassembly Checks

Depending on the transmission design, there are other checks that should be done before disassembly. An example of this is the intermediate-band apply pin check for a THM 125. This check requires a special tool that is attached to the transmission case and operated with a torque wrench (Fig. 12-33). This determines if the intermediate-band apply pin is the correct length for the band. If the check indicates that the apply pin is correct, there is probably not too much band wear. If the check indicates the apply pin is too short, either the pin, band, or both have to be replaced during the overhaul.

12.3.7 Pump Removal

On transmissions and many transaxles, the pump is the unit that holds the parts inside the case. Its removal allows the disassembly of the rest of the parts. The pump is held in place by a group of bolts that are relatively easy to remove. The close fit between the outer pump diameter and the case plus the rubber sealing ring makes pump removal a little difficult at times. Several different methods can be used for pump removal: using slide hammers or special screw-type pullers or prying the gear train. The use of slide hammers or prying the gear train is sometimes limited by transmission design.

323

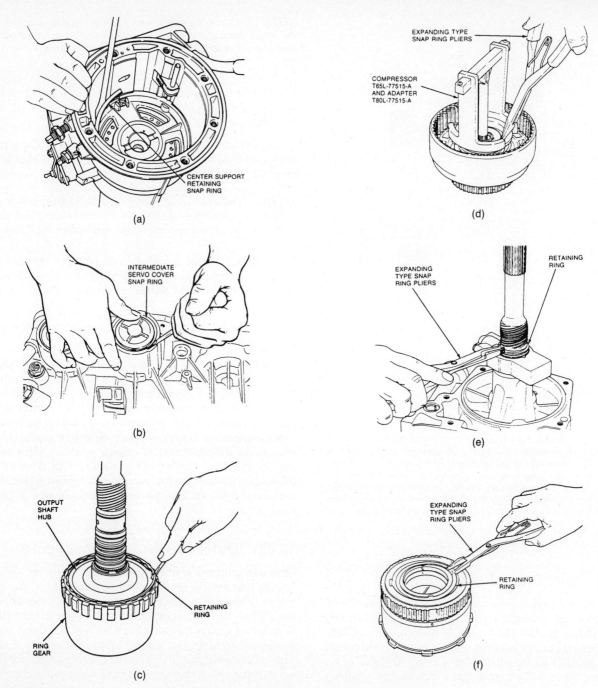

Fig. 12-23 Some uses for internal retaining rings are (a) to hold the center support in position, (b) to retain the intermediate servo cover, and (c) to keep the ring gear in the proper position on the output shaft. Some uses for external retaining rings are (d) to hold the clutch spring retaining plate, (e) to hold the governor, and (f) to hold planetary gear parts in the proper position. *(Courtesy of Ford Motor Company.)*

Fig. 12-24 The most common styles of retaining rings are (a) external, pin type; (b) internal, pin type; (c) external, plain type; (d) internal, plain type; and (e) E-clips.

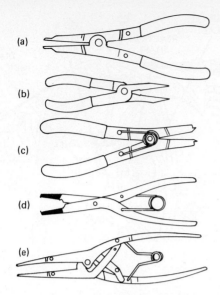

Fig. 12-25 Snap ring pliers for (a) external, pin type; (b) internal, pin type; and (c–e) three types for external, plain-type snap rings. The pliers in (e) have jaws that open in a parallel action. *(Courtesy of Snap-On Tools.)*

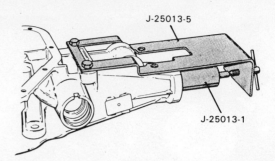

Fig. 12-27 Fixture J-25013-5 with part J-25013-1 is required to accurately check the end play on this THM 200 transmission. *(Copyrighted material reprinted with permission from Hydramatic Div., GM Corp.)*

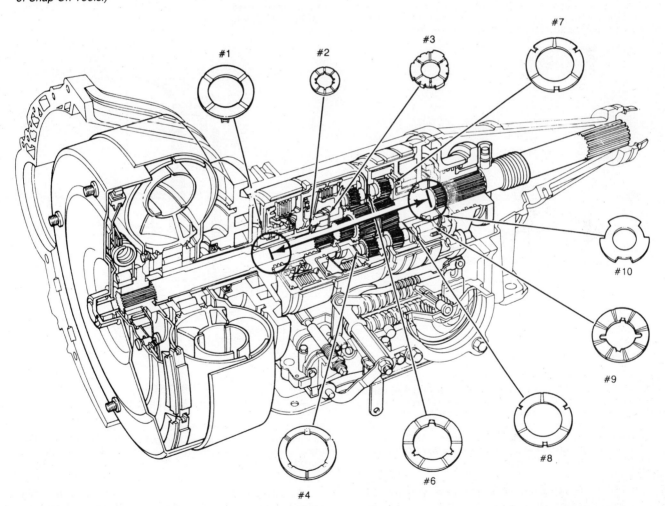

Fig. 12-26 Transmission end play occurs between thrust washer 1 at the front and 10 at the back of the case. *(Courtesy of Ford Motor Company.)*

NOTE: WASHER #5 IS PART OF SHELL AND SUN GEAR ASSEMBLY

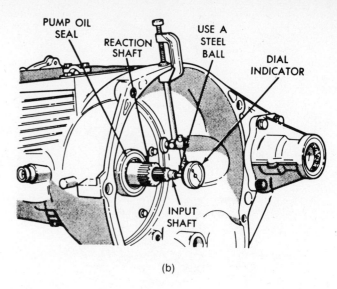

Fig. 12-28 (a) A dial indicator has been mounted on the turbine shaft with the indicator stem against the pump. (b) A dial indicator has been mounted to the transmission case with the indicator stem against the turbine shaft. The amount of end play can be read by moving the turbine shaft up and down. [(b) *Courtesy of Chrysler Corporation.*]

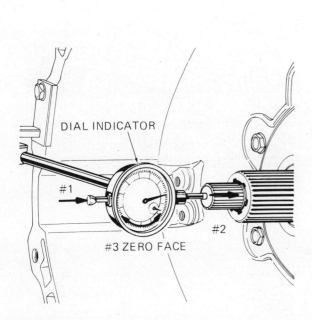

Fig. 12-29 A dial indicator should be set up with the indicator stem parallel to the turbine shaft so the stem is loaded into the indicator about one revolution of the needle (1). Next the turbine shaft should be moved completely inward (2), and (3) the dial rotated so zero aligns with the needle. The amount of end play is read when the turbine shaft is moved outward to its limit. (*Courtesy of Ford Motor Company.*)

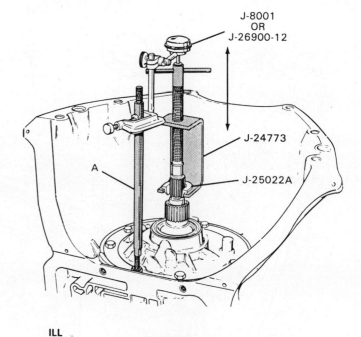

ILL NO.	DESCRIPTION
A	2.78mm (11″) BOLT & LOCKNUT OR J-25025-7A

Fig. 12-30 With the puller assembly shown (J-24773), end play can be read by lifting upward on the pulling tool. (*Copyrighted material reprinted with permission from Hydra-matic Div., GM Corp.*)

To pull a pump using slide hammers, you should:

1. Remove the pump-retaining bolts.

2. Check the pump body bolt holes for threads (Fig. 12-34). These threads are one size larger than the pump-retaining bolts, and only two holes on opposite sides of the pump are threaded.

3. Thread the slide hammer bolts into the two threaded holes (Fig. 12-35). Adapters are available to thread the slide hammer bolts into different sizes of pump threads.

4. Using the slide hammers along with a lifting action, remove the pump assembly.

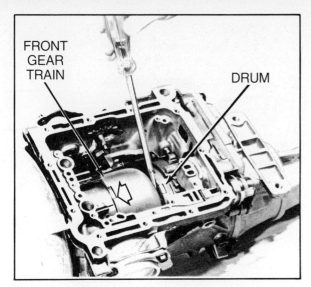

Fig. 12-31 A screwdriver is being used to pry forward on the sun gear shell of this C5 transmission so the end play can be read. *(Courtesy of Ford Motor Company.)*

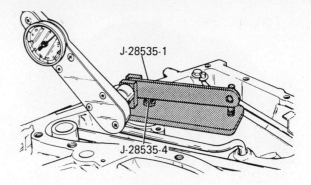

INTERMEDIATE BAND APPLY PIN

LENGTH	IDENTIFICATION
Short	2 Grooves
Medium	1 Groove
Long	No Grooves

Fig. 12-33 Tool J-28535-1 along with a torque wrench checking whether the correct intermediate-band apply pin is being used. The length of the pin can be identified by the number of grooves. *(Copyrighted material reprinted with permission from Hydramatic Div., GM Corp.)*

THICKNESS		IDENTIFICATION NUMBER AND/OR COLOR
1.66 - 1.77mm	(0.065" - 0.070")	1 - —
1.79 - 1.90mm	(0.070" - 0.075")	2 - —
1.92 - 2.03mm	(0.076" - 0.080")	3 - BLACK
2.05 - 2.16mm	(0.081" - 0.085")	4 - LIGHT GREEN
2.18 - 2.29mm	(0.086" - 0.090")	5 - SCARLET
2.31 - 2.42mm	(0.091" - 0.095")	6 - PURPLE
2.44 - 2.55mm	(0.096" - 0.100")	7 - COCOA BROWN
2.57 - 2.68mm	(0.101" - 0.106")	8 - ORANGE
2.70 - 2.81mm	(0.106" - 0.111")	9 - YELLOW
2.83 - 2.94mm	(0.111" - 0.116")	10 - LIGHT BLUE
2.96 - 3.07mm	(0.117" - 0.121")	11 - BLUE
3.09 - 3.20mm	(0.122" - 0.126")	12 - —
3.22 - 3.33mm	(0.127" - 0.131")	13 - PINK
3.35 - 3.46mm	(0.132" - 0.136")	14 - GREEN
3.48 - 3.59mm	(0.137" - 0.141")	15 - GRAY
		HH0033-200C

Fig. 12-32 Available thrust washers for a THM 200C transmission. The sizes and method of identification are shown. *(Copyrighted material reprinted with permission from Hydra-matic Div., GM Corp.)*

Fig. 12-34 Portion of a pump assembly showing hole threaded to accept a slide hammer. Note that the other hole is not threaded.

To pull a pump using a screw-type puller you should:

1. Check the end of the input shaft for a check valve that can be damaged by the puller screw. If you find one, a protection device must be positioned between the puller and the end of the shaft.

2. Remove the pump-retaining bolts.

3. Attach the puller to the stator support shaft. Some pullers merely slide over the unsplined portion of the shaft; others clamp onto the shaft (Fig. 12-36).

4. Turn the puller screw inward to pull the pump assembly.

To pull a pump by prying, you should:

1. Lay the transmission so the top of the case is down.

2. Remove the pump-retaining bolts.

3. Place a large screwdriver between the sun gear input shell and the reaction carrier, and pry the input shell forward to remove the pump (Fig. 12-37). In some cases, you can pry directly on the pump.

In the THM 125, THM 440, and AXOD transaxles, the main gear train is behind the valve body; case cover, channel plate or chain cover; and drive chain and sprockets (Fig. 12-38). Removal of these parts provides access to the driven sprocket support, which supports the input end of the gear train.

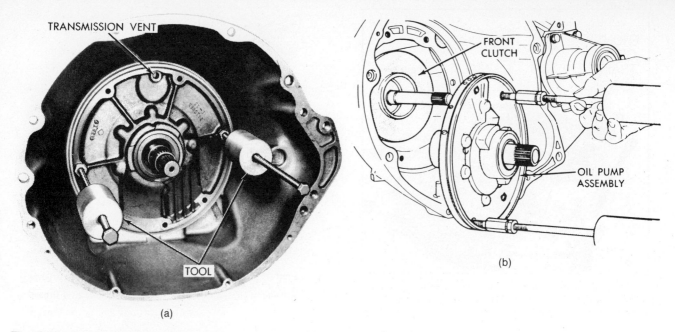

Fig. 12-35 (a) A pair of slide hammers have been threaded into the pump assembly. (b) After it is pulled free, the tools can be used to lift the pump assembly out of the transmission. *(Courtesy of Chrysler Corporation.)*

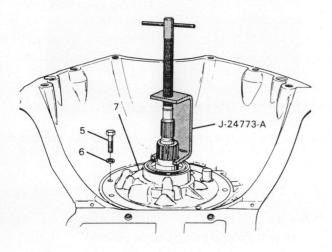

ILL.
NO.   DESCRIPTION

5 BOLT, PUMP TO CASE M8 X 1.25 X 35
6 WASHER, PUMP TO CASE BOLT
7 PUMP BODY ASSEMBLY

(a)

Fig. 12-36 (a) Tool J-24773-A is designed to pull the pump on some Hydramatic transmissions; note how it fits into a groove in the stator support. (b) Tool clamps onto the stator support so it can be used on a variety of transmissions. A valve protector (arrow) has been placed between the puller screw and the end of the turbine shaft to protect the check valve in the end of the shaft. *[(a) Copyrighted material reprinted with permission from Hydra-matic Div., GM Corp.)]*

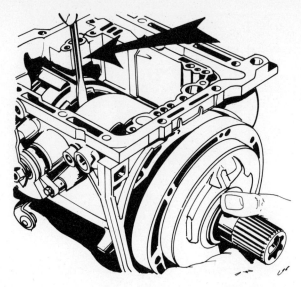

Fig. 12-37 The pump can be removed from a C5 transmission by prying forward on the gear train. *(Courtesy of Ford Motor Company.)*

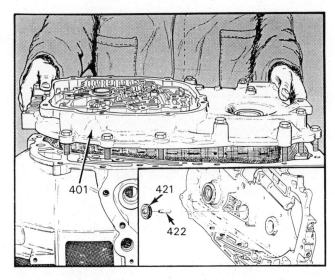

Fig. 12-38 Removing the case cover (401) from a THM 125C transaxle to gain access to the drive chain and sprockets. *(Copyrighted material reprinted with permission from Hydra-matic Div., GM Corp.)*

12.3.8 Major Disassembly

With the pump removed, some transmissions almost fall apart; others require disassembly in particular stages. As mentioned earlier, a service manual should be checked to determine the exact procedure.

After removal of the pump assembly, a typical disassembly is as follows:

1. Carefully note the position of the band struts, loosen the band adjusting screw, remove the servo cover and piston, and remove the band struts and band (Fig. 12-39).

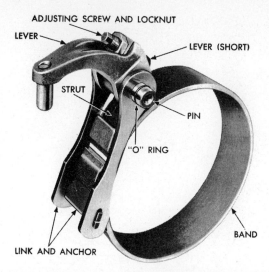

Fig. 12-39 Ends of a Torqueflite band fit between the strut at the top and the link and anchor. *(Courtesy of Chrysler Corporation.)*

On transmissions using a clutch mounted in the case next to the pump, remove the clutch friction, steel, and pressure plates being sure to note the position of the different plates (Fig. 12-40).

2. Remove the driving clutch assemblies that are next to the pump (Fig. 12-41).

3. Remove the extension housing. It should be noted that some transmissions require expansion of a snap ring during this operation (Fig. 12-42).

4. On transmissions with shaft-mounted governors, remove the governor and drive ball, if used, along with its support (Fig. 12-43).

5. On some transmissions, the planetary gear train can be slid out of the case as one assembly (Fig. 12-44).

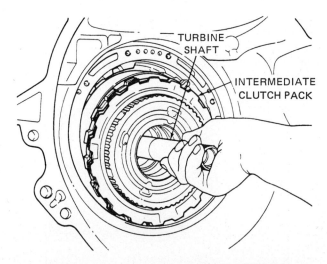

Fig. 12-40 After the pump assembly has been removed from an AOD transmission, the intermediate clutch can be removed by itself or along with the turbine shaft, reverse clutch, and forward clutch. *(Courtesy of Ford Motor Company.)*

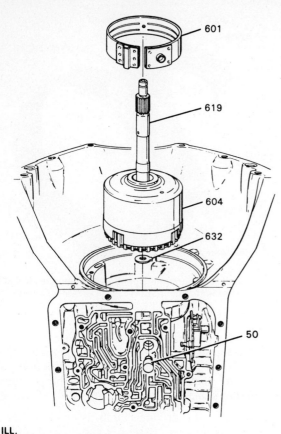

ILL. NO.	DESCRIPTION
50	PIN, BAND ANCHOR
601	BAND ASSEMBLY, INTERMEDIATE
604	HOUSING, DIRECT CLUTCH & SEAL
619	HOUSING ASSEMBLY, FORWARD CLUTCH
632	WASHER, THRUST SELECTIVE (FRONT)

Fig. 12-41 Lifting the turbine shaft (619) from a THM 200C transmission lifts the forward clutch and direct clutch housing (604) out of the case. *(Copyrighted material reprinted with permission from Hydra-matic Div., GM Corp.)*

On others, remove the necessary retaining rings to disassemble the gear train. A seal pick or snap ring pliers should be used to remove the retaining rings. You should note exactly where each retaining ring is located, and as a portion of the gear train is removed, you should note if a thrust washer is used and how it is positioned

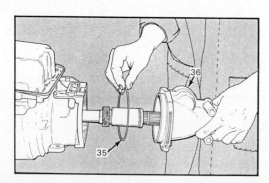

Fig. 12-42 Removing an extension housing (36) with the seal (35). *(Copyrighted material reprinted with permission from Hydra-matic Div., GM Corp.)*

(a)

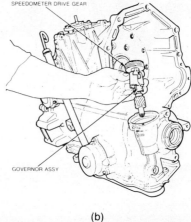

(b)

Fig. 12-43 (a) A C5 governor is removed from the support after the four bolts are removed; on the transmission shown, the governor support remains on the output shaft. (b) The governor is lifted out of an AXOD after the cover has been removed. *(Courtesy of Ford Motor Company.)*

(Fig. 12-45). The condition of these parts should be checked as they are removed.

6. On those transmissions using a center support, remove the retainer, usually a large snap ring, and lift the center support out of the case. The low–reverse clutch assembly in some transmissions acts as a center support. A special tool is often required to grab and lift out the center support, but in some cases, lifting the output shaft will push it along with the rest of the gear train and the center support (Fig. 12-46).

HINT Some technicians modify a large screwdriver, as shown in Fig. 12-47 to allow easy removal of large snap rings.

NOTE Some center supports use an "anticlunk" spring to load them in a counterclockwise direction (Fig. 12-48). Look for this spring and how it is positioned.

7. On those transmissions using a clutch assembly in the back of the case, remove the retaining ring, pressure plate, and clutch plates (Fig. 12-49). Carefully note the position of each of these parts and if a wave or Belleville plate is used. A compressor is usually required to compress the piston return springs to allow removal of the retaining ring, return springs, and clutch piston. The piston can often be removed by blowing air into the apply passage (Fig. 12-50).

(a)

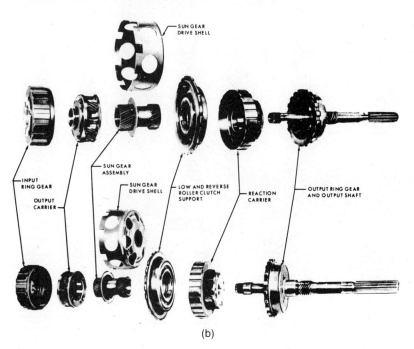

(b)

Fig. 12-44 (a) The entire gear train of a Torqueflite transmission can be removed as an assembly. (b) The gear train of a THM 250C transmission is removed one part at a time starting with the input ring gear and finishing with the output ring gear and output shaft. *[(b) is Copyrighted material reprinted with permission from Hydra-matic Div., GM Corp.]*

SAFETY NOTE
Whenever using a spring compressor, compress the springs just enough to allow the snap ring to be removed and be careful that the compressor is positioned correctly and will not slip and allow the springs to come loose. Also, make sure it has enough travel to allow the springs to extend completely and that the retainer does not catch on the snap ring groove during removal (Fig. 12-51).

8. On transmissions, remove any remaining parts as required by the manufacturer's instructions. On transaxles, remove the final drive gears and differential plus any other remaining parts as required by the manufacturer's instructions (Fig. 12-52).

9. As the park gear is removed, the gear and park pawl should be inspected for wear and damage. Also, the pawl return spring should be checked to ensure that the park pawl is moved completely away from the gear when released (Fig. 12-53).

12.4 COMPONENT CLEANING

Many of the internal parts along with the case and extension housing are cleaned right after removal. Others (e.g., a clutch assembly) are cleaned as they are disassembled further. Three procedures are commonly used: hot spray wash, cold dip, and solvent wash. After washing, the parts are air dried. Transmission parts should not be dried by wiping them with a shop cloth. More

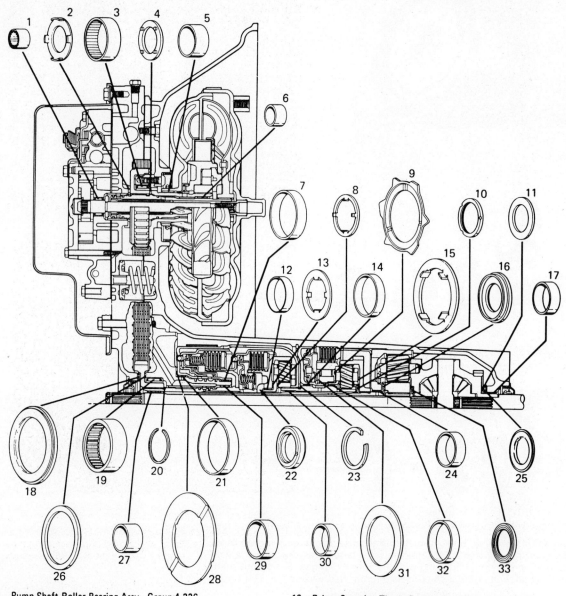

1. Pump Shaft Roller Bearing Assy. Group 4.226
2. Case Cover To Driven Sprocket Thrust Washer Group 4.131
3. Bearing Assembly Group 4.131
4. Case Cover To Drive Sprocket Thrust Washer Group 4.131
5. Converter Bushing Group 4.115
6. Drive Sprocket Support Bushing Group 4.226
7. Direct Clutch Drum Bushing Group 4.169
8. Input Carrier To Input Sun Gear Thrust Washer Group 4.159
9. Reaction Carrier To Lo Race Thrust Washer Group 4.180
10. Reaction Sun To Internal Gear Thrust Bearing Group 4.159
11. Differential Carrier To Case Sel. Thrust Washer Group 4.176
12. Input Internal Gear Bushing Group 4.158
13. Input Carrier To Input Int. Gear Thrust Washer Group 4.159
14. Lo And Reverse Clutch Housing Bushing Group 4.159
15. Reaction Carrier To Int. Gear Thrust Washer Group 4.180
16. Sun Gear To Internal Gear Thrust Bearing Group 4.178
17. Case Bushing Group 4.319
18. Driven Sprocket Thrust Bearing Assembly Group 4.131
19. Bearing Assembly Group 4.131
20. Selective Snap Ring Group 4.169
21. Direct Clutch Bushing Group 4.169
22. Input Shaft Thrust Washer Group 4.158
23. Selective Snap Ring Group 4.216
24. Final Drive Internal Gear Bushing Group 4.319
25. Differential Carrier To Case Thrust Brg. Assy. Group 4.176
26. Driven Sprocket Support Thrust Washer Group 4.131
27. Input Shaft Bushing Group 4.158
28. Thrust Washer Group 4.169
29. Driven Sprocket Support Bushing Group 4.226
30. Reaction Sun Gear Bushing Group 4.159
31. Reverse Housing To Lo Race Selective Washer Group 4.180
32. Reaction Carrier Bushing Group 4.159
33. Sun Gear To Carrier Thrust Bearing Group 4.159

Fig. 12-45 Location of bearings, bushings, and thrust washers in a THM 125C. Note how the thrust washers are different. *(Copyrighted material reprinted with permission from Hydra-matic Div., GM Corp.)*

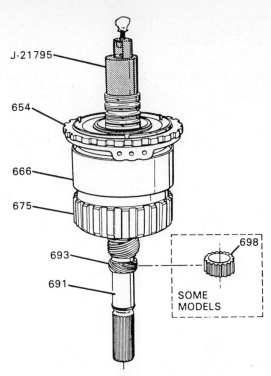

Fig. 12-46 Tool J-21795 is used to lift the center support and planetary gearset from a THM 400 transmission. Some technicians use a slide hammer in place of the thumb screw for a better handle. *(Copyrighted material reprinted with permission from Hydra-matic Div., GM Corp.)*

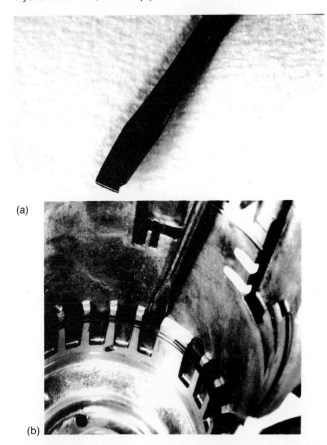

Fig. 12-47 (a) A special screwdriver is being used to pry and lift a large retaining ring out of its groove. (b) A small notch has been ground into the side of the blade so it hooks the retaining ring.

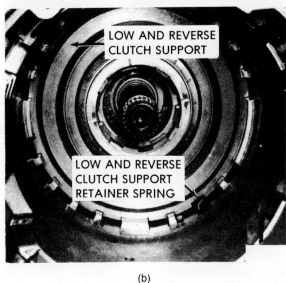

Fig. 12-48 (a) The anticlunk or (b) low–reverse clutch support spring is difficult to see when the clutch support is in place. *(Copyrighted material reprinted with permission from Hydramatic Div., GM Corp.)*

than one transmission has died from "red rag disease." Lint left on parts wiped by a common shop cloth can clog the filter and block the fluid flow into the pump.

With hot spray wash, the parts are placed inside a cabinet that resembles a large dishwasher (Fig. 12-54). The washer sprays the parts with a hot water–detergent solution as the parts rotate. This is a relatively expensive piece of equipment, and it is an effective and quick cleaner. Also, some units trap the sludge and dirt in a manner that makes their disposal as well as the disposal of the used cleaning solution relatively easy and inexpensive. This material is toxic and is carefully controlled in many areas.

Cold dipping soaks the parts in a strong cleaning agent similar to carburetor cleaner for a period of time. The only equipment required for this is a container, but the solution used is relatively expensive and is difficult to dispose of when spent. The cold dip cleaning solutions are also toxic.

333

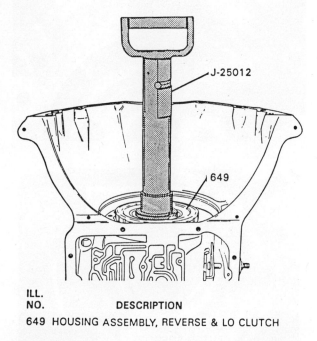

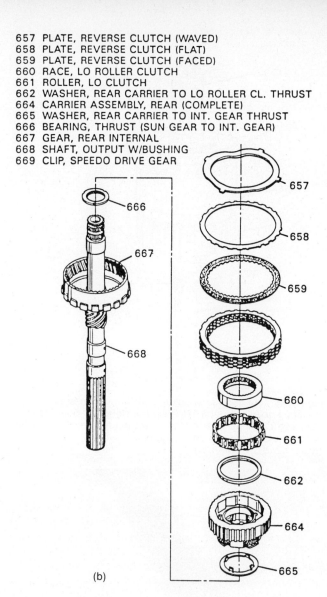

657 PLATE, REVERSE CLUTCH (WAVED)
658 PLATE, REVERSE CLUTCH (FLAT)
659 PLATE, REVERSE CLUTCH (FACED)
660 RACE, LO ROLLER CLUTCH
661 ROLLER, LO CLUTCH
662 WASHER, REAR CARRIER TO LO ROLLER CL. THRUST
664 CARRIER ASSEMBLY, REAR (COMPLETE)
665 WASHER, REAR CARRIER TO INT. GEAR THRUST
666 BEARING, THRUST (SUN GEAR TO INT. GEAR)
667 GEAR, REAR INTERNAL
668 SHAFT, OUTPUT W/BUSHING
669 CLIP, SPEEDO DRIVE GEAR

ILL.
NO.   DESCRIPTION
649 HOUSING ASSEMBLY, REVERSE & LO CLUTCH

(a)

(b)

Fig. 12-49 (a) Tool J-25012 used to lift the reverse–low clutch housing (649) out of a THM 200C transmission. (b) The clutch plates, low roller clutch, and rear carrier are under this clutch housing. *(Copyrighted material reprinted with permission from Hydra-matic Div., GM Corp.)*

Solvent wash is commonly used for small parts. A cleaning brush is used on the parts as they are dipped or sprayed with a petroleum- or water-based solvent. Petroleum solvents are flammable and cause redness along with soreness of the skin in many people. Petroleum solvents are also considered toxic. As the parts are blown dry, the solvent–air mixture is very flammable.

After cleaning with either the spray washer or cold dip, the parts should be rinsed with clear water and air dried to prevent rusting. Solvent wash also requires air drying to remove any residue. Be careful as you do this because of the flying debris that can enter your eyes or penetrate your skin.

12.5 CASE INSPECTION

After the case has been cleaned, several areas should be checked or serviced: the bushings, all fluid passages, the valve body worm tracks, all bolt threads, the clutch plate lugs, and the governor bore.

Fig. 12-50 After the return springs have been removed, the low–reverse clutch piston can be partially removed by blowing air into the passage indicated. *(Copyrighted material reprinted with permission from Hydra-matic Div., GM Corp.)*

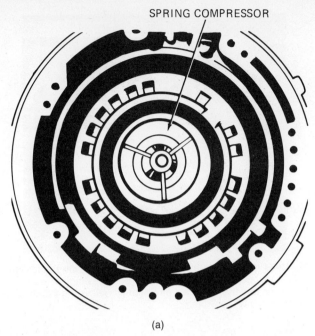

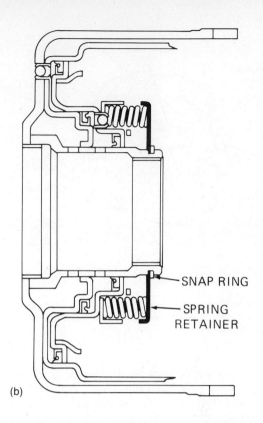

Fig. 12-51 (a) A clutch spring compressor should be set up to push against the spring retainer as indicated by the arrows and the spring compressed just enough to remove the retaining snap ring. (b) After the snap ring has been removed, the compressor should be slowly released making sure the retainer does not catch in the snap ring groove.

Some cases have a bushing that supports the output shaft, and this bushing should be replaced if it is worn or scored. This operation is described in the next section.

Every fluid passage in the case must be clean and open. It is a good practice to blow air into each passage and make sure that it comes out the other end. Next you should plug off the other end of the passage while the air is blowing to make sure that pressure has developed, which is a sign that no leaks are caused by cracks.

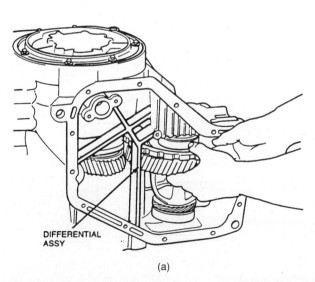

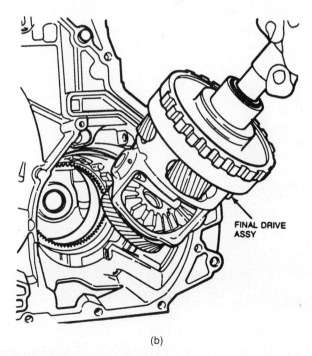

Fig. 12-52 After the bearing retainer has been removed, the differential assembly can be removed from an ATX transaxle. *(Courtesy of Ford Motor Company.)*

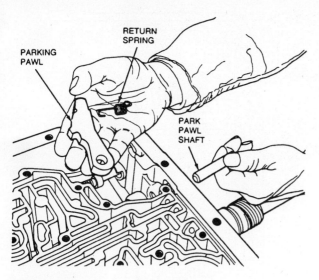

Fig. 12-53 Removal of the shaft allows removal of the park pawl. It should be checked for wear as it is removed. *(Courtesy of Ford Motor Company.)*

Fig. 12-55 A flat, smooth file is being drawn sideways across the valve body area of a case to smooth out any dings or warpage. It is drawn in a front-to-back direction as well as the side-to-side direction shown.

Check for warpage in the "worm track" (grooves for the valve body fluid flow) area. This can produce a cross leak—a leak from one passage to another. Some technicians draw file this area using a 16-in., single-cut file as standard practice. To do this, place the case with the worm tracks upward; grip each end of the file; and draw the file sideways across the case (Fig. 12-55). The file should be drawn in both a lengthwise and crosswise direction until a light cut has been made on all of the raised portions of the worm tracks. At this time, you can be sure that this area is flat.

If the transmission has failed because of a burned clutch or band you should check for pressure loss and cross leaks. (Cross leaks can cause an unwanted partial clutch application and drag.) You can check this by locating the passages for the two circuits; plugging the

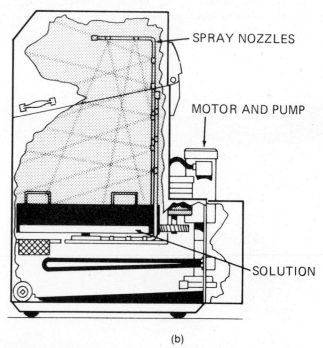

Fig. 12-54 (a) A clean engine block is being removed from a spray wash cabinet that also can be used for transmissions. (b) Note the spray nozzles inside the cabinet from which a hot cleaning solution will be pumped. *(Courtesy of Better Engineering.)*

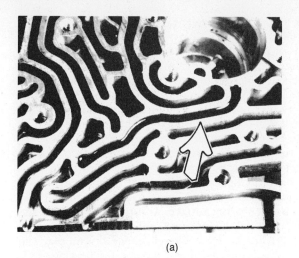
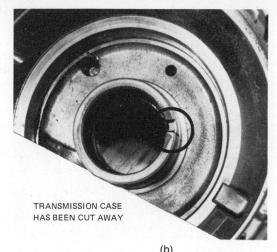

Fig. 12-56 (a) A worm track (arrow) has been closed off and filled with ATF. (b) The low–reverse clutch in a transmission failed because of a pressure loss through a crack (circled); it can also be checked by filling with ATF. Any leakage shows up as a loss of ATF with a red wetness appearing in the area of the leak.

openings in one of the passages with putty; and filling that passage with automatic transmission fluid (ATF) (Fig. 12-56). The cylinders for clutch and band servo pistons can also be checked for leakage by filling them with ATF. If fluid leaks from this passage to the other or out of the cylinder, there is a cross leak probably caused by a crack or porous casting.

Most faulty bolt threads are found during disassembly, when the bolts come out with aluminum on the threads. It is good practice to check all of the bolt threads visually to make sure they are in good shape. Damaged threads should be repaired as described in Chapter 11.

If the transmission has a holding clutch—for example, the low–reverse clutch or intermediate clutch in a THM 350—check the area of the case where the clutch lugs fit (Fig. 12-57). A badly worn case should be replaced.

Transmissions with gear-driven governors can have excessive wear in the governor bore, which can cause a loss in governor pressure. Minor scores can be cleaned using a brake wheel cylinder hone (Fig. 12-58). The bore size can be checked by placing the governor into its bore and filling the passages with ATF, as shown in Fig. 12-59. If the fluid leaks out too quickly (to point A in less than 30 seconds), the bore should be repaired. A worn bore can be repaired by installing a special reaming fixture; reaming the bore oversize; and installing a bushing to return the bore to the original size (Fig. 12-60).

The final service step for the case is to install a new seal for the manual shifter and, in some cases, the throttle valve linkage or rear bushing seal (Fig. 12-61). These operations are described in Chapter 11. After this is done, the case can be set aside until reassembly.

Fig. 12-57 Case showing wear in the area where the low–reverse clutch plates contact the lugs. The case is still usable, but the wear grooves might hamper proper clutch apply and release.

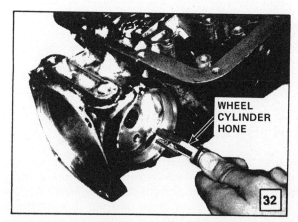

Fig. 12-58 A brake cylinder hone dipped in oil can be used to smooth out minor scores in a governor bore. It should be turned by hand; a power tool should not be used. (*Copyrighted material reprinted with permission from Hydra-matic Div., GM Corp.*)

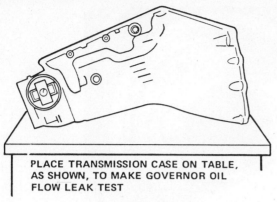

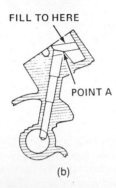

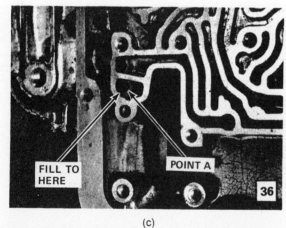

Fig. 12-59 (a) Governor bore of a THM 350 can be checked for excessive wear by filling the bore with ATF as shown. (b,c) If the fluid level drops from the fill point to point A in 30 seconds or less, a bushing should be installed in the governor bore. *(Copyrighted material reprinted with permission from Hydra-matic Div., GM Corp.)*

12.6 BUSHING, BEARING, AND THRUST WASHER SERVICE

Bushings and bearings are used to support rotating shafts, and thrust washers are used to separate rotating parts from each other or from stationary parts (Fig. 12-62). Bushings are plain metal bearings that require a lubricant to reduce friction, and bearings have much less friction because they have rolling members—either balls or rollers—as well as a lubricant to reduce friction. Bushing, bearing, or thrust washer failure causes wear as the hard parts turning at different speeds rub against each other (Fig. 12-63).

A bushing is a metal sleeve that is lined with a soft bearing material, usually bronze, tin, or both. Thrust washers are plastic, fiber, or bronze- or tin-lined iron. Where sideways positioning is critical or thrust loads are very high, a radial needle bearing commonly called a *Torrington* is used (Fig. 12-64). This bearing type uses needle bearings to absorb the end loads. The bearings must run against a very smooth, hard surface—either the face of a gear or a race. A bushing or thrust washer must have an operating clearance of about 0.003 to 0.005 inch (0.076 to 0.127 mm) to allow a good oil flow through the bearing surface. A Torrington bearing and a tapered roller bearing operate best with almost no clearance; in fact, excessive clearance or end play can cause a pounding that might damage the bearing.

An experienced technician judges bushing condition by appearance and feel. Any scoring, galling, flaking, excess wear, or rough operation is cause for replacement (Fig. 12-65). A bushing is relatively inexpensive, but the damage caused by a worn bushing is very expensive to repair. Most technicians will not gamble on a so-so bushing. If in doubt, replace it. Most manufacturers do not publish clearance or size specifications for bushings; some publish thrust washer thicknesses. If you want to measure bushing clearance to gain experience, use the rule of thumb that anything over 0.006 in. (0.152 mm) is excessive (Fig. 12-66).

12.6.1 Bushing Removal

Several methods can be used to remove a bushing depending on the size and how the bore is shaped. If the bore is straight, the bushing is normally pressed or driven out the other end of the bore (Fig. 12-67). The driving tool is usually a stepped device that fits into the bushing bore and has a raised shoulder to press against the end of the bushing. The shoulder should be slightly smaller than the bushing outer diameter to keep it from damaging the bushing bore.

If the bore is stepped so the bushing cannot be pressed through, there are removal tools that can pull the bushing out. These tools are often used with a slide hammer. They either enter the bushing and expand to secure a hold onto it or thread into the bushing (Fig. 12-68).

A commonly used method to remove a bushing in a stepped bore is to collapse or cut it. A bushing is collapsed by catching the bushing edge with a bushing cutter or sharp chisel and folding it inward (Fig. 12-69). This will be easier if you locate the seam in the bushing and work next to it. It is also possible to cut a groove in the bushing. But this is time consuming, and there is a possibility of damaging the outer bore.

Very small bushings are usually removed by cutting threads in them so a slide hammer can be threaded into the bushing (Fig. 12-70). For example, a THM 350 output shaft bushing allows $\frac{9}{16}$-in. threads to be cut into it. Use either coarse or fine threads to match those on your slide hammer. Some technicians remove this bushing using an impact wrench and a C6 band adjusting screw modified as described in Fig. 12-71.

FOLLOW STEPS 1-6 TO REPAIR THE GOVERNOR BORE

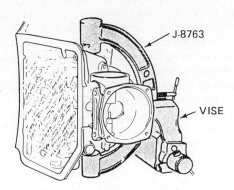

STEP 1 Install holding fixture J-8763 and mount in vise.

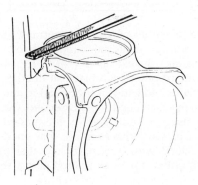

STEP 2 Remove (file) any excess material from the governor face.

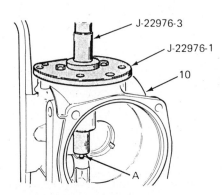

A PIN, GOVERNOR SUPPORT
10 CASE, TRANSMISSION

STEP 3 Install J-22976-3 and J-22976-1. Torque bolts to 13 N·m (10 ft.-lbs.). Make sure J-22976-3 rotates freely and then remove it.

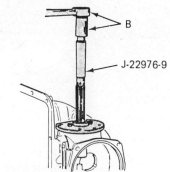

B RATCHET & SOCKET

STEP 4 Ream the governor bore as follows:
- Oil J-22976-9, J-22976-1 and the governor bore with transmission fluid.
- After each ten revolutions, remove the reamer and dip in transmission fluid to clean.
- After the reamer reaches the end of the bore and bottoms on the governor support pin, rotate the reamer ten additional revolutions.
- Remove the reamer. Be certain to rotate during removal to prevent scoring the bore.
- Remove the tools and thoroughly clean the case.

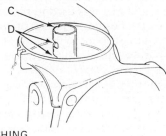

C BUSHING
D SLOTS

STEP 5 Align the slots in the bushing with the slots in the governor bore.

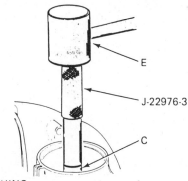

C BUSHING
E BRASS HAMMER

STEP 6 Install the bushing until the slots in the bushing align with the feed holes in the governor bore.

Fig. 12-60 (a) A THM 350/250 governor bore is repaired by installing a special fixture and reaming the bore oversize. (b) Then a governor bushing is driven into the bore to reduce it back to the correct size. *(Copyrighted material reprinted with permission from Hydramatic Div., GM Corp.)*

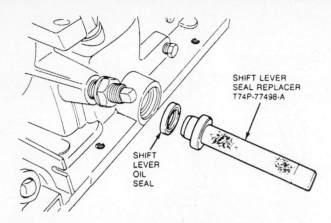

Fig. 12-61 Tool T74P-77498-A used to drive the shift lever seal into a case. *(Courtesy of Ford Motor Company.)*

12.6.3 Tapered Roller Bearing Service

Some transaxles use tapered roller bearings in their final drives (Fig. 12-74). These bearings are checked by visual inspection and by rotating the bearing with a pressure between the cup, bearing, and cone. Any scoring or flaking of the cup, cone, or roller surfaces or a rough feel is cause for replacement. Special tools are often required to press a bearing cone off of or onto a shaft or a cup out of or into a bore (Fig. 12-75).

Tapered roller bearings must be adjusted to get the correct end play or preload. This adjustment is normally accomplished by changing selective sized shims that can be positioned under the cup or cone (Fig. 12-76). The procedure for this adjustment varies between manufacturers (Fig. 12-77).

12.6.2 Bushing Installation

Replacement bushings are available as single items or as a set. From past experience, technicians automatically replace every bushing in certain transmissions.

A bushing installer should be used to push the new bushing into its bore to prevent damaging it or the bore. This is often the same tool that was used to remove the bushing. In most cases, the bushing is placed on the tool and pressed or driven into the bore to the correct depth (Fig. 12-72). Some bushing drivers have steps so they "bottom out" and stop at the correct depth.

In some cases a bushing is staked to lock it in place. Staking is usually done using a punch to bend bushing metal into a recess (Fig. 12-73). After staking a bushing, you should remove any raised metal with a scraping tool.

12.7 FRICTION MATERIAL SERVICE

There are three points of view concerning automatic transmission friction material (lined plates, unlined plates, and bands) during an overhaul: replace all of them as a standard practice, replace all lined plates and the unlined plates and bands if they need it, and replace only those items that need it. The second view is the one held by many technicians. As mentioned earlier, some states (e.g., California) require that lined friction plates be replaced with new parts and that bands be replaced with either a new or relined part if the transmission is being "overhauled" or "rebuilt."

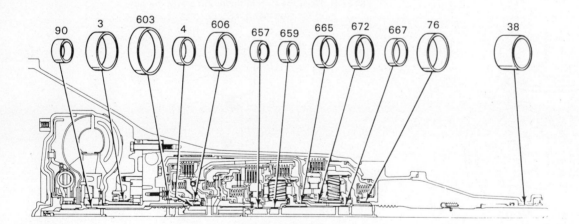

ILL. NO.	DESCRIPTION	ILL. NO.	DESCRIPTION
3	BUSHING, OIL PUMP BODY	657	BUSHING, INPUT SUN GEAR (FRONT)
4	BUSHING, STATOR SHAFT (REAR)	659	BUSHING, INPUT SUN GEAR (REAR)
38	BUSHING, CASE EXTENSION	665	BUSHING, REACTION SHAFT (FRONT)
76	BUSHING, CASE	667	BUSHING, REACTION SHAFT (REAR)
90	BUSHING, STATOR SHAFT (FRONT)	672	BUSHING, REACTION SUN GEAR
603	BUSHING, REVERSE INPUT CLUTCH (FRONT)		
606	BUSHING, REVERSE INPUT CLUTCH (REAR)		

Fig. 12-62 Location of the various bushings in a THM 700 transmission. *(Copyrighted material reprinted with permission from Hydra-matic Div., GM Corp.)*

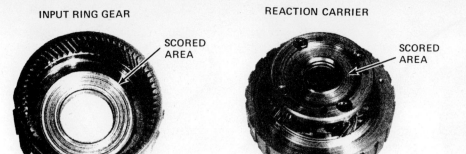

Fig. 12-63 The scored areas on the face of the carrier and ring gear are the result of thrust washer failure. *(Copyrighted material reprinted with permission from Hydra-matic Div., GM Corp.)*

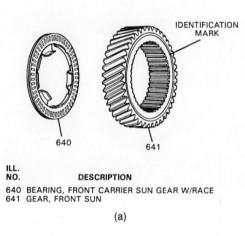

ILL. NO.	DESCRIPTION
640	BEARING, FRONT CARRIER SUN GEAR W/RACE
641	GEAR, FRONT SUN

(a)

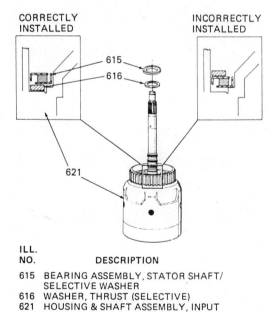

ILL. NO.	DESCRIPTION
615	BEARING ASSEMBLY, STATOR SHAFT/ SELECTIVE WASHER
616	WASHER, THRUST (SELECTIVE)
621	HOUSING & SHAFT ASSEMBLY, INPUT

(b)

Fig. 12-64 (a) Some roller bearing thrust washers use only one race with the other side running against the gear face; (b) others use two races. In either case, the races must be positioned correctly to prevent bearing damage. *(Copyrighted material reprinted with permission from Hydra-matic Div., GM Corp.)*

12.7.1 Lined Plate Service

If lined plates are to be reused, they should be carefully inspected (Fig. 12-78). The requirements for reuse of a lined plate are:

1. The lining thickness must be almost the same as a new plate.

2. There must be no breaking up or pock marks in the lining.

3. There must be no metal particles embedded in the lining.

4. The lining must not come apart when scraped with a coin, fingernail, or knife blade (Fig. 12-79).

5. The lining must not have a glazed, shiny appearance.

6. The plate must be flat.

7. The splined area must be flat and even.

Plates that are to be reused should not be washed. If they need cleanup, they should be quickly rinsed in solvent and then air dried.

12.7.2 Steel Plate Service

Unlined steel plates can often be reused, but they also have certain requirements (Fig. 12-80):

1. The plate must be flat (except for wave or Belleville plates).

2. There must be no sign of surface irregularities.

3. The splined area must be flat and even.

4. Shiny or slightly burned plates must be reconditioned.

A quick way to check the plates for flatness is to stack the plates and look for any gaps between the plates. Overheated plates tend to warp into either a slightly conical or a "potato chip" shape. Gaps indicate warped, unusable plates. Then restack the plates, turning every other one upside down, and recheck for gaps, which indicate faulty plates.

341

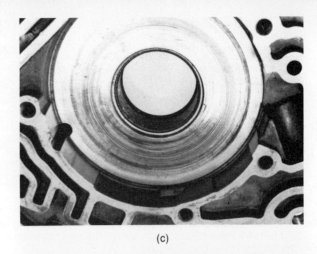

Fig. 12-65 (a) Bushing is used but appears in good condition. (b) Bushing is badly scored. (c) Bushing is badly pitted. (d) Bushing seized and spun in its bore. The clutch assembly in (d) and pump cover in (e) should be replaced.

Fig. 12-66 A quick and simple way to check bushing clearance is to place a strip of paper between the converter hub (in this case) and the bushing. The paper is about 0.003 in. thick so one should be able to remove it using a slight pull.

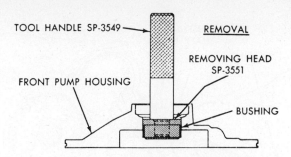

Fig. 12-67 Tool with a flange that catches the edge of a bushing so it can be driven out. *(Courtesy of Chrysler Corporation.)*

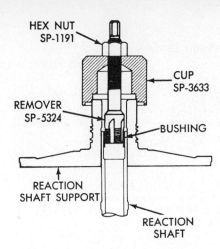

Fig. 12-70 Tool cuts its own threads as it is threaded into the bushing. After it has entered the bushing 3 or 4 turns, the hex nut at the top is tightened to lift the bushing out. *(Courtesy of Chrysler Corporation.)*

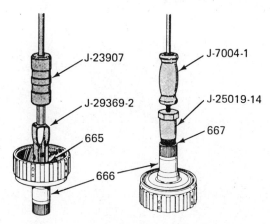

665 BUSHING, REACTION CARRIER SHAFT — FRONT
666 SHAFT, REACTION CARRIER
667 BUSHING, REACTION CARRIER SHAFT — REAR

Fig. 12-68 Tools J-29369-2 and J-25019-14 attach a slide hammer to a bushing so the bushing can be pulled out. *(Copyrighted material reprinted with permission from Hydra-matic Div., GM Corp.)*

Fig. 12-69 A chisel is used to cut a groove through the bushing so it can be collapsed. Care should be taken to ensure that the chisel does not damage the bushing bore.

Fig. 12-71 A band adjusting screw from a C6 has cutting grooves ground in the threads, and the nut has been welded in place. With the use of an air wrench, it cuts threads into the bushing in the front of a THM 250/350 main shaft, bottoms in the hole, and forces the bushing out.

343

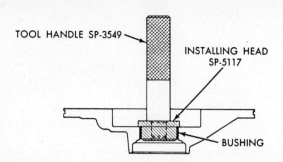

Fig. 12-72 Tool with a flange that extends over the edge of a bushing so it can be driven or pressed into its bore. Care should be taken to ensure that the bushing enters straight into the bore. *(Courtesy of Chrysler Corporation.)*

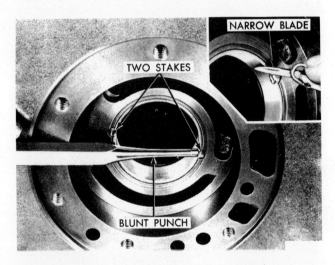

Fig. 12-73 Bushing should be staked to ensure that it does not spin. A hammer and punch is used for this as shown. After staking, a narrow scraping blade is used to remove any burs or high spots that might have resulted from the staking operation. *(Courtesy of Chrysler Corporation.)*

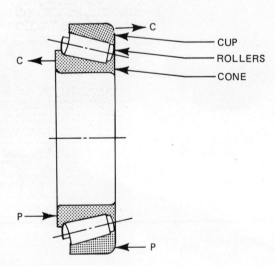

Fig. 12-74 The cup, rollers, and cone of a tapered roller bearing are ground at an angle as shown. Thus, they can resist a side thrust in the direction indicated by arrows *P*; the direction that preload is applied onto the bearing. A clearance results in the bearing if either race is moved in the direction indicated by arrows *C*.

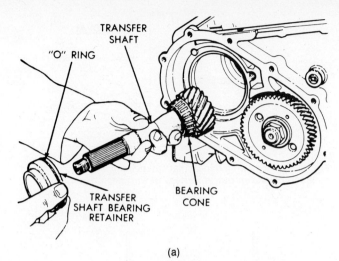

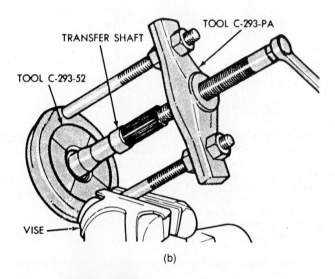

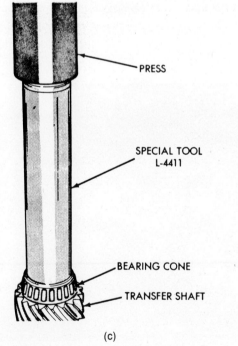

Fig. 12-75 (a) Bearing that requires (b) a special tool to remove it from the transfer shaft. The new bearing is pressed onto the shaft using (c) a special tool. *(Courtesy of Chrysler Corporation.)*

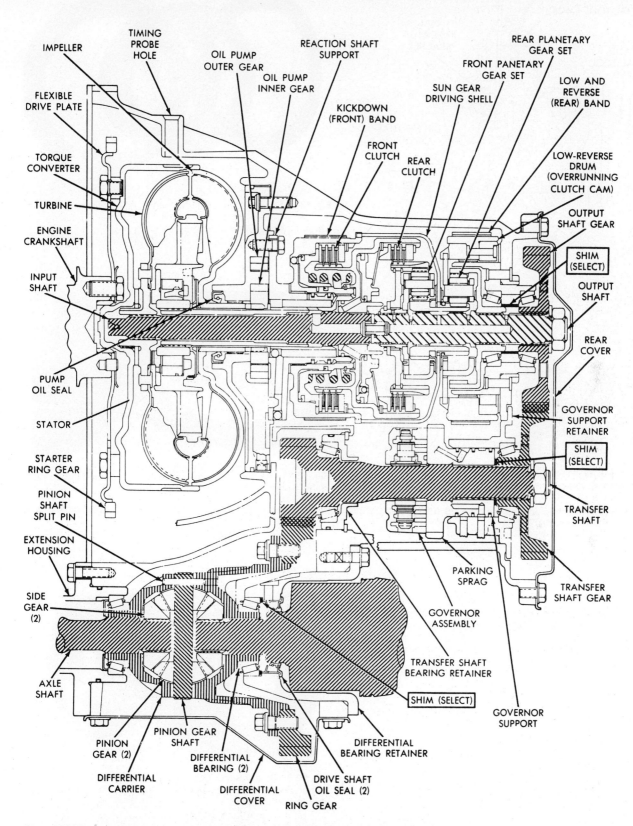

Fig. 12-76 Three selective sized shims are used to adjust the clearance of three pairs of tapered roller bearings in a Torqueflite transaxle. *(Courtesy of Chrysler Corporation.)*

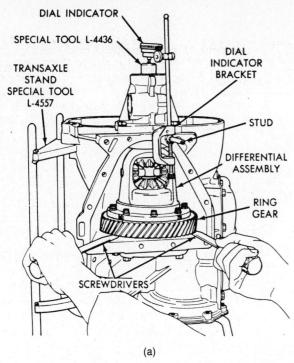

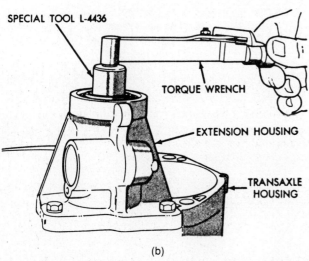

Fig. 12-77 (a) Bearing end play is checked using a dial indicator. End play is read as the differential assembly is pried upward and lowered. (b) Bearing preload is measured using a torque wrench as the shaft is rotated. *(Courtesy of Chrysler Corporation.)*

Fig. 12-78 Four faulty lined clutch plates: (a) glazed and polished; (b) losing lining material; (c) worn to the metal backing; (d) warped into a potato chip shape.

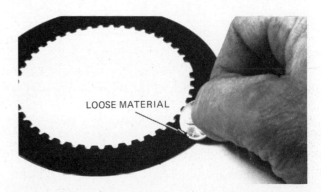

Fig. 12-79 The lining is easily scraped from a lined clutch plate using a coin; this disc should not be reused.

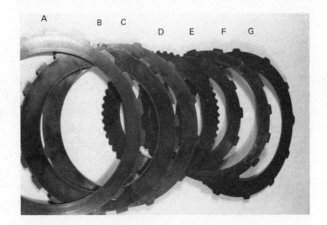

Fig. 12-80 Steel plates: (a,b) can be reused if they are flat; (b, c,e) show some burn marks from slippage; (f,g) warped; (c–f) should be replaced.

A steel plate must be slightly rough in order to produce proper break-in of the new lined plates. The friction surface of the lined plate has to wear slightly until it matches the surface of the unlined plate. Too smooth of a steel plate surface allows increased slippage, which produces heat and causes a glazing of the friction material. Too rough of a steel plate surface produces severe operation and rapid friction material wear. The ideal surface finish for a steel plate is a tumbled finish, like that of a new steel plate (Fig. 12-81). Some shops use a tumbler to recondition used steel plates (Fig. 12-82). This machine rotates a basket of plates along with a ceramic media and soap, and the tumbling action creates the small nicks and dents that produce the correct surface finish. Some sources recommend roughing up a steel plate by sanding or blasting with sand, grit, or glass beads, but these processes are considered too time consuming. Also, blasting produces a surface finish that is too rough and severe.

Fig. 12-81 A good steel plate should have a surface that is flat but has a series of nicks as shown. The slight roughness holds ATF and also tends to cause the new lined plate to wear slightly or break in.

Fig. 12-82 Parts tumbling machine. Steel plates are put in it with a ceramic or steel media and a cleaning agent. Rotation of the machine causes them to bump into each other as the cleaning action takes place. The result is a clean steel plate with the right surface texture. *(Courtesy of Intercont Products.)*

12.7.3 Band Service

If a band is to be reused, it should be checked to ensure that:

1. The lining material is sound with no breaking up or pock marks.

2. The lining material does not come apart when scraped with a thumbnail or knife blade.

3. The lining thickness is almost the same as a new band.

4. The lining material is not badly discolored or does not appear burned.

5. There are no metal particles embedded in the friction material.

6. The end lugs appear tight and unworn.

If the friction surface on a fabric-lined band has a dark, shiny appearance but is good otherwise, it can be scraped using a knife or bearing scraper to remove the glazed surface. A glazed paper-lined band should be replaced.

The drum surface for a band must also be in good condition, and like the surface of a steel plate, the drum surface should be slightly rough to help break in the new band (Fig. 12-83). A rough, badly scored drum should be replaced. If the scoring is light and the equipment is available, some technicians true up the drum surface using either an engine lathe or a brake drum lathe. This usually requires a band with an oversize lining to compensate for the smaller drum diameter. Some of the newer transmissions use a drum made from a steel stamping rather than from a forging or a casting. These units should be checked for dishing by laying a straight edge along the drum surface (Fig. 12-84).

If the drum surface is smooth but polished, it can be sanded to produce a rougher finish. If a paper-lined band is used, use 120- to 160-grit sandpaper, and sand in a direction that goes around the drum (Fig. 12-85). If a fabric-lined band is used, use coarser sandpaper (40 to 60 grit) and sand in a front-to-back direction.

12.7.4 New Friction Material Preparation

New friction material must never be used in a dry condition. Dry lining acts like an insulator. It easily overheats and burns from the friction heat created during the first shift. The lined material should be soaked in ATF for at least 30 minutes before installation; some shops soak the plates overnight.

When a clutch pack is assembled, the clearance should always be checked and adjusted if necessary. The clearance check ensures that the clutch is assembled correctly with parts that are the correct size—lined plates, unlined plates, pressure plate, apply ring, and snap ring. Depending on the transmission and the particular clutch, several variable-size parts can be used. The parts that can be of variable thickness are the piston, pressure plate, snap ring, apply ring, steel plate, and lined plate. Most manufacturers publish clearance specifications for some, if not all, of the clutches used in their transmissions. Sometimes, though, the only specification

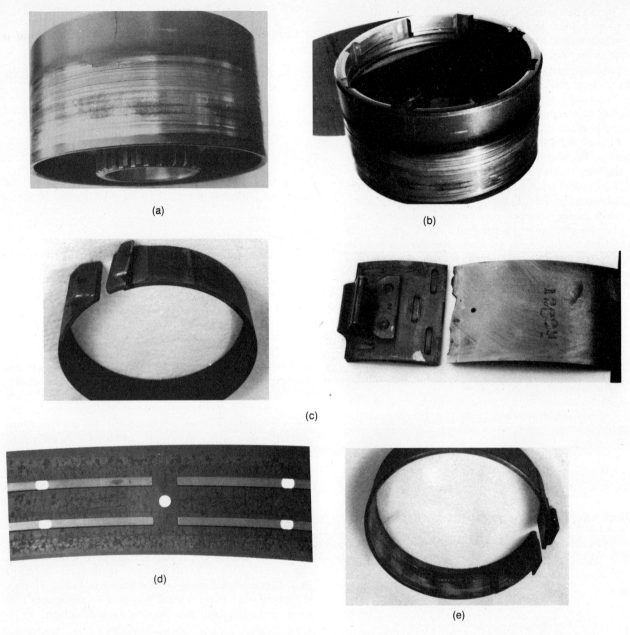

Fig. 12-83 Badly worn drums: (a) scored and polished; (b) badly scored and shows burn marks. Faulty bands: (c) broken; (d) lining flaked off; (e) worn.

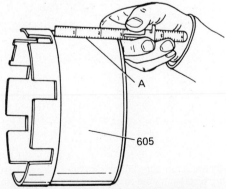

Fig. 12-84 Stamped steel drums should be checked for dishing by placing a straight edge along the drum surface. There should be no clearance at "A". *(Copyrighted material reprinted with permission from Hydra-matic Div., GM Corp.)*

A   CHECK FOR DISHING AT THIS POINT
605   HOUSING & DRUM ASSEMBLY, REVERSE INPUT CLUTCH

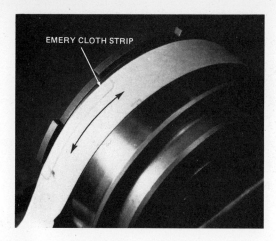

Fig. 12-85 If the drum surface is polished, it should be sanded to slightly roughen the surface. If a paper band is to be used, sand around the drum using 120- to 180-grit emery cloth or sand paper. If a fabric band is to be used, sand front to back on the drum using coarser 40 to 60 grit.

is to make sure the plates turn freely with the clutch in a normal position (Fig. 12-86). There should never be a drag between the plates. If no clearance specifications are available, use the rule of thumb that for a clutch that is applied in all forward gears the clearance should be about 0.010 in. (0.5 mm) or slightly less for each lined plate; if the clutch is released in any of the forward gears, use a clearance of 0.010 in. or slightly more for each lined plate.

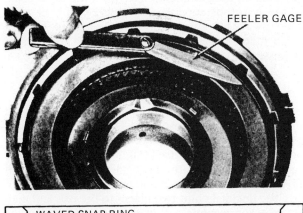
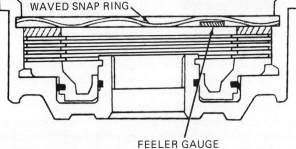

Fig. 12-86 Measuring clutch clearance using a feeler gauge. Note that the feeler gauge is inserted between the pressure plate and the waved snap ring. If a flat snap ring is used, measurement should be in the same place. *(Courtesy of Chrysler Corporation.)*

Clutch clearance, also called backing (pressure) plate travel, is normally measured using a feeler gauge placed between the pressure plate and the snap ring. If a waved snap ring is used, position the feeler gauge in the widest area under a wave. Clearance can also be measured using a dial indicator. Position the dial indicator as shown in Fig. 12-87, and raise and lower the backing plate to measure its vertical travel—the clutch clearance. If a wave or Belleville plate is used, the pressure plate should be pushed downward with a light, even pressure so the cushion plate(s) are not distorted. In some cases, the clutch clearance is measured from a point on the transmission case to the clutch pack (Fig. 12-88). In others, the clutch has to be assembled temporarily before installation in order to measure the stack height (Fig. 12-89). Measuring clutch clearance seems to be a lot of trouble to some, but if it saves a comeback from a burned-out clutch or a repeat teardown, it is well worth the time.

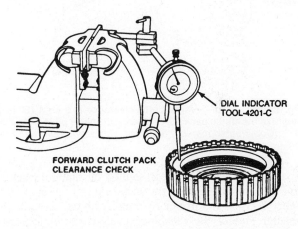

Fig. 12-87 Clutch clearance can be quickly measured after setting up a dial indicator as shown and lifting and dropping the pressure plate. The amount of clearance is indicated by the amount of dial movement. *(Courtesy of Ford Motor Company.)*

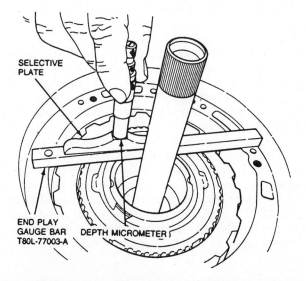

Fig. 12-88 When a clutch pack is installed in a case, the amount of clearance can be measured using a depth micrometer and a gauge bar. The depth to the top plate is compared to a specified dimension. *(Courtesy of Ford Motor Company.)*

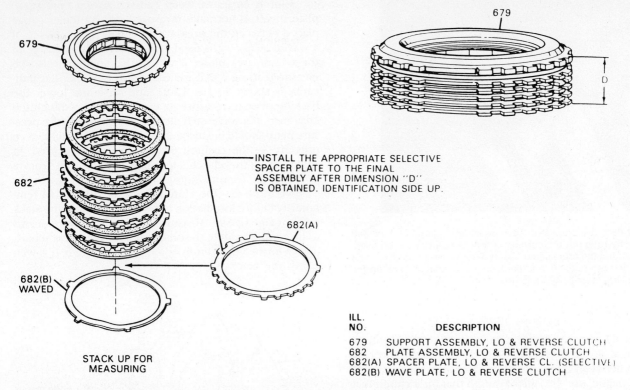

Fig. 12-89 The low–reverse clutch clearance in a THM 700 is checked by measuring the height of the clutch pack "D". The correct spacer plate should be selected from dimension "D". *(Copyrighted material reprinted with permission from Hydra-matic Div., GM Corp.)*

Clutch clearance is adjusted by changing the selective component, usually the snap ring or pressure plate. Many technicians prefer to build a clutch toward the smaller end of the clearance in order to reduce the amount of piston travel and produce crisper shifts. For example, if you measure an AOD forward clutch (five plates) with 0.092 in. (2.34 mm) of clearance and a 0.075-in. (1.9-mm) snap ring (Fig. 12-90), you can reduce the clearance to the 0.050 to 0.089 in. (1.27 to 2.26 mm) specification with a snap ring that is either 0.088 to 0.092 in. or 0.102 to 0.106 in. (2.24 to 2.34 mm or 2.59 to 2.69 mm). The larger snap ring (assume it is 0.104 in.) would be preferred because it would reduce the clearance by about 0.029 in. (0.104 − 0.075 = 0.029 in.) (0.74 mm, 2.64 − 1.9 = 0.74 mm) to a clearance of about 0.063 in. (0.092 − 0.029 = 0.063 in.). The snap ring that is 0.088 to 0.092 in. reduces the clearance by about 0.015 in. (0.38 mm) to about 0.077 in. (1.9 mm)—acceptable but not as good as 0.063 in.

If the selective parts do not correct the clearance or are not available, clutch clearance can be reduced by using extra thick steel plates or adding an extra unlined steel or lined friction plate right next to another one (Fig. 12-91). If an additional friction plate is too thick, one or both of the adjacent friction linings can be scraped off using a knife or gasket scraper.

12.8 INTERNAL SEAL SERVICE

An important part of a transmission overhaul is the correct replacement of the internal seals. These are the rubber and metal seals used on accumulator, servo, and clutch pistons and the rotating metal or Teflon sealing rings used on the pump clutch support, the input shaft, and the governor support (Fig. 12-92). Rubber seals should always be replaced during an overhaul. If metal seals or metal or Teflon sealing rings are in good shape, it is possible to reuse them.

12.8.1 Fitting Rubber Seals

A seal should always be checked in the bore where it will operate. In most cases, this takes very little time. An O-ring seal (round or square cut) is first checked by placing it in the bore by itself (Fig. 12-93). If it is slightly larger or smaller than the bore (3 percent or less), it will work. (To compute 3 percent, take the diameter, move the decimal point two places to the left, and multiply by 3. For example, a $2\frac{1}{4}$-in. servo is 2.250 in. Moving the decimal point gives us 0.0225; multiplying by 3 gives us 0.0675 in. The allowable over- or undersize of 3 percent of $2\frac{1}{4}$ in. is slightly more than $\frac{1}{16}$ in.)

Next the seals should be checked for drag, one at a time. Starting with the easiest one to remove and replace, place one seal on the piston, lubricate it with ATF, and work it into the bore. Now stroke the piston, and feel the drag. A round O-ring should produce some drag; a square-cut seal is okay if it produces a barely noticeable drag. Now remove the piston, remove the seal, install the next seal, and check its drag.

Lip seals are checked by placing them on the piston and checking the drag also, one at a time, just like O-ring seals. It must produce a drag on the way into the

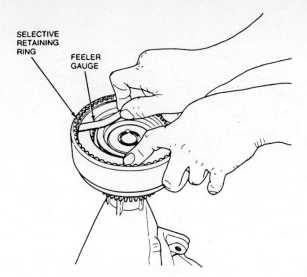

The clearance should be:

<u>3.8L (232 CID)</u>

1.02-1.80mm (0.040-0.071 inch)

<u>5.0L (302 CID) HO SEFI, 5.0L (302 CID) SEFI, 5.8L (351 CID)</u>

1.27-2.26mm (0.050-0.089 inch)

If the clearance is not within specification, selective snap rings are available in the following thicknesses:

- 0.060-0.064 inch
- 0.074-0.078 inch
- 0.088-0.092 inch
- 0.102-0.106 inch

Install the correct size snap ring and recheck

Fig. 12-90 The clearance in an AOD forward clutch is adjusted by selecting the snap ring of the correct thickness. *(Courtesy of Ford Motor Company.)*

drum, but it is okay if it falls outward when turned over. On those pistons that use three lip seals, you should check each one, one at a time.

12.8.2 Installing Pistons with Lip Seals

During installation, a lip seal often catches the edge of the bore and tries to turn outward. This probably cuts the seal lip and causes a fluid leak. Several things can be done to ease this installation and produce a reliable clutch: use a wax lubricant, an installing tool, a seal guide, and precompress the seal lip.

A seal should never be installed dry. Both the bore and the seals should be lubricated with ATF, petroleum jelly, transmission assembly lubricant, or wax lubricant. The wax lubricant is commonly used for car door or hood striker plates. Smearing a film of wax lubricant around the seal and at the edge of the bore is often just enough lubricant to allow the seal to slip into the bore (Fig. 12-94).

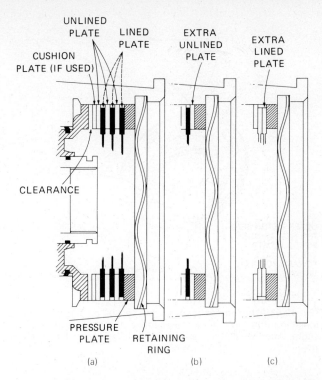

Fig. 12-91 (a) Clutch clearance can be reduced by adding an extra lined or (b) an unlined plate. (c) When two lined plates are next to each other [as in (c)], clearance can be increased by shaving the lining off of one or both adjacent sides of the two plates.

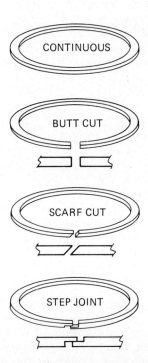

Fig. 12-92 Teflon sealing rings are supplied in four styles as shown here. The cut rings are easy to install by merely placing them in the proper position. The uncut, continuous ring requires special tools.

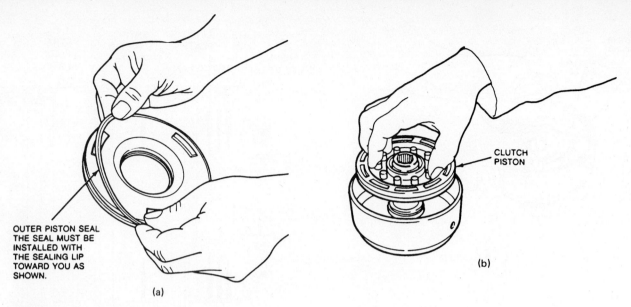

Fig. 12-93 A piston seal can be checked by placing it on the piston, lubricating it, and installing it in the bore so one can feel the amount of drag. It should be noticeable but slight. After checking the outer seal, it is removed so the inner seal can be checked. *(Courtesy of Ford Motor Company.)*

Fig. 12-94 Piston and seal installation can be made easier if a film of wax lubricant is smeared around the rubber seals and seal area.

Seal installation tools are commercially available or you can make one by rounding the edges of a feeler gauge that is 0.005 to 0.010 in. (0.12 to 0.25 mm) or squeezing a bent, thin piece of piano wire into the end of a small tubing (Fig. 12-95). As the piston is being installed, use this tool to coax the seal lip into the bore.

Seal guides are available for some clutch units. These are smooth steel bands with a slight funnel or cone shape (Fig. 12-96). They are placed in the drum and lubricated with ATF. As the piston is being installed, the guides prevent the seal lips from catching on the edge of the bore.

Some seals are on a piston with a construction that does not allow an installing tool or guide to be used and precompression is necessary. For example, the front clutch piston of an A727 has a ledge that extends out-

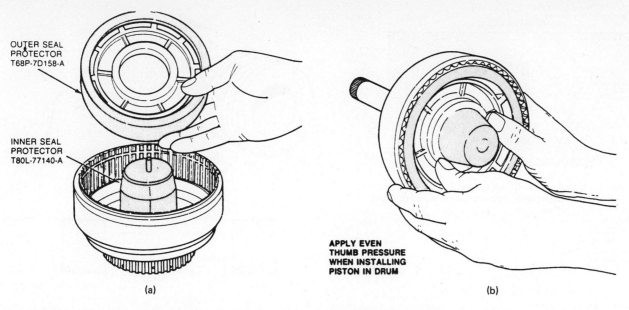

Fig. 12-95 Inner and outer seal protectors are used to guide the seal lips into place. With the seals and sealing areas lubricated with petroleum jelly, one should be able to slide the piston into position using thumb pressure. *(Courtesy of Ford Motor Company.)*

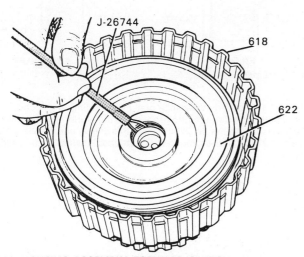

618 HOUSING ASSEMBLY, FORWARD CLUTCH
622 PISTON ASSEMBLY, FORWARD CLUTCH

Fig. 12-96 Tool J-26744 used to guide the lip of the sealing ring past any obstructions while the piston (622) is installed. *(Copyrighted material reprinted with permission from Hydra-matic Div., GM Corp.)*

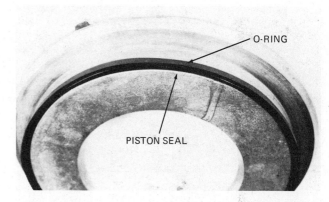

Fig. 12-97 An extension housing sealing O-ring (arrow) has been placed around the sealing lip of the outer sealing ring of a Torqueflite front clutch. It reshapes the sealing lip to make installation easier.

12.8.3 Fitting Sealing Rings

A sealing ring has to make a seal on one of its sides and at the outer diameter. Fluid pressure plus the elasticity of the ring pushes the ring outward where it engages the bore, and if the bore is rotating, the sealing ring rotates with the bore (Fig. 12-98). Rubbing action should always take place between one side of the sealing ring and the groove. Some sliding action takes place between the ring and the bore as end play allows the bore along with the drum to move forward and backward.

A metal sealing ring should be checked by placing it in its bore (Fig. 12-99). A hook- or interlock-type ring should be hooked to itself first. There should be a tight and close fit between the outer diameter of the ring and the bore. An undersized or odd-shaped ring should not be used. With metal rings, there should be a slight gap—about 0.002 to 0.012 in. (0.05 to 0.3 mm)—between the ends of the ring to allow for expansion of the ring

ward, well over the edge of the bore. The extension housing sealing O-ring from a Powerglide is slightly smaller than the A727 outer seal. If you place the powerglide seal over the Torqueflite lip seal, it pulls the seal lips tightly inward (Fig. 12-97). Leave the O-ring in place for a while (the longer the better); then remove the O-ring; and quickly install the piston into the drum. The colder the temperature, the more time you have to do this before the lip seal expands back to its normal size.

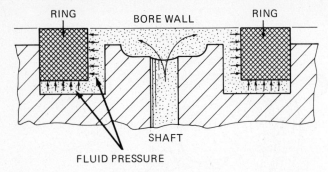

Fig. 12-98 A metal seal ring is forced outward and away from the pressurized oil. It tends to wear into the sides of the grooves.

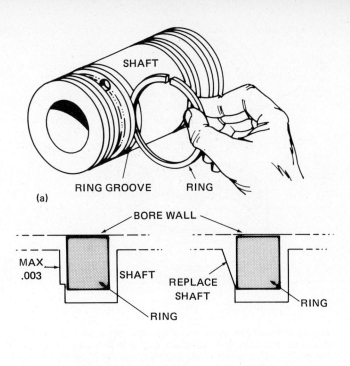

Fig. 12-100 (a) Side clearance of a metal sealing ring is checked by placing the ring into the groove and measuring the clearance with a feeler gauge. (b) While doing this check, one should make sure the side of the groove is not stepped, nicked, or tapered.

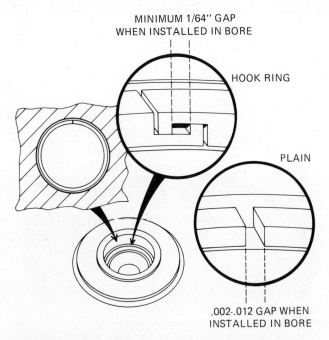

Fig. 12-99 A metal sealing ring has been hooked and placed into its bore. It should enter the bore with only a slight pressure and make full contact with the bore. A plain sealing ring should have a slight clearance as shown.

metal. Slight nicks or imperfections can be smoothed out using sandpaper.

Next check the ring in the groove (Fig. 12-100). There should be a maximum of about 0.003 in. (0.07 mm) of groove wear, and the sides of the groove should be smooth and straight. Small imperfections can be smoothed using a small file. Excessive or tapered wear requires shaft or clutch support replacement.

12.8.4 Installing Teflon Sealing Rings

Scarf-cut, Teflon sealing rings are easy to install. Merely place them in the groove with the ends lapped in the right direction (Fig. 12-101). Be careful when you install

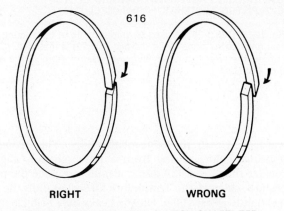

Fig. 12-101 When a scarf-cut Teflon ring is installed, the ends of the ring must lap on the correct sides. *(Copyrighted material reprinted with permission from Hydra-matic Div., GM Corp.)*

the next unit on top of the sealing ring. The ends tend to stick out, and they can get caught and cut during installation.

Uncut Teflon rings require two special tools for installation: an installing tool and a resizing tool. To install a Teflon ring, place the installing tool over the shaft; adjust it to the correct depth if necessary; lubricate the ring and the tool with ATF; and slide the ring over the tool and into its groove (Fig. 12-102). You can expect the ring to stretch to a size that is too large. Now lubricate the ring and resizing tool with ATF, and work the resizing tool over the ring, being sure the ring enters its groove correctly. The resizing tool should compress the ring down to the correct diameter. Once the transmission operates and the ring gets hot, it takes the shape of the bore around it.

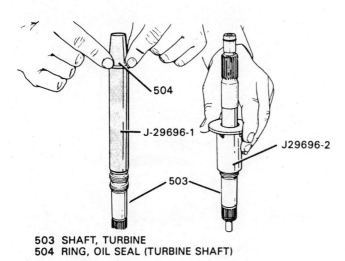

503 SHAFT, TURBINE
504 RING, OIL SEAL (TURBINE SHAFT)

Fig. 12-102 An uncut Teflon ring is installed by lubricating it and sliding it over a seal installer (J-29696-1). Then it must be resized smaller using tool J-29696-2. *(Copyrighted material reprinted with permission from Hydra-matic Div., GM Corp.)*

12.9 SUBASSEMBLY REPAIR

Subassembly operations are done to check and repair the various units inside the transmission. These are the pump, clutch assemblies, gearset, valve body, and governor. A technician normally disassembles the unit, makes any necessary checks and inspections, makes the needed repairs, and reassembles the unit with whatever new parts that are required. Rather than have an entire transmission disassembled, the technician usually works on one unit at a time. Some assemblies, (e.g., the clutch) take a small amount of time to disassemble, clean up, check, make any minor repairs to, and reassemble.

12.9.1 Pump Service

Service of most pumps consists of disassembly; inspection of the pumping members, stator support shaft and splines, front bushing, and clutch support surface with the sealing ring grooves; replacement of the front seal; checking of any valves; cleaning of all fluid passages and the drainback hole; and reassembly (Fig. 12-103).

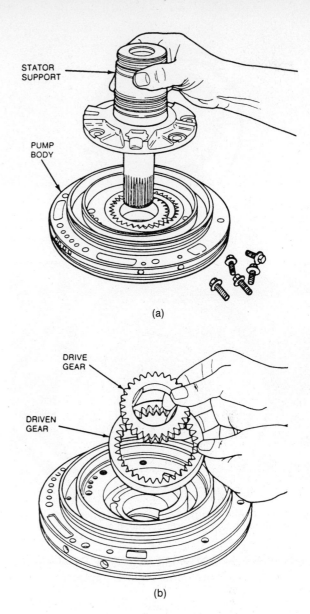

Fig. 12-103 Removing the pump cover/stator support exposes the inside of an internal–external gear pump and allows removal of the gears. *(Courtesy of Ford Motor Company.)*

Pump disassembly is fairly easy; simply remove the bolts that secure the cover onto the body. As you remove the gears or the rotor and slide, carefully note the front–back positioning; many gears are marked to make this easier.

Experienced technicians check a pump mostly by visual inspection. They carefully check the areas where wear normally occurs. Remember that the pump has a high-pressure area, and this high pressure tries to force the gears outward. A technician will inspect these areas, the sides of the gears or rotors, the areas of the body and cover where the sides of the gear run, and the flanks of the gear teeth/rotor lobes for score marks, which indicate a worn pump (Fig. 12-104). Also check for a worn pump bushing. Manufacturers sometimes publish clearance specifications for the pump wear locations, and

Fig. 12-104 (a) The rotor and slide of a vane pump are broken and (b) the pump body is badly scored. It must be replaced.

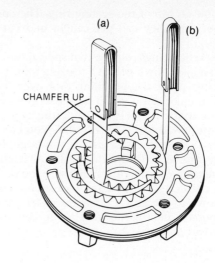

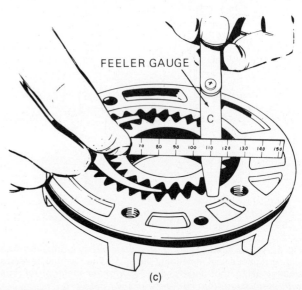

Fig. 12-105 An internal–external gear pump is checked for wear at three locations: (a) between the teeth and the crescent, (b) between the external gear and the body, and (c) at the side of the gears. *(Copyrighted material reprinted with permission from Hydra-matic Div., GM Corp.)*

these clearances can be checked using a feeler gauge (Fig. 12-105). A worn pump requires replacement with a new or rebuilt unit.

Vane-type pumps are also checked by visual inspection. Besides the inner slide area, be sure to check the pump guide rings, vanes, rotor, pump guide, slide, slide seals, seal support, slide pivot pin and spring, and slide sealing ring and backup seal (Fig. 12-106). If replacement is necessary, the thickness of the rotor and slide should be measured to ensure the correct replacement parts (Fig. 12-107).

With the pump disassembled, it is usually fairly easy to remove and replace the seal and also the bushing if necessary. A seal driver should be used when installing the new front seal (Fig. 12-108). If a seal driver is not available, a flat piece of $\frac{1}{8}$-in. (3-mm) or thicker metal should be placed over the seal to prevent damage while pressing or driving it into place. A damaged stator shaft can usually be pressed out and a new one pressed in at this time.

Metal caged lip seals are not designed to retain pressure. The fluid that is fed to lubricate the front bushing normally drains back into the inside of the transmission case through a drainback hole between the bushing and the seal (Fig. 12-109). If this hole plugs up, pressure can build up behind the seal, and this causes the seal to blow out. Make sure this passage is open.

On those pumps that include a clutch piston assembly, the piston seals should be serviced as described in the next section.

After the pump has been cleaned thoroughly, checked, and had the necessary bushings and seal replaced, it can be reassembled. The gears or rotor and slide assembly should be well lubricated with ATF and placed in the pump body in the correct position. Some technicians lubricate the gears with a heavy coating of petroleum jelly or transmission assembly lubricant to ensure that the pump will prime and pick up fluid immediately on start-up.

On some pumps, the cover has a much smaller diameter than the body; on these pumps, the cover is merely placed in position and the bolts are installed and tightened to the correct torque. Other pump covers and bodies have the same outer diameter, and these diameters are only slightly smaller than the bore in the transmission case. On these pumps, the two outer diameters

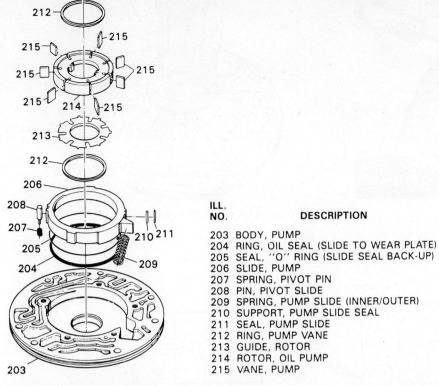

ILL. NO.	DESCRIPTION
203	BODY, PUMP
204	RING, OIL SEAL (SLIDE TO WEAR PLATE)
205	SEAL, "O" RING (SLIDE SEAL BACK-UP)
206	SLIDE, PUMP
207	SPRING, PIVOT PIN
208	PIN, PIVOT SLIDE
209	SPRING, PUMP SLIDE (INNER/OUTER)
210	SUPPORT, PUMP SLIDE SEAL
211	SEAL, PUMP SLIDE
212	RING, PUMP VANE
213	GUIDE, ROTOR
214	ROTOR, OIL PUMP
215	VANE, PUMP

Fig. 12-106 Exploded view of a THM 200-4R vane pump showing the relative position of its parts. *(Copyrighted material reprinted with permission from Hydra-matic Div., GM Corp.)*

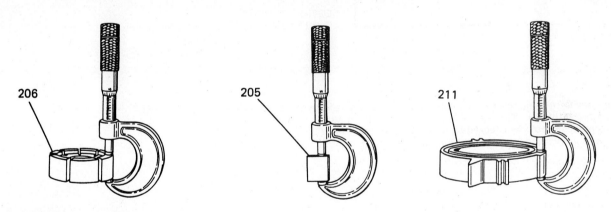

205 VANE, OIL PUMP
206 ROTOR, OIL PUMP
211 SLIDE, OIL PUMP

ROTOR SELECTION		VANE SELECTION		SLIDE SELECTION	
THICKNESS (mm)	THICKNESS (in.)	THICKNESS (mm)	THICKNESS (in.)	THICKNESS (mm)	THICKNESS (in.)
17.593 - 17.963	.7068 - .7072	17.943 - 17.961	.7064 - .7071	17.983 - 17.993	.7080 - .7084
17.963 - 17.973	.7072 - .7076	17.961 - 17.979	.7071 - .7078	17.993 - 18.003	.7084 - .7088
17.973 - 17.983	.7076 - .7080	17.979 - 17.997	.7078 - .7085	18.003 - 18.013	.7088 - .7092
17.983 - 17.993	.7080 - .7084			18.013 - 18.023	.7092 - .7096
17.993 - 18.003	.7084 - .7088				

Fig. 12-107 The pump used in a THM 440 has selective sized rotors, vanes, and slides. During replacement, the sizes should be checked using a micrometer and matched to ensure proper operation. *(Copyrighted material reprinted with permission from Hydra-matic Div., GM Corp.)*

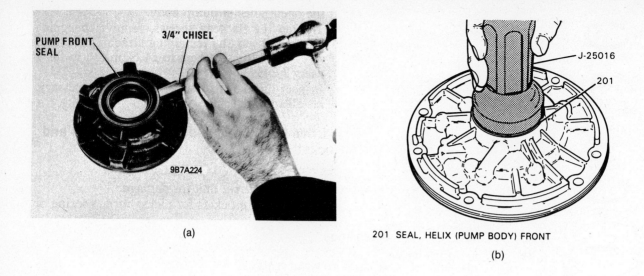

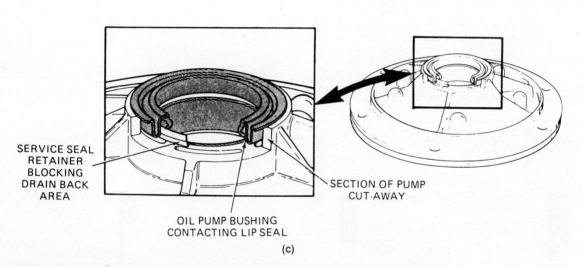

Fig. 12-108 (a) A front pump seal can be removed using a hammer and chisel. (b) A properly sized driver (J-25016) should be used to install the new seal. (c) After installation, seal-to-bushing clearance should be checked to ensure proper fluid flow. *(Copyrighted material reprinted with permission from Hydra-matic Div., GM Corp.)*

have to be exactly aligned before tightening the bolts. The two pump halves can be aligned by using the band shown in Fig. 12-110. This band is available commercially, or a very large screw-type hose clamp can be used in its place. An alternate aligning method is to place the pump upside down in an empty transmission case before tightening the bolts (Fig. 12-111).

Some pumps require alignment of the pump gears as they are installed. This requires a special alignment set that positions the drive gear properly to the pump housing and front bushing (Fig. 12-112). Improper alignment might cause breakage of the gears or housing or damage to the bushing or front seal. In all cases it is good practice to visually center the pump drive gear to the front bushing as the pump is assembled.

At this point, the pump is ready to be reinstalled.

12.9.2 Clutch Assembly Service

The service procedure for most clutch assemblies is to remove the clutch plates, disassemble the return spring(s) and piston; thoroughly clean the parts; inspect the drum, piston, and check ball as well as the bushing and seal ring area; install new seals on the piston; install the piston and return spring(s); install the clutch plates; and check the clutch clearance (Fig. 12-113).

Clutch plate removal is usually easy. A screwdriver or seal pick is used to work the snap ring out, and then the pressure plate and clutch plates can be removed (Fig. 12-114). Always note the position of the pressure plate, the number of lined and unlined plates, and if any cushion plates are used and their position. Clutch stacks will vary, and the easiest way to determine what is in a

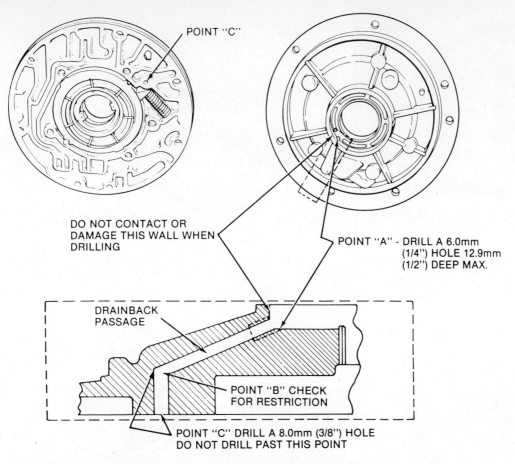

Fig. 12-109 The fluid flowing past the front pump bushing must be allowed to flow back to the pan through a drain-back passage. If this passage is blocked, a drill with the proper size bit should be used to open it. *(Copyrighted material reprinted with permission from Hydra-matic Div., GM Corp.)*

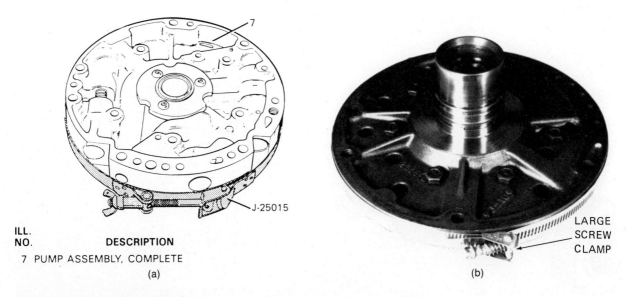

ILL. NO.	DESCRIPTION
7	PUMP ASSEMBLY, COMPLETE

(a)

(b)

Fig. 12-110 When the pump body and cover are the same diameter, they must line up exactly while the bolts are tightened. This can be done using (a) a special tool (J-25015) or (b) a long screw-type clamp. *[(a) Copyrighted material reprinted with permission from Hydra-matic Div., GM Corp.]*

Fig. 12-111 The body and cover of a pump can be aligned correctly by placing them into a case backward while the bolts are tightened.

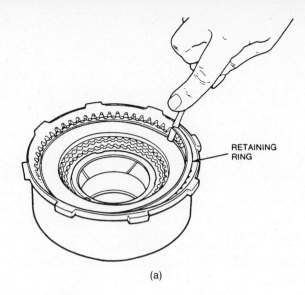

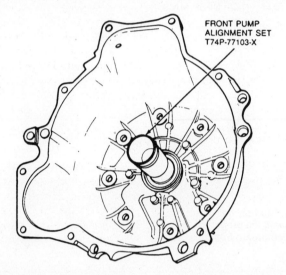

Fig. 12-112 Tool T74P-77103-X used to align the pump gears of an A4LD transmission while the bolts are tightened. *(Courtesy of Ford Motor Company.)*

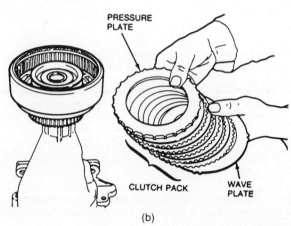

Fig. 12-114 (a) The clutch pack retaining ring is removed using a small screwdriver. (b) With the ring removed, the pressure plate and clutch plates can be lifted out. *(Courtesy of Ford Motor Company.)*

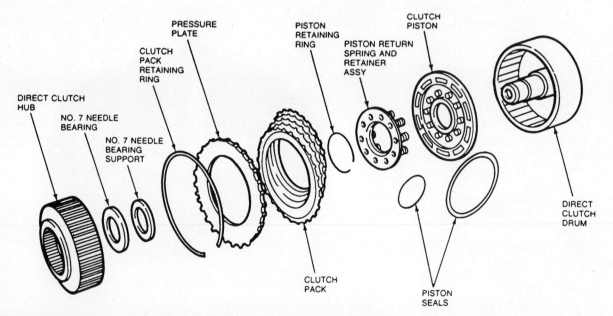

Fig. 12-113 Exploded view of an AOD direct clutch. It is disassembled, cleaned, and carefully inspected during a transmission rebuild. *(Courtesy of Ford Motor Company.)*

particular stack is to look at it as it comes apart. Also, as the plates come out, check them for wear as described earlier.

A spring compressor is required to remove the spring(s) and retainer so the piston can be removed. There is a large variety of spring compressors, and no one unit will work best in all clutches because of the variety of clutches—with a bore in the center, with a shaft in the center, or in a case (Fig. 12-115). Many shops use bench- or floor-mounted compressors because these are usually faster. Be careful when compressing the spring(s), removing the retainer, and allowing the spring(s) to extend. Some springs store quite a bit of energy—one clutch uses an 800-lb (364-kg) release spring—and are rather long. If they fly loose, they can cause injury. Review the safety note following Step 7 of Section 12.3.8. While the springs are being removed on some clutches, be sure to note the number and location of the springs if all of the pockets are not filled with springs (Fig. 12-116).

With the springs and retainer removed, some clutch pistons almost fall out of the bore. Other pistons have to be coaxed with air pressure in the clutch apply oil hole or by slamming the clutch drum onto a block of wood piston side down (Fig. 12-117). With the piston out, remove the old seals, wash the parts in solvent, and air dry them.

Some pistons are made from stamped steel parts that are pressed together, and in a few cases, these parts can shift position, which can cause a seal groove that is too wide or too narrow. Check this width; it should be slightly wider than the seal portion that fits into it. Also place a screwdriver into the groove, and try prying it open wider (Fig. 12-118). If the parts move easily, replace the piston.

Some clutch drums are made from stamped steel parts, and some of these can leak at a weld or a crack next to a weld and cause a pressure loss. If you have cause to suspect this, place the clutch with the bore upward, and fill the piston area with ATF or solvent. If fluid leaks out, you need to replace the drum. Another method is to spray the assembled clutch with a soapy water mix, and air check it using varied low and high pressures.

The check ball should be captured in its cage and still be moveable and free to rattle when shaken (Fig. 12-119). You should be able to move it with either an air blast or a seal pick. If you suspect that you have a leaky check ball, fill the drum with ATF as just described. It should not leak. The check ball assembly can be removed and replaced in stamped steel drums; in the heavier cast drums, the ball can be staked by striking it using a thin pin punch to make it seat tighter.

The bushing and seal ring bore in some clutches or the turbine shaft bushing and seal ring grooves should be smooth and free from scores and other damage (Fig. 12-120). Small imperfections can be removed with a file or by sanding. It should be noted that some turbine shafts have two or more fluid passages, and some passages have a cup plug in it for fluid flow control (Fig. 12-121). These passages should be clean and open and not leak into each other.

When the clutch parts check out and are thoroughly clean, install the new seals on the piston or in the bore; thoroughly lubricate the seals and bore with ATF; carefully install the piston completely into the bore; and replace the return springs and retainer using a spring compressor. If three piston seals are used, be sure the middle one faces in the proper direction. Next install the new lined friction plates and new or reconditioned unlined steel plates in the correct order. Remember that the lined plates must be soaked in ATF; plates and any cushion springs must be stacked in the correct order; and the plates must be thoroughly lubricated during assembly. Don't forget to check the clutch clearance after the snap ring has been replaced.

Most technicians air check a clutch as soon as it is assembled. This simple operation ensures that the clutch works properly. Air pressure is blown into the clutch apply hole; the piston should stroke and squeeze the clutch together and not leak much air (Fig. 12-122). Then air pressure is released, and the piston should release to provide clutch clearance. In some cases, a special air nozzle is necessary to reach the apply hole. Placing the clutch over the pump clutch support shaft allows the use of a standard rubber-tipped air gun. Do not air check a clutch before installing the clutch plates; the piston can travel too far, and the piston seal can catch at the end of its bore.

At this point, a clutch is ready for assembly.

12.9.3 Gearset Service

Servicing gearsets is primarily a visual inspection of the various gears and a side play and rotation check of the planet gears. In some cases there is also an end play check of the assembled gear train to ensure the thrust washers are not worn excessively and that there is the correct clearance between the components (Fig. 12-123).

In most transmissions, the gearset comes out one part at a time. A technician normally gives a quick check of each part as it is removed. After cleanup, the parts are carefully inspected.

All ring and sun gears should be checked for chipped or broken teeth and worn or stripped drive splines (Fig. 12-124). The thrust surfaces at the sides of the gears and any support bushings or bushing surfaces should be checked for scores, wear, or other damage (Fig. 12-125). Drive shells should be checked for stripped splines, damaged lugs, or cracks (Fig. 12-126).

When checking a carrier, the pinion gears must be undamaged and also turn freely. Check for worn or missing pinion thrust bushings (Fig. 12-127). If they appear sloppy, measure the amount of pinion gear end play and/or side clearance using a feeler gauge (Fig. 12-128). Some manufacturers provide pinion gear side clearance specifications, but if none are available, use the rule of thumb of about 0.005 to 0.025 in. (0.127 to 0.635 mm). In some cases, the pinion gear assembly can be removed from the carrier to replace the bearings, gear, or thrust washers, but the needle bearings commonly used can make this a tedious task (Fig. 12-129). Shims are available to tighten the side clearance on some carriers (Fig. 12-130). One variety of these shims is said to improve pinion gear and carrier life because of increased lubrication oil flow.

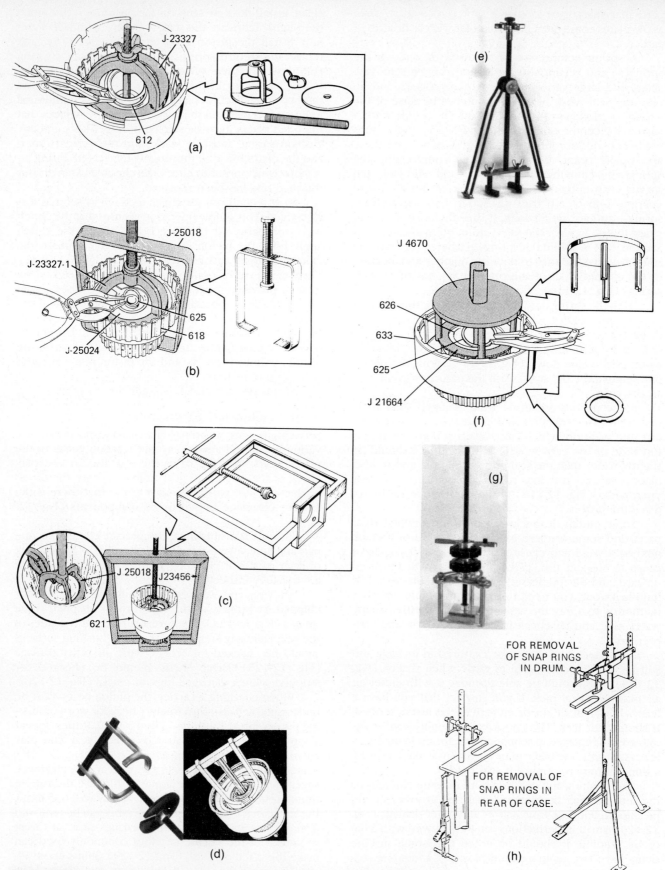

Fig. 12-115 The variety of clutch spring compressors include (a–e) styles that are completely portable, (f) a type to be used with a press, (g) bench-mounted types, and (h) floor-mounted types. *[(a1, b1, c1, f1) Courtesy of Kent-Moore. (a2, b2, c2, f2) Copyrighted material reprinted with permission from Hydra-matic Div., GM Corp.) (d) Courtesy of OTC Tool & Equipment Div. of SPX Corp. (e,g) Courtesy of K-D Tools.]*

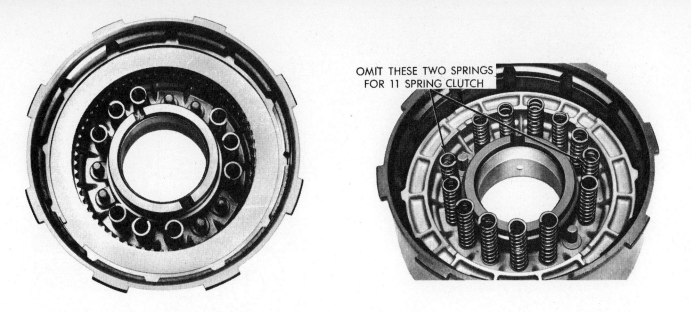

Fig. 12-116 Torqueflite front clutches using a different number of return springs from the number of spring pockets. *(Courtesy of Chrysler Corporation.)*

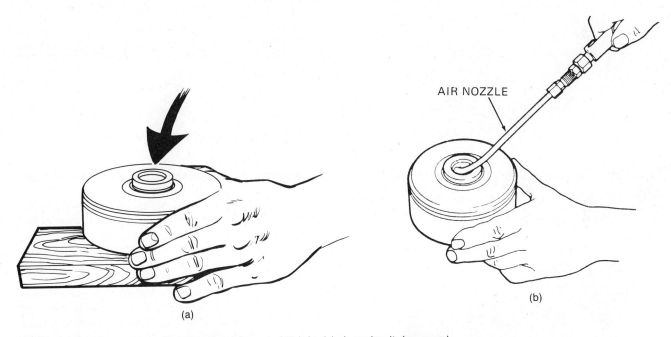

Fig. 12-117 A piston can usually be removed from a clutch by (a) slamming it downward onto a block of wood or (b) using air pressure. *[(b) Courtesy of Ford Motor Company.]*

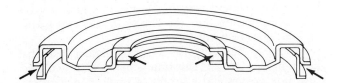

Fig. 12-118 Piston made from several stamped steel parts that can separate. Test this by inserting a screwdriver and trying to pry the groove open with a light pressure. *(Copyrighted material reprinted with permission from Hydra-matic Div., GM Corp.)*

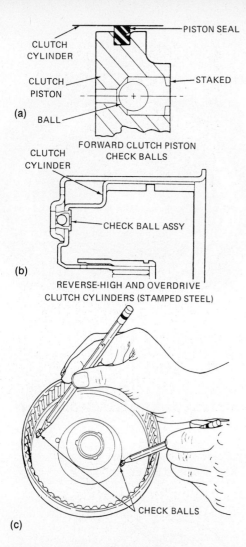

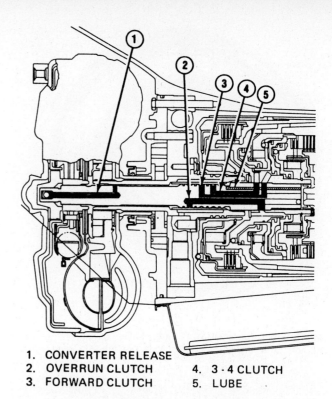

1. CONVERTER RELEASE
2. OVERRUN CLUTCH
3. FORWARD CLUTCH
4. 3-4 CLUTCH
5. LUBE

Fig. 12-121 The turbine shaft of a THM 700 has five fluid passages. Note that the converter release passage has a check ball in it. *(Copyrighted material reprinted with permission from Hydra-matic Div., GM Corp.)*

12.9.4 Valve Body Service

Despite its complexity, the valve body is one of the more reliable parts in a transmission. This is probably because the valves are so well lubricated, and in a way, they normally don't do much—only move slightly once in a while. Most valve body service operations consist of disassembly, cleaning, checking for free movement, reassembly, and in a few transmissions, adjusting the pressure regulator valve (Fig. 12-131). The valve(s) in a governor are serviced in the same manner.

The biggest problem during valve body repair is getting it back together with everything in the right order and location. Most technicians complete a valve body repair as quickly as possible; the longer it takes, the easier it is to forget which spring goes where or how it fits on a particular valve (Fig. 12-132). In many cases, the service manual is only an aid. Each transmission model and version is slightly different, and these differences are often in the valve body—often just a spring change. You should remember that in taking a valve body apart, you become the expert in how that particular unit goes back together.

Some technicians use a spring holder like that shown in Fig. 12-133. This unit is made from two electrical box covers and a group of small, 3-in.-long machine screws. Each screw location on the holder is numbered. As a valve and spring are removed, the spring is placed on a screw and the number is written next to the spring in the manual illustration. This does two

Fig. 12-119 (a) Check ball in a cast drum is placed into a drilled hole and locked in place by staked metal. (b) A stamped drum uses a cartridge check ball assembly that is pressed into the drum. (c) The balls should move when checked using a pencil point. *(Courtesy of Ford Motor Company.)*

Fig. 12-120 Bushing appears usable and the two sealing ring areas (arrows) show normal operation.

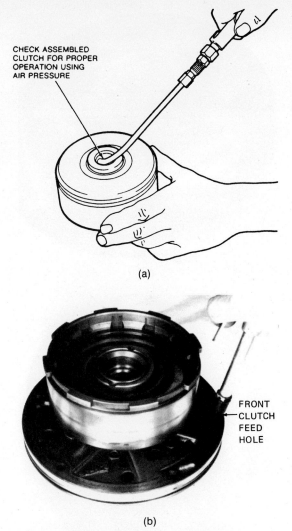

Fig. 12-122 (a) With the proper tip on the air gun, a clutch can be air checked to test the seals and clutch operation. (b) A clutch can also be checked by installing it over a pump assembly. *[(a) Courtesy of Ford Motor Company.]*

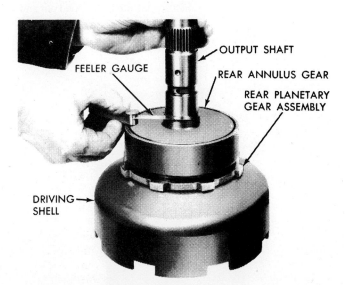

Fig. 12-123 The end play of a Torqueflite gear train is checked using a feeler gauge. Excessive end play indicates worn thrust washers and thrust surfaces. *(Courtesy of Chrysler Corporation.)*

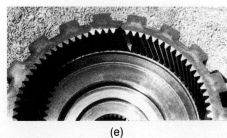

Fig. 12-124 Planetary gear damage shows up as (a–d) wear and (e) cracks.

365

Fig. 12-125 Gear train components show severe wear and scoring in the thrust surface areas.

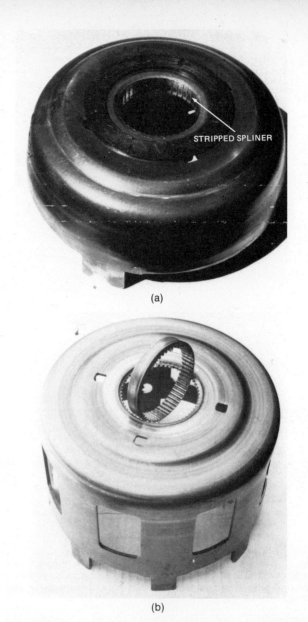

Fig. 12-126 Sun gear drive shell splines are (a) stripped out and (b) ripped out.

Fig. 12-127 Pinion gear has too much side clearance; it should be repaired or the carrier should be replaced.

things: It helps locate a certain spring and helps teach the technician the name of the spring and valve with which it works. As each spring comes out, a mental or written note should be made of how it fits the valve, how many coils it has, and if it has a particular color.

Several methods are used to retain the valve(s) in a bore. Many units use a cover plate that holds one or more valves (Fig. 12-134). Removal of the retaining screws allows removal of the plate, valve(s), and spring(s). Many valves use a plug or sleeve at the end

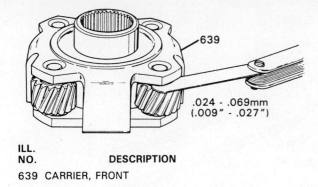

ILL. NO.	DESCRIPTION
639	CARRIER, FRONT

Fig. 12-128 Planet gear side clearance is measured using a feeler gauge. The clearance should be between 0.024 and 0.069 in. on the THM 200 front carrier shown. *(Copyrighted material reprinted with permission from Hydra-matic Div., GM Corp.)*

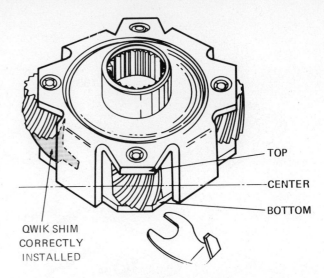

Fig. 12-130 Aftermarket shim developed to reduce the amount of side clearance in some Hydramatic carriers. It is installed by sliding it into position and bending a tab to lock it in place.

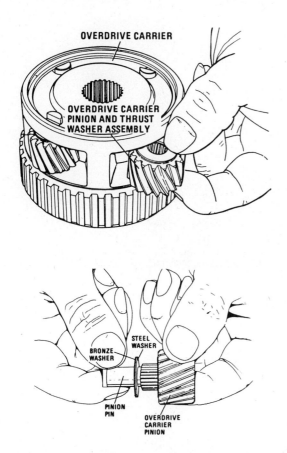

Fig. 12-129 Some carriers can be rebuilt so a new pinion pin, needle bearings, and thrust washers can be installed. This should correct the amount of gear side clearance. *(Copyrighted material reprinted with permission from Hydra-matic Div., GM Corp.)*

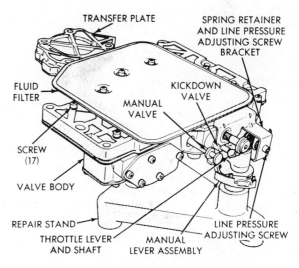

Fig. 12-131 Torqueflite valve body mounted on a repair stand prior to disassembly. Note that it has an adjustable pressure regulator valve. *(Courtesy of Chrysler Corporation.)*

of each bore and retain this plug/sleeve with a keeper—pin, plate, or key (Fig. 12-135). The keeper cannot come out while the valve body is in place on the transmission, and the valve's spring pressure keeps it in place while the valve body is removed. Pushing the plug/sleeve slightly inward allows the keeper to be pulled out easily, and then the plug/sleeve, valve(s), and spring(s) will come out. Some valve bodies use a coiled spring pin to hold the valve plug/sleeve in place (Fig. 12-136). In some cases, this coiled pin can be pulled out by gripping it with a pair of pliers; in other cases, you will need to make an extractor tool. Grind the flats of a small [about $\frac{3}{32}$-in. (3-mm)] Allen wrench or the end of a No. 49 drill bit so it tapers to a point, as shown in Fig. 12-137. Then tap this pointed end into the coiled pin; rotate the pin in a direction that wraps the pin tighter; and as the pin rotates, lift it out of the bore.

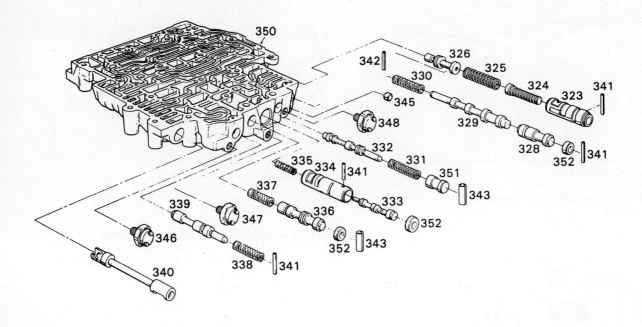

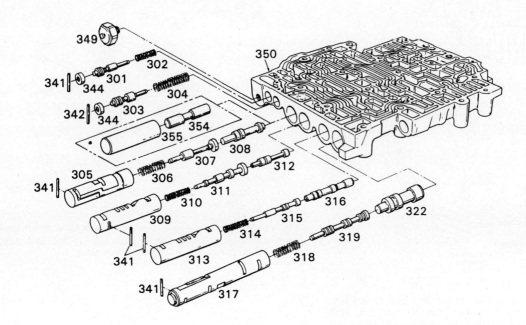

*USED ON VEHICLES HAVING ECM CONTROL OF T.C.C. APPLY

Fig. 12-132 A disassembled THM 700 valve body. The valves from one side are at the top, and the valves from the other side are at the bottom. *(Copyrighted material reprinted with permission from Hydra-matic Div., GM Corp.)*

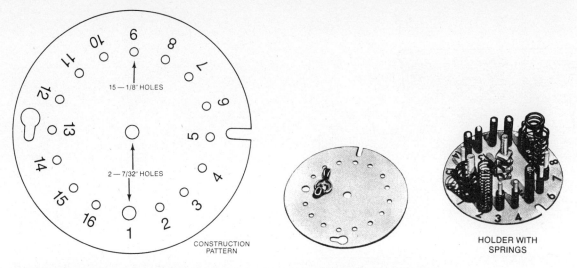

Fig. 12-133 The simple valve body spring holder shown can be made from two electrical box covers, 2¼ × 2-in. stove bolts with nuts and 15⅛ × 1½-in. stove bolts and nuts. As a valve is removed from the valve body, the spring is placed over a numbered peg, and the number of the spring is written on a blowup picture of the valve body. *(Courtesy of Ford Motor Company.)*

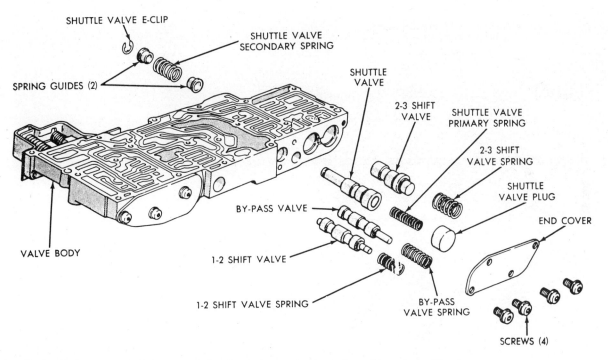

Fig. 12-134 Removal of four screws holding the end cover in place allows four valves with springs to be removed. *(Courtesy of Chrysler Corporation.)*

Most technicians place a lint-free shop cloth(s) or a carpet scrap under the valve body while disassembling it. The cloth helps keep the check balls, screws, and pins from rolling away and might prevent a nick or dent in a valve if one happens to drop.

The major reason for disassembling a valve body is so it can be cleaned thoroughly and to make sure that all valves are working freely and not sticking. The standard test for a sticking valve is to hold the valve body so the bore is vertical. In this position, a valve should fall freely from one end of the bore to the other—at least it should fall through the area of normal valve movement (Fig. 12-138). Any valve that doesn't fall freely is sticking, and this can be a fault of the valve, the bore, or both.

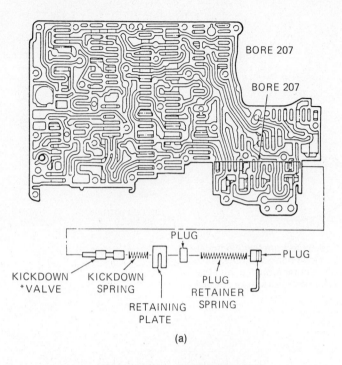
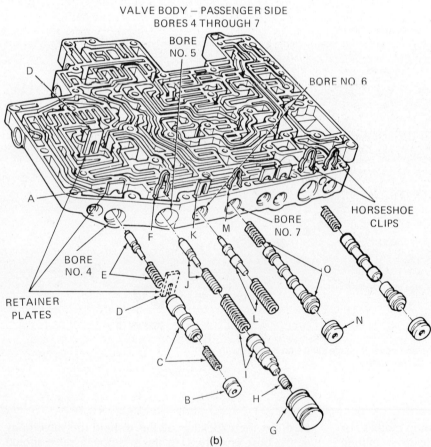

Fig. 12-135 (a) A4LD kickdown valve assembly is held into bore 207 by an L-shaped pin and a retaining plate. (b) Horseshoe clips and retainer plates are used to retain the valves. *(Courtesy of Ford Motor Company.)*

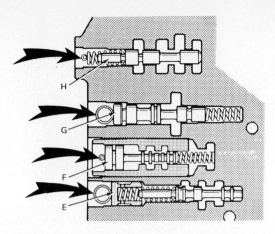

Fig. 12-136 Valves retained by coiled retaining pins (arrows). *(Copyrighted material reprinted with permission from Hydra-matic Div., GM Corp.)*

Fig. 12-138 The standard test for a valve is to hold the valve and bore in a vertical position. The valve should fall out of the bore so be ready to catch it.

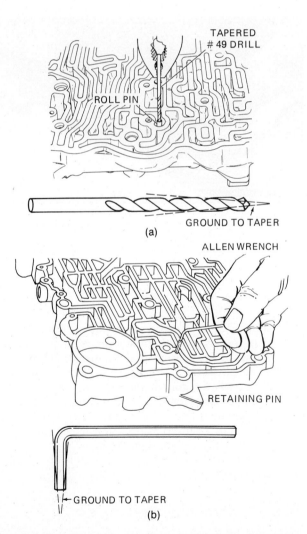

Fig. 12-137 (a) Drill bit or (b) Allen wrench can be ground to a taper so they enter a roll pin and can be turned to rotate the roll pin and lift it out. *(Copyrighted material reprinted with permission from Hydra-matic Div., GM Corp.)*

Carefully inspect the valve and valve body for varnish—a light brown or golden brown coating. It can be cleaned off with carburetor/automatic choke cleaner used either in aerosol spray can form or container with dip basket (Fig. 12-139). Soak the valves, springs, and valve body for 5 to 15 minutes, rinse them in hot water, and then dip the valves and springs in ATF and air dry the valve body.

If a valve is still sticky, inspect it for nicks or dents. The raised metal can be removed using a fine-grit knife-sharpening stone. Lay the valve on the stone and carefully rotate it until all traces of raised metal are gone (Fig. 12-140). Be careful not to make a flat or groove on the valve or round off the corners of a land.

If a valve is smooth but still sticks in the bore, carefully examine the bore for debris or nicks that might cause raised metal. Small metal particles tend to become embedded in the bores of aluminum valve bodies. One cure is to drive the metal completely into the aluminum as shown in Fig. 12-141. A few sharp taps on the valve is often all that is necessary. If this doesn't free up the valve, clean the bore using crocus cloth or a lapping compound. Crocus cloth is a very fine grit abrasive cloth. Wrap a section into a tube or roll it with the grit outward, slide this tube into the valve bore, and then rotate it in a direction that tries to unwrap the tube (Fig. 12-142). After turning the crocus cloth tube several revolutions, clean the bore and retry the valve fit. Some technicians make a lapping compound by mixing household scouring powder and ATF to make a thin paste. This paste is spread onto the valve and in the bore, and then the valve is stroked and rotated in the bore (Fig. 12-143). The valve and bore are then cleaned in solvent and air dried, and the valve fit is tried again.

After all the valves, springs, and valve body are cleaned and the valves move freely in their bores, the valve body can be reassembled. Each spring should be checked to make sure there are no damaged or distorted coils, and each valve should be dipped in ATF before

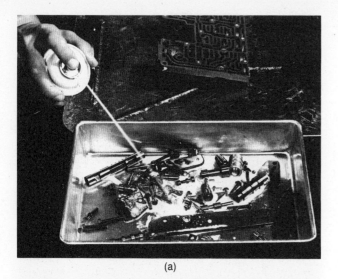

(a)

(b)

Fig. 12-139 (a) Valve body and its valves are being sprayed with a carburetor/choke cleaner; (b) a cleaning brush is being used to loosen any deposits in the body passages. After a soak period they will be rinsed in hot water and then solvent and air dried. After cleaning, they will be dipped in ATF and reassembled. *(Courtesy of Ford Motor Company.)*

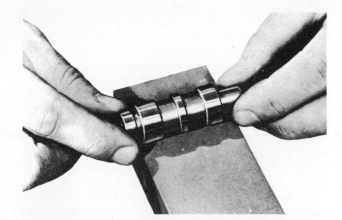

Fig. 12-140 A valve can be trued by placing it on a sharpening stone and rotating it. This method leaves the valve with sharp edges as expected. *(Courtesy of Ford Motor Company.)*

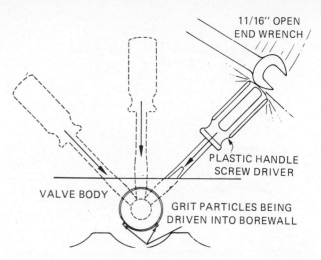

Fig. 12-141 (a) Sometimes a valve in an aluminum valve body will stick in its bore because grit will embed in the walls of the bore. (b) This valve can often be freed by striking it using a screwdriver and wrench as shown.

Fig. 12-142 A scored bore can be cleaned by inserting a small roll of crocus cloth into the bore and rotating it in a counterclockwise direction (in this particular case). *(Courtesy of Ford Motor Company.)*

installation. The reassembly procedure is generally the reverse of the disassembly procedure making sure each valve moves after its installation. Be sure to tighten each retainer screw to the correct torque (Fig. 12-144).

The pressure regulator valve spring is adjustable in most Chrysler Corporation transmissions, and changing its length changes line pressure (Fig. 12-145). The adjustment can be checked using a small ruler and, if necessary, changed using an Allen wrench.

The valve body should now be ready to install in the transmission.

12.9.5 Differential Service

Transaxle differentials should be checked to make sure that the differential gears, thrust washers, and differential pinion shaft are in good condition (Fig. 12-146). A differential can usually be disassembled by removing the

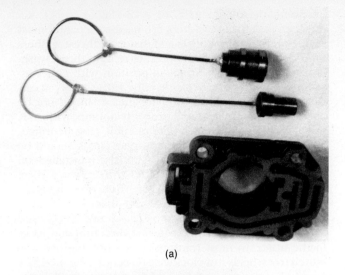

(a)

(b)

Fig. 12-143 (a) A wire handle has been attached to a valve so it can be used to lap the bore of the governor. (b) A mixture of ATF and scouring powder is used as the lapping compound.

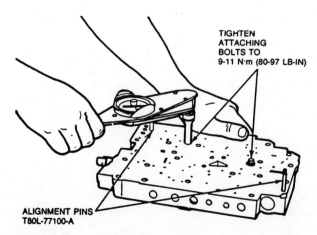

Fig. 12-144 As a valve body is assembled, the screws should be tightened to the correct torque. The alignment pins are being used to keep the separator plate in the correct position. *(Courtesy of Ford Motor Company.)*

Fig. 12-145 Torqueflite pressure regulator valve should be adjusted so there is $1\frac{5}{16}$ in. between the spring retainer and the valve body. *(Courtesy of Chrysler Corporation.)*

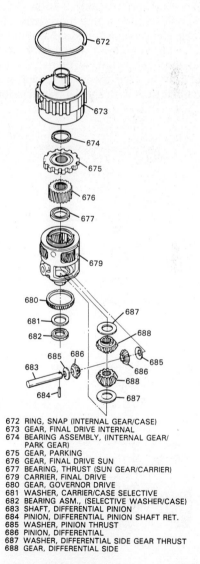

672 RING, SNAP (INTERNAL GEAR/CASE)
673 GEAR, FINAL DRIVE INTERNAL
674 BEARING ASSEMBLY, (INTERNAL GEAR/PARK GEAR)
675 GEAR, PARKING
676 GEAR, FINAL DRIVE SUN
677 BEARING, THRUST (SUN GEAR/CARRIER)
679 CARRIER, FINAL DRIVE
680 GEAR, GOVERNOR DRIVE
681 WASHER, CARRIER/CASE SELECTIVE
682 BEARING ASM., (SELECTIVE WASHER/CASE)
683 SHAFT, DIFFERENTIAL PINION
684 PINION, DIFFERENTIAL PINION SHAFT RET.
685 WASHER, PINION THRUST
686 PINION, DIFFERENTIAL
687 WASHER, DIFFERENTIAL SIDE GEAR THRUST
688 GEAR, DIFFERENTIAL SIDE

Fig. 12-146 A disassembled THM 440 final drive and differential assembly. *(Copyrighted material reprinted with permission from Hydra-matic Div., GM Corp.)*

373

lock pin and driving the differential pinion shaft out. This allows removal of the gears and thrust washers (Fig. 12-147). The gears should be inspected for chipped or broken teeth and scoring on the bearing surfaces. The thrust washers and differential pinion shaft should be checked for wear and scoring.

Some differentials are a combination planet carrier that includes a set of planet pinions along with the differential. Like those inside the transmission portion, these pinion gears must turn freely on their shafts and not have an excessive amount of end play (Fig. 12-148).

12.10 TRANSMISSION ASSEMBLY

The reassembly procedure for an automatic transmission varies depending on the make and model of the transmission. A service manual should be followed to make sure that the parts go together in the right order and position and that important checks like end play are not skipped (Figs. 12-149 and 12-150).

Reassembly begins with a thoroughly clean, reconditioned case and reconditioned subassemblies. All parts should be air dried and not wiped with cloth or paper towels (so there is no lint on them). As the parts are assembled into the case, the bushings and shaft surfaces should be lubricated with special transmission assembly lubricant, petroleum jelly, or ATF (Fig. 12-151). Remember that thrust washers are used to separate most of the internal components and that most thrust washers interlock. It is especially important to hold thrust washers and Torrington bearings in position with either assembly lubricant or petroleum jelly. These parts can easily slip out of position. Ordinary grease should not be used for this because it will probably not mix with ATF, might stay in a lump, and could possibly clog an orifice.

On some transaxles, an end play check should be made as the final drive and differential are installed; on others, the differential and transfer shaft bearings must be adjusted to the correct preload. These operations are described in the appropriate service manual.

As a roller clutch is assembled, it should be checked to make sure it freewheels in one direction and locks in the other. In most transmissions a holding one-way clutch freewheels in a clockwise direction (Fig. 12-152). Remember the power flows and the purpose for the one-way clutch. In some newer transmissions, the freewheel direction of a one-way clutch is hard to remember, but a service manual often illustrates how to check it (Fig. 12-153).

As a gearset is assembled, each part should turn freely without binding (except for one-way clutches in one direction). It is often possible, by turning the input/turbine shaft and holding the correct reaction member, to make the gearset produce each of its different gear ratios.

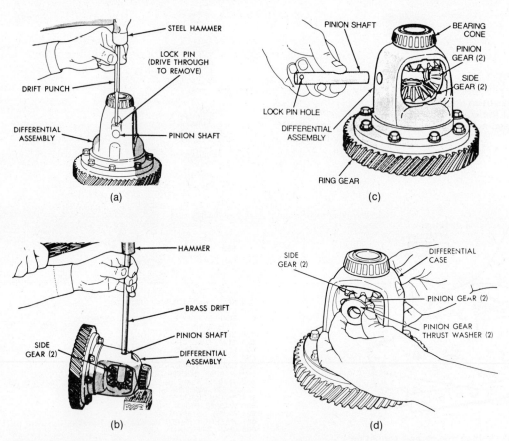

Fig. 12-147 The normal procedure to disassemble a differential is (a) remove the pinion shaft lock pin, (b & c) remove the pinion shaft, and (d) roll the differential gears around so they can be removed along with the side gears. *(Courtesy of Chrysler Corporation.)*

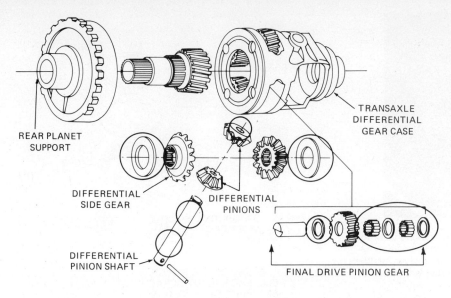

Fig. 12-148 Normally wear occurs in a differential at the circled areas. They should be inspected while the differential is disassembled. *(Courtesy of Ford Motor Company.)*

As a center support is installed, don't forget to install the anticlunk spring in the proper position if one is used. Also, usually a clearance or end play check ensures that the portion captured by the center support is free to operate but does not have excessive clearance (Fig. 12-154).

Some clutch assemblies are a little difficult to install because of the close fit of the plates over the hub. Swinging the clutch in a kind of spiral direction (as shown in Fig. 12-155) usually works the plates over the hub. Often you can count the plates as they line up and drop over the hub to determine when they are all on. You can also determine if a clutch is completely installed by measuring its position in the transmission case or by lifting the clutch straight up, about $\frac{1}{4}$ in. (6 mm), and then dropping it (Fig. 12-156). A solid "thunk" indicates the clutch is dropping onto its thrust washer, which is good. If the clutch is dropping onto the edge of a misaligned plate, it makes a quieter sound.

The final assembly point of most gear trains is the pump. A new pump gasket must be used, and many technicians use two guide pins to hold the gasket in place and align the pump as it is dropped into place (Fig. 12-157). The internal end play is always checked and adjusted at this time.

Two common methods are used to check this: a special H-shaped gauge or a dial indicator. The special

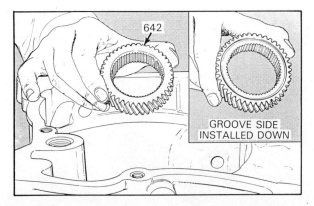

Fig. 12-149 Modern transmissions like the THM 125 shown have parts that must be positioned correctly. A service manual should be followed during assembly to ensure that they fit. In this case, the groove identifies the bottom of the gear. *(Copyrighted material reprinted with permission from Hydra-matic Div., GM Corp.)*

Fig. 12-150 One of several end play checks in a THM 125 is to determine whether the low roller clutch race has the correct thrust washer. *(Copyrighted material reprinted with permission from Hydra-matic Div., GM Corp.)*

375

Fig. 12-151 A generous layer of petroleum jelly is being used to place a thrust washer so it won't slip to the wrong position as the pump is installed. After the end play check, a layer for lubrication is spread on the other side of the thrust washer.

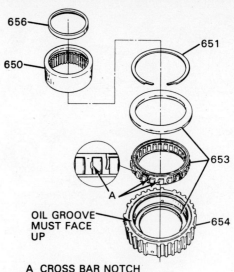

A CROSS BAR NOTCH
650 RACE, INPUT SPRAG INNER
651 RING, SNAP (SPRAG)
653 SPRAG ASSEMBLY, INPUT CLUTCH
654 RACE, INPUT SPRAG OUTER
656 DAM, INPUT SPRAG RACE LUBE

(a)

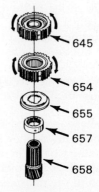

SPRAG & ROLLER CLUTCH ASSEMBLY

(b)

Fig. 12-153 (a) A sprag clutch should be disassembled for cleaning. (b) It must be reassembled so it operates in the proper direction. The THM 440 sprag and roller clutch assembly shown has two units that operate in opposite directions. (*Copyrighted material reprinted with permission from Hydra-matic Div., GM Corp.*)

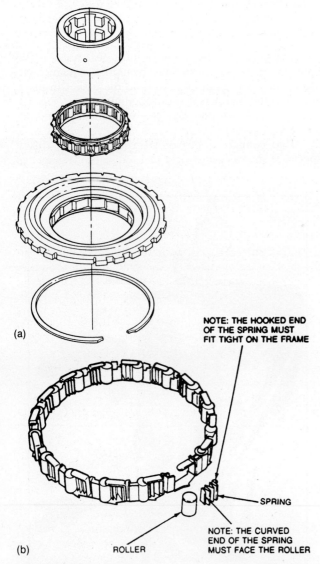

Fig. 12-152 (a) A one-way clutch should be disassembled for cleaning and inspection. (b) Note how the accordion spring should be positioned during reassembly. [(a) *Copyrighted material reprinted with permission from Hydra-matic Div., GM Corp.* (b) *Courtesy of Ford Motor Company.*]

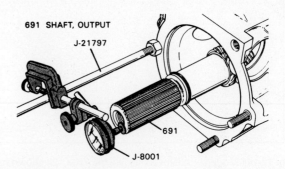

Fig. 12-154 As the center support is installed in a THM 400, the rear unit end play should be checked and corrected if necessary. (*Copyrighted material reprinted with permission from Hydra-matic Div., GM Corp.*)

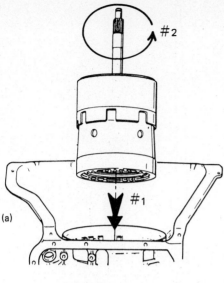

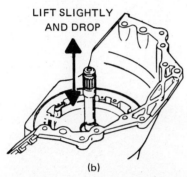

Fig. 12-155 The clutch shown is often difficult to install because the clutch plates have to fit over a hub. (a) Lower it so it rests on the plates and (b) rotate the turbine shaft in a spiral fashion. You should feel and hear it drop as each plate works into alignment. *(Copyrighted material reprinted with permission from Hydra-matic Div., GM Corp.)*

Fig. 12-156 Many manufacturers provide clutch-installed height specifications that make it easy to determine if a clutch is installed correctly. *(Copyrighted material reprinted with permission from Hydra-matic Div., GM Corp.)*

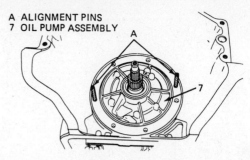

Fig. 12-157 Two alignment pins are being used to hold the gasket in position and align the pump as it is installed. *(Copyrighted material reprinted with permission from Hydra-matic Div., GM Corp.)*

gauge is simpler and faster. Place the front gasket in position; adjust the gauge to fit the pump mounting surface; and then adjust the gauge rod to touch the thrust surface on the clutch (Fig. 12-158). Now turn the gauge over, and place it onto the pump. The clearance between the gauge rod and the thrust surface of the pump is the room you have for the thrust washer plus the amount of clearance for end play. If the end play is measured using a dial indicator, install the pump using the gasket but without the O-ring, install and tighten two bolts, and measure the end play as described in section 12.3.5 (Fig. 12-159). Then remove the pump, and correct the end play by selecting the correct thrust washers or shims. The selective thrust washer is normally No. 1, the one in front. If there is no end play and the transmission is bound, there is a probability that a thrust washer has slipped out of position or a clutch is not completely installed.

Most technicians prefer to adjust the end play on the tight side of the specifications. This reduces the thrust direction motions of the gear train and tends to hold the clutch and gear train members straighter in the case. For example, let's say you measure 0.038 in. (0.96 mm) of end play, and there is a No. 69 selective thrust washer that is 0.087 to 0.091 in. (2.21 to 2.31 mm). The end play specification is 0.005 to 0.036 in. (0.13 to 0.92 mm) so the end play is slightly big (Fig. 12-160). According to most technicians, the transmission operates better if the end play is reduced—as much as 0.033 in. (0.83 mm) [0.038 − 0.005 = 0.033 (0.96 − 0.13 = 0.83 mm)]. The end play can best be reduced by using a No. 74 washer [0.122 − 0.089 = 0.033 and 0.038 (measured end play) − 0.033 = 0.005 inch] [3.09 − 2.26 = 0.83 mm and 0.96 (measured end play) − 0.83 = 0.13 mm]. Any of the thrust washers between Nos. 69 and 74 can be used because they reduce the amount of end play to within specifications.

The time to install the band servos and adjust or check the band clearance is after the pump is installed. If one of the ends of a band is heavier than the other, the heavy end is placed at the anchor. Once this is done, many technicians air check the clutch assemblies, servos, and governor as described in Chapter 10. On transmission reassembly, the lubrication passages can also be checked. With the extension housing removed, blow air into the cooler return port; you should be able to hear

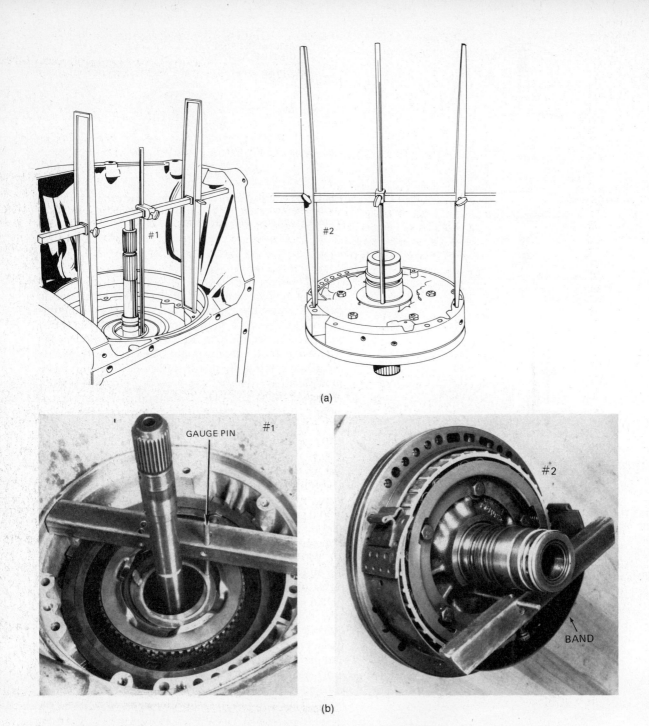

Fig. 12-158 (a) An H gauge is being used to check end play. (b,c) The shop-made gauge is similar to that shown in (a). It is first adjusted to the distance of the thrust washer surface in (b). Then this distance is transferred to the thrust washer on the pump in (b). The amount of clearance between the gauge rod and the thrust washer is the amount of end play. The gauge rod length compensates for the width of the bar and the C6 band.

air escape from the rear bushing. Each of the components should respond with the correct motion or sound and with a minimum of air leakage. If the air checks produce the correct results, the valve body, filter, and pan can be installed.

In review, some general rules for transmission assembly are:

1. New or reconditioned replacement components should always be used.

2. New rubber seals, new or very good sealing rings, and new gaskets and metal-clad seals should be used.

3. All friction elements should be soaked in ATF.

4. All clutches and bands should have the correct clearance.

5. All one-way clutches should freewheel in the proper direction.

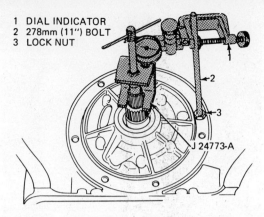

Fig. 12-159 A dial indicator set up to measure the end play of a THM 700 transmission. *(Copyrighted material reprinted with permission from Hydra-matic Div., GM Corp.)*

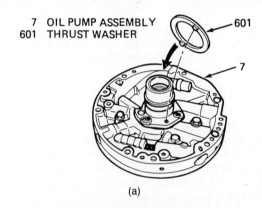

TRANSMISSION END PLAY WASHER SELECTION CHART

WASHER THICKNESS		I.D.
1.87 - 1.97 mm	(.074" - .078")	67
2.04 - 2.14 mm	(.080" - .084")	68
2.21 - 2.31 mm	(.087" - .091")	69
2.38 - 2.48 mm	(.094" - .098")	70
2.55 - 2.65 mm	(.100" - .104")	71
2.72 - 2.82 mm	(.107" - .111")	72
2.89 - 2.99 mm	(.113" - .118")	73
3.06 - 3.16 mm	(.120" - .124")	74

(b)

Fig. 12-160 (a) The size of thrust washer 1 (601) will be changed if the amount of end play is wrong in the THM 700. (b) One of eight different thickness washers can be used. *(Copyrighted material reprinted with permission from Hydra-matic Div., GM Corp.)*

6. All support bushings and thrust washers must be in good condition and assembled correctly.

7. All moving parts should be lubricated with ATF or transmission assembly lubricant.

8. All end plays should be correctly adjusted.

9. All snap rings must be installed completely into their grooves.

10. All bolts must be tightened to the correct torque.

11. All bolt threads should be dipped in ATF before installing to reduce thread galling.

12. Each hydraulic component should air check correctly.

13. The gear train should rotate without excessive drag.

12.10.1 Torque Converter Installation

With the transmission assembled, the torque converter can be installed. Remember that the converter should be checked and cleaned as described in Chapter 13 or rebuilt. Many technicians pour about 1 quart of ATF into the converter before installation. Remember that the converter must fill before fluid goes to the cooler and back to the transmission lubrication passages.

The converter must align at three points: turbine spline, stator spline, and pump. Most converters use two flats or notches that engage the pump drive gear, and these notches should be positioned to align with the pump drive gear before installation (Fig. 12-161). If the drive gears do not line up, they will be "stacked" on top of the pump gear. If the transmission is installed with a stacked converter and the engine is started, the pump and possibly the converter hub will be ruined.

Make sure that the converter is completely installed before connecting the converter to the engine. Some manufacturers provide a dimension for checking this (Fig. 12-162). If no dimension is available, pump tang engagement can be checked by noting how deep the converter is in the housing, rotating the converter a turn or so, lifting the converter back out, and noting if the drive tangs are aligned with the location of the converter hub (Fig. 12-163). If they are not aligned, the converter was not completely installed and will need to go deeper into the transmission. A more common method is to check the converter when it is connected to the flexplate, as described in the next section.

12.11 TRANSMISSION INSTALLATION

Replacing a transmission is another operation that is the reverse of the removal procedure. The last operation in the removal is usually the first step in the installation. This is normally to lift the transmission in place and bolt it to the engine, being sure to tighten the bolts to the correct torque. Next the supports and mounts are placed in position and the bolts are installed and tightened so the transmission and engine are safely supported.

The final proof of correct converter installation is if you can move the converter forward and rotate it to align with the flexplate. If the converter is jammed in place, it is stacked, and the transmission must be removed so the converter can be aligned and installed correctly (Fig. 12-164). Don't forget to align the converter to the flexplate using the marks you made during removal and to tighten the mounting bolts to the correct torque.

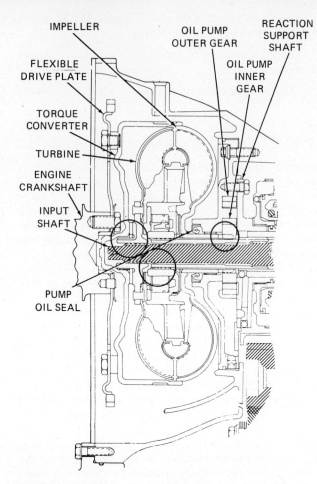

Fig. 12-161 A torque converter must align with the pump, stator support, and turbine shaft (circled) after it is installed into the transmission. It should be possible to rotate it for connection with the flexplate after the transmission has been bolted onto the engine. *(Courtesy of Chrysler Corporation.)*

Fig. 12-163 The converter must engage the pump drive tangs (circled). The position of the drive tang can be marked on the case so the converter can be properly aligned as it is installed.

The rest of the transmission installation is rather routine—replacing the cooler lines, shift linkage, TV linkage, electrical connectors, speedometer cable, and drive shaft(s). When this is completed, the transmission should be filled with the correct amount of ATF and the engine started. If you are unsure of the correct amount of fluid, add enough fluid to cause it to read slightly high on the dipstick, and adjust it to the correct level after starting the engine and operating the transmission in the different gear positions. Next complete your transmission overhaul with a road test to ensure that it operates correctly.

Shops that specialize in rebuilding units that are sold to other shops for installation often use a transmission test machine for their quality-control checks (Fig. 12-165). These devices use an engine to drive the transmission while a power absorber provides a load on the drive shaft(s). They are made so a realistic test of the transmission can be made in a short time period.

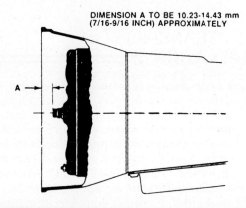

Fig. 12-162 Some manufacturers provide the converter-installed dimension so it is easy to determine if the converter is installed correctly. *(Courtesy of Ford Motor Company.)*

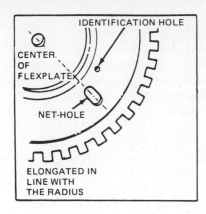
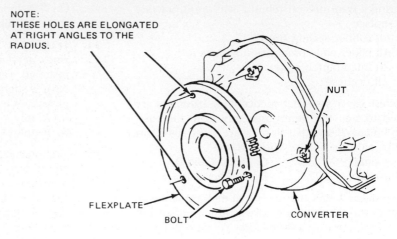

1. MOVE CONVERTER FORWARD TO CONTACT ATTACHING SURFACE ON FLEXPLATE PRIOR TO TIGHTENING BOLTS.

2. ALIGN SLOT IN FLEXPLATE THAT HAS AN IDENTIFICATION HOLE NEAR IT, WITH AN ATTACHMENT HOLE IN CONVERTER. INSTALL BOLT AND NUT AND TIGHTEN TO SPECIFIED TORQUE. TIGHTEN ALL REMAINING BOLTS TO SPECIFIED TORQUE AS THEY ARE INSTALLED.

Fig. 12-164 Some manufacturers recommend that the flexplate-to-converter bolts be tightened in a particular order. *(Copyrighted material reprinted with permission from Hydramatic Div., GM Corp.)*

Fig. 12-165 Transmission test machines are available for thoroughly testing a unit before it is installed or sold to someone else for installation. The transmission is driven by an automotive engine (left) while a power absorber (right) is used to simulate the load of the car. *(Courtesy of Intercont Products.)*

REVIEW QUESTIONS

The following questions are provided so you can check the facts you have just learned. Select the response that best completes each statement.

1. Technician A says that the torque converter is usually left on the engine to save time when removing a transmission. Technician B says that it is more difficult to remove a transaxle than a transmission. Who is right?
   A. A only
   B. B only
   C. Both A and B
   D. Neither A nor B

2. Technician A says that the transmission should be thoroughly cleaned right after removal. Technician B says that most transmission-holding fixtures allow the technician to move the transmission to the best working position. Who is right?
   A. A only
   B. B only
   C. Both A and B
   D. Neither A nor B

3. Technician A says that some servos and accumulators require special holding tools or spring compressors to hold the internal spring captive during disassembly. Technician B says that an end play check should be made before removing the transmission pump. Who is right?
   A. A only
   B. B only
   C. Both A and B
   D. Neither A nor B

4. Technician A says that excessive transmission end play is usually caused by a stretched case. Technician B says that loose pump bolts are the major cause of excess end play. Who is right?
   A. A only
   B. B only
   C. Both A and B
   D. Neither A nor B

5. During a transmission rebuild, it is good practice to replace
   A. All rubber sealing rings
   B. All paper and composition gaskets
   C. All lined clutch plates
   D. All of these

6. Technician A says that special slide hammers are used to pull the pump assembly from the case. Technician B says that air pressure in the right port helps pull the pump. Who is right?
   A. A only   C. Both A and B
   B. B only   D. Neither A nor B

7. Technician A says that the usual method of cleaning transmission parts is to wash them in solvent and dry them with a shop cloth. Technician B says that spray washers are fast and effective ways of cleaning parts. Who is right?
   A. A only   C. Both A and B
   B. B only   D. Neither A nor B

8. After the case has been cleaned it should be checked for
   A. Damage, warpage, and cross leaks at the worm tracks
   B. A worn governor bore
   C. Damaged bolt threads
   D. All of these

9. Technician A says that worn or damaged bushings are repaired by driving them out and replacing them with new bushings. Technician B says that some bushings are removed by cutting threads in them so a puller bolt can be used. Who is right?
   A. A only   C. Both A and B
   B. B only   D. Neither A nor B

10. Technician A says that it is good practice to replace all of the friction materials (lined plates, unlined plates, and bands) when rebuilding a transmission. Technician B says that lined friction material must be soaked in ATF before it is installed. Who is right?
    A. A only   C. Both A and B
    B. B only   D. Neither A nor B

11. Technician A says that an unlined steel clutch plate can be reused if it is smooth and shiny and only has a little bit of "potato chip" warpage. Technician B says that a lined plate that shows any darkening should not be reused. Who is right?
    A. A only   C. Both A and B
    B. B only   D. Neither A nor B

12. Clearance in a multiple-disc clutch is adjusted using any of these methods except
    A. Selective sized clutch pressure plates
    B. Selective sized pressure plate snap rings
    C. Selective sized clutch housings
    D. Selective sized steel plates

13. Working the lip of a new sealing ring into a clutch bore can be made easier by using any of these methods except
    A. Seal installers
    B. A lubricant such as chassis grease
    C. Wax lubricant
    D. A wirelike seal-installing tool

14. Technician A says that full-circle Teflon sealing rings need to be cut at an angle so they can be installed. Technician B says that the seal ring must have a slight clearance when it is installed in its groove. Who is right?
    A. A only   C. Both A and B
    B. B only   D. Neither A nor B

15. Technician A says it is good practice to install clutch seals, one at a time, onto the piston and test their fit into the bore. Technician B says that there should be no seal drag at all when this is done. Who is right?
    A. A only   C. Both A and B
    B. B only   D. Neither A nor B

16. Technician A says that incomplete installation of a clutch assembly can cause a loss of end play and a bound transmission after the pump has been installed. Technician B says that it is possible to tell if a clutch assembly is completely installed by noting its depth in the case or by lifting it slightly and dropping it. Who is right?
    A. A only   C. Both A and B
    B. B only   D. Neither A nor B

17. Technician A says that you should ensure the cleanness of all internal parts by wiping them with a clean shop cloth just before installation. Technician B says that petroleum jelly is a good stickum for thrust washers as well as being a good assembly lubricant. Who is right?
    A. A only   C. Both A and B
    B. B only   D. Neither A nor B

18. Technician A says that transmission end play is normally measured using a dial indicator mounted at the turbine shaft. Technician B says that transmission end play is always corrected at the thrust washer number 1 next to the pump. Who is right?
    A. A only   C. Both A and B
    B. B only   D. Neither A nor B

19. Technician A says that all the clutches and servos should be air checked right after the valve body has been installed. Technician B says that all transmission bolts and especially the valve body bolts should be tightened to the correct torque. Who is right?
    A. A only   C. Both A and B
    B. B only   D. Neither A nor B

20. Technician A says that you should be able to rotate the converter to align the bolt holes with the flexplate after the transmission has been bolted to the engine. Technician B says that a transmission overhaul is not complete until it passes a road test. Who is right?
    A. A only   C. Both A and B
    B. B only   D. Neither A nor B

CHAPTER 13

TORQUE CONVERTER SERVICE

OBJECTIVES

After completing this chapter, you should:

- Be able to make the checks needed to determine whether a converter is usable.
- Be able to recommend whether a converter should be reused or rebuilt.
- Be familiar with the procedure used to rebuild a converter.

13.1 INTRODUCTION

If a transmission problem is caused by a bad torque converter, it should be serviced, probably by removing and replacing it. Also, the converter should be checked during every transmission overhaul. This is especially true with lockup converters, which provide a few more opportunities for problems. If a torque converter has an internal failure, it is normally replaced with a new or rebuilt unit. Unless there is some rather specialized and expensive equipment, most shops do not rebuild converters.

A converter does not necessarily wear out and require a rebuild during a transmission overhaul. However, converters tend to collect the metal, dirt, and other debris that enters with the fluid, and it is impossible to thoroughly check them from the outside. Because of their internal shape and the high amount of centrifugal force generated in them, this debris packs around the outer diameter and is extremely difficult to remove from the outside (Fig. 13-1). Lockup converters make this process even harder because of the added parts. Many transmission rebuilders fear that removing the converter for a few days during an overhaul plus the unusual motions and agitations involved in the process can loosen the internal debris. If the converter is not serviced, this debris can leave the converter, pass through or possibly clog the cooler, and return to the lubrication passages of the transmission. It can start the wear process in the transmission as soon as the engine is started after the transmission overhaul.

Many shops flush out and thoroughly check the converter during every transmission overhaul. Other shops install a rebuilt converter as standard practice, especially on high-mileage transmissions, ones that show much metal wear or aluminized oil, or units with lockup converters. Installing a rebuilt converter is more expensive than servicing the existing one. Cleaning and reinstalling a good converter is an adequate and safe repair in cases where the transmission is relatively clean.

13.2 CONVERTER CHECKS

If converter problems are suspected or before one is used on an overhauled transmission, it should be

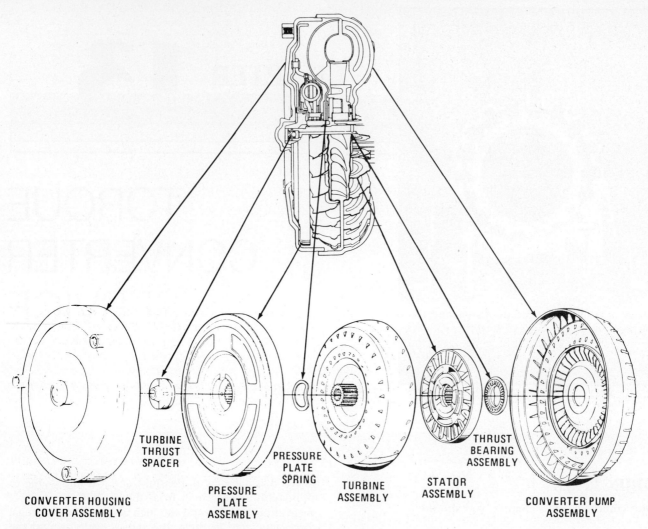

Fig. 13-1 Exploded views of a torque converter with a lockup clutch. *(Copyrighted material reprinted with permission from Hydra-matic Div., GM Corp.)*

checked to make sure it is in good condition. These checks include the visual condition, stator one-way clutch operation, turbine end play, internal interference, lockup clutch operation, and external leakage. A converter that checks out is usable. One that fails one or more of the checks should be replaced with a new or rebuilt unit.

13.2.1 Visual Checks

The first checks performed are visual, often as the converter is removed from the transmission. A technician inspects the following:

1. The outside (especially at all welds) for wetness, which might indicate a leak.

2. The drive lugs or studs for physical damage (Fig. 13-2).

3. The pilot for damage and signs of motion.

4. The hub and pump drive tangs for signs of seal or bushing area wear.

5. The starter ring gear, on some converters, for wear or damage (Fig. 13-3).

Small imperfections on the converter hub can be cleaned and smoothed off using crocus cloth so it won't damage the new front seal. Stripped or damaged female threads in the drive lugs can often be repaired by installing a thread insert as described in Chapter 11.

13.2.2 Stator One-Way Clutch Check

There are several ways to check a stator one-way clutch. Remember that it should lock in one direction and slip or freewheel in the other. The easiest way to check it is to reach into the converter hub so the tip of your index finger contacts the splines of the one-way clutch (Fig. 13-4). The converter should be laying flat on a bench top or floor. Now try rotating the splines. You should be able to turn them in one direction (clockwise) but not the other if the clutch is good. If you can turn the splines in both directions, the clutch is probably slipping; this would go along with a complaint of no

Fig. 13-2 Converter checks should begin with visual checks to ensure that the pilot, drive lugs, starter ring gear (if used), hub, and pump drive tangs are in good shape.

power or poor low-speed performance from the car. If you cannot turn it in either direction, the clutch is locked or bound; this would go along with a complaint of poor cruising ability because of the fluid drag in the converter.

Some technicians grip the stator splines using long, thin, jaw snap ring pliers in order to rotate the stator (Fig. 13-5). The added leverage of the pliers allows you to turn the splines in both directions. But in one direction (clockwise), only the inner race and splines revolve; in the other direction, the entire stator and clutch rotate. If you feel the stator carefully, you can note the difference, especially if you turn it rapidly or snap it.

A stator-holding tool can be made as shown in Fig. 13-6. This tool can be inserted into a groove in the thrust washer on some stators to keep it and the stator from rotating. Next a special one-way clutch torquing tool is inserted into the stator splines, and a torque wrench is used to try to turn the torquing tool and one-way clutch inner race. The one-way clutch should turn freely in a clockwise direction, and it should lock and hold at least 10 ft-lb (14 N·m) of torque in a counterclockwise direction. Don't exert any more torque than this because you can break the holding tool.

A converter with a faulty one-way clutch must be replaced or rebuilt.

13.2.3 End Play Check

The force forward and backward on the front cover and rear impeller tries to make the converter longer—deeper from front to back. This action plus any wear of

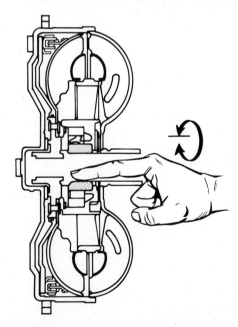

Fig. 13-3 The starter ring gear can be changed in the field by cutting welds using a hack saw or grinder and driving the gear off. The new gear must be welded in place following the procedure recommended by the manufacturer. *(Courtesy of Chrysler Corporation.)*

Fig. 13-4 A stator clutch can be checked by reaching into the hub so your index finger touches the splines. You should be able to rotate the splines in one direction but not the other. *(Copyrighted material reprinted with permission from Hydra-matic Div., GM Corp.)*

385

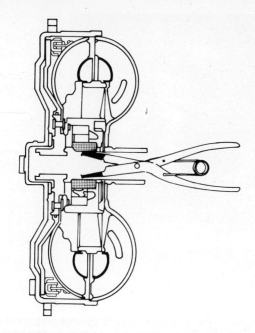

Fig. 13-5 A stator clutch can be checked by reaching into the hub and gripping the splines with snap ring pliers. You should feel the weight of the stator in only one direction if you try rotating it rapidly.

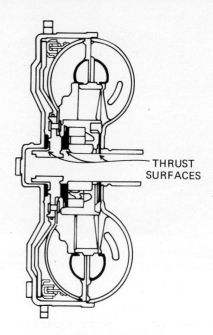

Fig. 13-7 The thrust surfaces are the major wear points inside a converter; if these thrust surfaces wear, the internal end play increases.

the thrust washers allows more front-to-back movement, and this is called end play (Fig. 13-7). The pressure inside a converter also tends to "balloon" it, and this might increase end play even more.

End play is normally measured using a dial indicator, and two styles of devices are commonly used for this. One tool uses an expandable stem that fits into the turbine splines and is expanded to lock into the splines (Fig. 13-8). The dial indicator is then clamped onto the stem and adjusted so the measuring stylus is against the converter and the dial reads zero. Then the stem along with the turbine and dial indicator is lifted through the end play as far as it will go, and the amount of travel is read on the dial indicator.

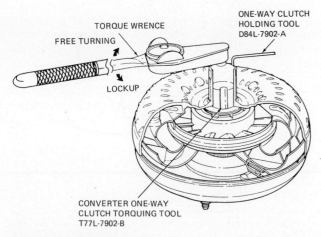

Fig. 13-6 The stator-holding tool keeps the stator from turning while the torquing tool connects the torque wrench to the stator clutch hub. The clutch should turn freely in a clockwise direction and hold up to 10 ft-lb in the other direction. *(Courtesy of Ford Motor Company.)*

The second tool is designed so the converter sits on top of it (Fig. 13-9). The measuring stem is then moved upward to contact the turbine splines; the dial indicator is adjusted to zero; and then the turbine is lifted as far as possible. The amount of end play is read on the dial indicator.

A rough end play check can be made by gripping the turbine or stator clutch splines with a pair of snap ring pliers or your fingers and trying to lift them (Fig. 13-10). You can feel the amount of end play as the turbine or stator is lifted and then dropped, and with experience, you can get a pretty good idea of how much end play there is.

Some manufacturers publish torque converter end play specifications. If none are available, use the rule of thumb that 0.030 in. (0.76 mm) is normal and 0.050 in. (1.27 mm) is the maximum amount of end play allowable. A converter with excess end play should be replaced.

13.2.4 Internal Interference Checks

The thrust washers inside a converter can wear to the point where the impeller, turbine, or stator can rub against each other. The normal wear pattern caused by the dynamic fluid pressure moves the turbine toward the front of the converter and the stator toward the rear. Converter interference should be checked twice—with the turbine and stator toward the front and toward the rear. This is controlled by gravity and how the converter is placed while checking.

Lay the converter on a work area with the hub upward, and gravity will move the turbine and stator toward the front of the converter. Next insert the stator support of the pump into the converter so the support splines align with the stator clutch splines, and then in-

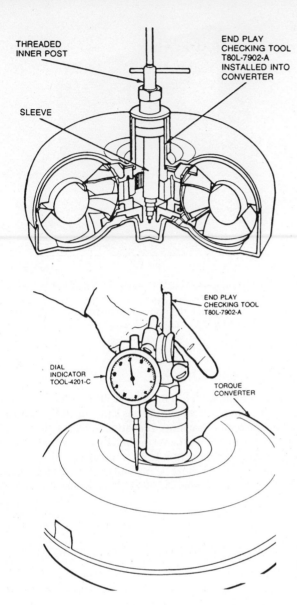

Fig. 13-8 The checking tool is installed so that it locks into the turbine hub. After the dial indicator is set up, the tool is raised and lowered, and the amount of end play is read on the dial indicator. *(Courtesy of Ford Motor Company.)*

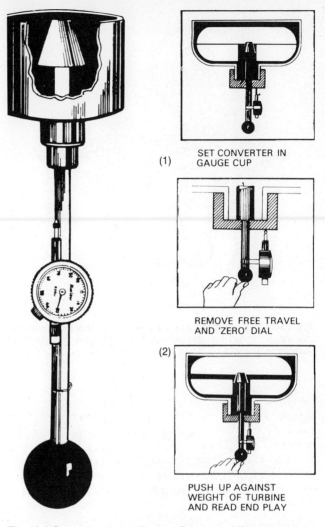

Fig. 13-9 Converter gauge sits on a fixture, and the converter is placed over it (1). After the dial indicator is adjusted (2), the turbine is lifted so the amount of end play can be read. *(Courtesy of Better Engineering.)*

sert the input shaft into the converter so its splines enter the turbine splines (Fig. 13-11). Rotate the pump and input shaft in both clockwise and counterclockwise directions, one at a time and together. If there is any sign of contact or rubbing—either a rubbing or grating sound or feel—the converter needs to be replaced.

Next turn the whole assembly over so the turbine and stator move toward the rear of the converter, and repeat the check by rotating the converter while holding the other units. Again any sign of internal contact indicates a converter that should be replaced.

13.2.5 Converter Clutch Checks

The clutch portion of a lockup converter can fail in several ways: the friction lining can break up or wear out, the seals at pressure plate can fail, or the damper assembly can fail. The damper assembly and lining material cannot be checked on the bench; the only way to check them is to check their operation on a road test or to cut the converter open. Lining breakup often shows up as a plugged oil cooler or lining material in the cooler line, and occasionally, a fragment of a broken damper spring can be found in the cooler line. These are definite signs of a faulty lockup converter.

Two styles of testers are available for checking converter clutches. One type uses adapters that replace the turbine shaft that allow a vacuum to be exerted on the front side of the clutch plate assembly. If this chamber can hold a vacuum, the center seal and the clutch lining (which forms the outer seal) must be good. The second tester style uses adapters that attach to the turbine and provide a means of applying the clutch using air pressure (Fig. 13-12). With the clutch applied, torque is exerted to try to turn the turbine. A good converter clutch locks the turbine and prevents it from turning.

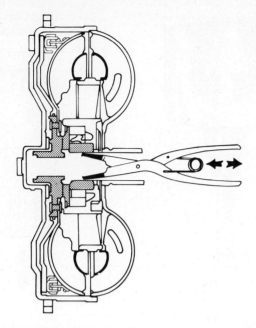

Fig. 13-10 You can make a quick check for end play by gripping the stator or turbine splines and trying to move the stator or turbine in and out.

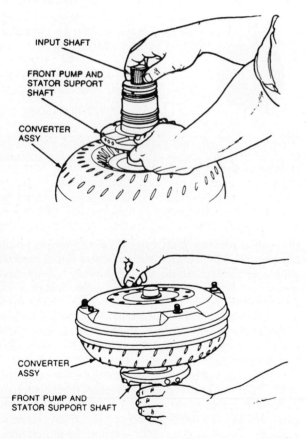

Fig. 13-11 Internal interference check. (a) The stator support and turbine shaft have been installed into a converter and are rotated to make sure there is no internal interference. (b) Everything is upside down and the internal parts are rotated. Again, there should be no interference when the stator support or turbine shaft are rotated. *(Courtesy of Ford Motor Company.)*

Fig. 13-12 A lockup converter is placed in a test fixture and air pressure is applied to apply the clutch. At this time the turbine shaft should be locked if the converter clutch is good. *(Courtesy of Intercont Products.)*

13.2.6 Leak Checks

A converter can develop fluid leaks at the various welds, fittings, or plugs. These fluid leaks show up at the bottom of the converter housing and appear like a front seal leak. They cause a wetness on the converter that can be used as an indication of a leaky converter.

The converter must be pressurized to test for a leak, and this requires a special piece of equipment (Fig. 13-13). This normally consists of an expandable plug that fits into the hub and a device to keep the plug from blowing out of the hub. The plug is equipped with an air chuck to allow pressure to be added or adjusted.

The following description of the procedure to check a converter is very general; you should always follow the exact procedure for the equipment you are using. In general, to check a converter for leaks, you should:

1. Drain as much fluid out of the converter as possible.

2. Select the correct size adapter or sleeve that fits into the converter hub.

3. Place the converter and adapter into the holding fixture (Fig. 13-14).

> *CAUTION:*
> Be sure the assembly is mounted so the adapter cannot be blown out of the hub when it is pressurized.

4. Tighten the adapter as required, and apply an air pressure of 30 to 40 psi (207 to 275 kPa).

5. Spray a soapy water solution over all of the areas where you suspect a leak. Foamy bubbles indicate a leak.

6. When finished, completely release the air pressure before removing the test pieces.

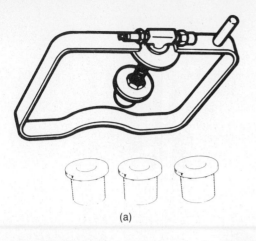

Fig. 13-13 (a) Torque converter leak test fixture with adapters needed to fit hubs of different sizes. A slightly different leak tester is shown (b) with its adapters and installed into its fixture. *[(a) Courtesy of Kent-Moore.]*

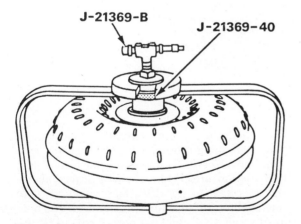

Fig. 13-14 Torque converter installed in a test fixture. The converter should be filled with an air pressure of 80 psi and the whole assembly submerged in water or sprayed with soapy water. Leaks show up as a series of bubbles. *(Copyrighted material reprinted with permission from Hydra-matic Div., GM Corp.)*

Fig. 13-15 Pinhole leaks in a converter can be sealed using a punch and hammer. A series of punch marks around the hole should displace metal to close off the hole.

Very small pinhole leaks can be sealed by peening the hole closed. This is done using a punch and hammer and striking the area immediately around the hole (Fig. 13-15). The malleable iron material of the converter or weld is forced over into the hole. Larger holes are sealed by welding, which requires rebalancing the converter, or by installing a new or rebuilt converter.

13.3 CONVERTER CLEANING

Because of their internal shapes, converters are very difficult to clean. It does no good to run solvent or cleaners into them; in fact, it may even do more harm than good if all the solvent isn't removed. A torque converter cleaner pumps cleaning solvent through the converter in a backward direction: At the same time, the device rotates the turbine in order to create fluid flows that loosen the debris so it can be flushed out (Fig. 13-16). The cleaning operation usually lasts for 5 to 15 minutes, until the solvent leaving the converter runs clear.

After cleaning, the internal end play should be rechecked to ensure that it is not excessive. Occasionally end play increases as varnish or debris is removed. All the solvent should be removed also. If the converter has a drain plug, this is fairly easy. If there is no drain, drain as much solvent as possible out through the hub; pour 2 quarts (1.9 L) of clean ATF into the hub, agitate the converter to mix the ATF and solvent; and drain as much of the ATF–solvent mixture out through the hub. The small amount of solvent remaining should cause no problem.

Exactly how much of the packed dirt and metal debris is removed from a really dirty converter can be questioned, and any remaining debris can damage the transmission. Many technicians prefer to rebuild a dirty converter to ensure complete cleaning and a thorough inspection of the inside.

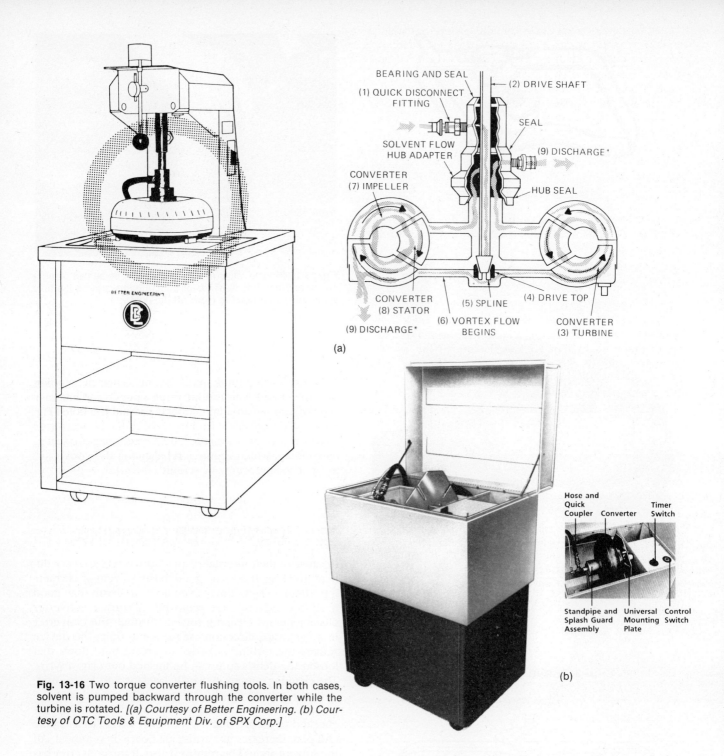

Fig. 13-16 Two torque converter flushing tools. In both cases, solvent is pumped backward through the converter while the turbine is rotated. [(a) Courtesy of Better Engineering. (b) Courtesy of OTC Tools & Equipment Div. of SPX Corp.]

13.3.1 Drilling a Converter

It is possible to drill a converter to provide a drain opening, but this requires a plug for the opening. Drilling a converter is not recommended by some technicians.

Normally the plug used to fill the hole is a special, closed-end pop rivet or a $\frac{1}{8}$-in. NPT (national pipe thread) plug. The pop rivet is the quickest, but the pipe plug is the most secure. If a pipe plug is used, a recessed head type should be used to reduce the plug's weight and the possibility of disturbing the balance of the converter. If balance is a problem, two plugs spaced exactly opposite each other can be used.

When drilling a converter, it is important not to drill into anything inside (only the housing), and with lockup converters, it is important not to drill into the clutch friction area. On some converters, there is a raised area next to the drive lugs; in others, it is recommended to drill into the impeller area. The hole size is usually $\frac{1}{8}$-in. (3.5 mm) if a pop rivet is to be used or $\frac{11}{32}$ in (8.7 mm) if a $\frac{1}{8}$-in. pipe plug is to be used. Before drilling, place a depth stop or a piece of tape on the drill bit so you can drill to a maximum depth of $\frac{1}{4}$ in. (6.3 mm). This is usually just deep enough to penetrate the outer housing and allow the converter to drain.

After the converter has drained, if you are using a pop rivet, simply remove any burrs from the outside of the hole; apply a small amount of sealant on the rivet; place the rivet into the hole; and using a pop rivet tool, expand the rivet to lock it in place. If you are using a pipe plug, apply grease to the flutes of a $\frac{1}{8}$-in. NPT tap to catch the metal chips, and cut threads in the hole opening. While tapping, make sure not to run the tap into the turbine. Also periodically back the tap up to break the chips and cut a cleaner hole. Then apply sealing tape or sealant to the threads of the pipe plug and tighten it into the opening to a torque of 8 ft-lb (10.8 N-m).

It is good practice to pressure test the plug before installing the converter in a transmission if this was done on the bench.

13.3.2 Cooler Cleaning

Some torque converter cleaners are also used to clean oil coolers. This can also be done with a simple fluid pump (Fig. 13-17). Cleaning a cooler is normally a matter of pumping solvent backward through the cooler lines and cooler. The cleaner or fluid pump is connected to the transmission end of the lubrication cooler return line, and the transmission end of the cooler feed line is opened to return the solvent to the cleaner or to a catch bucket (Fig. 13-18). Fluid is then pumped through the circuit until it runs clear at the return. Some technicians alternate fluid and air flows and periodically reverse the flow direction if they feel debris might be caught in the cooler fins. Be careful if using compressed air because excessive pressure can rupture the cooler assembly. One commercial cooler cleaning device produces a pulsating solvent flow intended to loosen and remove packed debris. Don't forget that any remaining debris in the converter and cooler will find its way into the lubrication circuit of the transmission. All of the solvent should be blown out of the cooler and lines using clean compressed air.

13.4 CONVERTER REBUILDING

Because it is a sealed unit, a converter must be cut apart and welded back together to be rebuilt. These operations require some rather expensive equipment and some machining and metal working skills, and because of this, converter rebuilding is normally done by converter rebuilding specialty shops or the very large transmission shops.

If possible, a transmission shop rebuilds its own converters to ensure a top-quality unit and also reduce the problem of getting an exact replacement for a particular car make or transmission. Remember that converters for the same transmission can have different stall speeds, and installing the wrong converter can cause drastic changes in vehicle performance. Eight different converters are used with one of the common transaxles. They look about the same and fit any of the flexplates and transaxles. They are easily interchanged, but this creates improper operation. A poor-quality rebuilt converter can ruin an otherwise perfect transmission. It takes about 20 to 30 minutes to rebuild a converter, and

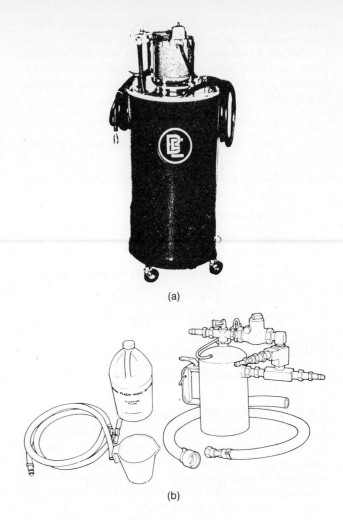

Fig. 13-17 Oil cooler flusher (a) pumps solvent through the cooler and (b) uses shop air pressure. *[(a) Courtesy of Better Engineering. (b) Courtesy of Kent-Moore.]*

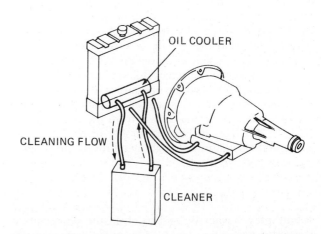

Fig. 13-18 The cooler flushing unit is connected to the cooler lines so a cleaning agent is forced backward or forward (depending on the line connections) through the cooler unit until it is clean.

this is often not much more than the time spent to check, clean and flush, or to get an exchange unit.

When a converter is rebuilt, all damaged or worn parts are either repaired or replaced as necessary. Rebuilding a torque converter consists of the following operations:

- Inspection of the exterior for damage.
- Cutting the unit open on a specially equipped lathe.
- Inspection of the interior components for wear or damage.
- Thoroughly cleaning the interior parts.
- Disassembling and inspecting the stator one-way clutch.
- Prelubrication of the internal bearing surfaces.
- Adjustment of the internal end play.
- Careful reassembly and rewelding of the two housing sections.
- Leak checking of all welds.
- Rebalancing the assembly.

13.4.1 Disassembling a Converter

Since a converter is welded together, the weld must be cut to allow the separation of the front and rear housings and access to the inside. To do this, the converter is mounted in a metal-cutting lathe. To ensure that it is mounted true, the converter is centered to the lathe axis using an adapter that matches the crankshaft pilot and is bolted to a face plate at the drive lugs. Duplicating the mounting points that are used to secure the converter to the engine should give true, wobble-free rotation for accurate machining.

When the lathe is started and the converter is spun, runout of the hub can be observed to determine how well the converter was assembled previously. Excessive hub runout causes wear at the front pump, bushing, and seal. It is corrected by careful alignment during reassembly of the converter housings. A faulty hub will be removed and replaced at this time (Fig. 13-19). The weld is cut by using a hardened steel alloy cutter bit to remove the weld metal. The new replacement hub is welded in place using an automatic wire welder.

The converter housing weld is cut using a hardened carbide cutter bit. A skilled operator will remove a minimum amount of metal—just enough to allow separating the two parts, the front "bowl" from the rear impeller (Fig. 13-20). This makes reassembly of the converter easier and also allows enough metal for future rebuilding.

13.4.2 Cleaning

A hot spray washer is used to clean the disassembled converter parts (Fig. 13-21). This is the same cleaning process and equipment that is used for the transmission parts during a rebuild. Some of the parts are cleaned several times as they go through different repair pro-

(a)

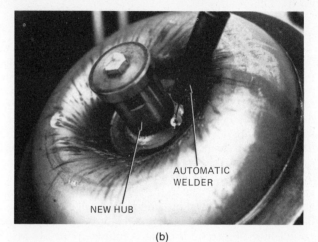

(b)

Fig. 13-19 (a) Converter hub showing roughness in the seal-bushing area and hub with a bushing seized onto it with major damage in the pump drive area. These were cut out of converters. (b) Both hubs were replaced with new hubs that were welded into the converter using a special welder.

cedures. For example, the front bowl and rear impeller must have the packed-in debris removed and be deburred from the lathe cut and have the edges beveled slightly to ease reassembly. They are rewashed after these operations to ensure complete cleanliness.

13.4.3 Subassembly Repair

With the converter disassembled, the following items are checked and repaired as necessary:

- Converter Hub—Removed and replaced.
- Starter Ring Gear—Removed and replaced (Fig. 13-22).
- Crankshaft Pilot—Welded to build up material and machined back to original size.
- Drive Lugs/Studs—Removed and replaced.
- Turbine Drive Hub—Removed and replaced.
- Thrust Washers—Removed and replaced; can be upgraded to Torrington bearings.

(a)

(b)

Fig. 13-20 For rebuilding, (a) a converter is cut apart on a machine lathe and (b) a carbide tool is used to cut away the weld so the front cover can be separated from the rear impeller section. Once the weld is cut, the converter is split apart for inspection.

Fig. 13-21 After being cut, the usable components of the converter are placed in a spray washer for a thorough cleaning.

Fig. 13-22 A new starter ring gear is tack welded onto a converter bowl. A straight, true installation is ensured by doing this in a lathe.

- Support Bushings—Removed and replaced.
- Impeller Fins—Spot welded to lock in place.
- Stator Clutch—Disassembled, inspected, and reassembled.
- Stator Clutch Outer Race—Pinned to lock into the stator.
- Clutch Lining—Relined.
- Damper Springs or Assemblies—Removed and replaced.

With the exception of the stator one-way clutch, which is always disassembled and inspected, the preceding operations are done on an as-needed basis, so a rebuild varies from one that involves only disassembling, cleaning, and inspecting to one with many services required.

13.4.4 Converter Reassembly

Converter reassembly requires several operations:

- Check the height of the stacked components to ensure that the assembled unit is not too long or too short (Fig. 13-23).
- Check and adjust the internal clearance to ensure there is correct end play. Adjustment is done by changing or shimming the thrust washers.
- Make sure the bowl and impeller housings are correctly aligned so there is a minimum of hub runout and tack weld them securely into position (Fig. 13-24).

Fig. 13-23 Fixture used to ensure that the converter has the correct overall length when it is assembled. Shims are added internally if necessary.

- Weld the housing. This is done in an automatic wire welder to provide a continuous, leak-free weld (Fig. 13-25).

13.4.5 Postassembly Checks

After the converter has been welded back together, it is pressure checked to ensure that there are no leaks (Fig. 13-26). This operation is similar to that described in an earlier section. Small leaks can often be repaired

Fig. 13-25 Converter that has just been welded back together on an automatic welder. After two more checks to make sure it is good, it is ready for installation.

by peening the hole closed using a punch and hammer. Larger leaks require rewelding.

The final converter check is to rebalance the assembled unit (Fig. 13-27). This compensates for changes in the amount of weld metal and the possible repositioning of the two housing sections between disassem-

(a)

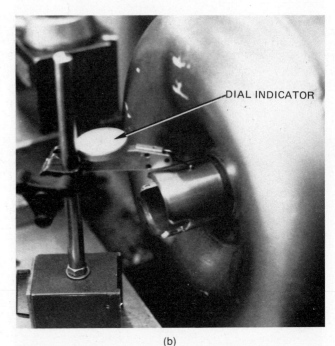
(b)

Fig. 13-24 (a) Converter has the front bowl attached to the lathe and the hub attached to a special fixture adjusted to give the correct internal end play. Note the tack welds being used to hold the two halves together. (b) Checks are made at this time to ensure that the hub is centered and true with the front of the converter.

Fig. 13-26 Converter in a special pressure test fixture. After being pressurized, it is sprayed with soap solution to expose any voids or pinholes in the welds.

bly and reassembly. During the balancing operation, the turbine and stator must be locked in a centered position to prevent them from affecting the overall balance. Balancing a converter is made more difficult by the three units that can revolve together as one unit or as three separate units.

Fig. 13-27 Converter being balanced on a special balancer. If necessary, small weights are spot welded onto the converter to correct any unbalance.

REVIEW QUESTIONS

The following questions are provided so you can check the facts you have just learned. Select the response that best completes each statement.

1. Technician A says that a dirty torque converter can ruin a transmission rebuild. Technician B says that most used torque converters are worn-out. Who is right?
   A. A only
   B. B only
   C. Both A and B
   D. Neither A nor B

2. The visual checks to be made on a converter include the condition of
   A. The drive lugs or studs
   B. The hub and pump drive tangs
   C. The pilot
   D. All of these

3. Technician A says that it should not be possible to turn the stator splines in either direction with one's finger. Technician B says that normally these splines can be turned both ways by gripping them with a pair of snap ring pliers. Who is right?
   A. A only
   B. B only
   C. Both A and B
   D. Neither A nor B

4. Technician A says that a converter with a bad stator one-way clutch should be replaced with a new converter. Technician B says a converter can be rebuilt and the one-way clutch repaired. Who is right?
   A. A only
   B. B only
   C. Either A or B
   D. Neither A nor B

5. Technician A says that excessive end play in a converter is caused by a worn-out stator. Technician B says one can perform an accurate converter end play check with a pair of snap ring pliers. Who is right?
   A. A only
   B. B only
   C. Both A and B
   D. Neither A nor B

6. Technician A says that a good rule of thumb for the maximum allowable amount of converter end play is 0.050 in. Technician B says that any converter with less than 0.040 in of end play should be replaced or rebuilt. Who is right?
   A. A only
   B. B only
   C. Both A and B
   D. Neither A nor B

7. Technician A says that a stator support and a turbine shaft are needed in order to check a converter for internal interference. Technician B says that this is a simple check in which one merely rotates parts in the converter while listening and feeling for internal rubbing. Who is right?
   A. A only
   B. B only
   C. Both A and B
   D. Neither A nor B

8. Technician A says that if the converter clutch lining breaks up, it probably plugs the cooler. Technician B says that a failed converter clutch causes too little internal end play. Who is right?
   A. A only
   B. B only
   C. Both A and B
   D. Neither A nor B

9. Technician A says that during a leak check the converter is filled with oil to a pressure of about 80 psi. Technician B says that a very small leak at a converter weld can be stopped using a hammer and punch. Who is right?
   A. A only
   B. B only
   C. Both A and B
   D. Neither A nor B

10. When a transmission is rebuilt the torque converter should be
    A. Balanced
    B. Cleaned
    C. Painted
    D. All of these

11. Technician A says that a converter can be drained adequately by merely turning it so the hub is downward over a container. Technician B says that one can clean a converter quite well by running a solvent hose into it and turning on the parts washer. Who is right?
    A. A only
    B. B only
    C. Both A and B
    D. Neither A nor B

12. Technician B says that a converter flushed with solvent should be filled with ATF and then as much of the ATF/solvent mix as possible drained. Technician B says that it is possible to completely drain the converter by drilling a hole in it. Who is right?
    A. A only
    B. B only
    C. Both A and B
    D. Neither A nor B

13. Technician A says a transmission cooler is cleaned by flushing it with solvent in both directions. Technician B says that a dirty cooler can cause a transmission burnout. Who is right?
    A. A only
    B. B only
    C. Both A and B
    D. Neither A nor B

14. During a rebuild, a converter
    A. Is taken apart and cleaned
    B. Has the stator clutch disassembled and checked
    C. Is adjusted to the correct internal end play
    D. All of these

15. Technician A says that fairly expensive, special equipment is required to rebuild a converter. Technician B says that the only sure way to get all of the dirt and debris out of a converter is to rebuild it. Who is right?
    A. A only
    B. B only
    C. Both A and B
    D. Neither A nor B

CHAPTER 14

AUTOMATIC TRANSMISSION MODIFICATIONS

OBJECTIVES

After completing this chapter, you should:

- Be familiar with the modifications that can be made to improve the reliability or change the operating characteristics of a transmission.

14.1 INTRODUCTION

There are several reasons for modifying an automatic transmission, and most involve matching the transmission to the driver's or vehicle's particular driving characteristics or conditions. Most modifications are done to improve shift qualities and are relatively mild and inexpensive. Other changes, such as those done to produce an all-out race transmission, can be very expensive.

When considering transmission modifications, it should be remembered that any transmission work that does not follow a manufacturer's recommendations affects the operating characteristics, transmission life, and possibly safety. These changes probably void any manufacturer's warranty, and the technician making the changes is responsible for any adverse results.

When planning transmission modifications we should first establish the different stages or levels of the modifications:

1. Street Usage—Changes to improve shift quality.

2. Street Usage—Changes to improve fuel economy.

3. Street Usage—Changes to improve strength and longevity.

4. Street Strip—Changes to improve vehicle acceleration rate.

5. Racing—Numerous changes for strictly racing purposes.

Most transmissions are designed and programmed to suit the average new car buyer. At one time, a major design criteria was to produce a transmission in which the shifts were very smooth—to the point where they could not be felt (Fig. 14-1). This type of shift has a certain amount of slippage, and the slippage produces wear and heat. It is possible to change the shift characteristics of many transmissions to the other extreme, a shift that jerks the car and produces tire chirp.

We will discuss several types of modifications. Some are inexpensive—a shift kit or, more simply, a hole size increase in a transfer plate or a spring removal from an accumulator. Other changes are expensive, such as a special-purpose gearset or a highly modified torque converter. Table 14-1 lists various modifications and the ones often used for the modification stages. Each modification will be discussed in more detail.

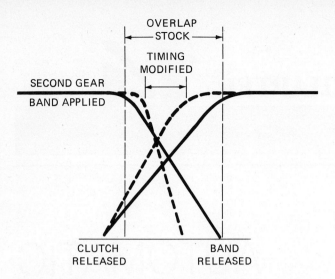

Fig. 14-1 With stock shift programming, the overlap period between two gears is relatively long. This is usually reduced to a minimum when transmissions are modified.

Most of the modifications that require specially made, nonproduction parts are very expensive and are only made on all-out competition units. Although they might provide a benefit for ordinary operation, they are usually not worth the cost. Some of these parts for modifying a transmission are available from the car dealership or the original equipment manufacturer (OEM). These components are designed for the largest engine/heaviest or sportiest vehicle using that transmission and usually handle more torque or provide firmer, tighter operating characteristics. These are the safest, most secure parts for units that require a warranty. Aftermarket suppliers that service the transmission repair industry are a good source for shift kits and heavy-duty components and friction materials. Better suppliers pride themselves in providing parts that cure many transmission weaknesses. All-out racing components are available from about 40 companies that specialize in high-performance transmissions and are available through most speed shops.

TABLE 14.1 Modifications to Produce Desired Operating Characteristics

Stage	A	B	C	D	E	F	G	H	I	J	K	L	M	N	O
1	X														
2	X														X
3	X	X	X	X	X	X	?		?						X
4	X	X	X	X	X	X	?			X					X
5		X	X	X	X	X	X	X	X	X	?	X	?		X

Symbols: ?, might be made depending on exact usage and other variables; A, shift kit; B, increase clutch pack size; C, improve friction material quality; D, added fluid/deep pan; E, added cooling; F, improve lube flow; G, low-ratio gearset; H, lightweight drums; I, high-strength components; J, heavy-duty sprag races; K, low-drag components; L, low-volume oil pump; M, special valve body; N, internal transmission brake; O, converter stall speed change;

Transmission modifications can be a trial-and-error operation. The changes made often depend on the particular transmission design and are determined after studying the operation, hydraulic circuit, and physical construction of that unit. Most technicians use a kit that has already been worked out and tested by a supplier. Those developing their own modifications normally work in such a way that it is possible to easily go back to where they started. It is then possible to make smaller or larger changes until the desired operation is obtained. Also, some of these changes will interact and work together to produce greater effects.

14.2 SHIFT KITS

Shift improver kits are the most popular transmission modification. On certain transmission models, some transmission shops install a shift kit on each rebuild as a standard practice. This is done to increase the life of the rebuild and to make a noticeable firmness in the shift quality, thus increasing customer satisfaction. The contents of a shift kit vary depending on the transmission and the supplier. It can be a drill bit and direction sheet for increasing an orifice size, a replacement transfer plate, a valve body spring, a spacer block for an accumulator, an orifice plug, or any combination of these (Fig. 14-2). Some shift kits include the ability to provide

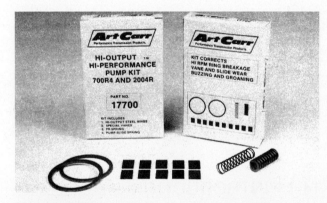

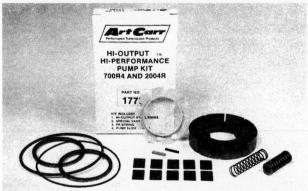

Fig. 14-2 Shift improver kits vary from the minor ones that firm up the shifts to major ones that involve significant internal transmission changes. *(Courtesy of Art Carr Performance Transmission Products.)*

different stages of performance changes from the barely noticeable to the rather severe.

14.2.1 Orifice Changes

Because the size of an orifice in the feed passage is used as a major control for the quality of a shift, the speed and/or quality of that shift can be easily altered by changing the size of the orifice. The orifice is normally located in the transfer plate, a part that can be changed easily (Fig. 14-3). As described earlier, the speed at which a clutch or band servo strokes to take up a clearance and apply the clutch or band is directly related to the size of the servo piston and the fluid flow rate to the piston. An orifice reduces the flow rate and therefore increases the time for the piston stroke. If we were to increase the area of the orifice by 10 percent, we would increase the flow rate by 10 percent and reduce the time by 10 percent. This would speed up the shift and reduce the amount of slippage that takes place during the shift.

It should be remembered that we are referring to the area of the orifice, not the diameter. The area of an orifice that is a circle is the product of:

$$A = \pi r^2 \quad \text{or} \quad A = 0.785 d^2$$

where

π (pi) = 3.1416

r = radius, one-half the hole diameter

d = hole diameter

For example, assume there is an orifice with a diameter of 0.180 in., and we want to increase its size (area) by 10 percent. The area of a 0.180-in. hole is $0.180 \div 2 = 0.090$ (radius), $0.090 \times 0.090 = 0.008$ (radius squared), and $0.008 \times 3.1416 = 0.025$ in.^{2} (hole area). We need to make the area of the hole 0.0025 larger (0.025×0.10) for a total of 0.0275 if we want to increase it by 10 percent. To determine the diameter of a hole that is 0.0275 in.^{2}, we work through the formula backward: $0.0275 \div 3.1416 = 0.00875$ (radius squared), and the square foot of this is 0.0936 (radius). Multiplying this by 2 gives us a hole diameter of 0.187 in. Increasing an orifice diameter from 0.180 to 0.187 in. increases the flow rate about 10 percent. Although it appears rather complicated, a simple pocket calculator makes the mathematics rather easy.

When drilling an orifice, finish the hole so there are no burrs or rough hole edges. These can create fluid turbulence that reduces the flow. Normally the hole is drilled using a sharp drill bit to the desired size, and then a larger drill or a scraper is used to debur or chamfer each end of the hole (Fig. 14-4).

If a hole is too large, the diameter can be reduced by soldering the hole closed and then drilling it to the desired size. It is good practice to counterbore each end of the hole before soldering it to give the solder a good surface to which to adhere.

The flow through a round passage can be closed by driving a slightly larger cup plug into the passage (Fig. 14-5). The cup plug is then drilled to provide the desired orifice.

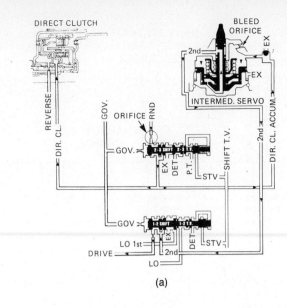

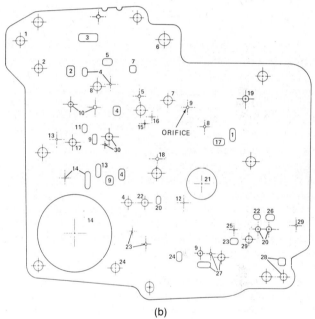

Fig. 14-3 (a) The orifice leading to the 2–3 shift valve slows direct-clutch fluid flow and increases the time period for the 2–3 shift. (b) Orifice marked with an arrow. *(Copyrighted material reprinted with permission from Hydra-matic Div., GM Corp.)*

14.2.2 Accumulator Changes

An accumulator is a relatively easy device to alter. In some cases, it is mounted at the outside of the case, and this gives quick access. The most common change made with accumulators is to block its motion. If the piston cannot stroke, it will not take fluid away from the clutch or band operation and the application will be faster and firmer. Like modifying an orifice size in a transfer plate, blocking an accumulator is fairly easy to change back if it does not produce the desired results.

In a few cases an accumulator can be blocked by turning the piston upside down relative to the bore and

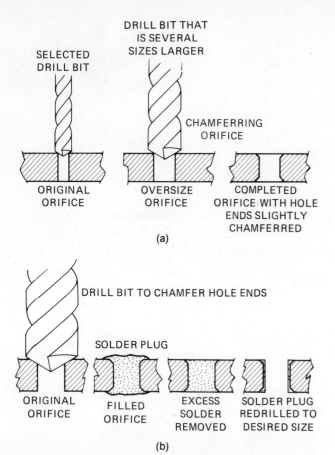

Fig. 14-4 (a) After an orifice in a transfer plate has been drilled to a larger diameter, the edges of the hole should have a slight chamfer (using an oversize drill bit). This can usually be done by hand. (b) If an orifice needs to be made smaller, it should be chamferred, soldered closed, filed flat, and drilled to the proper diameter. Don't forget to chamfer the ends of the hole.

guide rod. Another way to block one is to cut a piece of tubing to a length slightly shorter than the accumulator piston stroke (Fig. 14-6). The tubing diameter should be smaller than the piston diameter so it fits against a strong portion of the piston.

14.2.3 Other Programming Changes

Depending on the intended operation, there are additional modifications that can be made to the hydraulic circuits to reprogram the automatic operation. These include:

Fig. 14-5 This cup plug can be driven into a passage to close it off. Normally it is drilled with the proper size drill to make an orifice of the correct diameter.

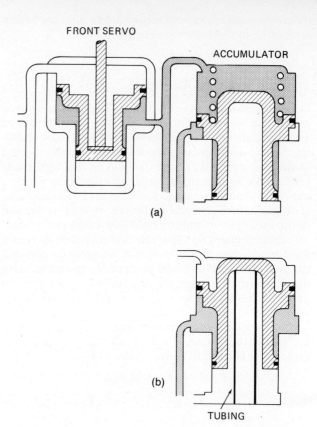

Fig. 14-6 (a) The action of a Torqueflite accumulator can be eliminated by cutting a piece of thin wall conduit to the right length and (b) installing it. The spring is left out.

- Modulator or modulator rod length to change shift quality.
- Shift valve spring to change shift points.
- Governor weights or springs to change shift points.
- Line pressure to apply more pressure on the friction material.
- Removal of center seal from clutches using three seals to increase clutch apply force.

Shift quality can be changed—softened or made firm—by several changes in the modulator as described in an earlier chapter. Many transmission shops install a firmer modulator (one with a high spring pressure or a smaller diaphragm area) or a longer modulator pin to obtain firmer upshifts (Fig. 14-7). A softer modulator or shorter pin is used for softer, smoother shifts. Also, the adjustable modulators can be adjusted to make minor changes in shift quality. In addition to these changes, there are two styles of kits available—one driver adjusted and the other technician adjusted—to vary the vacuum signal to the modulator to produce the desired shift quality (Fig. 14-8).

You can see how this works if you road test a car using a transmission test package like that illustrated in Chapter 10. Make a moderate-to-high throttle acceleration from a stop to the 2–3 shift with things normal,

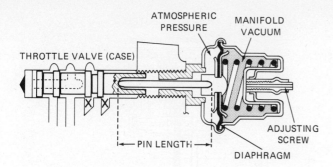

Fig. 14-7 Line pressure can be adjusted to a higher pressure through the modulator by making the pin longer, turning the adjuster screw inward, or using a modulator with a stronger spring pressure. *(Courtesy of Ford Motor Company.)*

and note the shift quality. Now have an assistant generate a vacuum leak at the gauge—you can note the leak effect on the vacuum and line pressure gauges—and repeat the baseline test. Depending on the amount of vacuum leak, you should be able to notice a definite change in shift quality. It should be noted that reducing the amount of the vacuum signal will not affect full-throttle shifts because the vacuum is zero at this time.

Many shift kits contain one or more replacement shift valve springs. The strength of these springs is altered from stock to produce different shift timing for that particular upshift or downshift.

Shift timing for all of the shifts is altered by changing the governor weights or springs. Some manufacturers have governor weights available that weigh different amounts. Drilling a hole or cutting away some of the metal in the weights reduces their centrifugal force and produces lower governor pressure and higher speed shifts (Fig. 14-9). Placing a bolt and nut in a drilled hole or welding additional metal onto the weights produces earlier and higher governor pressure and lower speed shifts. Changing the springs has a similar effect. Stronger springs produce later, higher speed shifts, and weaker springs produce earlier, lower speed shifts.

An individual usually has difficulty locating a spring of a particular diameter, length, and strength for these changes. A good, complete spring selection is not commonly available. The strength effect of a spring can be increased by stretching it slightly or by placing a shim under one end of it. If tinkering with springs, remember that all straight-wound springs have a constant rate of compression. Even though valve body springs are relatively small, they follow the normal rules for springs. Spring rate refers to the rate of compression relative to the load, and this means that if it takes 1 ounce (28.4 g) of pressure to compress the spring $\frac{1}{4}$ in. (6.35 mm), it takes 2 ounces (56.8 g) of force to compress it $\frac{1}{2}$ in. (12.7 mm), and 3 ounces (85.2 g) of force to compress it $\frac{3}{4}$ in. (19 mm). Shimming or stretching a spring causes it to enter further into its rate increase and does so sooner. If a spring can be shortened by squeezing it, it enters the rate slower and is weaker. The actual strength of a spring is determined by the diameter of the wire that it is wound from and how much wire is used (the diameter and number of coils). The thicker the wire, the stronger the spring.

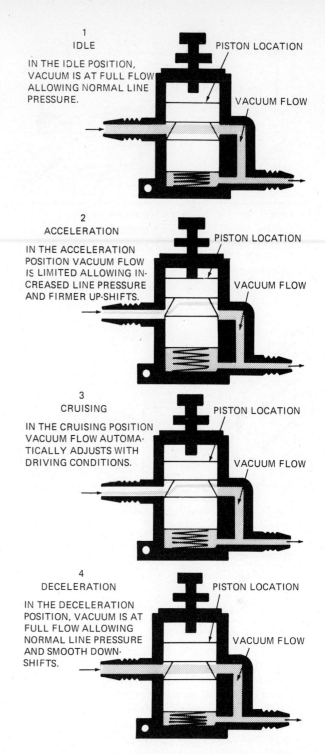

Fig. 14-8 Device designed to be mounted in the vacuum line to a modulator. It reduces the vacuum signal to the modulator during acceleration to produce firmer upshifts. It provides normal modulator operation and line pressure during other phases of operation.

Many people automatically shim the pressure regulator spring in order to raise line pressure when reprogramming a transmission. This produces firmer and stronger shifts. But it also increases the horsepower loss in the transmission and produces more heat in the fluid.

401

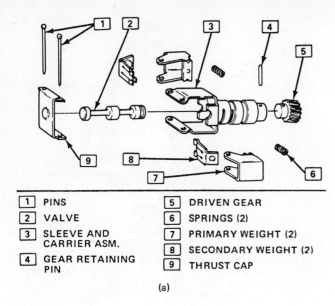

1. PINS
2. VALVE
3. SLEEVE AND CARRIER ASM.
4. GEAR RETAINING PIN
5. DRIVEN GEAR
6. SPRINGS (2)
7. PRIMARY WEIGHT (2)
8. SECONDARY WEIGHT (2)
9. THRUST CAP

(a)

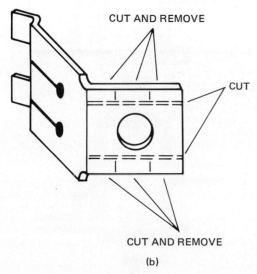

(b)

Fig. 14-9 Grinding material off the secondary weights reduces the centrifugal force of the weights and reduces governor pressure. The result is later upshifts.

A knowledgeable transmission tuner uses the other methods described in this chapter to produce stronger and firmer shifts and uses line pressure increase as the last resort. Some tuners can reduce line pressure. These tuners have made other changes for the needed strength so line pressure can be dropped to improve transmission efficiency.

Most General Motors three-speed rear-wheel drive (RWD) transmissions use outer, center, and inner seals in the direct-clutch pistons and some forward-clutch pistons (Fig. 14-10). The center seal divides the piston chamber into two separate areas to produce smoother neutral–first and neutral–reverse driveway shifts and the ability to have a strong direct clutch in reverse. Removing the center seal produces a stronger forward-clutch application; on most street-driven vehicles, it will be considered harsh. Removing the center seal from the direct clutch produces a severe leak into the reverse passages in third gear. If the passages are altered significantly, this seal can be omitted to produce a stronger

DIRECT CLUTCH

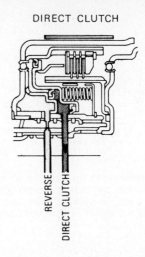

Fig. 14-10 Direct clutch with two fluid chambers separated by the center piston seal. The inner chamber is used in third gear, and both inner and outer chambers are used in reverse.

direct clutch in third gear. This is only done on racing transmissions.

14.3 CLUTCH PACKS

For the most part, changes to the clutch pack are used to improve the torque-carrying capacity and the longevity of the transmission. Drilling the clutch hub and drum increases the shift speed and the longevity. Modifications to a clutch pack include using the best friction material available, increasing the number of clutch plates to the maximum, and drilling the hub and drum for improved fluid flow in and out of the plate area.

Friction material quality is important in any transmission, but high torque or extended life demands the best material available. This is often a top-quality OEM material with new steel plates. Some shops grit blast the steel plates. As explained earlier, this usually produces a firmer shift, but friction material loss during shifts reduces the life.

14.3.1 Increasing Pack Size

Adding one more friction and steel plate to a four-plate clutch increases the torque capacity by 25 percent. The longevity or number of shifts that the clutch can make also increases a like amount.

Most manufacturers produce different clutch assemblies with various amounts of plates depending on the engine torque and body weight of the car. The differences that determine the number of plates are usually the piston thickness, pressure plate thickness, drum depth, and hub length (Fig. 14-11). The piston from some four-plate clutch is the thickness of a three-plate clutch piston minus the thickness of one friction plate, one steel plate, and a clearance of about 0.010 in. (0.25 mm) (Fig. 14-12). Using the four-plate piston in a three-plate clutch usually allows enough room for an additional friction and steel plate. When doing this, you should always compare the length of the clutch hubs to ensure that the added plate is positioned correctly on the hub. In some transmissions, a clutch with more plates has a longer hub.

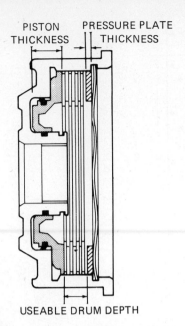

Fig. 14-11 The number of plates that can be used in a clutch assembly is determined by the length of the drum as well as the thickness of the pressure plate, piston, lined plates, and unlined plates.

14.3.2 Drilling a Hub or Drum

When a clutch is released, fluid must enter between the clutch plates to reduce lining heat, friction, and wear. When a clutch applies, this fluid must leave (Fig. 14-13). The clutch hubs in many transmissions have holes drilled in them to allow this fluid flow; others do not (Fig. 14-14). But these holes can be drilled.

If drilling a hub, use 8 to 12 holes about $\frac{1}{8}$ in. (3.17 mm) in diameter. The holes should be spaced equally around the hub so that pairs of holes are directly opposite (Fig. 14-15). Also the hole pattern should be staggered inside the area where the plates are located so oil is directed to all of the clutch plates. After the holes have been drilled, check to remove any burrs from the drilling. Carefully spaced holes should not disrupt the balance of the hub.

If the clutch has a band fitted around it like the direct clutch in many three-speed Simpson gear train units, the band must release as the clutch is applied. Clutch apply and band release can be speeded up slightly by routing the fluid from between the clutch plates under the band. This is done by drilling a series of holes through the band friction surface of the drum. Drill 8 to 12 equally spaced $\frac{1}{8}$-in.-diameter holes through the drum. Again, the holes should be staggered across the friction area. The holes should align with grooves for the clutch plate splines (Fig. 14-16). The sharp edges of these holes tend to cause band wear so it is good practice to chamfer their outer edge using a $\frac{3}{16}$-in. (4.76-mm) or $\frac{1}{4}$-in. (6.35-mm) drill bit.

14.3.3 Band Changes

Changes to increase the strength of a band are limited to drilling holes in the drum surface as just described

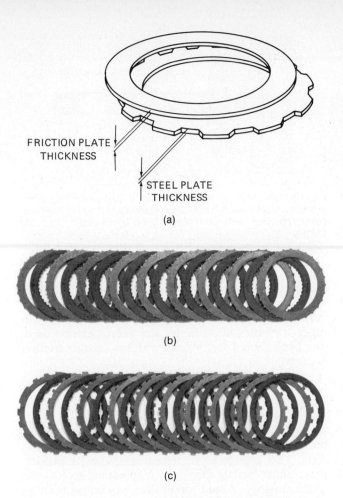

Fig. 14-12 Measuring the thickness of a lined and unlined clutch plate plus the clearance of 0.010 inch is normally the amount of room that is required in order to add an additional plate (a). This clutch pack (b) with slightly thinner friction plates will allow 8 friction plates to be used in an AOD direct clutch which normally uses 6 frictions. The 9 clutch pack shown in C replaces the 6 plate 3–4 clutch pack of a THM 700. *[(b) and (c) are courtesy of Alto Products Corp.]*

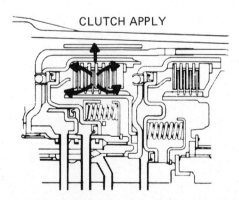

Fig. 14-13 When a clutch applies, fluid must leave the area between the plates. Normally it flows to the ends of the clutch assembly; if the drum is drilled, it can also move outward. When the clutch releases, fluid must flow back between the plates. It normally flows from the center.

403

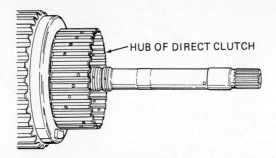

Fig. 14-14 A THM 700 clutch hub was drilled at the factory to provide a fluid flow into the clutch plate area. *(Copyrighted material reprinted with permission from Hydra-matic Div., GM Corp.)*

and using the various parts available from the manufacturer. In some transmissions, different band levers are available that increase the apply force on the band. In other transmissions, different-size servos can be used. The band should also be one with best-quality friction material.

14.4 IMPROVING LUBRICATION

As mentioned several times, the transmission's life can be severely reduced if the fluid becomes too hot or too dirty. This applies to all transmissions and especially those under heavier-than-normal operation, such as RV, trailer towing, off-road, street-strip, or competition operation.

Extra deep oil pans are available from several aftermarket sources, and these hold one to four more quarts of fluid than the stock pan (Fig. 14-17). The increased surface area, which allows more air contact, provides added cooling, and the increases amount of oil

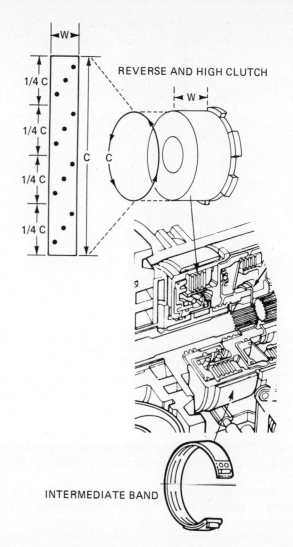

Fig. 14-16 If drilling a clutch drum, lay out the holes in the same manner as that used for the hub. The hole location should be underneath the band if possible. After drilling the drum, chamfer the outer edge of the holes using a slightly oversize drill bit.

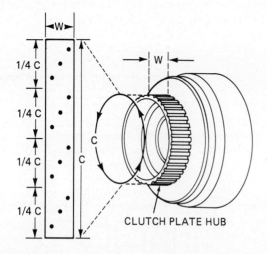

Fig. 14-15 If drilling a clutch hub, cut a strip of paper to the width and circumference of the hub; divide the circumference into fourths; lay out four or six sets of equally spaced holes; wrap the marked paper around the hub; transfer the hole locations; drill 8 to 12 holes about $\frac{1}{8}$ in. in diameter; debur the holes; and clean the hub to remove any metal chips.

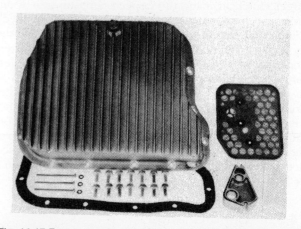

Fig. 14-17 Extra-capacity pans that hold about 2 more quarts of ATF than a stock pan and are finned to provide extra cooling ability. *(Courtesy of A-1 Automatic Transmissions.)*

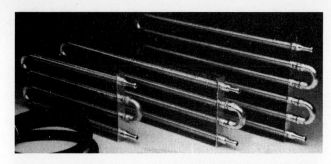

Fig. 14-18 A supplementary oil cooler can be added to provide more cooling capacity. *(Courtesy of A-1 Automatic Transmissions.)*

offers more of a heat sink, which reduces the rate of heat buildup. Some deep pans are made from cast aluminum with fins to increase the amount of metal-to-air even more. Some deep pans have air tubes passing through the sump area to provide an increased oil-to-air cooling. Most deep pans are used with an extension to place the filter lower, in the same proximity to the bottom of the pan to ensure fluid pickup under all operating conditions. A possible drawback with a deep pan is reduced ground clearance, which might cause the pan to run into road obstructions.

Add-on, oil-to-air supplementary coolers are available from many aftermarket sources. These come in kits to make installation quite easy (Fig. 14-18). Coolers are available in different sizes to suit the cooling needs of different size vehicles. Larger vehicles or heavier loads generate more transmission heat and require more cooling. Cooler setups for street-driven vehicles normally use rubber hoses with slip-on connections secured by common hose clamps. The coolers for racing/competition vehicles normally use more positive, threaded connections. The coolers using threaded connections also use larger line fittings to ensure adequate oil flow.

For transmissions where increased longevity is a concern, remote filters can be added to the cooler line (Fig. 14-19). Some filters are rather small, in-line devices; others use a standard, spin-on filter cartridge. These filters provide positive filtering of the fluid returning to the transmission to become lubrication oil.

The valve body or pressure regulator can be modified to provide a full-time fluid flow into the torque converter, cooler, and lubrication circuit. Remember that the pressure regulator valve shuts off this flow whenever line pressure falls below the pressure regulator setting. The modification is to grind a portion of land #2 (or whichever land separates torque converter feed pressure from line pressure) from the pressure regulator valve or drill a small feed hole (0.060 in., 1.5 mm) in the passage wall between torque converter feed and line or supply pressure. This change might drop line pressure in some gear positions slightly, but it ensures a converter and lubrication flow at all times.

For competition transmissions, lubrication oil quantity is a concern. Some types of competition require transmission operation in first and second gear at extremely high engine rpm. Some transmissions have rather tortuous and restricted lubrication oil passages; others have restricters in the passages to limit lubrication oil flow. Under these operating conditions, the gearset and the released apply devices must have an adequate fluid flow to reduce friction and remove heat. When building up a transmission for racing, it is good practice to trace out the cooler–lubrication circuit and study the path that the fluid must follow. Restrictions can often be reduced by drilling the passages to a larger diameter and improving any sharp corners that might limit fluid flow if there is enough metal surrounding them. This should be done with discretion, but the lubrication flows can be increased significantly in some transmissions.

14.5 GEAR TRAIN CHANGES

Several aftermarket sources provide low-ratio gearsets for the more popular transmission models (Fig. 14-20). These are relatively expensive, as are other special-purpose components. They provide additional torque and pulling power in first and second gears. For example, the replacement gearset for a THM 400 lowers the first-gear ratio from 2.48:1 to 2.75:1. If this is used in a vehicle with a 4:1 rear-axle ratio, the overall ratio is reduced from a stock ratio of 2.48×4, or 9.92:1, to a lower ratio of 1.75×4, or 11.1:1. The modified ratio is over 10 percent lower so it should increase the torque and pulling power by more than 10 percent.

A possible disadvantage with the lower gear ratio is that the engine rpm in first gear increases by 10 percent, and the top speed is reduced by 10 percent. There is also a greater rpm drop during the shift to the next higher gear.

Low-ratio gearsets are designed primarily for competition vehicles, but they are also somewhat popular with trailer towing, off-road, and RV vehicles, which need more pulling power.

Another gear ratio change is the addition of an overdrive gearset, which provides one overdrive gear ratio. The most common sets are units placed behind the transmission. These require a shorter drive shaft plus

Fig. 14-19 A filter can be added to the cooler line to remove any contaminants that might clog the cooler or cause wear in the transmission. *(Courtesy of A-1 Automatic Transmissions.)*

Fig. 14-20 Low-ratio gearsets for THM 400, C6, and Torqueflite. They provide a low-gear ratio that is about 11 percent lower and a second-gear ratio that is about 6 percent lower than the stock ratios. *(Courtesy of A-1 Automatic Transmissions.)*

mounting brackets to support the unit and absorb any torque reactions. Overdrive units are available that adapt to the extension housing area of some of the more popular transmissions. This installation, like the addition of a low-ratio gearset, requires complete transmission teardown and rebuilding.

14.6 COMPETITION TRANSMISSION: STRENGTHENING AND LIGHTENING CHANGES

If the engine's horsepower is increased, the tire size is increased to improve traction, or the vehicle's load is increased, some stock automatic transmission parts will fail. They are not designed to handle the added torque requirements. Several extra-duty or high-performance parts are available from aftermarket sources to cure the weaknesses in the more popular transmissions. Examples of these parts are:

- The outer sprag race for a THM 350—special sprag races are over three times as strong as the stock unit.

- The intermediate sprag for a THM 400—special sprags more than double the strength of the stock unit.

- The Powerglide high clutch hub—the special unit is significantly stronger and more reliable than the stock unit.

- Input shafts for several different transmissions are made from stronger steel alloys than the original ones.

For competition transmissions, some companies market special parts that allow the use of needle or roller bearings to support the rotating parts and Torrington bearings to replace the plastic or metal thrust washers. These reduce the friction loss and some of the heat that is generated in the transmission. An alternative to the expensive special bearings is to use wider-than-stock replacement bushings or two stock bushings (right next to each other) whenever possible. Increasing the bushing support area often reduces the amount of wobble and misalignment that a part can produce when spinning at high speed.

Lightweight clutch drums, hubs, and planetary gearset parts are also available for some transmissions. Some clutch drums are rather heavy, and the high–reverse clutch of a three-speed Simpson gear train spins about $2\frac{1}{2}$ times faster than the engine in first gear. During a fast acceleration, it takes a significant amount of power to speed up this heavy drum, and bringing the drum to a stop in second gear puts a fairly heavy load on the band. These lightweight units are made from high-strength aluminum alloys and are rather expensive. In racing situations that demand rapid acceleration, they offer the optimum performance. Some lightweight gearset components are drilled with rather large holes to reduce the amount of metal and weight.

14.7 MANUAL SHIFT VALVE BODIES

Manual shift valve bodies eliminate automatic upshifts and downshifts. The shifter has a gear position for each gear, and the transmission immediately shifts into whatever gear position the selector is placed. Most manual shift valve bodies are designed for drag racing and use a reverse shift pattern. The gear positions will be in the order: park, reverse, neutral, first, second, and third. Manual shift valve bodies are also designed for pulling (tractor or truck), off-road, and rally racing vehicles.

For the most part, most manual shift valve bodies look like stock valve bodies on the outside. The internal fluid passages and valves have been significantly changed. They are designed so the shifts occur instantly with no lag.

Some drag racing valve bodies have the added feature of an internal transmission brake. These units use an electric solenoid that opens passages to shift the transmission into both first and reverse at the same time (Fig. 14-21). When the solenoid is deactivated by a driver-controlled switch, the transmission is instantly in first gear only, and the car is free to accelerate. The purpose of this is to produce an instant response so the car will "launch" instantly, much like a car using a clutch and standard transmission. If used excessively with the

Fig. 14-21 Valve body for a THM 350 with a solenoid valve and some tubing added for the transmission brake, which allows the transmission to totally lock and hold the drag race vehicle stationary while it is staged. *(Courtesy of A-1 Automatic Transmissions.)*

converter at full stall, a transmission brake can significantly shorten the life of the converter.

14.8 HIGH-STALL CONVERTERS

It should be remembered that stall speed is the maximum speed that an engine can run against a stalled, stationary transmission input shaft and torque converter turbine, and this speed is determined by how strong the engine is and how tight or loose the converter is. A high-stall converter is one that allows the engine to reach a higher than normal speed. As described in Chapter 7, the stall speed is primarily controlled by the following factors:

- Converter Diameter—The smaller the converter, the higher the stall speed, about 30 percent difference in speed per inch in diameter (Fig. 14-22).
- Impeller Fin Angle—Negative fin angle produces a higher stall speed.
- Stator Fin Angle—The greater the fin angle, the higher the stall speed and the greater the torque multiplication ratio.
- Impeller to Turbine Fin Clearance—The greater the clearance, the higher the stall speed and the lower the converter efficiency.

High-stall converters are commonly used in drag racing and truck- or tractor-pulling vehicles. On these vehicles, it is important that the engine be allowed to quickly reach the best rpm for producing maximum power. For example, a typical 300-in.^{3} (5-L) engine that is built up for racing produces about 90 horsepower at 2000 rpm and about 200 horsepower at 4500 rpm. Acceleration from a dead stop with a stock (about 2000 rpm stall speed) converter begins with an engine turning 2000 rpm and producing 90 horsepower. A higher stall speed converter allows the engine speed to increase, and the vehicle's acceleration begins with an engine that is producing more power, as much as 110 horsepower more depending on the converter. After the two converters reach their coupling speeds, there is very little difference in their performances.

In some cases, the added horsepower can be a disadvantage if it produces too much wheel spin or breaks drive train parts. Another possible disadvantage that can be created by excessive stall speed is poor fuel mileage and noise. Imagine driving a car around town that leaves from stops at engine speeds of over 4000 rpm, and compare this with the average car, which begins accelerating at 1000 to 2000 rpm. High-stall converters often prove to be noisy fuel wasters during normal driving.

Most high-performance torque converter manufacturers market a variety of converters with very high stall competition converters at one extreme and high torque-improved gas mileage units at the other. This second converter is popular with RV- and trailer-towing vehicles as well as other vehicles where low-end pulling power and fuel mileage are important. These converters have a lower than stock stall speed and often have a greater torque multiplication ratio.

All-out competition converters are usually precision remanufactured stock converters that have the fins welded or brazed in place for improved strength; have

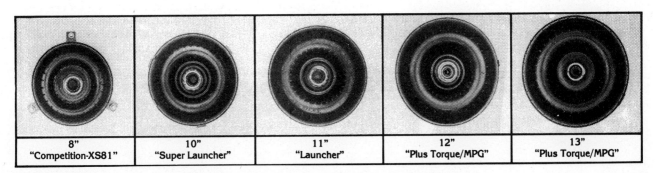

Fig. 14-22 Five converters built especially for five different purposes. Note the different diameters that help determine the different stall speeds. *(Courtesy of A-1 Automatic Transmissions.)*

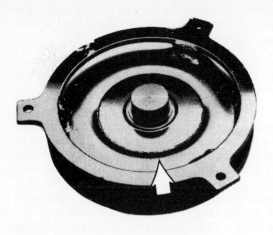
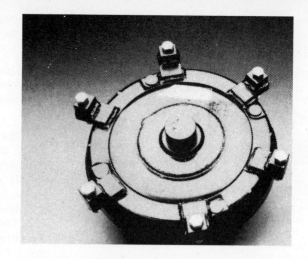

Fig. 14-23 Competition converters with extra reinforcement and mounting mounts in the drive area. *(Courtesy of A-1 Automatic Transmissions.)*

Torrington thrust bearings, steel turbine splines, and heavy-duty stator clutches installed for improved durability; and are assembled in a precision manner. In some cases the drive lugs are reinforced for added strength (Fig. 14-23).

14.8.1 Direct Couplers

In some cases, automatic transmissions are used without torque converters. The transmission input shaft and hydraulic pump are driven by a shaft connected directly to the crankshaft or flywheel (Fig. 14-24). This allows the length and weight of the transmission as well as the rotating mass to be reduced significantly. Direct couplers are used in boats and in some circle track racing cars. The transmission becomes essentially an in-and-out box with two or three forward-gear ratios plus reverse.

14.9 CONCLUSION

It has been the intent of this author to introduce you to the aspects, possibilities, and some of the excitement of automatic transmission tuning. It is impossible to cover all of the information needed to fully develop a transmission for all of the various vehicles (racing and production) in a book of this type. More than a few knowledgeable manufacturers of specialized transmission components can provide information concerning transmission modifications for those who want to carry this farther.

Some transmission tuning, such as shift kit installation and torque converter swapping, is done in the average transmission shop to cure minor transmission problems and improve customer satisfaction. The more specialized work involved in producing a custom racing transmission is not. It is considered too time consuming and raises problems of warranty if something goes wrong. This type of work is usually left to the highly specialized shop, race car builder, and highly knowledgeable hobbyist.

Fig. 14-24 Coupler attaches directly to the flexplate and is used to drive the transmission pump and input shaft. *(Courtesy of A-1 Automatic Transmissions.)*

APPENDIX 1

ENGLISH-METRIC-ENGLISH CONVERSION TABLE

MULTIPLY	BY	TO GET/ MULTIPLY	BY	TO GET
LENGTH				
inch (")	25.4	millimeter (mm)	0.3939	inch
mile	1.609	kilometer	0.621	
AREA				
inch^2	645.2	millimeter^2	0.0015	inch^2
PRESSURE				
pounds/in.^{2}	6.895	kilopascals (kPa)	0.145	psi
VOLUME				
inch^3	16,387	millimeter^3	0.00006	inch^3
inch^3	6.45	centimeter^3	0.061	inch^3
inch^3	0.016	liter	61.024	inch^3
quart	0.946	liter	1.057	quart
gallon	3.785	liter	0.264	gallon
WEIGHT				
ounce	28.35	gram (g)	0.035	ounce
pound	0.453	kilogram (kg)	2.205	pound
TORQUE				
pound-inch	0.113	newton-meter (N·m)	8.851	pound-inch
pound-foot	1.356	newton-meter	0.738	pound-foot
VELOCITY				
miles/hour	1.609	kilometer/hour	0.6214	miles/hour

TEMPERATURE
(degree Fahrenheit − 32) × 0.556 = degree Celsius
(degree Celsius × 1.8) + 32 = degree Fahrenheit

APPENDIX 2

DESCRIPTION OF GENERAL MOTORS TRANSMISSIONS/TRANSAXLES

	No Upshift	Two-speed	Three-speed	Four-speed	RWD	RWD Transaxle	FWD	FWD Transaxle	Cast-Iron Case	Fluid Coupling	Multistage Converter	Lockup Converter	Semiautomatic	Adjustable Band	Buick	Cadillac	Chevrolet	Oldsmobile	Pontiac
Hydramatic				X	X				X	X				X		X		X	X
Roto-Hydramatic, "Slim Jim"			X		X				X	X								X	X
Controlled Coupling, "Jetaway"				X	X				X	X						X		X	X
Dynaflow	X				X				X	X	X				X				
Powerglide (cast-iron case)		X			X				X						X		X		
Turboglide	X				X						X						X		
Powerglide (Corvair)		X				X											X		
Hydramatic, F-85			X		X				X									X	
Dual-Path		X			X										X				
Powerglide (aluminum case)		X			X												X		
Tempest Torque		X				X													X
THM 300		X			X										X			X	X
THM 400			X		X								S		X	X	X	X	X
THM 425			X				X						S		X			X	
Torquedrive	X				X									X			X		
THM 350			X		X							X			X		X	X	X
THM 250			X		X							X		X	X		X	X	X
THM 200			X		X							X			X	X	X	X	X
THM 180			X		X							X					X		
THM 325			X				X					X			X			X	
THM 125			X					X				X			X		X	X	X
THM 200-4R				X	X							X			X	X		X	X
THM 700-R4				X	X							X					X		X
THM 325-4L				X			X					X			X			X	
THM 440				X				X				X			X	X	X	X	X
THM F-7				X				X								X			
4L30-E				X	X											X			
4T60-E				X				X											

Abbreviations: RWD, rear-wheel drive; FWD, front-wheel drive.
S indicates that some models used a variable-pitch stator in the torque converter.
Lockup torque converters were used after 1980 in those models indicated.
These descriptions apply to transmissions and transaxles used in domestic passenger cars.

APPENDIX 3

CHRONOLOGY OF GENERAL MOTORS AUTOMATIC TRANSMISSIONS/ TRANSAXLES

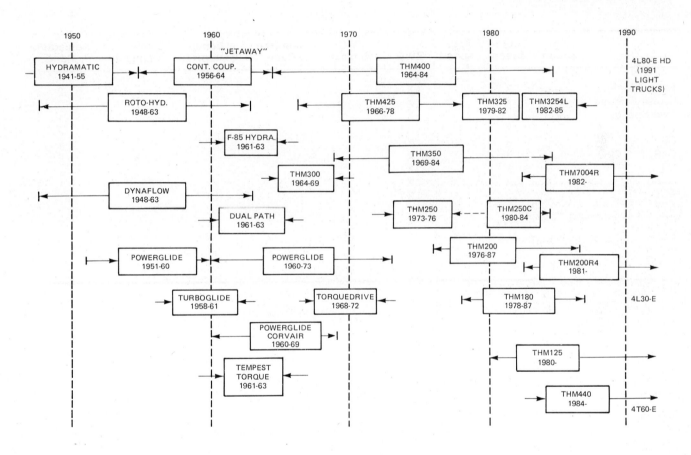

APPENDIX 4

DESCRIPTION OF FORD MOTOR COMPANY TRANSMISSIONS/ TRANSAXLES

	Two-speed	Three-speed	Four-speed	RWD	FWD Transaxle	Cast-iron Case	Lockup Converter	Semi-automatic	Adjustable Band
Ford-O-Matic		X		X		X			2
Cruise-O-Matic		X		X		X			2
Two-Speed	X			X					1
C4		X		X					2
C6		X		X					1
C4S		X		X				X	2
C3		X		X					2
Jatco		X		X					1
AOD			X	X			DD		
ATX		X			X		S		
C5		X		X			X		2
A4LD			X	X			X		
AXOD			X		X		X		
E4OD			X	X			X		

DD indicates direct drive, S indicates some.

These descriptions apply to transmissions and transaxles used in domestic passenger cars.

APPENDIX 5

CHRONOLOGY OF FORD MOTOR COMPANY AUTOMATIC TRANSMISSIONS/TRANSAXLES

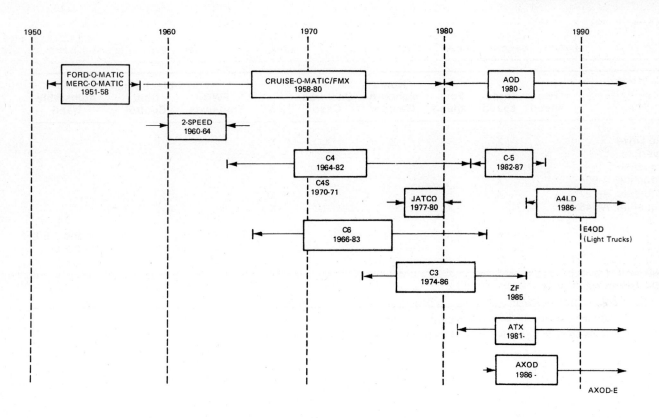

APPENDIX 6

DESCRIPTION OF CHRYSLER CORPORATION TRANSMISSIONS/ TRANSAXLES

	Two-speed	Three-speed	Four-speed	Non-planetary Gearset	Cast-iron Case	RWD	FWD Transaxle	Electronic Shift Control	Adjustable Band
Fluid Drive		X		X	X	X			
Powerflite	X				X	X			
Torqueflite		X			X	X			
Torqueflite, A-904		X				X			
Torqueflite, A-727		X				X			
Torqueflite, A-345			X			X			
A-404		X					X		1.7 L
A-413		X					X		1.4 and 1.6 L
A-415		X					X		2.2 L
A-470		X					X		2.6 L
A-500			X			X		X	
A-604, Ultradrive			X				X	X	

Numbers in right margin indicate engine size used with transaxle.
These descriptions apply to transmissions and transaxles used in domestic passenger cars.

414

APPENDIX 7

CHRONOLOGY OF CHRYSLER CORPORATION AUTOMATIC TRANSMISSIONS/ TRANSAXLES

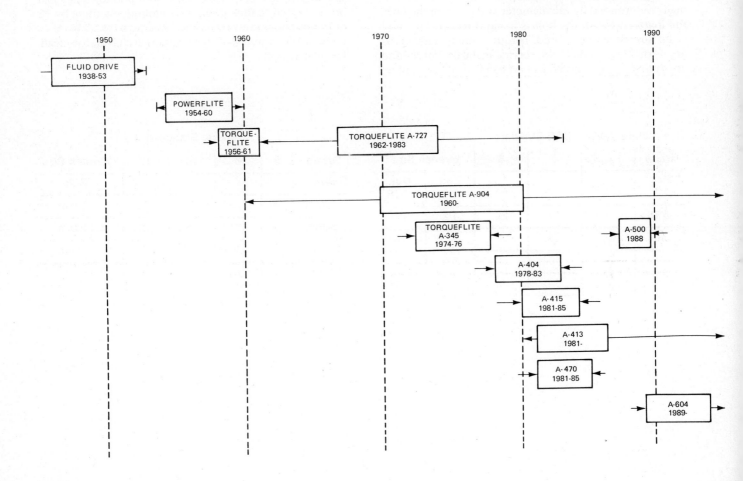

APPENDIX 8

BOLT TORQUE TIGHTENING CHART

The tightening value for a particular nut or bolt is normally determined by the diameter and grade of the bolt, the material which the bolt is being threaded into, and if the threads are lubricated. Torque value specifications are normally printed in the service manuals, but if none are available, the following chart can be used as a guide.

The following values are given in foot-pounds. They can be converted to inch-pounds by multiplying them by 12 or to newton-meters by multiplying them by 1.356. Also, it should be noted that these values are given for clean, lubricated bolts.

SAE Standard

Grade	1 & 2	5	8	Wrench Size	
Size				Bolt	Nut
1/4	5	7	10	3/8	7/16
5/16	9	14	22	1/2	9/16
3/8	15	25	37	9/16	5/8
7/16	24	40	60	5/8	3/4
1/2	37	60	92	3/4	13/16

Metric Standard

Grade	5	8	10	12	Wrench Size
Size					Bolt
6 mm	5	9	11	12	10 mm
8 mm	12	21	26	32	14 mm
10 mm	23	40	50	60	17 mm
12 mm	40	70	87	105	19 mm
14 mm	65	110	135	160	22 mm

GLOSSARY

Accelerate: To increase speed.

Accumulator: A device used to dampen fluid apply pressure so as to cushion or soften a shift.

Additives: Chemicals added to automatic transmission fluid (ATF) to improve the operating characteristics.

Aerated fluid: Fluid of a foamy nature that has been whipped to the point where it contains air bubbles.

Annulus gear: Ring gear, internal gear.

Apply device: Hydraulically operated clutches and bands and mechanical one-way clutches that drive or hold portions of the planetary gearset.

Atmospheric pressure: Pressure of the atmosphere around us; generally considered to be 14.7 psi at sea level.

Automatic transmission: A transmission that changes forward-gear ratios automatically.

Automatic transmission fluid (ATF): The oil used in automatic transmissions.

Balanced valve: A valve that is in a balanced position between two separate opposing forces.

Ballooning: The undesirable expansion of a component such as a torque converter because of excessive internal pressure.

Band: A lined metal strap that wraps around a drum; used to stop or hold the drum stationary.

Belleville spring: A conical steel ring that gives a spring action because of its resistance to forces that try to flatten it.

Bevel gear: A gear with teeth that are cut at an angle so it can transmit power between shafts that are not parallel.

Bump: A harsh shift condition produced by a sudden and forceful application of a clutch or band.

Case: The rigid housing for a transmission.

Centrifugal force: The force on a revolving object that tries to push it away from the center of revolution.

Chuggle: A bucking or jerking condition usually most noticeable during acceleration and when the torque converter clutch is engaged. Chuggle is often engine related.

Clutch: A device that controls the power transfer between two points. It can allow or stop the transfer.

Coastdown: A complete release of the throttle so the moving vehicle slows down; sometimes called "zero-throttle coastdown."

Coefficient of friction: A reference to the amount of friction between two surfaces.

Coupling phase: A condition where the torque converter turbine speed is almost equal to that of the impeller.

Damper: A device that reduces torsional vibrations between the engine and transmissions.

Decelerate: Reduce speed.

Delayed: When a shift occurs later than expected. Also called "late" or "extended."

Detent: A spring-loaded device used to position the manual valve correctly.

Detent (throttle position): A movement of the throttle to the maximum. *See* Wide-open throttle.

Differential: A gear arrangement that allows the drive wheels to be driven at different speeds.

Direct drive: A 1:1 gear ratio.

Double bump: Two sudden and forceful clutch or band applications.

Dynamic friction: The relative amount of friction between two surfaces that have different speeds. *See* Static friction.

Early: When a shift occurs before the correct speed is reached and causes a lugging or laboring of the engine.

End bump: A condition that produces a firmer feel at the end of a shift. Also called "end feel" or "slip bump."

End play: The amount of motion a gearset has in a direction that is parallel to the input shaft.

Energy: The ability to do work.

Engine braking: A condition where engine drag is used to slow the vehicle down.

Exhaust: The release of fluid pressure back to the sump.

Final drive: The last set of reduction gears before the power flows to the differential and drive axles.

Firm: A noticeable quick application of a clutch or band that is considered normal.

Flare: A condition that produces an increase in engine rpm with a loss of torque during a shift. Also called "slipping."

Flexplate: The thin drive plate that connects the torque converter to the crankshaft of the engine.

Fluid coupling: A device that transfers power through the fluid.

Flywheel: The rotating metal mass attached to the crankshaft that helps even out power surges.

Force: A push or pull measured in units of weights, usually pounds.

Friction: The resistance to motion between two bodies in contact with each other.

Friction modifier: An additive that changes the lubricity of a fluid.

Fulcrum: The pivot/supporting point for a lever.

Gear reduction: A condition in which the driving gear is smaller than the driven gear. This is normally used to produce a torque increase.

Governor: A hydraulic device that provides hydraulic pressure that is relative to road speed.

Harsh: A very noticeable and unpleasant application of a clutch or band. Also called "rough."

Heavy throttle: An acceleration with the throttle opened about three-fourths of the pedal travel.

Helical gear: A gear that has the teeth cut at an angle.

Hunting: A series of repeating upshifts and downshifts that cause noticeable engine rpm changes. Also called "busyness."

Hydraulics: A branch of science dealing with the transfer of power through fluids under pressure.

Hydrodynamics: A branch of science dealing with the action of fluids in motion.

Hypoid gear: A special form of a bevel gear that has the teeth cut in a curvature and that positions the gear on nonintersecting planes; commonly used in rear-wheel drive (RWD) final drives.

Idler gear: A gear positioned between two other gears in the gear that is used to reverse the direction of rotation but does not change the ratio.

Impeller: Also called converter pump, the input member of the torque converter.

Inertia: The physical property that a body at rest remains at rest and a body in motion remains in motion and travels in a straight line.

Initial feel: A distinct feel that is firmer at the start of a shift than at the end.

Internal gear: A gear with the teeth pointing inward toward the center of the gear.

Kickdown: An arrangement that produces a downshift when the driver pushes the accelerator pedal all the way down.

Land: The large-diameter portion of a spool valve.

Late: A shift that occurs above the correct speed that causes overreving of the engine.

Light throttle: An acceleration with the throttle opened about half of the pedal travel.

Manual valve: The valve operated by the gear selector that directs pressure to the apply devices needed to put a transmission in gear.

Medium throttle: An acceleration with the throttle opened about half of the pedal travel.

Minimum throttle: The least amount of throttle opening that can accelerate a car and produce an upshift.

Modulator: A vacuum device that senses engine load and causes a hydraulic valve to change pressure relative to that load.

Multiple-disc clutch: A clutch that uses more than one friction disc.

Needle bearing: A very thin roller bearing.

Oil cooler: A radiatorlike component that reduces the fluid temperature.

Oil filter: The part that removes dirt and debris from oil that is circulated through it.

One-way clutch: An overrunning clutch that locks in one direction and overruns or freewheels in the other.

Orifice: A restricted opening in a fluid passage designed to reduce fluid pressure while flow is occurring.

O-ring: A sealing ring made from rubberlike material.

Overdrive: A gear arrangement that causes the output shaft to turn faster than the input shaft.

Parking gear: A transmission lock that prevents the output shaft from turning.

Passing gear: A downshift to the next lower gear.

Pawl: A locking device that fits into a gear tooth to hold the gear stationary.

Pinion gear: A small gear that meshes with a larger gear.

Piston: The moveable portion of a hydraulic cylinder or servo.

Planetary gearset: A gear system that uses a particular gear arrangement to produce one or more gear ratios.

Power: The rate at which work is done.

Pressure: A force per unit area; generally measured in pounds per square inch (psi) or units of atmospheric pressure (bars or kilopascals).

Pump: A device that transfers fluid from one point to another and is used to create hydraulic pressure.

Reaction member: The portion of the planetary gearset that is held stationary in order to produce a reduction or overdrive.

Reservoir: A tank or pan where fluid is kept.

Roller clutch: A one-way clutch that uses a series of rollers positioned in a special cam for the locking elements.

Rotary flow: The fluid motion inside a torque converter in the same direction as the impeller and turbine.

Servo: A hydraulic device that changes fluid pressure into mechanical motion or force.

Shift feel: A description of clutch or band application or release that is usually described, for example, as firm or soft.

Shudder: An easily noticed jerking sensation that is a more severe form of chuggle.

Slipping: A loss in torque with a noticeable increase in engine rpm.

Soft: A slow and almost unnoticeable clutch or band apply.

Spline: A groove or slot cut into a shaft or bore that is used to connect a matching spline.

Sprag: The locking element in a one-way sprag clutch.

Stall: A condition where the engine is running but the transmission input shaft is not rotating.

Static friction: The relative amount of friction between two stationary surfaces or two surfaces that are turning at the same speed. *See* Dynamic friction.

Stator: A component in the torque converter that is used to change the direction of fluid motion.

Sump: The fluid storage point or reservoir; usually the transmission pan.

Surge: A barely noticeable engine feel that is similar to chuggle.

Thrust washers: Bearings that separate rotating parts that turn against each other.

Tie-up: A condition where two opposing clutches or bands are attempting to apply at the wrong time. This tends to slow down the engine or vehicle. Also called "fight."

Torque: Turning or twisting effort; usually measured in foot-pounds or newton-meters.

Torque converter: A fluid coupling that transfers power from the engine to the transmission and can produce a torque increase.

Transaxle: A transmission that is combined with the final drive assembly; normally used in front-wheel drive cars.

Transmission: A device in the power train that provides different forward-gear ratios as well as neutral and reverse.

Turbine: The output member of a torque converter.

Valley: The area of a spool valve between the lands.

Valve: A hydraulic device that controls fluid flow.

Viscosity: The resistance to flow in a fluid.

Vortex flow: A recirculating fluid flow in the converter that is outward in the impeller and inward in the turbine.

Wide-open throttle (WOT): Full travel of the pedal to produce maximum power from the engine and full travel of the throttle or detent valve in the transmission.

Work: The result of force that changes speed or direction of motion.

INDEX

A

Accumulator cover seal, 304
Accumulators, 124, 399–400
Additives, 127–128
Air tests, 259, 283–285
Aisin-Warner (AW-4) transmission, 236, 238
All-wheel drive (AWD), 21
Amperes, 165
Annulus gear (*see* Ring gear)
Antiwear agents, 127
Apply devices, 16, 32–47 (*see also* Bands)
Automatic Transmission Rebuilder's Association (ATRA), 315
Automatic Transmission Service Group (ATSG), 315
Automatic transmission fluid (ATF), 124, 126

B

Ballooning, 137
Band adjustments, 256
Bands, 16, 40–41, 42–43, 45, 347, 403–404
Bearings, 338, 340
Belleville plate, 40
Belleville spring, 37
Bolt torque tightening table, 416
Boost, 119, 135, 157
Brake switch, 167, 175–176
Breakaway first, 52
Buick, Dual-Path transmission, 192
Bushings, 338, 340

C

Center support, 73
Checks, torque converter, 384–389
Chevrolet, Turboglide, 49
Chrysler Corporation:
  A-404, 52, 201–203
  A-413, 52, 201–203
  A-415, 52, 201–203
  A-470, 52, 201–203
  A-500, 62, 164, 200–201
  A-604, 81, 86–87, 164, 169, 203, 205
  A-727, 199
  A-904, 52
  A-904, 199
  automatic transmissions/transaxles, chronology of, 415
  automatic transmissions/transaxles description, of, 414
  detent valve, 143
  Fluid Drive, 2
  front clutch, 34, 60

Chrysler Corporation (*contd.*)
  mechanical throttle valves, 143
  Powerflites, 49, 66
  rear clutch, 54
  switch valve, 137
  Torqueflites, 52, 198–199
Circuit, 165
Cleaning:
  cooler, 391
  torque converters, 389–391, 392
  transmissions, 319, 333–334
Clutch:
  assembly, service, 358, 361
  band application, 142
  clearance, 349–350
  cone, 41
  control, torque converter, 137, 142
  converter, 137, 188–190, 386–387
  direct, 34, 60, 75
  forward, 54, 73, 81, 84
  front, 34, 60
  high, 71
  high-reverse, 34, 60, 73
  intermediate, 56, 75
  intermediate roller, 56
  low and reverse, 81, 84
  low-reverse, 56
  multiple disc, 16, 40, 42
  one-way, 16, 41, 45
  operation, 38–39
  overrun, 45, 81, 84
  packs, modifications, 402–405
  plates, 37
  rear, 54
  roller, 45
  second, 75
  sprag, 45
  stator one-way, check, 384–385
  third, 75
  torque converter, tests, 259, 266–267
  viscous converter, 190, 192
Coasting downshift, 159
Comfort power switch, 174
Compound planetary gearsets, 15–16
Cone clutch, 41
Continuously variable transmission (CVT), 25–26
Control devices (*see* Apply devices)
Control pressure, 113, 132
Conversion table, English-metric, 409
Converter clutch, 188–190
  checks, 387
  control valve, 137
  interference, 386–387

Converters:
   high-stall, 407–408
   rebuilding, 391
Cooler flow, 139, 141
Copper-to-copper contact, 177
Corrosion inhibitors, 128
Crescent pump, 109
Cushion plates, 40

D

Default mode, 174
Depth filter, 113
Detent cam, 147
Detent downshift, 159
Detent valve, 143, 147
Detergents-dispersants, 127
Diagnostic tools and procedures, 261, 288–290
Diaphragm spring (*see* Belleville spring)
Differential, 21
Direct clutch, 34, 60, 75
Direct couplers, 408
Disassembly, 314
Downshifts, 159
Downshift valve, 143
Driving members, 32, 34
Drum, 35
Dynamic friction, 39
Dynamic seal, 120
Dynamometer, 3

E

Elastomers, 122
Electrical circuit testing, 177–179
Electrical system checks, 259, 274–275
   computer-controlled circuits, 275–276
   solenoids, 277
   speed sensors, 279
   switches, 278–279
   throttle position sensor, 279–280
Electronic transmission control unit (ETCU), 166, 170, 174–176
Electronic transmission controls, 164–179
End play, 323, 385–386
Engine Control Module (ECM), 116–117, 137, 166
Engine coolant temperature sensor, 176
Exhaust, 132
Exhaust gas recirculation (EGR) valve, 146
Extension housing, 306–308
External retaining rings, 322

F

Filters, 111–113
First gear, 52
Flexible band, 42
Flexplate, 182
Fluid:
   changes, 244–247
   checks, 242–244, 258

Fluid (*contd.*)
   manual valves, 147
   power (*see* Hydraulic systems)
Foam inhibitors, 128
Ford Motor Company:
   A4LD, 62, 208
   AOD, 71, 79, 81, 213–214
   ATX, 71, 74–75, 77, 79, 192, 194, 212–213
   AW4, 66
   automatic transmissions/transaxles, chronology of, 413
   automatic transmissions/transaxles, description, of, 412
   AXOD, 81, 93, 96, 97, 214–216, 327
   C3, C4, C5, 52, 56, 205, 207
   C6, 52, 207
   E4OD, 164, 208
   FMX, 66, 73, 79, 211
   forward clutch, 54
   high-reverse clutch, 34, 60
   Jatco, 52, 207
   transmission fluids, 124
   two-speed transmission, 49, 66
   vacuum modulators, 145
   ZF-4HP-22, 216
Forward clutch, 54, 73, 81, 84
Forward sprag, 81, 84
Four-wheel drive (4WD), 21–25
Friction members (*see* Control devices)
Friction modifiers, 128
Friction plates (*see* Lined plates)
Front clutch, 34, 60
Front-wheel drive (FWD), 1

G

Gears:
   bevel, 5
   external, 7
   first, 52
   helical, 5
   internal, 7
   planetary, 7, 192, 193
   ratios, 3–5, 12–14
   reduction, 3
   ring, 7
   service, 361
   spur, 5
   sun, 7
   third, 60
Gear trains:
   modifications, 405–406
   types, 52–101
General Motors Corporation:
   automatic transmissions/transaxles description and chronology of, 410–411
   direct clutch, 34, 60
   Dynaflows, 66
   F7, 4L30-E, 4T60-E, 164
   forward clutch, 54
   Hydra-matic transmission, 217
   Powerglide, 49, 66
   THM 125, THM 125C, 52, 56, 220, 222–224, 323, 327
   THM 180 (Hydra-matic 3L30), 66, 231

General Motors Corporation (*contd.*)
    THM 180, 74–75, 79
    THM 200, THM 250, 52, 56, 219–220
    THM 200-4R, 62, 66, 225
    THM 300, 66, 230–231
    THM 325, 52, 56, 219–220
    THM 325-4L, 62, 66, 225
    THM 350, 52, 56, 225, 337
    THM 350, 60, 406
    THM 375, 52
    THM 400 (Hydra-matic 3L80), THM 425, 52, 224–225
    THM 400, 406
    THM 440, 81, 93, 96, 327
    THM 440, 327
    THM 440-T4 (4T60), 231, 236
    THM 700, 81
    THM 700-R4 (Hydramatic 4L60), 231
    THM 700-R4, 35, 41, 86–87
    Turbo Hydra-matic transmission, 217
    vacuum modulators, 145
Gerotor pump, 109
Governor, 17
Governor valve assemblies, 154–155
Grooves, 113
Ground circuit, 165, 176–177

H

Hard parts, 314
Heat exchanger, 139
High clutch, 71
High-reverse clutch, 34, 60, 73
High-stall converters, 407
Holding members, 40–41
Hondamatic, 25
Hook ring, 121
Horsepower, 3
Housing, 35
Hydra-matic 300, 49
Hydraulic pressure tests, 259, 269–270
    checking governor pressure, 272, 274
    fluid flow diagrams, 271–272
    interpreting pressure readings, 270–271
Hydraulic systems:
    automatic transmission fluid (ATF), 124–129
    basics, 108–109
    components of, 107–108
    flow and pressure, 109–113
    flow control, 113–118
    modifying flow and pressure, 124
    operation, 131–132
        governor valve, 153–155
        manual valve, 147
        pressure development and control, 132–135
        shift modifiers, 161
        shift valves, 155–159
        throttle pressure, 141–147
        torque converter, 135–141
    pressure control, 118–119
    principles, 104–107
    sealing pressure, 119–124

I

Impeller, 18, 182
Import transmissions, 238
Intermediate band, 56, 73
Intermediate clutch, 56, 75
Intermediate roller clutch, 56
Internal retaining rings, 322

J

Jatco, 52, 207
Jeep, 236, 238

K

Kickdown band, 56
Kickdown valve, 143, 147
Kits, 315

L

Lands, 113
Lathe-cut seal, 121
Leak checks, 388–389
Limp-in mode, 174
Line pressure, 113, 132
Lined plates, 37, 341, 346
Linkage checks, 247–255
Lip seals, 350, 351
    replacement, 294
    shape, 121
Lockup converters, 188–192
Low and reverse band, 56
Low and reverse clutch, 81, 84
Low band, 71
Low-reverse band, 56
Low-reverse clutch, 56
Lubrication:
    flow, 141
    improvement, 404–405

M

Mainline pressure, 109, 132
Maintenance (*see* Service and maintenance)
Manual downshift, 159
Manual linkage checks, 247–250
Manual shift valve bodies, 406–407
Manual valves, 17, 109, 113–114, 147
Metal deactivators, 128
Modifications (*see* Transmissions, modifications)
Modulators:
    checks, 258
        bellows comparison check, 264
        diaphragm leakage, 261–262
        load check, 264
        spring pressure, 263–264
        stem alignment, 263
        vacuum supply, 262–263

Modulators (*contd.*)
   pressure, 114
   removal and replacement, 294
Modulator valve, 17
Multiple-disc clutches, 16, 40, 42
Multiple-disc driving clutches, 35, 37

N

Neutral start switch, 166, 174–175
Neutral, 52
Noise and vibration checks, 259, 285, 288

O

O-ring seal, 121, 350, 353
Ohms, 165
Oil cooler, 139, 141
Oil leak checks, 259, 285
Oil pan debris check, 259, 282
One-way clutches, 16, 41, 45
Open circuit, 176–177
Orifice, 124, 399
Overdrive, 62
Overhaul procedures, 315–316
   disassembly, 319–331
   removal, 316–319
Overlap control, 157
Overrun clutch, 45, 81, 84
Oxidation inhibitors, 127

P

Park pawl, 51
Part-throttle downshift, 159
Pascal's law, 104
Pinions, 7
Planet carrier, 7
Planetary gear ratios, 12–14
Planetary gear torque converters, 192, 194
Planetary gearsets, 7, 15–16
Pollution control devices, 146
Power flow, 49–101
Power-economy switch, 167
Powerglide, 73
Pressure, 105–107, 132–135
Pressure regulator valve, 108–109, 118–119
Pumps, 18, 182, 108–109, 111, 355–358

Q, R

Quad-ring seal, 122
Ravigneaux gearsets, 66–81
Reaction carrier, 52
Reaction members, 16
Reaction surfaces, 113
Rear clutch, 54
Rear-wheel drive (RWD), 1
Reassembly, 374–379
Red rag disease, 333

Repair, in-car:
   aluminum thread, 298
   clogged cooler, 300
   cooler line and fitting, 299–300
   extension housing seal and bushing, 304–308
   governor, 308–309, 311
   lip seal replacement, 294
   manual shift linkage and seal, 294
   neutral start switch, 301
   servo and accumulator piston, cover, and seal, 303–304
   throttle valve (TV) linkage and seal, 294
   vacuum modulator removal and replacement, 294
   valve body, 302–303
Restricter plate, 117
Retaining rings, 322
Reverse, 60
Reverse band, 71
Reverse clutch, 71, 75
Rigid band, 42
Ring gear, 7
Road tests, 258–266
Roller clutches, 45
Rooster comb, 147
Rotary floe, 184
Rotor pump, 109
Rust inhibitors, 128

S

Seals:
   quality, 122, 124
   service, 350–353
Seal swellers, 127
Second clutch, 75
Selector valve, 147
Semiautomatic transmissions, 2
Sensors, 166–167, 174–176, 279, 280
Separator plate (*see* Unlined plates)
Service and maintenance:
   band adjustments, 256
   fluid changes, 244–247
   fluid checks, 242–244
   manual linkage checks, 247–250
   throttle linkage checks, 250–255
Service manuals, 315
Servo, 16, 43, 45, 303–304
Shift feel, 124, 142
Shift improver kits, 398
Shift modifier valves, 17
Shift modifiers, 161
Shift overlap, 157
Shift quality, 39–40, 124, 400
Shifts, abnormal, 264–266
Shift timing, 142
Shift valves, 17, 155–160
Shorted circuits, 176–177
Short-to-ground (*see* Grounded circuits)
Shuttle valve, 117
Simple machines, 3
Simpson gear train, 15, 51–62
Simpson gear train plus additional gearset, 62–66
Soft parts, 314

Solenoids, 116, 167–170, 277
Solid-state electronics (*see* Electronic transmission controls)
Speed sensors, 166, 174, 279
Spool valves, 113
Sprag clutches, 45
Square cut seal, 122
Stall speed, 187–188, 407
Stall tests, 188, 259, 280–282
Static friction, 39
Static seal, 119–120
Stator, 18, 182, 184, 186
Stator one-way clutch check, 384–385
Steel plate (*see* Unlined plates)
Strut, 42
Subassembly repair, 355–374
Sun gear, 7
Supply pressure, 109, 132
Surface filter, 112
Switch valve, 137
Switches, 165, 278–279

T

Tapered roller bearings, 340
Teardown, 314
Teflon rings, 121, 350, 351, 354–355
Tests and checks, 258–259
Third clutch, 75
Third gear, 60
Throttle linkage checks, 250–255
Throttle position sensors, 166, 174, 279–280
Throttle pressure (TP), 141–143, 157
Throttle valves (TVs), 17, 114, 294
   mechanical, 143, 145
   vacuum, 145–147
Thrust washers, 338, 340
Torque, 2–3
Torque converter clutch control, 137, 142
Torque converter clutch tests, 259, 266–267
Torque converter clutch valve, 17
Torque converter pressure control, 135, 137
Torque converters:
   checks for, 384–389
   construction of, 182–183
   definition of, 19–20
   fluid temperatures, 245
   installation of, 379–380
   lockup, 118, 188–192
   oil cooler, and lubrication circuit, 135–141
   operation of, 183–188
   planetary gear, 192, 193
   purpose of, 181
   rebuilding, 391–394
Torqueflites, 57, 198–199
Torque multiplication, 185–186
Torrington bearing, 338
Tractive force, 5
Transaxles, 11, 20–21
   Chrysler, 198
   Ford Motor Company ATX, 198
   Ford Motor Company AXOD, 198
   GM, 198

Transfer plate, 117
Transmission control unit (TCU), 16
Transmissions:
   Aisin-Warner, 236–237
   Chrysler, 198–205
   competition, 406
   components, 2–29
   Ford Motor Company, 205–216
   GM, 217–236
   imports, 238
   in-car repair, 293–311
   modifications, 397–408
   overhaul, 314–380
   problem solving and diagnosis, 258–290
   purpose, 1–2
   service and maintenance, 242–256
   standard, 7
Troubleshooting charts (*see* Diagnostic tools and procedures)
Turbines, 18, 182, 186–187
Turbulator, 138
Two-4 band, 81, 84
Two-clutch transmission, 25–26
Two-way check valve, 117

U

Unlined plates, 37, 117, 341, 346

V

Vacuum modulators, 145
Valleys, 113
Valves:
   governor, 153–155
   manual, 147
   service, 364
   shift modifiers, 161
   shift, 155–160
   throttle, 141–147
   types, 17
Vane pump, 109
Viscosity index improvers, 127
Viscous converter clutch, 190, 192
Visual converter check, 384
Volkswagen, Automatic Stick Shift, 2
Volt-ohmmeters, 178
Voltage, 165
Vortex flow, 183–184

W

Walking, 7
Watts, 165
Wave plate, 40
Wide-open throttle (WOT), 116
Wide-open throttle (WOT) kickdown valve, 147
Worm tracks, 119, 336